Ecological Studies, Vol. 139

Analysis and Synthesis

Edited by

M.M. Caldwell, Logan, USA
G. Heldmaier, Marburg, Germany
O.L. Lange, Würzburg, Germany
H.A. Mooney, Stanford, USA
E.-D. Schulze, Jena, Germany
U. Sommer, Kiel, Germany

Ecological Studies

Volumes published since 1992 are listed at the end of this book.

Springer
New York
Berlin
Heidelberg
Barcelona
Hong Kong
London
Milan
Paris
Singapore
Tokyo

Robert A. Mickler   Richard A. Birdsey
John Hom
Editors

Responses of Northern U.S. Forests to Environmental Change

With 172 Figures, 78 in Color

 Springer

Robert A. Mickler
ManTech Environmental
Technology, Inc.
1509 Varsity Drive
Raleigh, NC 27606
USA

Richard A. Birdsey
USDA Forest Service
Northeastern Forest Experiment Station
11 Campus Boulevard
Newton Square, PA 19073
USA

John Hom
USDA Forest Service
Northeastern Forest Experiment Station
11 Campus Boulevard
Newton Square, PA 19073
USA

QK
115
.R46
2000

Library of Congress Cataloging-in-Publication Data
Responses of northern U.S. forests to environmental change/edited by
  Robert Mickler, Richard A. Birdsey, John Hom.
    p. cm.—(Ecological studies; 139)
    Includes bibliographical references and index.
    ISBN 0-387-98900-5 (hardcover: alk. paper)
    1. Forest ecology—United States. 2. Global environmental change.
  I. Mickler, Robert A. II. Birdsey, Richard A. III. Hom, John Lun.
  IV. Title: Responses of northern US forests to environmental change.
  V. Title: Responses of northern United States forests to
  environmental change. VI. Series: Ecological studies; v. 139.
  QK115.R46 2000
  577.3'0973—dc21   99-40960

Printed on acid-free paper.

Production coordinated by Chernow Editorial Services, Inc., and managed by Tim Taylor;
manufacturing supervised by Jacqui Ashri.
Typeset by Scientific Publishing Services (P) Ltd., Madras, India.
Printed and bound by Sheridan Books, Ann Arbor, MI.
Printed in the United States of America.

9 8 7 6 5 4 3 2 1

ISBN 0-387-98900-5 Springer-Verlag New York Berlin Heidelberg   SPIN 10737845

Preface

In the Global Change Research Act of 1990, "global change" is defined as "changes in the global environment (including alterations in climate, land productivity, oceans or other water resources, atmospheric chemistry, and ecological systems) that may alter the capacity of the Earth to sustain life." For the purposes of this book, we interpret the definition of global change broadly to include physical and chemical environmental changes that are likely to affect the productivity and health of forest ecosystems over the long term. Important environmental changes in the Northern United States include steadily increasing atmospheric carbon dioxide, tropospheric ozone, wet and dry deposition of nitrogen and sulfur compounds, acidic precipitation and clouds, and climate variability. These environmental factors interact in complex ways to affect plant physiological functions and soil processes in the context of forest landscapes derived from centuries of intensive land use and natural disturbances.

Research in the North has begun to unravel some key questions about how environmental changes will impact the productivity and health of forest ecosystems, species distributions and abundance, and associations of people and forests. Initial research sponsored by the USDA Forest Service under the United States Global Change Research Program (USGCRP) was focused on basic process-level understanding of tree species and forest ecosystem responses to environmental stress. Chemical pollution stresses received equal emphasis with climate change concerns.

At the most basic plant level, research has highlighted some of the mechanisms that determine how physiological processes respond to combinations of factors that affect northern forest trees. Investigators conducted experiments on the impacts of increases in carbon dioxide (CO_2) and ozone (O_3), and the impacts of nitrogen (N) and acidic deposition. These environmental factors are expected to have continuing effects on forests. Less certain but still of concern are anticipated changes in temperature and precipitation that may be induced by increasing concentrations of greenhouse gases. Experimental soil warming and modeling studies are examples of how the potential effects of climate change on ecosystems are investigated.

Many past experiments involved seedlings or small trees exposed to gases in chambers. Current research projects involve scaling up experiments from highly controlled chamber studies to more realistic experiments and observations on whole ecosystems using open-air exposure systems and gas exchange measurements. Basic physiological research, combined with experimental and observational research at the ecosystem level, will lead to understanding of the causes of observed changes in forest health and productivity of northern forests, including prospective changes in growth and biomass, species composition, pest outbreaks and mortality, carbon allocation and storage, water quality and yield, and wildlife habitat.

Because of these potential changes, there is a need to develop effective management practices to protect forest health and productivity on both public and private lands. Landscape-scale studies have an important integrating function directed at understanding how changes in the physical and chemical climate affect the abundance, distribution, and dynamics of species, populations, and communities. Social interactions and economics research are directed at understanding how the use of trees and forests by people will be influenced by potential changes in forest ecosystems because of global change (adaptation), and how human activity can initiate or alter the processes of change (mitigation). Modeling is used to integrate study results, to provide understanding and prediction of global change effects on forest ecosystems, and—along with landscape-scale studies—to provide an important bridge to assessment, resource management, and policy. Assessment and policy activities ensure that research results and assessments are transferred to sound management practices and interpreted into policy options.

Global change researchers in the North are active participants in regional and national assessments of resource conditions and trends, with a focus on how forest health and productivity may be affected by global change. A large part of this effort requires development and application of the modeling tools needed to make such assessments, to provide scientific input for national assessment efforts, and to develop and analyze policy options for local, regional, and national decision makers.

In this book we report progress in understanding how multiple interacting stresses are affecting or are likely to affect forest ecosystems at multiple spatial and temporal scales. Although there has been much progress under the sponsorship of the USGCRP, we are only beginning to understand how our forest ecosystems are likely to evolve over the next century. Global change research in the northern United States is built on a solid foundation of long-term ecological research and a decade of air pollution studies sponsored by the National Acid Precipitation Assessment Program. The USGCRP introduced a heightened awareness of the potential of climate change and climate variability to affect ecosystems, which can only be understood in the context of widespread chemical stresses. Now that the USGCRP is nearing the end of its first decade, it is timely to assemble the available knowledge as a basis for targeting future global change research and for transferring information to land managers and policy makers through syntheses such as this book, and through participatory assessments.

The U.S. Department of Agriculture (USDA) Forest Service Global Change Research Program (FSGCRP) has been a key player in global change research in the Northeast and North Central United States. Through full or partial sponsorship of more than 100 research projects in the region, the FSGCRP established linkages with most of the regional networks and teams of scientists studying global change issues. Vigorous partnerships among scientists sponsored by various funding institutions foster the interdisciplinary research approaches that are essential for understanding the complex impacts of environmental change. The USDA Forest Service has a unique role as a land management agency working with both public and private landowners throughout the region's forested lands. Thus our research has been regionally dispersed to teams that address specific issues at temporal and spatial scales that are relevant to land managers. Parallel efforts are aimed at aggregating our understanding to landscape and larger domains so that both land managers and policy makers are aware of both the local and global effects of environmental change, whether positive or negative, in the Northeastern and North Central United States.

Robert A. Mickler
Richard A. Birdsey
John Hom

Acknowledgments

Many of the studies described in this book were partially or fully sponsored by the Northern Global Change Program (NGCP), one of five regional research cooperatives that comprise the United States Department of Agriculture (USDA) Forest Service Global Change Research Program (FSGCRP). These programs are designed to provide a sound scientific basis for making regional, national, and international management and policy decisions regarding forest ecosystems in the context of global change. The FSGCRP is part of the U.S. Government's Global Change Research Program, developed under the direction of the Office of Science and Technology Policy in the Executive Office of the President, through the Federal Coordination Council on Science, Engineering, and Technology and its Committee on Earth and Natural Resources.

The FSGCRP is directed by William Sommers and coordinated by Elvia Niebla. Their encouragement and support over the years is greatly appreciated. We also thank Deputy Chiefs Jerry Sesco and Robert Lewis, Staff Director Richard Smyth, and Research Station Directors Ron Lindmark, Linda Donoghue, and Bov Eav for their support. Key supporters in the USDA included Gray Evans, Carol Whitman, and Margot Anderson.

There are many additional sponsors of the research reported in this book. We are especially grateful for the support provided by the U.S. Department of Energy, the U.S. Environmental Protection Agency, the U.S. National Science Foundation, and the National Council for Air and Stream Improvement.

Over the past decade, the NGCP has received technical guidance from many individuals. We are grateful for the insight provided by Phil Wargo, Chip Scott, Keith Jensen, Robert Long, Joanne Rebbeck, Kevin Smith, Jud Isebrands, Lew Ohmann, Rolfe Leary, Mike Vasievich, John Zasada, Alan Lucier, Eric Vance, Alan Ek, David Shriner, Steve Rawlins, Doug Ryan, Paul Van Deusen, Larry Hartmann, Darrell Williams, Boyd Strain, Keith Van Cleve, and Jerry Melillo. We also thank the many scientists who provide peer reviews of research proposals and manuscripts.

This book has not been subject to policy review by the USDA Forest Service or any other U.S. government sponsors and, therefore, does not represent the policies of any agency. The research findings reported in this book have not been subject to scientific review by the National Council for Air and Stream Improvement.

Robert A. Mickler
Richard A. Birdsey
John Hom

Contents

Contributors

John D. Aber

University of New Hampshire
Institute for the Study of Earth,
Oceans, and Space, Durham,
NH 03824, USA

Mary Beth Adams

USDA Forest Service
Northeastern Forest Experiment
Station, Parsons, WV 26287, USA

Allan N.D. Auclair

Rand, Washington, DC 20005, USA

Scott W. Bailey

USDA Forest Service
Northeastern Forest Experiment
Station, Campton, NH 03223, USA

Richard A. Birdsey

USDA Forest Service
Northeastern Forest Experiment
Station, Newton Square, PA 19073,
USA

Bruce Bongarten

University of Georgia
School of Forest Resources
Athens, GA 30602, USA

Donald H. DeHayes

University of Vermont
School of Nature Resources
Burlington, VT 05405, USA

Richard E. Dickson

USDA Forest Service
North Central Forest Experiment
Station, Rhinelander,
WI 54501, USA

Ann C. Dieffenbacher-Krall

University of Maine
Institute for Quaternary Studies
Orono, ME 04469, USA

Ivan J. Fernandez

University of Maine
Department of Plant, Soil and
Environmental Sciences
Orono, ME 04469, USA

Linda S. Heath

USDA Forest Service
Northeastern Forest Experiment
Station, Durham, NH 03824, USA

Warren E. Heilman

USDA Forest Service
North Central Forest Experiment
Station, East Lansing,
MI 48823, USA

John Hom

USDA Forest Service
Northeastern Forest Experiment
Station, Newton Square, PA 19073,
USA

Judson G. Isebrands

USDA Forest Service
North Central Forest Experiment
Station, Rhinelander, WI 54501, USA

Louis Iverson

USDA Forest Service
North Central Forest Experiment
Station, Delaware, OH 43015, USA

George L. Jacobson Jr.

University of Maine
Institute for Quaternary Studies
Orono, ME 04469, USA

Jennifer C. Jenkins

USDA Forest Service
Northeastern Forest Experiment
Station, Burlington, VT 05405, USA

David F. Karnosky

Michigan Technology University
Department of Forestry
Houghton, MI 49931, USA

David W. Kicklighter

Marine Biological Laboratory
The Ecosystems Center
Woods Hole, MA 02543,
USA

John A. Laurence

Cornell University
Boyce Thompson Institute
Ithaca, NY 14853, USA

Gregory B. Lawrence

U.S. Geological Survey
Troy, NY 12180, USA

Andrew M. Liebhold

USDA Forest Service
Northeastern Forest Experiment
Station, Morgantown, WV 26505,
USA

Robert P. Long

USDA Forest Service
Northeastern Forest Experiment
Station, Delaware, OH 43015, USA

Patrick J. McHale

State University of New York
College of Environmental Science
and Forestry, Syracuse, NY 13210,
USA

William H. McWilliams

USDA Forest Service
Northeastern Forest Experiment
Station, Newton Square, PA 19073,
USA

Jerry M. Melillo

Marine Biological Laboratory
The Ecosystems Center
Woods Hole, MA 02543, USA

Robert A. Mickler

ManTech Environmental
Technology, Inc. 1509 Varsity Drive
Raleigh, NC 27606, USA

Rakesh Minocha

USDA Forest Service
Northeastern Forest Experiment
Station, Durham, NH 03824, USA

Subhash Minocha   University of New Hampshire,
                         Department of Plant Biology,
                         Durham, NH 03824, USA

Myron J. Mitchell   State University of New York
                         College of Environmental Science
                         and Forestry, Syracuse,
                         NY 13210, USA

Scott V. Ollinger   University of New Hampshire
                         Complex Systems Research Center
                         Durham, NH 03824, USA

Brian E. Potter   USDA Forest Service
                         North Central Forest Experiment
                         Station, East Lansing, MI 48823,
                         USA

Joanne Rebbeck   USDA Forest Service
                         Northeastern Forest Experiment
                         Station, Delaware, OH 43015,
                         USA

Gordon C. Reese   USDA Forest Service
                         Northeastern Forest Experiment
                         Station, Newton Square, PA 19073,
                         USA

Lindsey E. Rustad   University of Maine
                         Department of Plant, Soil and
                         Environmental Sciences, Orono,
                         ME 04469, USA

Paul G. Schaberg   USDA Forest Service
                         Northeastern Forest Experiment
                         Station, Burlington, VT 05402,
                         USA

Thomas L. Schmidt   USDA Forest Service
                         North Central Forest Experiment
                         Station, St. Paul, MN 55108,
                         USA

Walter C. Shortle   USDA Forest Service
                         Northeastern Forest Experiment
                         Station, Durham, NH 03824,
                         USA

Kevin T. Smith

USDA Forest Service
Northeastern Forest Experiment
Station, Durham, NH 03824, USA

Paul A. Steudler

Marine Biological Laboratory
The Ecosystems Center
Woods Hole, MA 02543, USA

Joel P. Tilley

Yale University
School of Forestry and Environmental
Studies, New Haven, CT 06511,
USA

Margaret Tyrrell

Yale University
School of Forestry and Environmental
Studies, New Haven, CT 06511,
USA

Daniel J. Vogt

Yale University
School of Forestry and Environmental
Studies, New Haven, CT 06511, USA

Kristiina A. Vogt

Yale University
School of Forestry and Environmental
Studies, New Haven, CT 06511, USA

Philip M. Wargo

USDA Forest Service
Northeastern Forest Experiment
Station, Hamden, CT 06514, USA

David W. Williams

USDA Forest Service
Northeastern Forest Experiment
Station, Newton Square, PA 19073,
USA

Peter B. Woodbury

Cornell University
Boyce Thompson Institute
Ithaca, NY 14853, USA

1. An Introduction to Northern U.S. Forest Ecosystems

1. Forest Resources and Conditions

William H. McWilliams, Linda S. Heath, Gordon C. Reese, and Thomas L. Schmidt

The forests of the northern United States support a rich mix of floral and faunal communities that provide inestimable benefits to society. Today's forests face a range of biotic and abiotic stressors, not the least of which may be environmental change. This chapter reviews the compositional traits of presettlement forests and traces the major land use patterns that led to the development of contemporary forested ecosystems. Human impacts have dominated forest development over the last 150 years, so considerable attention is paid to current compositional, structural, and successional traits resulting from these impacts. Estimates of forest carbon storage set the stage for later chapters dealing with environmental factors affecting forest health and resiliency. Resource sustainability is addressed by examining productive capacity (growth) in relation to forest drain components: mortality and removals. Historical information has been gleaned from the literature. Source material for recent trends in timberland area, species composition, stand structure, and net forest drain is from the successive state-level forest inventories conducted by the United States Department of Agriculture (USDA) Forest Service's Forest Inventory and Analysis (FIA) unit. Much of this information was compiled from the FIA's Eastwide Database (Hansen et al., 1992).

Northern Forests in Perspective

The Northern Region covers a wide range of physiographic and climatic regimes, spanning 12 degrees latitudinally and 30 degrees longitudinally, or from Minnesota to Maine in the north and from Missouri to Maryland in the south. The region lies wholly within the Humid Temperate Domain described by Bailey (1994), which includes the Warm Continental Division, portions of the Prairie and Hot Continental Divisions, and a small portion of the Subtropical Division in southeastern Missouri and southern Maryland. Major geophysical relief is found in the Superior Uplands and Central Lowlands of the Lake and Central States; the Ridge and Valley, Appalachian Plateaus, and Adirondacks of the Mid-Atlantic States; and the Green Mountains and White Mountains of New England. For this analysis, the Northern Region is divided into four subregions: Lake States (Michigan, Minnesota, and Wisconsin), Central States (Illinois, Indiana, Iowa, and Missouri), New England (Connecticut, Massachusetts, Maine, New Hampshire, Rhode Island, and Vermont), and Mid-Atlantic (Delaware, Maryland, New Jersey, New York, Ohio, Pennsylvania, and West Virginia).

Of the four regions that make up the United States, the Northern Region comprises 18% of the land area and 23% of the nation's forest area (Table 1.1). What distinguishes this region from the others is its high concentration of people. The Northern Region contains 45% of the population of the United States. Heavily forested areas are common in regions that are not far removed from urban areas with high population densities, such as Minneapolis and Milwaukee in the Lake States, and the urban corridor from Baltimore to Boston along the Atlantic coast. Still, the region's population has a significant impact on forestland for recreational use, timber utilization, and land conversion. Forest fragmentation has become a major concern as urban centers expand into traditionally forest areas. Forest policy of the Northern Region is strongly influenced by the fact that private forestland owners control 8 of every 10 ha of timberland (Powell et al., 1993). The primary reasons cited for private forestland ownership are recreation and esthetic enjoyment. These objectives account for 29% of the region's 3.9 million private owners or 26% of the private forestland (Birch, 1996). Only 1% of the private owners, controlling 19% of the private forest, cite timber production as the primary reason for ownership. These objectives do not necessarily conflict with intentions to harvest, as 35% of the private owners, controlling 61% of the private forest, have expressed intentions to conduct some form of harvest in the next 10 years.

The Northern Region plays an important role in national resource supply, with deciduous forests containing an abundance of high-value oak (*Quercus* spp.), black cherry (*Prunus serotina* Ehrh.), ash (*Fraxinus* spp.),

Table 1.1. Total Land Area, Area of Forestland (Thousands of Hectares), Volume of Growing Stock, Average Annual Gross Growth, Average Annual Mortality, and Average Annual Removals of Growing Stock (Millions of Cubic Meters) by Region and Species Group, Northern U.S., 1997 (Source: Data for Eastern States were Compiled from the USDA Forest Service, Eastwide Database. Data for Western States are from USDA Forest Service, 1992)

Region	Total Land Area	Area of Forestland	Volume of Growing Stock		
			Total	Softwood	Hardwood
North	165,968.3	67,261.0	5,827.7	1,336.5	4,491.2
South	216,314.5	82,802.6	7,117.6	2,933.7	4,184.0
Rocky Mountains	300,366.0	56,547.3	3,121.6	2,873.8	247.8
Pacific Coast	231,896.6	87,664.5	6,163.7	5,507.8	655.8
United States Total	914,545.4	294,275.4	22,230.6	12,651.8	9,578.8

	Average Annual Gross Growth			Average Annual Mortality			Average Annual Removals		
	Total	Softwood	Hardwood	Total	Softwood	Hardwood	Total	Softwood	Hardwood
North	177.2	39.1	138.1	40.1	11.0	29.1	64.0	18.5	45.5
South	357.8	190.6	167.1	62.9	29.1	33.7	260.0	165.2	94.7
Rocky Mountains	81.0	73.3	7.7	19.5	17.1	2.4	23.4	22.5	0.8
Pacific Coast	149.4	128.7	20.6	28.8	24.7	4.0	106.0	101.9	4.0
United States Total	765.4	431.7	333.5	151.3	81.9	69.2	453.4	308.1	145.0

Note: Data may not add to totals due to rounding. Computation of volume, growth, mortality, and removals is based on the merchantable stem of growing-stock trees.

and maple (*Acer* spp.); and coniferous forests of spruce (*Picea* spp.), fir (*Abies* spp.), and white pine (*Pinus strobus* L.) supplying the pulp, paper, and lumber industries. In merchantable volume, the Northern Region has 47% of the nation's hardwood inventory volume and 11% of the softwood inventory. Using gross growth (exclusive of removals) as a measure of productivity, this region averages $2.6\,\mathrm{m^3\,ha^{-1}\,yr^{-1}}$ and ranks second behind the Southern Region ($4.3\,\mathrm{m^3\,ha^{-1}\,yr^{-1}}$). Compared with other regions, cutting activity is having a relatively minor impact across much of the Northern Region. The ratio of net growth (gross growth minus mortality) to removals is one measure of overall resource sustainability. The Northern Region's ratio of 2.1 ranks second to the Rocky Mountains' ratio of 2.6 with respect to conditions that favor resource sustainability. Some subregions of the Northern Region have more active timber harvesting and are associated with tighter relationships. The ratio for both the Southern and Pacific Coast Regions is 1.1:1.0.

Presettlement Forest

Today's northern forests have evolved in relation to a complex set of natural and anthropogenic influences. Prior to settlement, the major influences on forest character were wind, pathogens, and fire originating from lightning strikes and Native American activity (Seischab and Orwig, 1991). Thus, the presettlement forest was not a continuous stand of immense trees of uniform composition as some envision (Cline and Spurr, 1942). Rather, it contained compositional and structural diversity over a wide range of physiographic and climatic gradients found in this region. Little is known about the extent and composition of the presettlement forest, though some information has been developed from pollen data, land survey records, and studies of old growth forests. Pollen data suggest that prior to settlement, composition was changing toward increasing dominance of pine (*Pinus* spp.) and spruce, and decreasing dominance of birch (*Betula* spp.) and beech (*Fagus grandifolia* Ehrh.) (Russell et al., 1993). The increasing trend for pine and spruce was strongest in the northern hardwood forest. Russell et al.'s finding corroborates Gajewski's (1987) observation that from about A.D. 1450 to the time of European settlement, the abundance of northern tree species was increasing while the more southern species were decreasing, perhaps due to a cooling climate during the period known as the Little Ice Age.

Heavy logging in the Lake States all but obliterated the original "climax" pine–hemlock (*Tsuga canadensis* [L.] Carr.) forests of the region (Weaver and Clements, 1938). Using land office records for counties representative of northern lower Michigan, Whitney (1987) was able to estimate the distribution and composition of precolonial forests. Mixed pine forests containing red pine (*Pinus resinosa* Ait.), white pine, and jack pine (*Pinus banksiana* Lamb.) comprised about 45% of the area represented by the study area. Oak species were rarely found in the main canopy of mixed pine stands but were common as a suppressed understory component. Pine's predominance was largely due to drought-prone, coarse-textured soils and periodic natural fires. The pure jack pine type, a particularly fire-dependent type, likely accounted for 15% of the original forest. Beech, sugar maple (*Acer saccharum* Marsh.), hemlock, and an associated mix of hemlock, white pine, and northern hardwoods accounted for about 20% of the forest. The remaining 20% was comprised of swamp conifers, including northern white cedar (*Thuja occidentalis* L.), tamarack (*Larix laricina* [Du Roi] K. Koch), and spruce.

Some information on original forest vegetation of the Central States can be gleaned from the work of Blewett and Potzger (1950) and Potzger et al. (1956) in Indiana. Using land survey records, they found that beech and sugar maple constituted at least half of the stems in most of Indiana's townships. Beech often was twice as abundant as sugar maple. Other species of importance were oak, hickory (*Carya* spp.), and ash. Beech–maple forests were common on north-facing slopes and moist upland sites. Beech–maple forests in Indiana were considered a subset of the "mixed mesophytic association" in which beech and maple were the most prominent (Potzger and Friesner, 1940; Braun, 1938). Oak–hickory stands typically occupied south-facing slopes and ridgetops. The composition of oak–hickory forests varied but oaks often were the dominant species. Intermediate slope exposures supported a mixture of species. Bottomland and transient forests occurred infrequently. Although not completely representative of the entire Central States subregion, Indiana contained examples of the major formations of the subregion (Braun, 1950).

In New England, Siccama (1971) found that original forests of northern Vermont contained the same species as today, though in markedly different proportions. The most striking example is beech, which comprised 40% of presettlement composition compared with less than 10% today. In southern New England, Bromley (1935) cited fire as a major factor in restricting the abundance of fire-sensitive species such as hemlock, white pine, and beech. Fire also was important in Maine, where numerous, widespread natural fires played a role in the development of northern conifer, hardwood, and mixed forests. However, the interval between severe disturbances was longer than needed to attain a climax, all-aged structure (Lorimer, 1977). A comparison of Lorimer's estimate of composition with the recent forest inventory for Maine (Griffith and Alerich, 1996) suggests the increased importance of northern white cedar and maple, especially red maple (*Acer rubrum* L.), and the decreased importance of birch and beech. Using numbers of stems as the importance

Table 1.2. New England Presettlement and Current Species Composition as a Percent of Total Forest Composition

Presettlement (%)		Current (%)	
Spruce	21	Spruce	21
Birch	17	Cedar	17
Beech	15	Maple	15
Balsam fir	14	Balsam fir	15
Cedar	12	Birch	10

Note: The current inventory data for Maine was compiled for growing-stock trees at least 12.5 cm in diameter in Aroostok, Penobscot, and Piscataquis Counties—the counties where Lorimer's study sites were located.

value, six genera accounted for at least 10% of total composition (Table 1.2).

Studies of the presettlement forest in the Mid-Atlantic States also have documented differences between precolonial and current composition. Hough and Forbes (1943) found that beech was ubiquitous throughout the Allegheny Plateaus of Pennsylvania. Hemlock–beech and beech–maple were considered as climax associations on most sites. Red maple, yellow birch (*Betula alleghaniensis* Britton), sweet birch (*Betula lenta* L.), white ash (*Fraxinus americana* L.), and black cherry—species of considerable abundance today—were found only locally. Lutz (1930) presented similar findings, citing the low frequency of black cherry and yellow poplar (*Liriodendron tulipifera* L.). Beech and sugar maple were the most widely distributed species in western New York (Seischab, 1990). Seischab and Orwig (1991) supported the view that the presettlement forest contained a large component of "steady state" communities, primarily due to a lack of catastrophic disturbance other than occasional windthrow events.

Land Use History

The process of settlement began along the Atlantic seaboard in the mid-1600s. Colonization began slowly and accelerated gradually, with settlers moving westward and southward. The human population of the Northern Region increased relatively slowly until the mid-1800s when a period of rapid growth began. At first, forests were cleared for cropland to satisfy increased demand for food. On average, every person added to the U.S. population during the 1800s was matched by 1.2 to 1.6 ha of cropland and even more pasture and hayland (MacCleery, 1992). As the population continued to increase, so did demand for wood to build houses and railroads, provide fuel and chemicals for tanning, supply mining props, and support other industries. The advent of rail transportation made vast tracts available, while earlier timbering had concentrated on areas accessible by water and horse. By the 1920s, conversion of forest to cropland had nearly ceased. Most of the softwood stands were exhausted and the only large reservoirs of virgin hardwood timber in the East were in the Lake States, the southern Appalachians, and the Lower Mississippi Valley (USDA Forest Service, 1920). By this time, most of the forestland in the Northern Region was heavily cut over.

Following the 1920s, northern forests began a period of reestablishment and growth. During the 1930s and 1940s, abandoned farmland provided the source for a new generation of forests. In the 1930s, the Forest Service began conducting forest inventories to track the extent and condition of the nation's forests. Most inventories conducted since that time have documented the steady expansion and improvement of northern forests. The demise of American chestnut (*Castanea dentata* [Marsh.] Borkh.) was one of the major events that occurred during the period of forest regrowth. First documented in 1904, the chestnut blight disease (*Endothia parasitica*) virtually eliminated the species by the 1950s (Harlow et al., 1979).

Today's Northern Forest

The Forestland Base

Today's northern forest encompasses 67.3 million ha (Table 1.3). The Lake States and Mid-Atlantic subregions account for two-thirds of the forestland in the Northern Region. In New England, 81% of the total land base is classified as forest, making it the nation's most heavily forested region. Ninety-five percent of the forestland in the Northern Region is classified as timberland, that is, forestland capable of producing timber crops and which is not reserved from harvesting activity. The timberland estimate is the most useful for analyzing trends because, over time, most FIA inventories have focused on measuring trees on timberland. Some

Table 1.3. Area of Forestland (Thousands of Hectares), Forestland as a Percent of Total Land Area, Area of Timberland, and Percent Change in Timberland Area by Subregion, Northern U.S., 1997 (Source: USDA Forest Service, Eastwide Database)

Subregion	Area of Forestland	Forestland as a Percent of Total Land Area	Area of Timberland	Percent Change in Area of Timberland
Lake States	20,765.8	42	19,464.8	+5
Central States	10,017.3	18	9,566.8	+10
New England	13,078.7	81	12,677.3	no change
Mid-Atlantic	23,399.2	53	22,427.7	+3
North Total	67,261.0	41	64,136.7	+4

Note: Data may not add to totals due to rounding.

reserved forestland in parks, wilderness areas, and urban settings has not been routinely inventoried. The most recent inventories of northern forests indicate that the area of timberland has increased by 4% despite conversion to other land uses. The major source of new forest has been the reversion of retired agricultural land, primarily in rural areas. In areas with relatively high population densities, reversions to forest are offset by conversion of forest to urban, suburban, industrial, and other land uses. The most significant increases have occurred in the Lake and Central States. Timberland in the Mid-Atlantic States increased slightly while there was no change in New England. It is likely that future inventories will show decreases in timberland area as new sources of forestland are scarce and conversion continues.

Ownership

Private forestland ownership is fundamental to understanding forest policy, economics, and management decisions in the Northern Region. Here 80% of the timberland is in the hands of 3.9 million private owners (Birch, 1996). Only 10% of the privately owned timberland is owned by forest industry (defined as those that own and operate a wood-using plant), making the "other private" group the dominant owner in the region (Table 1.4). Often referred to as the nonindustrial private forestland (NIPF) owner, this group is composed of individuals, partnerships, and miscellaneous corporations. Thus, a NIPF owner might be a pension fund investment firm, a retired grandmother, an Indian tribe, a timberland management corporation, or a farmer. A key to understanding NIPF owners is that large numbers of owners control small tracts and a relatively few owners control large tracts. For example, an estimated 3.7 million private owners, or 94% of the private owners, own forest tracts that are smaller than 40 ha; these tracts make up 44% of the private timberland. The remaining 6% of private owners, with tracts larger than 40 ha, control 56% of the private timberland.

Table 1.4. Area of Timberland (Thousands of Hectares) by Ownership Class and Subregion Northern U.S., 1997 (Source: USDA Forest Service, Eastwide Database)

Subregion	All Owners	National Forest	Other Public	Forest Industry	Other Private
Lake States	19,464.8	2,268.6	5,148.6	1,384.7	10,663.0
Central States	9,566.8	693.1	557.9	103.8	8,212.0
New England	12,677.3	407.2	832.4	3,410.3	8,027.3
Mid-Atlantic	22,427.8	594.1	2,327.4	1,240.5	18,265.8
North Total	64,136.7	3,963.0	8,866.2	6,139.3	45,168.1

Note: Data may not add to totals due to rounding.

Although owner objectives are many and complex, private forestland management tends to be custodial, with a minimum of capital investment. Although most native forest types regenerate naturally, it can be difficult to regenerate some preferred species. Also, forest planting is relatively rare in the Northern Region. In 1996, 42,400 ha were planted in the North compared with 709,100 ha in the Southern Region.

In contrast to the Western United States, where public owners control 61% of the timberland, only 20% of the timberland in the Northern Region is in public ownership (Powell et al., 1993). The "other public" class contributes 69% of the public timberland, primarily due to significant state and county ownership in the Lake States. Although the 15 national forests within the Northern Region contribute only 6% of its timberland, they are major suppliers of recreation, timber, and other forest-related benefits.

Forest Type Groups

Data on forest composition, structure, and stage of stand development provide insights into landscape diversity and potential vulnerability to climate change. The FIA reports estimates of timberland area by forest type group, which is an assemblage of specific forest types. For example, the white–red–jack pine group includes the white pine, red pine, white pine–hemlock, hemlock, and jack pine forest types. Although the FIA does not measure stage of stand development directly, the forest type group and stand size class variables are useful indicators. Forest type groups imply some information on successional stage, as pioneer species merge with later successional species and stands convert to other types. Stand size class is a coarse measure of stand structure, as well as a surrogate for stage of stand development. Stand size class is assigned according to the dominance of sample trees by size class, including seedling–sapling (early successional stands), poletimber (midsuccessional), and sawtimber (mid- to late successional). The use of the term "late successional" is somewhat misleading, as stands in the 60- to 80-year range may be of sawtimber size, but are still young with respect to life expectancy. Examples of old growth forest conditions can be found in the North, but are extremely rare.

The major forest type groups of the Northern Region are oak–hickory and maple–beech–birch, accounting for 31 and 30% of total timberland, respectively (Table 1.5 on page 14). Oak–hickory forests are located primarily across the southern tier of the region (Fig. 1.1a in color insert). Oak–hickory is by far the dominant forest type group in the Central States, southeastern Ohio, West Virginia, and the Appalachian Mountains of Pennsylvania. Oak–hickory stands contain some of the highest value hardwood timber in the world. In general, large tracts of oak–hickory have recovered since the logging boom of the turn of the century and are reaching financial maturity. Sawtimber stands predominate across most of the oak–hickory region (Fig. 1.1b in color insert). Seedling–sapling stands occur infrequently, accounting for only 14% of the oak–hickory timberland. There is concern that a lack of oak regeneration following harvest will result in gradual, long-term declines in oak abundance (Southeastern Forest Experiment Station, 1993).

Timber harvesting and natural disturbances play a role in forest succession. Recent FIA data for West Virginia show that for the 10-year period between inventories, harvesting affected only 24% of the state's timberland (Birch et al., 1992). Clearcut harvests occurred on only 2% of the timberland. Most of the harvesting removed less than 40% of the stand's existing basal area. A preference for harvest of select species also was evident. A study of harvest activity in Pennsylvania produced similar findings (Gansner et al., 1993a). Another concern has been the impact of the gypsy moth (*Lymantria dispar* L.), which has caused significant oak mortality in areas where outbreaks have reached epidemic levels. Chestnut oak (*Quercus prinus* L.) and white oak (*Quercus alba* L.) have been affected more than other species (Gansner et al., 1993b). Most mortality has been in smaller, poorer quality trees. In hard-hit areas such as central Pennsylvania, the entire overstory was killed and young stands of black cherry, red maple, sweet birch, and other species have replaced oak. McWilliams et al. (1995) found that 92% of mixed oak stands in Pennsylvania were adequately stocked with woody species following major disturbance, but that oak stocking was far below predisturbance (including harvesting) levels. Only 16% of the stands were adequately stocked with oak. Yet another concern in Pennsylvania and other areas has been the impact of white-tailed deer (*Odocoileus virginianus borealis* Miller). Populations often exceed 20 per square mile, a threshold that indicates serious impacts on the forest understory. Excessive deer browsing has reduced both the density and diversity of woody understory species (Tilghman, 1989).

Maple–beech–birch, or northern hardwoods, are most common in the northern half of the study region (Fig. 1.2a in color insert). The maple–beech–birch group predominates in northern Pennsylvania, New York, and New England, and is found on cooler, high-elevation sites in the mountains of West Virginia. The Mid-Atlantic and New England subregions account for nearly two-thirds of the northern hardwood stands in the Northern Region. Most of the remaining stands are found in northern Wisconsin and Michigan. As with oak–hickory forests, northern hardwood forests are dominated by older, sawtimber-size stands (Fig. 1.2b in color insert), which make up 51% of the area for this group. Studies of harvest activity for New England and New York indicate many of the same harvest impacts as those found in oak–hickory states. Only about one-third of the timberland studied showed evidence of harvest between inventories (Gansner et al., 1990; McWilliams et al., 1996a).

Color Plate I

(a)

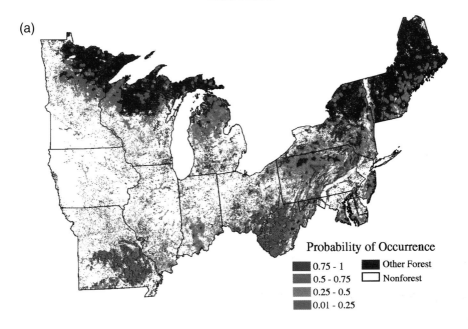

Probability of Occurrence

- 0.75 - 1
- 0.5 - 0.75
- 0.25 - 0.5
- 0.01 - 0.25
- Other Forest
- Nonforest

(b)

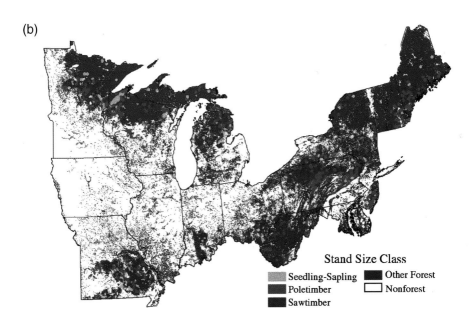

Stand Size Class

- Seedling-Sapling
- Poletimber
- Sawtimber
- Other Forest
- Nonforest

Figure 1.1. Location of forestland, estimated probability of occurrence for oak–hickory forestland (a), and probable distribution of oak–hickory forestland by stand size class (b), Northern Region (excluding southern New England), 1997.

Color Plate II

(a)

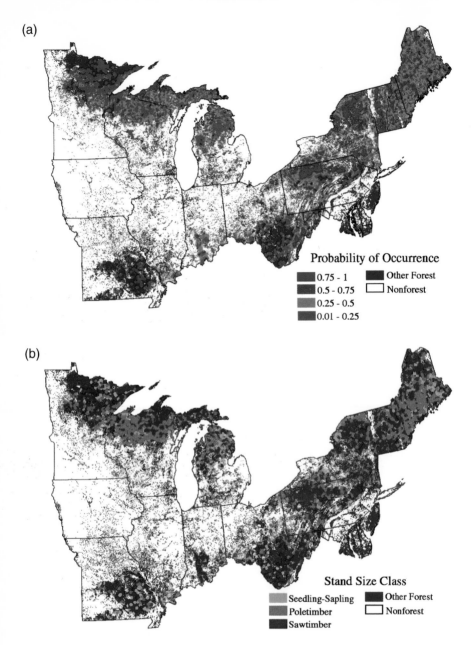

Probability of Occurrence

0.75 - 1   Other Forest
0.5 - 0.75   Nonforest
0.25 - 0.5
0.01 - 0.25

(b)

Stand Size Class

Seedling-Sapling   Other Forest
Poletimber   Nonforest
Sawtimber

Figure 1.2. Location of forestland, estimated probability of occurrence for maple–beech–birch forestland (a), and probable distribution of maple–beech–birch forestland by stand size class (b), Northern Region (excluding southern New England), 1997.

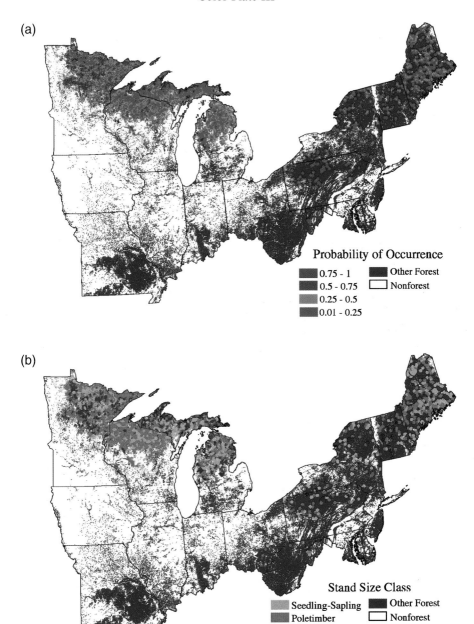

Figure 1.3. Location of forestland, estimated probability of occurrence for aspen–birch forest land (a), and probable distribution of aspen–birch forestland by stand size class (b), Northern Region (excluding southern New England), 1997.

(a)

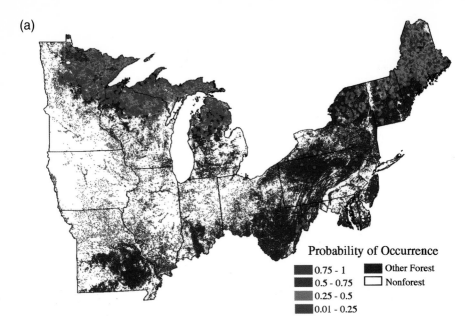

(b)

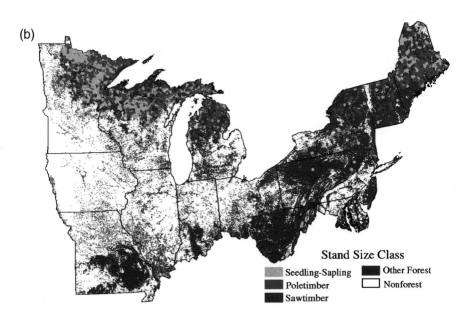

Figure 1.4. Location of forestland, estimated probability of occurrence for spruce–fir forest land (a), and probable distribution of spruce–fir forestland by stand size class (b), Northern Region (excluding southern New England), 1997.

Clearcut harvests were relatively rare compared with partial cutting, and a preference for selecting larger trees of preferred species was shown. These results are supported by a study of harvest disturbance in the Lake States. As northern hardwood stands continue to increase in stocking and size, susceptibility to a range of pests and pathogens also increases. In the Allegheny Plateau region of Pennsylvania, numerous damaging agents have taken a heavy toll in aging northern hardwood stands, particularly sugar maple (McWilliams et al., 1996b).

Aspen–birch is the third most prevalent group in the Northern Region, with 11% of the timberland. More than three-fourths of the aspen–birch forest is found in the northern Lake States, particularly Minnesota (Fig. 1.3a in color insert). The other significant amount of aspen–birch forest is in Maine. Unlike most timberland in the region, aspen–birch is distributed relatively evenly by stand size class with 36% seedling–sapling, 42% poletimber, and 22% sawtimber. Aspen–birch tends to be an early successional type characterized by trees that typically do not become large. In stands that are actively managed, clearcut harvests often perpetuate aspen–birch stands. Aspen–birch also is a common pioneer following major disturbance in other forest type groups. Younger stands of aspen–birch are common throughout the area where aspen–birch is prevalent (Fig. 1.3b in color insert).

Spruce–fir is the fourth most important group, with 10% of total timberland. The spruce–fir group comprises a number of specific forest types, including balsam fir (*Abies balsamea* [L.] Miller), red spruce (*Picea rubens* Sarg.), red spruce–balsam fir, white spruce (*Picea glauca* [Moench] Voss), black spruce (*Picea mariana* [Mill.] B.S.P.), northern white cedar, and tamarack. The group is found primarily in the northernmost areas of the region (Fig. 1.4a in color insert). In Minnesota, spruce–fir is dominated by swamp conifers, which generally are associated with lower stocking levels, growth rates, and disturbance. In Maine, balsam fir, red spruce, red spruce–balsam fir, and northern white cedar are the most common types. The recent inventory of Maine revealed tremendous changes in spruce–fir forests due to infestations of spruce budworm (*Choristoneura fumiferana* Clem.), associated salvage harvesting, and other harvesting to satisfy expanded timber markets. The result was a 20% decrease in spruce–fir timberland (Griffith and Alerich, 1996). Much of this decrease resulted from conversion of spruce–fir stands to northern hardwoods, white–red–jack pine, and aspen–birch. Young stands of spruce–fir are common throughout Maine and the Lake States (Fig. 1.4b in color insert). The distribution of spruce–fir timberland by stand size class is 32% seedling–sapling, 38% poletimber, and 30% sawtimber. A small portion of the red spruce forest type that occurs at high elevations (above 900 m) has been affected by acidic deposition.

The top four forest type groups account for more than 8 of every 10 ha of the timberland in the Northern Region. The remaining timberland is

Table 1.5. Area of Timberland (Thousands of Hectares) by Forest Type Group, Stand Size Class, and Subregion, Northern U.S., 1997 (Source: USDA Forest Service, Eastwide Database)

Forest Type Group	Stand Size Class	North	Lake States	Central States	New England	Mid-Atlantic
Oak–Hickory	Seedling–Sapling	2,838.8	352.3	1,069.7	119.3	1,297.4
	Poletimber	5,791.6	647.1	1,543.2	639.6	2,961.6
	Sawtimber	11,327.7	1,431.5	3,312.4	516.7	6,067.0
	Total	19,958.0	2,430.8	5,925.4	1,275.7	10,326.0
Maple–Beech–Birch	Seedling–Sapling	3,402.4	923.1	455.6	742.0	1,281.8
	Poletimber	6,087.9	1,665.1	321.9	1,920.5	2,180.3
	Sawtimber	9,953.8	2,491.2	908.1	2,295.0	4,259.5
	Total	19,444.2	5,079.3	1,685.6	4,957.5	7,721.6
Aspen–Birch	Seedling–Sapling	2,475.8	1,864.6	3.0	423.9	184.4
	Poletimber	2,956.2	2,208.2	—	549.0	199.1
	Sawtimber	1,535.6	1,340.8	—	132.8	61.9
	Total	6,967.6	5,413.6	3.0	1,105.7	445.4
Spruce–Fir	Seedling–Sapling	2,021.6	1,145.4	—	826.3	49.9
	Poletimber	2,435.6	1,167.0	—	1,163.3	105.3
	Sawtimber	1,879.6	753.5	—	1,000.7	125.5
	Total	6,336.8	3,065.9	—	2,990.2	280.5
White–Red–Jack Pine	Seedling–Sapling	594.4	376.6	9.5	89.3	119.0
	Poletimber	1,115.8	602.8	11.5	266.2	235.5
	Sawtimber	2,774.9	690.6	12.1	1,185.2	887.1
	Total	4,485.1	1,670.0	32.9	1,540.7	1,241.5
Elm–Ash–Cottonwood	Seedling–Sapling	931.0	437.5	150.0	97.3	246.2
	Poletimber	1,210.3	607.1	209.7	146.7	246.7
	Sawtimber	1,873.4	637.8	725.9	86.7	422.9
	Total	4,014.6	1,682.4	1,085.5	330.8	915.8

Oak–Pine					
Seedling–Sapling	230.0	—	108.3	25.8	95.8
Poletimber	517.2	—	129.5	148.2	239.5
Sawtimber	760.0	—	167.3	214.9	377.7
Total	1,507.2	—	405.1	388.9	713.1
Loblolly–Shortleaf					
Seedling–Sapling	221.0	—	69.8	14.1	137.0
Poletimber	290.4	—	75.3	27.0	188.1
Sawtimber	401.9	—	106.4	29.9	265.6
Total	913.3	—	251.5	71.0	590.7
Oak–Gum–Cypress					
Seedling–Sapling	47.8	—	13.7	—	34.2
Poletimber	68.2	—	20.5	—	47.7
Sawtimber	192.3	—	101.2	5.9	85.1
Total	308.3	—	135.4	5.9	166.9
All Forest Type Groups					
Seedling–Sapling	12,762.7	5,099.4	1,879.6	2,338.0	3,445.6
Poletimber	20,473.2	6,897.3	2,311.4	4,860.5	6,403.9
Sawtimber	30,699.3	7,345.5	5,333.4	5,468.0	12,552.2
Total	63,935.1	19,342.2	9,524.3	12,666.6	22,401.7

Note: Data may not add to totals due to rounding. Excludes 201.5 thousand hectares of nontyped timberland.

split among the white–red–jack pine, elm (*Ulmus* spp.)–ash (*Fraxinus* spp.)–cottonwood (*Populus* spp.), oak–pine, loblolly (*Pinus taeda* L.)–shortleaf pine (*Pinus echinata* Mill.), and oak–gum (*Liquidambar styraciflua* L.)–cypress (*Taxodium* spp.) groups. These groups often are of local economic importance and valuable for the species diversity and unique habitat they provide. For example, oak–gum–cypress forests are rare in the north (1% of timberland) but represent the northernmost extension of the group, provide critical wetland forest habitat, and contain high-value species such as cherrybark oak (*Quercus falcata* var. *pagodafolia* Ell.).

Biomass and Carbon

Growing concern over increasing concentrations of greenhouse gases in the atmosphere has drawn attention to the impact of forested ecosystems on global climate dynamics. The role of forested ecosystems in accumulating and storing carbon (C) has important ramifications for policies directed toward ameliorating the impact of increasing levels of atmospheric carbon dioxide (CO_2). The most direct method of tracking the amount of C in forested ecosystems is the conversion of existing biomass inventory data. Biomass on northern timberland totals 7.2 billion metric tons (Table 1.6), or roughly 112 tons ha^{-1}. (This estimate excludes the biomass of dead saplings and seedlings.) Nearly 60% of the total biomass is comprised of the nonmerchantable portion of the forest: branches, foliage, stumps, roots, bark, and small and dead trees.

Converting total biomass into C yields an estimate of 3.6 million tons (Table 1.7), or about 56 tons ha^{-1}. The average C content per hectare is relatively constant among Northern subregions. Thus, each subregion's contribution to total C is roughly comparable to the respective area of timberland. The Mid-Atlantic subregion accounts for the highest percentage of total C with 36%, followed by the Lake States (30%), New England (19%), and the Central States (15%).

Of particular interest to policymakers is how current estimates of C compare with historical trends and projections of future changes. Birdsey and Heath (1995) converted historical estimates of biomass to C and projected 1992 biomass estimates 50 years into the future. These estimates included the relative contribution of soils, the forest floor, and existing vegetation (Fig. 1.5). By far, the largest percentage of C stored in northern forests is contributed by forest soils (62%). Changes in total C over the last 50 years have been driven by change in land use from agricultural uses to forest. Although the rate of C flux is projected to decrease in the future, positive fluxes are expected over the next 50 years. As new sources of forest have dwindled over much of the Northern Region, future increases in C are expected to come from the continued growth and maturation of forest vegetation.

Table 1.6. Total Tree Dry Weight Content (Millions of Tons) on Timberland by Subregion, Forest Component, and Species Group, Northern U.S., 1997

	Total	Live Trees of Merchantable Size				Saplings[b]	Dead Trees of Merchantable Size[c]
		Main Stem	Branches and Foliage	Stump and Roots[a]	Bark		
Lake States							
Softwood	475.3	173.2	68.7	73.9	29.7	86.3	43.5
Hardwood	1,650.2	628.2	230.0	227.2	134.9	222.0	207.9
Total	2,125.5	801.4	298.7	301.1	164.6	308.3	251.4
Central States							
Softwood	39.3	16.6	4.8	7.9	2.0	6.2	1.8
Hardwood	1,015.0	410.9	144.7	151.8	77.4	144.4	85.8
Total	1,054.3	427.5	149.5	159.7	79.4	150.6	87.6
New England							
Softwood	518.5	199.0	67.5	91.1	42.9	80.7	37.3
Hardwood	841.6	381.8	71.4	140.1	58.8	154.5	35.0
Total	1,360.1	580.8	138.9	231.2	101.7	235.2	72.3
Mid-Atlantic							
Softwood	284.4	116.3	43.2	53.7	25.9	27.6	17.7
Hardwood	2,331.9	995.2	219.3	420.3	159.9	352.2	185.0
Total	2,616.3	1,111.5	262.5	474.0	185.8	379.8	202.7
North							
Softwood	1,317.5	505.1	184.2	226.6	100.5	200.8	100.3
Hardwood	5,838.7	2,416.1	665.4	939.4	431.0	873.1	513.7
Total	7,156.2	2,921.2	849.6	1,166.0	531.5	1,073.9	614.0

Note: Data may not add to totals due to rounding.
[a] Height of stumps assumed equal to 0.3 m. Stump bark is included.
[b] Includes stem, branches, bark, foliage, stump, and roots.
[c] Includes stem, branches, bark, stump, and roots.

Table 1.7. Total Tree Carbon Content (Millions of Tons) on Timberland by Subregion, Forest Component, and Species Group, Northern U.S., 1997

	Total	Live Trees of Merchantable Size				Saplings[b]	Dead Trees of Merchantable Size[c]
		Main Stem	Branches and Foliage	Stump and Roots[a]	Bark		
Lake States							
Softwood	250.2	90.2	38.7	38.5	15.7	44.4	22.7
Hardwood	839.0	312.8	117.1	113.2	81.8	110.6	103.5
Total	1,089.2	403.0	155.8	151.7	97.5	155.0	126.2
Central States							
Softwood	20.6	8.6	2.7	4.1	1.1	3.2	0.9
Hardwood	507.1	204.6	73.5	75.6	38.8	71.9	42.7
Total	527.7	213.2	76.2	79.7	39.9	75.1	43.6
New England							
Softwood	272.0	103.7	38.0	47.5	22.7	42.0	18.1
Hardwood	421.3	190.1	36.5	69.8	29.5	76.9	18.5
Total	693.3	293.8	74.5	117.3	52.2	118.9	36.6
Mid-Atlantic							
Softwood	150.0	60.6	24.1	28.0	13.7	14.4	9.2
Hardwood	1,163.9	495.6	111.8	209.3	80.1	175.4	91.7
Total	1,313.9	556.2	135.9	237.3	93.8	189.8	100.9
North							
Softwood	692.8	263.1	103.5	118.1	53.2	104.0	50.9
Hardwood	2,931.3	1,203.1	338.9	467.9	230.2	434.8	256.4
Total	3,624.1	1,466.2	442.4	586.0	283.4	538.8	307.3

Note: Data may not add to totals due to rounding.
[a] Height of stumps assumed equal to 0.3 m. Stump bark is included.
[b] Includes stem, branches, bark, foliage, stump, and roots.
[c] Includes stem, branches, bark, stump, and roots.

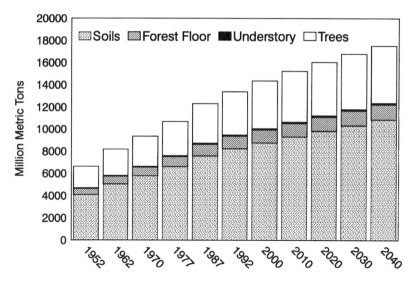

Figure 1.5. Historical and projected estimates of carbon storage (millions of tons) on forestland by forest component, Northern Region, 1997. (Source: Birdsey and Heath, 1995.)

Tree Species Importance

Information on biomass also is useful for describing tree-species richness and heterogeneity across the study region. The FIA inventories identified 143 tree species in the Northern Region, though only 19 species contribute more than 2% of the total biomass (Table 1.8). This diversity has positive implications for the forest's ability to respond to species-specific health concerns and changes in climate. The replacement of gypsy moth–killed white and chestnut oak by red maple and other species is a classic example from recent history. Red maple and sugar maple currently share dominance in the Northern Region, each with 10% of total regionwide biomass. Other important species are northern red oak (*Quercus rubra* L.) with 7% of total biomass, white oak with 6%, and quaking aspen (*Populus tremuloides* Michx.), black oak (*Quercus velutina* Lam.), hickory, and beech with 4% each. Recent inventories within red maple's native range have consistently documented the expansion of this species. Red maples' ability to thrive on a variety of sites (from hydric to xeric) and to regenerate prolifically from sprouts and wind-borne seed has contributed to its expansion. Red maple is now the dominant species in New England and the Mid-Atlantic States, and the third most dominant species in the Lake States.

The Mid-Atlantic subregion had the largest number of tree species tallied (127). The Lake States had the fewest species tallied but the highest number of species with at least 2%, a general indication of an even

Table 1.8. Total Live Tree Dry Weight (Millions of Tons) on Timberland by Species and Subregion, Northern U.S., 1997 (Source: USDA Forest Service, Eastwide Database)

North Species	Weight	Lake States Species	Weight	Central States Species	Weight	New England Species	Weight	Mid-Atlantic Species	Weight
Red maple (Acer rubrum L.)	669.2	Sugar maple	242.5	White oak	164.4	Red maple	174.1	Red maple	309.0
Sugar maple (A. saccharum Marsh)	654.9	Quaking aspen	199.4	Black oak	118.2	Sugar maple	136.8	Sugar maple	240.6
Northern red oak (Quercus rubra L.)	431.6	Red maple	176.8	Hickory spp.	106.6	Red spruce	107.3	Northern red oak	178.8
White oak (Q. alba L.)	369.2	Northern red oak	137.7	Post oak (Q. stellata Wangenh.)	68.2	Eastern white pine	105.5	White oak	133.8
Quaking aspen (Populus tremuloides Michx.)	242.4	Paper birch	114.3	Northern red oak	45.3	Balsam fir	99.6	American beech	130.6
Black oak (Quercus velutina Lam.)	239.6	Balsam fir	79.2	Sugar maple	35.0	Eastern hemlock	87.2	Black cherry	127.6
Hickory spp. (Carya spp.)	234.1	Northern white-cedar	77.6	American elm	26.4	Yellow birch	80.5	Chestnut oak	122.0
American beech (Fagus grandifolia Ehrh.)	232.9	Red pine (P. resinosa Ait.)	62.0	White ash	24.3	American beech	75.8	White ash	116.1
Eastern hemlock (Tsuga canadensis (L.) Carr.)	209.8	Black ash (F. nigra Marsh.)	59.6	Scarlet oak (Q. coccinea Muenchh.)	24.2	Northern red oak	69.8	Yellow-poplar	106.1
White ash (Fraxinus americana L.)	203.4	American basswood (Tilia americana L.)	55.6	Silver maple (A. saccharinum L.)	22.5	Paper birch	64.4	Hickory spp.	102.4
Eastern white pine (Pinus strobus L.)	198.1	White oak	54.1	Black walnut (Juglans nigra L.)	19.5	Northern white-cedar	42.4	Eastern hemlock	96.4
		Black spruce [Picea mariana (Mill.) B.S.P.]	51.7	Shortleaf pine (P. echinata Mill.)	19.1	White ash	35.8	Black oak	74.5
		Bigtooth aspen (Populus grandi-dentata Michx)	51.3	Bur oak (Cercis canadensis L.)	14.4	Quaking aspen	24.4	Eastern white pine	56.1
		Edropline		Eastern redcedar (Juniperus virginia L.)	14.2			Sweet birch (B. lenta L.)	54.6
		Jack pine (P. banksiana Lamb.)	41.8					Scarlet oak	38.4
		Eastern white pine	35.0						

Species		Value
Balsam fir [*Abies balsamea* (L.) Mill.]		185.9
Paper birch (*Betula papyrifera* Marsh.)		185.8
Black cherry (*Prunus serotina* Ehrh.)		177.6
Yellow birch (*B. alleghaniensis* Britton)		148.1
Chestnut oak (*Q. prinus* L.)		128.1
Northern white-cedar (*Thuja occidentalis* L.)		123.5
Red spruce (*Picea rubens* Sarg.)		119.9
Yellow-poplar (*Liriodendron tulipifera* L.)		119.7
Other species (124) ~		1,669.3
All species		6,543.1
Bur oak (*Q. macrocarpa* Michx.)		33.4
Yellow birch		33.1
American elm (*Ulmus americana* L.)		31.9
Black oak		29.3
Tamarack [*Larix larcina* (Du Roi) K. Koch]		29.0
Black cherry		28.4
Other species (61)		250.4
All species		1,874.1
Other species (87)		264.5
All species		966.8
Other species (74)		184.3
All species		1,287.9
Other species (112)		527.4
All species		2,414.4

Note: Species with less than 2% of total live tree dry weight are included as "other species."

distribution. New England had the fewest species tallied (87). The abundance of oak in the Central States is clearly evident, as oaks comprised 45% of the tree biomass in this subregion.

Components of Resource Change

The current growing-stock inventory in the Northern Region totals 5.8 billion m^3, of which more than 75% is in hardwood species (Table 1.9). The Mid-Atlantic contains 41% of the region's inventory, followed by the Lake States (28%), New England (21%), and the Central States (10%). Both timber removals and mortality represent drains on inventory levels. As used here, net growth is equal to gross growth minus mortality.

   The growth/removals ratio indicates that northern forests are growing at more than twice the rate of current removals. Positive ratios are apparent for all subregions except New England, where the ratio is 0.8:1.0. The negative drain for New England is driven by the decreases in the softwood resource in Maine. The most favorable conditions for expanding inventories are in the Central States, which has a ratio of 3.1:1.0, followed

Table 1.9. Volume of Growing-Stock, Average Annual Net Growth, Average Annual Removals, and Average Annual Mortality (Millions of Cubic Meters) on Timberland by Subregion and Species Group, Northern U.S., 1997 (Source: USDA Forest Service, Eastwide Database)

Subregion and Species Group	Total	Pulpwood	Sawlogs and Veneer	Other Products
Lake States				
Softwood	449.3	13.8	5.3	4.3
Hardwood	1,173.6	32.4	14.8	11.3
Total	1,622.9	46.1	20.1	15.6
Central States				
Softwood	33.8	1.1	0.2	0.2
Hardwood	552.1	15.2	5.1	4.3
Total	585.9	16.3	5.3	4.5
New England				
Softwood	546.1	5.1	10.1	4.7
Hardwood	673.0	6.2	4.0	1.7
Total	1,219.0	11.4	14.1	6.3
Mid-Atlantic				
Softwood	307.3	8.1	2.9	1.9
Hardwood	2,092.5	55.2	21.7	11.8
Total	2,399.9	63.3	24.6	13.7
North				
Softwood	1,336.5	28.1	18.5	11.0
Hardwood	4,491.2	109.1	45.5	29.1
Total	5,827.7	137.1	64.0	40.1

Note: Data may not add to totals due to rounding.

by the Mid-Atlantic States (2.6:1.0), and the Lake States (2.3:1.0). These findings are not surprising. Except for Maine, all of the recent inventories of northern states have shown increases in inventory volumes. In general, the Northern Region's forests have been undergoing increases in stocking levels and expansion of the number of medium and large trees in the inventory. This is consistent with the notion that these forests are maturing (Schmidt et al., 1996).

Timber Products

Trees that are removed from the existing inventory and converted to wood products provide another form of C storage. The Northern Region traditionally has been a strong supplier of wood products. The value of harvested trees provides incentive for adopting forest practices that enhance a multitude of resources. Projections indicate that demand for wood fiber will increase dramatically in the future.

In 1997, the Northern Region provided 71.6 million m^3 of wood in the form of timber products (Table 1.10). Hardwood species comprised 70%

Table 1.10. Timber Products Output (Thousands of Cubic Meters) by Subregion and Species Group, Northern U.S., 1997 (Source: Miscellaneous USDA Forest Service Timber Product Reports)

Subregion and Species Group	Total	Pulpwood	Sawlogs and Veneer	Other Products
Lake States				
Softwood	7,614.5	5,200.3	2,138.3	275.9
Hardwood	24,537.6	14,443.0	9,732.4	362.2
Total	32,152.1	19,643.3	11,870.7	638.1
Central States				
Softwood	113.1	80.4	18.0	14.7
Hardwood	5,948.6	1,004.9	4,819.4	124.3
Total	6,061.8	1,085.3	4,837.5	139.0
New England				
Softwood	10,969.6	5,998.8	4,671.4	299.4
Hardwood	5,474.3	3,752.7	1,205.2	516.4
Total	16,443.9	9,751.5	5,876.6	815.8
Mid-Atlantic				
Softwood	2,908.1	1,348.3	1,463.3	96.5
Hardwood	13,985.9	3,238.9	10,432.9	314.1
Total	16,894.0	4,587.2	11,896.2	410.6
North				
Softwood	21,605.3	12,627.8	8,291.0	686.5
Hardwood	49,946.5	22,439.6	26,190.0	1,316.9
Total	71,551.8	35,067.4	34,481.0	2,003.4

Note: Data may not add to totals due to rounding.

of the product output. Timber product categories include pulpwood, sawlogs–veneer, and other products. Pulpwood includes woody material that is converted to pulp at the point of first processing. In the northern part of the region, pulpwood is derived primarily from harvests of soft-wood species and aspen. In the hardwood region to the south, pulpwood is primarily a byproduct of harvesting quality sawlogs and the use of small and poorly formed trees and nonsawlog portions of sawtimber-size trees. Sawlogs and veneer logs are the highest value products derived from northern forests. Some of the other products include fuelwood, pallet wood, and railroad ties.

Nearly all of the wood harvested in the Northern Region is converted to pulpwood, sawlogs, and veneer. Total production is essentially split between these two categories. New England and the Lake States contribute most of the region's softwood products (86%), as well as most of the pulpwood production (84%). Aspen is a major source of pulpwood in the Lake States. The Mid-Atlantic and Lakes States are the primary sources of sawlogs and veneer, each with roughly one-third of total production. Hardwood species make up about three-fourths of total sawlog and veneer products.

References

Bailey RG (1994) *Ecoregions of the United States* (*map*). Rev ed. Scale 1:7,500,000, colored. USDA Forest Service, Washington, DC.

Birch TW (1996) *Private Forest-Land Owners of the Northern United States, 1994.* Resour Bull NE-136. USDA Forest Service, Northeastern Forest Experiment Station, Radnor, PA.

Birch TW, Gansner DA, Arner SL, Widmann RH (1992) Cutting activity on West Virginia timberlands. North J Appl For 9(4):146–148.

Birdsey RA, Heath LS (1995) Carbon changes in U.S. forests. In: Joyce LA (ed) *Productivity of America's Forests and Climate Change.* Gen Tech Rep RM-271. USDA Forest Service, Rocky Mountain Forest and Range Experiment Station, Ft Collins, CO, pp 56–70.

Blewett MB, Potzger JE (1950) The forest primeval of Marion and Johnson Counties, Indiana, in 1819. Butler Univ Bot Stud 10:40–52.

Braun EL (1938) Deciduous forest climaxes. Ecology 19:515–522.

Braun EL (1950) *Deciduous Forests of Eastern North America.* Hafner, New York.

Bromley SW (1935) The original forest types of southern New England. Ecol Monogr 5:61–89.

Cline AC, Spurr SH (1942) *The Virgin Upland Forest of Central New England: A Study of Old Growth Stands in the Pisgah Mountain Section of Southwestern New Hampshire.* Harvard For Bull 21.

Gajewski K (1987) Climatic impacts on the vegetation of eastern North America during the past 2000 years. Vegetation 68:179–190.

Gansner DA, Arner SL, Widmann RH, Alerich CL (1993a) Cutting disturbance in Pennsylvania: how much, where, and what. In: *Penn's Woods—change and challenge. Proceedings of 1993 Penn State Forest Resource Issues Conference,*

April 1–2, 1993, University Park, PA. Pennsylvania State University, State College, PA, pp 37–41.

Gansner DA, Arner SL, Widmann RH (1993b) After two decades of gypsy moth, is there any oak left? North J Appl For 10(4):184–186.

Gansner DA, Birch TW, Arner SL, Zarnoch SJ (1990) Cutting disturbance on New England timberlands. North J Appl For 7(3):118–120.

Griffith DM, Alerich CL (1996) *Forest Statistics for Maine, 1995.* Resour Bull NE-135. USDA Forest Service, Northeastern Forest Experiment Station, Radnor, PA.

Hansen MH, Frieswyk T, Glover JF, Kelly JF (1992) *The Eastwide Forest Inventory Data Base: User's Manual.* Gen Tech Rep NC-151. USDA Forest Service, North Central Forest Experiment Station, St Paul, MN.

Harlow WM, Harrar ES, White FM (1979) *Textbook of Dendrology.* 6th ed. McGraw-Hill, New York.

Hough AF, Forbes RD (1943). The ecology and silvics of forests in the high plateaus of Pennsylvania. Ecol Monogr 13:299–320.

Lorimer CG (1977) The presettlement forest and natural disturbance cycle of northeastern Maine. Ecology 58:139–148.

Lutz HJ (1930) Original forest composition in northeastern Pennsylvania as indicated by early land survey notes. J For 28:1098–1103.

MacCleery DW (1992) *American Forests: A History of Resiliency and Recovery.* Publ FS-540. USDA Forest Service, Washington, DC.

McWilliams WH, Bowersox TW, Gansner DA, McCormick LH, Stout SL (1995) Landscape-level regeneration adequacy for native hardwood forests of Pennsylvania. In: Gottschalk K, Fosbroke W, Sandra LC (eds) *Proceedings of the 10th Central Hardwood Conference, March 5–8, 1995, Morgantown, WV.* Gen Tech Rep NE-197. USDA Forest Service, Northeastern Forest Experiment Station, Radnor, PA, pp 196–202.

McWilliams WH, Arner SL, Birch TW, Widmann RH (1996a) Cutting activity in New York's forests. In: *Proceedings of The Empire Forest: Changes and Challenges, November 13–14, 1995, Syracuse, NY.* State University of New York, College of Environmental Science and Forestry, Syracuse, NY, pp 33–40.

McWilliams WH, White R, Arner SL, Nowak CA, Stout SL (1996b) *Characteristics of Declining Forest Stands on the Allegheny National Forest.* Res Note NE-360. USDA Forest Service, Northeastern Forest Experiment Station, Radnor, PA.

Potzger JE, Friesner RC (1940) What is climax in central Indiana? Butler Univ Bot Stud 5:81–195.

Potzger JE, Potzger ME, McCormick J (1956) The forest primeval of Indiana as recorded in the original U.S. land surveys and an evaluation of previous interpretations of Indiana vegetation. Butler Univ Bot Stud 13:95–111.

Powell DS, Faulkner JS, Darr DR, Zhu Z, MacCleery DW (1993) *Forest Resources of the United States, 1992.* Gen Tech Rep RM-234. USDA Forest Service, Rocky Mountain Forest and Range Experiment Station, Ft Collins, CO.

Russell EWB, Davis RB, Anderson RS, Rhodes TE, Anderson DS (1993) Recent centuries of vegetational change in the glaciated northeastern United States. J Ecol 81:647–664.

Schmidt TL, Spencer JS, Hansen MH (1996) Old and potential old growth forest in the Lake States, USA. For Ecol and Manag 86:81–96.

Seischab FK (1990) Presettlement forests of the Phelps and Gorham Purchase in western New York. Bull Torrey Bot Club 117(1):27–38.

Seischab FK, Orwig D (1991) Catastrophic disturbances in the presettlement forests of western New York. Bull Torrey Bot Club 118(2):117–122.

Siccama TG (1971) Presettlement and present forest vegetation in northern Vermont with special emphasis on Chittenden County. Am Midland Naturalist 85:153–172.

Southeastern Forest Experiment Station (1993) *Oak Regeneration: Serious Problems, Practical Recommendations*; *Symposium Proceedings, September 8–10, 1992, Knoxville, TN*. Gen Tech Rep SE-84. USDA Forest Service, Southeastern Forest Experiment Station, Asheville, NC.

Tilghman NG (1989) Impacts of white-tailed deer on forest regeneration in northwestern Pennsylvania. J Wildlife Manage 53:524–532.

USDA Forest Service (1920) *Timber Depletion, Lumber Prices, Lumber Exports, and Concentration of Timber Ownership. Report on Senate Resolution 311, 66th Congress, 2nd Session.*, USDA Forest Service, Washington, DC.

Weaver JE, Clements FE (1938) *Plant Ecology*. 2d ed. McGraw-Hill, New York.

Whitney GG (1987) An ecological history of the Great Lakes forest of Michigan. J Ecol 75:667–684.

2. Geologic and Edaphic Factors Influencing Susceptibility of Forest Soils to Environmental Change

Scott W. Bailey

There is great diversity in the structure and function of the northern forest across the 20-state portion of the United States considered in this book. The interplay of many factors accounts for the mosaic of ecological regimes across the region. In particular, climate, physiography, geology, and soils influence dominance and distribution of vegetation communities across the region. This chapter provides a review of the ecology of the northern forest, emphasizing the role of geology and soils.

The chapter begins with descriptive material reviewing the physiography, bedrock geology, and soils of the various provinces that constitute the northern forest. The distribution of vegetation communities and the role of climate, while of prime importance in defining the ecology of the region, are given limited coverage here as these are discussed more thoroughly elsewhere in this volume (see Chapters 1 and 3). However, to the extent that climate and vegetation are important soil-forming factors (Jenny, 1941), their characteristics in each province are summarized here. As the historical vegetation, which developed prior to large-scale anthropogenic alterations of the landscape in the last 150 years, is more germane to the distribution of soil types than current vegetation patterns, long-term climax vegetation or potential natural vegetation (Kuchler, 1964) is listed here.

In the second portion of the chapter, particular attention is paid to the important resource of the soils, upon which our forests grow. Perhaps the single most important factor in determining the health and productivity of the forest, soils integrate many of the same influences that result in distribution of a variety of ecological types across the region. Distribution of major soil types is reviewed, highlighting the important differences in soil-forming factors and processes among soil taxonomic types. In particular, characteristics of each soil type that most affect forest nutrient cycling, and which might be most dynamic in a changing environment, are highlighted.

Finally, efforts to predict distribution of soils susceptible to nutrient depletion are reviewed. Nutrient depletion is perhaps the aspect of environmental change that is most associated with soil processes and is the subject of much recent study. Opportunities for advancement of the methods used to evaluate this phenomenon are suggested.

Ecological Regions

The United States Department of Agriculture (USDA) Forest Service has adopted a hierarchical classification system for mapping ecological regions at multiple scales (McNab and Avers, 1994). At the highest level of classification, the domain, are broad climatic regions, such as Polar, Humid Temperate, Humid Tropical, and Dry. The entire northern forest region lies within the Humid Temperate Domain. At the second level of classification, four divisions occur in the northern forest. The Warm Continental, Hot Continental, Subtropical, and Prairie Divisions delineate broad differences in climate, primarily temperature and precipitation. Two divisions in this region are in mountainous terrain, characterized by altitudinal zonation of vegetation; the letter "M" designates these units (see Fig. 2.1 in the color insert). Broad distinctions within the northern forest region are best illustrated at the third level of the classification system, the province. The following is a breakdown of the study region at this level of detail, highlighting factors responsible for the distinction of ecological units at this scale.

Laurentian Mixed Forest Province

The Laurentian Mixed Forest occurs on flat to moderately hilly areas in the northern part of the region (see Fig. 2.1 in color insert). This province, which constitutes 22% of the region (Table 2.1), is characterized by

Table 2.1. Area of Ecological Provinces in the Northern Forest Region

Symbol	Province	Area (ha)	Area (%)
212	Laurentian Mixed Forest	37,656,021	22
M212	New England–Adirondack Mixed–Coniferous Forest–Alpine Meadow	10,938,240	6
221	Eastern Broadleaf Forest (Oceanic)	19,567,877	11
M221	Central Appalachian Broadleaf–Coniferous Forest–Meadow	8,426,163	5
222	Eastern Broadleaf Forest (Continental)	55,365,988	32
231	Southeastern Mixed Forest	79,560	0.05
232	Outer Coastal Plain Mixed Forest	3,475,754	2
234	Lower Mississippi Riverine Forest	1,054,295	1
251	Prairie Parkland (Temperate)	34,702,306	20
	Total	171,266,204	100

northern hardwood forests (American beech [*Fagus grandifolia* Ehrh.], yellow birch [*Betula alleghaniensis* Britton], sugar maple [*Acer saccharum* L.]). Lesser forest types include spruce (*Picea* spp.)–fir (*Abies* spp.) in Maine and Minnesota, Appalachian oak (*Quercus* spp.) forests in southern New York and adjacent Pennsylvania, and aspen (*Populus* spp.)–birch (*Betula* spp.) and mixed pine (*Pinus* spp.) forests in the upper Midwest. Mean annual precipitation ranges from 530 mm in northern Minnesota to 1270 mm in coastal Maine and portions of the Allegheny Plateau of Pennsylvania and New York. The frost-free growing season is relatively short, ranging from 100 to 160 days.

The age, structure, and composition of bedrock, as well as its influence on forests, is quite varied within the Laurentian Mixed Forest. In Maine, bedrock ranges from deformed but unmetamorphosed sedimentary rocks in the northeast to high-grade metasedimentary rocks in the southwest. Variable contributions of plutonic rocks, mostly granitic, as well as volcanic rocks also occur throughout Maine. The northern New York and Vermont portion is underlain by an assortment of unmetamorphosed sedimentary rocks, including carbonates, shale, and sandstone, their metamorphosed equivalents of marble, schist, and quartzite, as well as gneiss and amphibolite.

The influence of Appalachian deformation, metamorphism and igneous activity is less prevalent in the Allegheny Plateau of southern New York and adjacent Pennsylvania. Here, bedrock is slightly deformed sandstone, siltstone, and shale, with lesser amounts of limestone, conglomerate, and coal. Slightly deformed sandstone, limestone, and dolomite characterize the Michigan portion of the forest. In western sections of the province, highly deformed and metamorphosed rocks, associated with the Canadian Shield, dominate, including felsic to mafic plutonic and volcanic rocks, their metamorphic equivalents gneiss and amphibolite, as well as quartzites, and banded iron formation.

Direct influence of bedrock on forest processes such as nutrient- and water-cycling, is minimal in portions of the Laurentian Mixed Forest where thick surficial deposits blanket the bedrock surface. However, in areas of shallow surficial deposits, bedrock influence may be great, depending on topography, bedrock composition, and degree of water flow through bedrock pores or fractures. In the only unglaciated portion of this province, in southwestern New York and adjacent Pennsylvania, bedrock influences the texture and mineralogy of surficial deposits ultimately derived from saprolite. Indirect influence of the bedrock is important in all glaciated areas, as primary mineralogy, and to a certain extent texture, of surficial deposits is dependent on the bedrock origin of the sediments.

Surficial deposits in this province are nearly all of glacial origin, deposited during the final retreat of the continental glacier at the close of the Wisconsinan Stage, approximately 10,000 to 15,000 years ago. Minor areas of glacial deposits from earlier stages, up to 550,000 years old, are found in southern portions of the province adjacent to the limit of the Wisconsinan advance. Glacial deposits result from a variety of depositional modes, resulting in a variety of configurations and textures. Unsorted, unstratified ground moraine, or till, is the most common glacial deposit in this region. The till tends to be relatively thin (meters to tens of meters thick) in eastern portions, while it is up to hundreds of meters thick in some portions of the upper Midwest. Stratified drift, including outwash plains, kames, and eskers of lesser areal extent are common in most regions, while large sandy outwash deposits dominate some parts of Michigan. Lacustrine deposits, ranging from clays to stratified silts and sands, are common in some low-lying portions, for example, east of Lake Champlain in Vermont and in northern Minnesota. Marine clays are found in east central Maine and the Saint Lawrence Valley. These were deposited when sea level was higher, due to land subsidence under the weight of the continental glaciers. Recent surficial deposits include alluvium in river valleys and organic accumulations common in wetland basins throughout the region. These range greatly in size and frequency, with large peat deposits common in extreme eastern Maine and especially in northern Minnesota.

The only unglaciated portion of the Laurentian Mixed Forest occurs in southwestern New York and adjacent portions of Pennsylvania. This area is underlain by residuum in the most stable landscape positions, primarily gently sloping plateau tops. Colluvium dominates steeper slopes, whereas alluvium is found in lower slope positions.

The Laurentian Mixed Forest and its mountain analog, the New England–Adirondack Province, are the only forest provinces in the study region with a predominance of Spodosols (Fig. 2.2 in color insert; Table 2.2). In this province, these are found in a large variety of parent materials and landscape positions. Several other soil orders dominate certain parent materials or more limited portions of the landscape.

Table 2.2. Distribution of Dominant Soil Orders by Ecological Province (Areas in Hectares)

Soil Order	Laurentian Mixed Forest	New England–Adirondack Mixed–Coniferous Forest–Alpine Meadow	Eastern Broadleaf Forest (Oceanic)	Central Appalachian Broadleaf–Coniferous Forest–Meadow	Eastern Broadleaf Forest (Continental)	Southeastern Mixed Forest	Outer Coastal Plain Mixed Forest	Lower Mississippi Riverine Forest	Prairie Parkland (Temperate)
Alfisols	9,940,628	89,007	7,075,665	947,209	33,010,170	26,146	6,838	404,972	7,941,931
Entisols	2,371,920	29,167	1,478,551	216,489	3,589,220	0	772,070	407,287	1,176,208
Histosols	3,533,429	14,490	84,582	0	812,166	0	113,549	0	30,564
Inceptisols	7,788,578	2,238,438	5,638,106	4,726,155	3,212,236	26,177	121,577	18,683	84,500
Mollisols	389,422	0	7,691	0	9,089,137	0	0	0	24,579,910
Rock	47,066	0	0	0	64,544	0	0	0	0
Spodosols	11,466,970	8,365,341	796,962	0	714,190	0	179,717	0	0
Ultisols	1,129,595	0	4,286,735	2,510,427	4,302,733	24,995	22,166,920	2,124	219,431
Vertisols	0	0	0	0	56,990	0	0	209,192	473,069
Water	988,413	201,797	199,586	25,884	514,601	2,242	65,313	12,036	196,694
Total	37,656,021	10,938,240	19,567,877	8,426,163	55,365,988	79,560	3,475,754	1,054,295	34,702,306

Inceptisols are common, especially in wetter portions of the landscape and in finer parent materials. For example, while sandy to sandy loam parent materials are most prevalent, silt loams are common in some northeastern areas where lower metamorphic grade has resulted in large areas underlain by phyllites. Entisols are found in sandy stratified drift and alluvium. In low-lying basins in the cooler and wetter parts of the province, in Maine and Minnesota, large deposits of decaying organic matter accumulate, resulting in formation of large areas of Histosols. Most of the province is in the frigid temperature regime, with cryic soils in the Aroostook Hills of northern Maine and mesic soils at lower elevations of the Allegheny Plateau of New York and Pennsylvania. Soil moisture regime is xeric, udic, or aquic, largely dependent on parent material texture and landscape position.

New England–Adirondack Mixed Forest–Coniferous Forest–Alpine Meadow Province

This mountainous province is characterized by vegetational zonation, primarily with altitude, but also with latitude. Northern hardwood forests dominate lower elevations and latitudes, whereas spruce–fir forests are found at higher elevations and northern reaches of the province. Small areas of alpine tundra are found above timberline on the highest mountains of Maine, New Hampshire, Vermont, and New York. The climate is characterized by an abundance of precipitation, which is evenly spaced throughout the year. Frost-free growing season is the shortest in the northern forest, ranging from 80 to 150 days.

Most of the province is underlain by an extremely complex assortment of igneous and highly deformed metamorphic rocks. A large variety of compositions are represented, including slates, phyllites, schist, gneiss, marble, quartzite, granitic plutons, rhyolite, and amphibolite. In northwestern Maine, metasedimentary rocks are less deformed and at lower grade. Anorthosite, a relatively uncommon lithology on Earth, underlies large areas in the central Adirondacks. The Catskills and Tug Hill Plateau in New York differ from the rest of the province in that bedrock consists of only slightly deformed sandstone, conglomerate, siltstone, and shale.

Because surficial deposits are thin and composed primarily of material of local origin, bedrock has a relatively large influence on forest processes in this province. Where topography and fracture patterns are favorable, large amounts of water may exchange between bedrock and surficial deposits, yielding an influence of bedrock on composition of groundwater and surface water. Whereas bedrock type varies greatly over short distances, the mineralogic composition of glacial deposits may also vary greatly, reflecting differences in directions and efficacy of glacial erosion and deposition, as well as differences in lithology of bedrock source areas (Hornbeck et al., 1997).

Physiography includes glacially scoured, maturely dissected mountains and peneplains with scattered monadnocks. Surficial deposits are predominantly thin, stony till with some stratified drift and lacustrine deposits, primarily in the larger valleys. Spodosols underlie 76% of the province (Fig. 2.2; Table 2.2), a larger proportion than any soil order in any of the provinces. Much of the remainder is underlain by Inceptisols, which are more common further south, at lower elevations, and on finer, less acidic parent materials. Most of the province is in the frigid temperature regime, with cryic soils at the highest elevations and some mesic soils in the larger valleys in the southern portion. Soil moisture regimes are udic and aquic.

Eastern Broadleaf Forest (Oceanic) Province

Eastern Broadleaf Forest provinces cover 48% of the study region. This extensive type is subdivided into three provinces based on topography and climate. The eastern portions have an oceanic influenced climate, whereas a continental influenced climate dominates western and interior portions. In between, in the central Appalachians, lies a mountainous province characterized by altitudinal zonation of vegetation communities.

The Oceanic Eastern Broadleaf province is a transitional forest with northern hardwoods dominant in northern portions and oak–hickory (*Carya* spp.) dominant in the south. Oak–pine types are scattered throughout, especially in areas dominated by Entisols. The mixed mesophytic community is a component of the glaciated portion of the Allegheny Plateau in northwestern Pennsylvania and northeastern Ohio. Abundant precipitation is spaced evenly throughout the year, with winter snowfall ranging from none in the south to 2.5 m in the north. The frost-free growing season ranges from 120 to 250 days.

Bedrock geology is varied and complex in this province, which was heavily influenced by metamorphism, igneous activity, and subsequent sedimentation associated with closing of the Iapetus Ocean and opening of the Atlantic basin. The region from southern New England, south through eastern Pennsylvania and Maryland, is characterized by a highly deformed assemblage of metamorphic rocks with varying amounts of igneous rocks, primarily granites. Sandstones, siltstones, arkose, and basalt fill Mesozoic basins, such as the Connecticut Valley and the Newark Basin. The Hudson Valley includes a variety of siliciclastic rocks and carbonates in the valley, with metasediments and metavolcanics in eastern portions, bordering the Taconic Mountains. Toward the western part of the province, in western Pennsylvania, Ohio, and West Virginia, are slightly deformed sandstones, siltstones, and shales, with lesser amounts of limestone and coal.

There is a broad range of surficial deposits underlying this province, reflecting differing processes of coastal and inland settings, as well as differing glacial history. Approximately the northern half of this province is glaciated, whereas sections in southern Pennsylvania, Maryland, West Virginia, and southeastern Ohio are not. Thin till is the dominant surficial deposit in glaciated portions, although there are large areas of stratified drift, composed of sands and gravels, as well as marine clays, especially in southeastern New England. Glaciolacustrine deposits and alluvium of recent origin dominate the Hudson Valley of New York. Unglaciated portions are underlain by residuum in the most stable landscape positions, primarily gently sloping plateau tops. Colluvium dominates steeper slopes while alluvium is found in lower slope positions.

Dominant soil taxa reflect a diversity of parent materials as well as the broad range in climate and vegetation types (Table 2.2). Inceptisols are dominant in glaciated areas, with Entisols in alluvium and some stratified deposits, and lesser areas of Alfisols in parent materials influenced by carbonates and lacustrine deposits. Ultisols are common in unglaciated portions, especially in more stable landscape positions. Alfisols are found in lower landscape positions and areas influenced by base-rich parent materials. Inceptisols are found in more acidic, steeper portions of the unglaciated Piedmont and Allegheny Plateau. Soil temperature regime throughout the province is predominantly mesic, with udic and aquic moisture regimes.

Central Appalachian Broadleaf–Coniferous Forest–Meadow Province

Forest vegetation varies with altitude in this province. Oak forests and mixed oak–hickory–pine types occupy the lowest elevations. Northern hardwood forests are at moderate elevations, whereas the southern extension of the spruce–fir forest is found at the highest elevations. Precipitation is abundant and well spaced throughout the year. On average, the eastern Valley and Ridge is noticeably drier than western portions, as it lies in the rain shadow of the Allegheny Mountains. Twenty to thirty percent of the annual precipitation falls as snow. The frost-free growing season varies from 120 to 180 days.

Parallel narrow ridges and valleys characterize the eastern portion of the province, yielding to maturely dissected mountains and plateaus of the western and southern portions. Bedrock consists of folded sedimentary rocks, including shale, siltstone, sandstone, chert, limestone, and coal.

The bedrock has an important influence in this province by determining the chemistry, mineralogy, and texture of soil parent materials. Surficial materials are derived from saprolite in this unglaciated region. Residuum is found in the most stable landscape positions, primarily gently sloping plateau tops. Inceptisols are primarily found on steeper slopes and are the most common soil order in the province (Table 2.2). Substantial areas are underlain by Ultisols, primarily in residuum on plateau summits. Lower landscape positions develop Inceptisols on more acidic parent materials and Alfisols in calcareous parent materials. Soil temperature regime is mesic, with udic and aquic moisture regimes.

Continental Eastern Broadleaf Forest Province

Within the Continental Eastern Broadleaf Forest province, there is an east to west transition in vegetation types that corresponds to a longitudinal gradient in moisture excess (precipitation minus evapotranspiration). Northern hardwood forests are found at the eastern end of this province, in New York, northwestern Ohio, and southeastern Michigan. In the central parts of the province, in Michigan, Ohio, and Indiana, there is a transition from northern hardwood forest to oak–hickory forest with some oak savanna and minor bluestem prairie. Continuing west, oak savanna dominates southern Wisconsin with bluestem prairie becoming dominant further west, in the Minnesota portions of the province. The southwestern portions of the province, in southern Illinois and Missouri, are dominated by oak–hickory with some pine, oak–gum (*Liquidambar* spp.)–cypress (*Taxodium* spp.) in lowlands, and increasing proportions of prairie toward the west in Missouri.

Average annual precipitation varies from 635 mm in northern Minnesota to 1270 mm in the southern portions of the province. Moisture excess is less in this province than in other provinces in the subject region, approaching zero in the west. Frost-free growing season varies from 120 days in northern Minnesota to 200 days in southern Missouri.

Bedrock geology may have little direct influence on the forests of this province due to the great thickness of surficial deposits in most regions. However, an important indirect influence is on the mineralogy of surficial deposits. Most of the province is underlain by slightly to undeformed sedimentary rocks, including limestone, dolomite, sandstone, and shale. Igneous rocks, primarily granite, are found in northern Minnesota. The Ozark Highlands of Missouri are underlain by a variety of igneous rocks, including granite, gabbro, rhyolite, and andesite. Metamorphosed volcanics and sediments, associated with the Canadian Shield, are found only in the northern Minnesota portion.

Most of the province is level to gently rolling till plains, with till thickness up to 100 m or more. The till is mantled by loess in portions of the province, especially southern Wisconsin and southern Illinois. A lake plain, underlain by glaciolacustrine deposits from Lake Agassiz is found in northern Minnesota. Outwash deposits, some of great extent, are common, especially in the northern parts of the province. The portions in Missouri and southernmost Illinois, Indiana, and Ohio are unglaciated, maturely dissected plateaus, with steep to rolling hills. Sandstone bluffs are characteristic of the unglaciated portion of southern Illinois. Soils in the unglaciated portion are derived from residuum, colluvium, and alluvium, typical of unglaciated regions. However, owing to the proximity to the southern extent of glaciation, loess deposits are common in the unglaciated portion. In this region, loess is characteristically cherty, especially in Missouri.

With a wide variety of climate, parent materials, vegetation, and age characterizing the soils of this province, it is not surprising that a large variety of soil orders are found, each dominating different conditions. Alfisols are the dominant soil order, with substantial areas underlain by Mollisols. These two orders are characteristic of portions of the province with till or lacustrine deposits influenced by calcareous parent materials. Entisols are found on outwash and alluvial deposits. Ultisols are typical in more mature portions of the unglaciated terrain. Orders of lesser extent in this province include Spodosols, primarily in Michigan, Inceptisols, common in the Erie and Ontario lake plains, and Histosols in northern Minnesota. Soil temperature regime is mesic with a udic moisture regime.

Temperate Prairie Parkland Province

The Temperate Prairie Parkland in the western portion of the subject region represents a transition from the deciduous forests of the east to the open prairie of the Great Plains to the west. Bluestem prairie is the dominant vegetation with interspersed forested tracts. Oak–hickory forests are common along stream channels. Northern floodplain forests are found along a few rivers, notably the Minnesota and Red. Mean annual precipitation varies from 460 to 1015 mm, with minimal excess over potential evapotranspiration. Frost-free growing season varies from 111 to 235 days.

The Prairie Parkland, with the exception of the Osage Plains in southwestern Missouri, is glaciated, level to gently rolling till plain. Bedrock geology includes granite, gneiss, and metavolcanic rocks in Minnesota, with sedimentary rocks, including sandstone, shale, limestone, dolomite, and some coal in the remainder of the province. Till deposits are quite thick, blanketed with loess in some areas, and interspersed with lesser areas of glaciolacustrine and recent alluvial deposits. Loess and residuum are the dominant soil parent materials in the Osage Plains, similar to the unglaciated portions of the Continental Broadleaf Forest.

Mollisols, reflecting the dominance of prairie vegetation, are dominant, with substantial areas of Alfisols. Entisols are found primarily in alluvial positions, whereas Vertisols enter the study region primarily in this province, in westernmost Minnesota. Soil temperature regimes range from frigid in northern Minnesota, to mesic in most of the province, and thermic in southwestern Missouri. Soil moisture regimes include ustic, udic, and aquic.

Southern Ecological Provinces

Three ecological provinces, which together only account for 3% of the study area, are found along the southern border (Table 1.1). These provinces, which include the Coastal Plain Mixed Forest, Southeastern Mixed Forest, and Lower Mississippi Riverine Forest, are of much greater extent to the south. Climate is warm and moist, moderated by the marine influence, with a frost-free growing season ranging from 185 to over 250 days.

The Coastal Plain Mixed Forest, dominated by oak, hickory and pine, is found on Long Island, New York, southern New Jersey and Maryland, and Delaware. With the exception of Long Island, which is composed of sands at the terminal moraine, this province is an unglaciated, flat to weakly dissected alluvial plain interspersed with marine terraces and dunes. Ultisols and Entisols are the dominant soil orders (Table 2.2), with mesic to thermic temperature regimes, and xeric, udic, and aquic moisture regimes. These soils formed in unconsolidated marine silts, sands, and gravel.

The Southeastern Mixed Forest barely extends into the study region in central Maryland. This province, dominated by southern species of oak, hickory, and pine, is rooted in thick saprolite deposits developed from schists, phyllite, and gneiss. Soils include nearly equal proportions of Ultisols, Inceptisols, and Alfisols in the thermic temperature regime. In extreme southeastern Missouri, the Lower Mississippi Riverine Forest also barely extends into the study region. Oak–hickory and southern floodplain forest, composed of oak, gum, and cypress, grow in a dissected alluvial plain of marine and alluvial sediments. Entisols are the dominant soil order, with some Alfisols and Vertisols. Temperature regime is thermic, with udic to aquic moisture regime.

Soils

Although susceptibility to environmental change is not a criterion of soil taxonomic systems, taxonomic units provide a convenient framework for examination of the potential effects of global change issues on the soil resource. The American taxonomic system is based on differences in dominant genetic processes (Soil Survey Staff, 1996). These distinctions result from fundamental differences in parent material age, texture, and composition. These factors, along with climate, topography, and vegetation, result in differences in how water and nutrients move through the soil profile, as well as on the development and distribution of organic matter and cation exchange sites. In sum, these influences and processes result in general differences in nutrient content and flux rates between taxonomic units. Following is a review of this subject by taxonomic order, highlighting potential for environmental change in each.

Alfisols

Accumulation of translocated clay in a subsurface horizon (the argillic horizon) and increasing base saturation in the subsoil are the defining characteristics of an Alfisol. These soils form under many climatic regimes, but are most extensive in humid and subhumid temperate regions on relatively young, stable surfaces. Till, loess, and alluvium are typical parent materials. In humid temperate climates, Alfisols are found on most landscape positions except very steep slopes, alluvial floodplains, and very poorly drained depressions.

A prerequisite to Alfisol development is leaching of carbonates, which act as a flocculent, preventing clay translocation. Braunification, the release of the milder flocculent, iron, results in deposition of clay in the B-horizon. Deposition may also result from depletion of percolating waters as they are soaked up by peds, swelling of voids, slowing of percolating waters, sieve action of clogging fine pores, and by higher base saturation lower in the solum. Conditions for translocation may be relatively rare, occurring only during intense rains following prolonged drought (Buol et al., 1997).

In general, Alfisols are considered to be more highly developed than Inceptisols but less developed and less weathered than Ultisols. Ciolkosz et al. (1989) highlight the apparent anomoly that in central New York, ~12,000-year-old Alfisols are found, which are younger than the ~18,000-year-old Inceptisols of southern New York and northeastern Pennsylvania. Carbonates in glacial deposits have enhanced development of Alfisols compared with the coarser, carbonate-free deposits to the south.

With higher base saturation at depth, Alfisols are generally more nutrient rich than Ultisols. In New York, Cline (1949) considered forest soils to represent a chronolithosequence from unleached Mollisols (Udolls) to moderately leached Alfisols (Udalfs) to most leached Spodosols (Haplorthods). Organic matter loss from clearing or farming may degrade Mollisols to Alfisols. Changes in climatic patterns might alter clay translocation and deposition processes. Acidification of soil waters, resulting from atmospheric acid deposition or intensive forest management might accelerate soil development or even speed the transition of Alfisols toward Ultisols.

Entisols

Entisols, which exhibit the least soil development of any of the soil orders, are characteristic of mountainous and sandy regions. They form on young surfaces in recent deposits such as alluvial floodplains, deltas, and areas of active loess or dune sand deposition. They are also typical of recently exposed surfaces, such as on steep slopes, where mass wasting or other forms of rapid erosion have removed surficial material faster than pedogenic horizons can form. Entisols are also found in older deposits, such as sand dominated by the mineral quartz, where minerals exceptionally resistant to weathering have limited development.

Nutrient content of Entisols may vary widely. Many Entisols have very low cation exchange capacity due to low clay and organic matter content. On the other hand, high cation exchange capacity is typical of alluvial deposits. Thus, the ability of an Entisol to meet nutrient demands of various forest types is largely dependent on the parent material and mode of deposition.

Forest vegetation is important in stabilizing landforms; removal may accelerate erosional processes, resulting in Entisol formation on landscapes that otherwise are characterized by other soil types. Artificial drainage or drying climatic trends may promote oxidation of water-saturated soils, accelerating pedogenic development in floodplains. Forest management may also promote Entisols through changes in vegetative cover. In Wisconsin, conversion of stands from hemlock (*Tsuga* spp.) to aspen resulted in degradation of the spodic horizon in less than one hundred years, converting Spodosols to Entisols (Hole, 1976).

Histosols

Overall, Histosols comprise a small portion of the region's soils (Table 2.2), yet they are very widely distributed. Histosols form where production of organic matter exceeds mineralization, usually under saturated water conditions, which impedes decomposition. They develop in a variety of climates and substrates, although maritime climates and relatively impermeable substrates favor their formation. They are especially typical of glaciated regions, where glacial deposits include depressions and blocked drainage ways, and of the lower coastal plain where high water table and tidal inundation promote development. Although generally confined to depressions and low-lying areas, in extreme eastern Maine, where precipitation is high and frequent fog limits evapotranspiration, Histosols extend above depressional basins and even climb gentle slopes. Also, cooler temperatures and high precipitation at high elevations promotes Histosol development in mountainous regions.

Cation exchange capacity, derived from carboxyl, phenolic, and other functional groups in organic matter, may be quite high in Histosols. Exchange capacity may be quite sensitive to changes in pH due to the variable charge nature of the organic matter. Depending on hydrologic position, some Histosols are located in zones of groundwater discharge, where dissolved and exchangeable cations reflect the weathering regime of underlying unconsolidated mineral deposits or bedrock. At the opposite extreme, Histosols in ombrotrophic bogs may be completely dependent on atmospheric deposition for water inputs, in which case, pore waters and exchangeable cations are likely to be dominated by hydrogen ion.

Decomposition of organic matter is controlled by a number of interrelated factors, such as moisture content, temperature, composition of organic matter, acidity, and microbial activity. Therefore, changes in climate, hydrologic flow patterns, or vegetation communities might be expected to change the nature of Histosols.

Inceptisols

Inceptisols are generally considered to be immature soils, not having formed diagnostic subsurface horizons necessary for other orders. On the other hand, horizonation is too advanced to qualify as Entisols. Subsurface horizons of Inceptisols closely resemble parent material, with some development of structure and deposition of illuvial organic matter to distinguish B-horizons from similar C-horizons. These soils occur on young geomorphic surfaces, steep slopes subject to deposition, churning, and erosion, and depressions where water saturation limits development of spodic or argillic horizons. They are found in a wide range of climates and parent materials. Although many pedogenic processes may be at work, none predominates in an Inceptisol.

In acid parent material, Inceptisols in depressions tend to be more leached, with lower base content and higher exchangeable aluminum than soils in surrounding areas. In base-rich landscapes, Inceptisols in depressions tend to have higher base status than surrounding landscape positions (Buol et al., 1997). In some regions, Inceptisols are on unusually resistant parent materials; low weathering rates limit the amount of clay produced, impeding formation of an argillic horizon. In regions where Inceptisols are interspersed with other soil orders, they are generally more productive than the others.

With a wide range of parent materials, climates, and landscape positions, it is particularly difficult to generalize about the susceptibility of this soil order to environmental change. Base-poor Inceptisols, such as Dystrochrepts may be relatively sensitive to acidification and nutrient depletion, whereas Eutrochrepts may be relatively insensitive. Taxa with high organic matter (Haplumbrepts) and wetter moisture regimes (Aquepts) may be more sensitive to climatic or vegetative changes that alter organic matter decomposition or hydrologic regime.

Mollisols

Mollisols are defined by a deep, dark, relatively fertile A-horizon and higher base saturation in the subsoil. Most are found under grassland vegetation, while forested examples include poorly drained Mollisols of lowland hardwoods and relatively uncommon well-drained examples within the udic moisture regime (Udolls). Severe, relatively dry winters, a moist spring, and droughty summers with occasional thunderstorms and tornadoes typify the climate where Mollisols dominate the landscape.

Large annual inputs of organic matter through root turnover of deep-rooting prairie vegetation, partial decay of litter inputs during the dry growing season, reworking of organic matter by active soil faunal communities, and illuviation of organic colloids are among the processes responsible for development of thick, dark A-horizons. Subsoils are slightly leached, with a high base status, resulting from younger or more base-rich parent materials. Mollisols on older surfaces and in moister climates show deposition of clay coatings on ped faces lower in the B-horizon. At the forest–prairie boundary in Wisconsin, Mollisols occur on topographic positions that favor spread of fire, namely ridgetops and windward slopes (Buol et al., 1997).

Forests were much more widespread in the Midwest until about 5000 years ago, when climate changes brought a great expansion of the prairie and likely also of Mollisols (Ruhe, 1969). As this soil order is dependent on a relatively narrow range of climate and vegetation, changes in Mollisol quality and distribution might be expected in the event of future changes in climate and in response to present and future changes in vegetative cover brought about by agricultural and silvicultural management.

Spodosols

Spodosols are defined by eluviation of organic matter and iron, with or without aluminum from surface horizons and accumulation (illuviation) in the subsoil, the spodic horizon. The variably expressed albic or E-horizon is a zone of accumulation of resistant minerals and insoluble products of decomposition. Spodosols characteristically form in cool, humid climates under forest vegetation. Coniferous trees, notably hemlock are known for their capability of promoting spodic development (Buol et al., 1997). Leaching of carbonates and dominance of exchangeable cations by hydrogen and aluminum in the A-horizon are prerequisite to mobilization of organic matter and, with it, of iron and aluminum. Illuviation of clay may be a precursor to podzolization in finer parent materials.

Throughfall may be the major source of mobile organic matter for mobilization of aluminum and iron (Malcolm and McCracken, 1968). As throughfall quality is highly variable, dependent on species composition, this may explain the relative intensity of Spodosol formation with differing vegetative communities. Removal of hemlock from mixed hemlock–hardwood forests in northern Wisconsin resulted in fading of the spodic horizon (Milfred et al., 1967). The half-life of the spodic horizon after removal of hemlock may be a short as 100 years (Hole, 1975). Management activities that change species composition might result in changes in the intensity of podzolization processes.

In a transect of Spodosols and Inceptisols in Wisconsin and Michigan, Schaetzl and Isard (1996) found that Spodosols dominate the landscape in areas where the snowpack is thickest, limiting soil frost and allowing large snowmelt runoff events to infiltrate the soil. They hypothesize that the bulk of podzolization occurs during the snowmelt period. Thus, climatic changes that affect patterns of snow accumulation and melt might result in changes in Spodosol development.

Ultisols

Worldwide, areas between the limit of glaciation and the equator, in humid temperate to tropical climates, are dominated by Ultisols. Parent materials and landscapes are older than in glaciated areas, resulting in a longer period of weathering and soil development. A long frost-free season and an abundance of rain also contribute to deep weathering by promoting leaching over a long portion of the year. These soils are acidic to great depths, where the parent material is siliceous crystalline or sedimentary rock relatively poor in bases. On older surfaces, even soils developed in carbonate parent materials may be acidic to great depths. Weatherable minerals in the solum have been converted to secondary clays and oxides.

Ultisols are defined by low base saturation at depth, in association with pronounced clay accumulation in subsurface argillic horizons. There is more emphasis of in situ clay formation through advanced weathering in Ultisols as compared with argillic horizons in Alfisols that are more the result of clay translocation. Ciolkosz et al. (1989) emphasize the distinction between parent material Ultisols and genetic Ultisols. Parent material Ultisols have low base status that is inherited through low base parent materials and show less development than genetic Ultisols. As in Alfisols, clay translocation is the dominant process in parent material Ultisols. Parent material Ultisols typify many of the Ultisols found just south of the limit of glaciation, formed in unglaciated parent materials that were affected by severe erosion and frost-churning during periglacial periods. These soils contrast with well-developed Ultisols of the Piedmont, for example, which formed due to longer periods of weathering in older, more stable positions. Clay accumulation in these soils occurs deeper in the profile, in a thicker horizon, and in amounts that cannot be accounted for by translocation alone.

Although extensive leaching leads to a severe loss of bases in Ultisols, the pattern of base concentrations decreasing with depth suggests that biocycling successfully counters the leaching process. These soils of low fertility may be dependent on uninterrupted biocycling by forest vegetation for maintenance of organic matter and nutrient content. Cutting of native forests often leads to a major loss of fertility in Ultisols of tropical climates. To a lesser extent, the same concern holds for temperate region Ultisols.

Vertisols

Vertisols form through seasonal drying of a soil profile that is rich in clay, predominantly 2:1 expanding clays. The resultant seasonal shrink–swell cycles lead to alternating periods during which large vertical cracks characterize the upper portion of the soil profile. Vertisols occur in climates with a dry season of variable timing and intensity. They are characteristically alkaline, resulting from parent materials such as calcareous sediments and basic igneous. Small areas in the northwest and southwest portions of the study area exhibit Vertisols. Overall, this is not an important soil to the northern forest.

Sensitivity to Soil Nutrient Depletion

A particular global change issue involving dynamics of forest soils, that potentially could alter the health and productivity of the northern forest, is base cation nutrient depletion. In recent years, atmospheric acid deposition and intensive forest management practices have led to concerns about potential for nutrient depletion from forest soils (Federer et al., 1989; see Chapter 8). The theoretical basis for these changes is well founded. Empirical studies have documented these processes in laboratory settings. However, extreme spatial variability in forest soil quality and difficulty in maintaining consistency in collection techniques limit the ability of sampling programs to detect temporal changes.

A primary issue in determining the regional extent of this problem is whether mineral weathering rates, which are notoriously difficult to measure in the field, are high enough to replace base cations lost to forest regrowth and leaching (Federer et al., 1989). Mass balance studies, where weathering rates have been relatively well constrained, have documented net depletion of base cations, presumably derived from exchangeable soil pools, at the small watershed scale (Bailey et al., 1996; Likens et al., 1996). Several studies have documented decreases in soil base nutrient pools (Johnson and Todd, 1990; Joslin et al., 1992; Knoepp and Swank, 1994). However the causes and extent of these changes on a landscape basis are subject to great debate. Landscape and regional estimates of the extent of this phenomenon must be based on models of spatial patterns in deposition and soil processes.

Methods to Rate Sensitivity to Nutrient Depletion

The distribution of bedrock lithologies can tell much about the spatial distribution of areas sensitive to nutrient depletion. Table 2.3 lists the rock types found in the study region by generalized lithology. Within each

Table 2.3. General Lithology and Characteristics of Bedrock in the Northern Forest Region and Generalized Influence on the Acidity, Nutrient Content, and Texture of Soil Parent Materials

General Type	Lithology	Characteristics	Acidity	Nutrients	Texture
Sedimentary	Sandstone	Cemented sand particles, typically mostly quartz	Acidic	Low	Coarse
Sedimentary	Pelite	Cemented silt and clay particles	Acidic	Low to moderate	Fine to medium
Sedimentary	Limestone/marble	Cemented calcium and magnesium carbonate particles; marble is recrystallized	Basic	High	Fine to medium
Igneous	Mafic plutonic	Dark-colored rock crystallized from a melt; contains amphiboles, pyroxenes or olivine	Slightly acidic	Moderate	Medium to coarse
Igneous	Granitoid	Light-colored rock crystallized from a melt; composed primarily of quartz and feldspars	Acidic	Low	Coarse
Igneous	Syenite	Light-colored rock crystallized from a melt; composed primarily of feldspars with little or no quartz	Acidic	Low to moderate	Coarse
Igneous	Anorthosite	Light-colored rock crystallized from a melt; composed primarily of plagioclase; may be recrystallized	Slightly acidic	Moderate	Medium to coarse
Metamorphic	Slate	Metamorphic equivalent to pelite; mineral crystals too small to be seen with naked eye	Acidic	Low to moderate	Fine to medium
Metamorphic	Phyllite	Metamorphic equivalent to pelite; micaceous minerals barely large enough to be seen with naked eye	Acidic	Low	Fine to medium
Metamorphic	Mica schist	Coarse-grained metamorphic equivalent to phyllite with muscovite and/or biotite mica	Acidic	Low	Medium to coarse
Metamorphic	Sulfidic schist	Coarse-grained metamorphic equivalent to phyllite with iron sulfides	Very acidic	Low	Medium to coarse
Metamorphic	Calcareous schist	Coarse-grained metamorphic rock with carbonates at lower grades to calcium silicates at higher grades	Neutral	Moderate to high	Medium to coarse
Metamorphic	Metasandstone	Recrystallized sandstones; quartz-rich varieties referred to as quartzite	Acidic	Low	Coarse
Metamorphic	Gneiss	Granitoids and sediments in which minerals have separated into distinct bands during metamorphism	Acidic	Low	Coarse
Metamorphic	Amphibolite	Recrystallized mafic rocks composed primarily of hornblende and plagioclase	Slightly acidic	Moderate to high	Medium to coarse
Metamorphic	Ultramafic	Primarily of iron and magnesium silicates such as talc and serpentine	Basic	Low to high	Fine to medium

Color Plate V

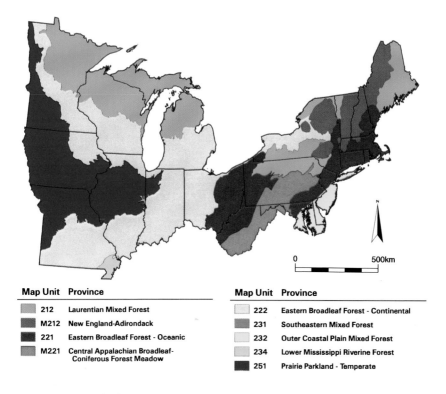

Map Unit	Province		Map Unit	Province
212	Laurentian Mixed Forest		222	Eastern Broadleaf Forest - Continental
M212	New England-Adirondack		231	Southeastern Mixed Forest
221	Eastern Broadleaf Forest - Oceanic		232	Outer Coastal Plain Mixed Forest
M221	Central Appalachian Broadleaf-Coniferous Forest Meadow		234	Lower Mississippi Riverine Forest
			251	Prairie Parkland - Temperate

Figure 2.1. Ecological provinces of the northern forest region (after Keys et al., 1995).

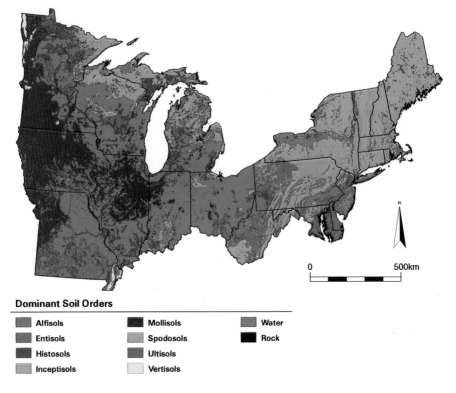

Figure 2.2. Dominant soil orders of the northern forest region (after Quandt and Waltman, 1997).

category, based on typical mineralogy and fabric, the general tendencies of the bedrock's influence on acidity, nutrient content, and texture of soil parent materials can be predicted. Norton (1980) categorized susceptibility to acidification based on generalized bedrock type. While this method is useful for comparing regions, it is of limited usefulness at finer scales. Local variation may be expected due to the transported nature of many parent materials, which may not reflect local bedrock; the highly variable age of landscape surfaces, resulting in parent materials of greatly differing age and degree of alteration from bedrock sources, and the variability in topography, thickness, and permeability of surficial deposits which affect the efficacy of weathering reactions. Models useful for predictions at scales from the landscape to the management unit must account for these factors.

An important limitation to classification schemes based on lithology, which has received relatively little attention, is that variability in mineralogic content of even narrowly defined lithologies may vary greatly. As an example, granites, coarse-grained igneous rocks composed of quartz, feldspar, and lesser amounts of amphibole and mica, are considered to form acidic, coarse-grained parent materials (Table 2.3) with low weathering rates and relative sensitivity to nutrient depletion. The Conway granite, which underlies large areas in central New Hampshire, is a good example. It is composed of quartz, microcline, and biotite (Billings, 1956), relatively slowly weathering silicate minerals (Table 2.4). Soils derived from this rock would be expected to be relatively

Table 2.4. Range of Base Cation Nutrient Content and Weathering Rate for Primary Minerals Commonly Found in Forest Soils. Nutrient Contents Are Midrange Examples for Minerals that Exhibit Solid Solution. Weathering Rates, Based on Laboratory Studies at pH 5, in Log Units (keq[m^2s]$^{-1}$), Were Compiled by Sverdrup and Warfinge (1995)

Mineral	Ca (%)	Mg (%)	K (%)	Weathering Rate
Calcite	40			−8 to −10
Dolomite	22	13		−8 to −10
Olivine		14		−12 to −13.5
Garnet	8	5		−12 to −13.5
Diopside	19	11		−12 to −13.5
Chlorite		11		−12 to −13.5
Epidote	18			−12 to −13.5
Plagioclase	7			−13.5 to −15
Hypersthene		11		−13.5 to −15
Augite	17	5		−13.5 to −15
Hornblende	5	3		−13.5 to −15
Actinolite	9	5		−13.5 to −15
Serpentine		26		−13.5 to −15
Biotite		8	8	−13.5 to −15
Microcline			14	−15 to −16
Muscovite			10	−15 to −16

poor in base cations and sensitive to acidification, true to the general prediction. On the other hand, granitic rocks of the nearby Highlandcroft Pluton typically contain about 25% plagioclase, 8% hornblende, and 3% calcite (Billings, 1956), faster weathering silicates and a very fast weathering carbonate (Table 2.4). Forested sites influenced by this rock might be relatively base-rich and well buffered from acidification.

Advancement of models to determine sensitivity was made by Warfinge and Sverdrup (1992) with development of PROFILE, a soil chemistry model which takes into account major nutrient-cycling processes and uses information about soil mineralogy to estimate weathering rates. This method has been used to map critical loads of atmospheric pollutants relative to soil and surface water acidification in several European countries. It has also been used to evaluate discrepancies between lab-based and field estimates of soil weathering rates (Sverdrup and Warfinge, 1995). A disadvantage of this method is its extensive site-specific data requirements. Additionally, soil mineral content is modeled based on a bulk element analysis. This does not take into account variations in mineral chemistry, inherent in soil parent materials, which may be critical in determining weathering stoichiometry and rates. Many minerals are solid solutions of two to four end-members; weathering rates can vary by several orders of magnitude depending on chemical composition (Table 2.4) (White and Brantley, 1995).

Potential for the Next Generation of Models

Opportunities to improve models of susceptibility to nutrient depletion will come through incorporation of a better understanding of spatial patterns of mineralogic composition of soil parent materials and better knowledge of mechanisms and locations where weathering reactions occur in the landscape. Extensive soil surveys to determine spatial patterns in mineralogy are extremely expensive, while estimates of mineralogic composition based on bulk element composition may be misleading. However, mineralogic composition of bedrock formations is often relatively uniform within map units. Furthermore, there is a wealth of data on bedrock chemical and mineralogic composition in the geologic literature. By better understanding how bedrock sources are sampled and incorporated by surficial deposits, we might be able to best predict spatial pattern in parent material composition. One effort using this approach is in progress for soils developed in glacial till (Hornbeck et al., 1997).

Further progress might be accomplished by development of methods which consider the properties of a larger portion of the soil than is traditionally studied. Soil mineralogy and chemistry protocols are performed on only the fine-earth fraction of the soil—that portion that passes a 2 mm sieve. These techniques have largely been developed for study of agricultural soils, which, due to the limitations of cultural practices, are centered mostly in relatively rock-free portions of the landscape. In contrast, forested soils, typically located on steeper upland portions of the landscape unsuitable for agriculture, frequently contain high amounts of rock fragments, termed "skeleton".

A further reason for ignoring these coarse fragments is that the rates of many reactions, including weathering, are dependent on surface area. Thus, fine particles are considered to be more reactive than coarse particles. For example, the specific surface area of colloidal clay ranges from about 10 to 1000 $m^2 g^{-1}$ compared with 0.1 $m^2 g^{-1}$ for fine sand. The specific surface area of gravel and pebbles, even where they make up the majority of a soil, is negligible. Yet, Bailey and Hornbeck (1992) found that pebbles in forest soils derived from glacial till contained highly weathered interiors. Fractures and zones of secondary alteration allow water to percolate into seemingly impervious pebbles. Ugolini et al. (1996) found that the skeleton of forest soils in Tuscany, derived from sandstone, contained more weatherable minerals than the fine-earth fraction, and contributed significantly to the soils' overall cation exchange capacity as well as available and total nutrient content for a number of nutrients, including nitrogen.

Yet another reason why specific surface area may not be important is that biologically mediated weathering may not differentiate between particle size, rather focusing on nutrient content of minerals and their susceptibiltiy to direct weathering by exuded organic acids. April and Keller (1990) documented extensive alteration of primary minerals in the rhizosphere compared with bulk soil. Minerals in direct contact with roots were most highly altered. Jongmans et al. (1997) documented direct weathering of feldspar and hornblende grains by mycorrhizal hyphae.

Finally, a better understanding of the role of bedrock weathering must be incorporated into models. In many regions, as stated previously in the review of ecological provinces, reactions in bedrock may influence forest nutrient cycles. For example, in areas of crystalline bedrock, such as granite–schist terrain in central New Hampshire, effective hydraulic conductivity of bedrock may be in the same order of magnitude as that of glacial till. Local zones of concentrated bedrock fractures may have conductivity a few orders of magnitude higher than till (Tiedeman et al., 1997). In this case, in landscape positions where groundwater flows from bedrock into surficial deposits, weathering products derived from mineral decomposition within bedrock may be delivered to soil and the rooting zone of the forest. Johnson and Todd (1998) found that nutrient flux measurements overpredicted calcium depletion from forest harvesting for an unglaciated site on Ultisols developed in calcareous bedrock. They suggest that deep rooting and bedrock weathering could account for this discrepancy.

References

April R, Keller D (1990) Mineralogy of the rhizosphere in forest soils of the eastern United States. Biogeochem 9:1–18.

Bailey SW, Hornbeck JW (1992) *Lithologic Composition and Rock Weathering Potential of Forested Glacial–Till Soils.* USDA Forest Service, Northeastern Forest Experiment Station, Radnor, PA.

Bailey SW, Hornbeck JW, Driscoll CT, Gaudette HE (1996) Calcium inputs and transport in a base-poor forest ecosystem as interpreted by Sr isotopes. Water Resour Res 32:707–719.

Billings MP (1956) *The Geology of New Hampshire Part II—Bedrock Geology.* Concord: New Hampshire State Planning and Development Commission, Concord, NH.

Buol SW, Hole FD, McCracken RJ, Southard RJ (1997) *Soil Genesis and Classification.* 4th ed. Iowa State University Press, Ames, IA.

Ciolkosz EJ, Waltman WJ, Simpson TW, Dobos RR (1989) Distribution and genesis of soils of the northeastern United States. Geomorph 2:285–302.

Cline MG (1949) Profile studies of normal soils of New York. I. Soil profile sequences involving Brown Forest, Gray–Brown Podzolic, and Brown Podzolic soils. Soil Sci 68:259–272.

Federer CA, Hornbeck JW, Tritton LM, Martin CW, Pierce RS, Smith CT (1989) Long-term depletion of calcium and other nutrients in eastern US Forests. Environ Mgt 13:593–601.

Hole FD (1975) Some relationships between forest vegetation and Podzol B horizons in soils of Menominee tribal lands, Wisconsin, USA. Soviet Soil Sci 7:714–723.

Hole FD (1976) *Soils of Wisconsin.* Soil Series 62. Wis Geol Nat Hist Surv Bull 87, University of Wisconsin Press, Madison, WI.

Hornbeck JW, Bailey SW, Buso DC, Shanley JB (1997) Streamwater chemistry and nutrient budgets for forested watersheds in New England: variability and management implications. Forest Ecol Mgt 93:73–89.

Jenny H (1941) *Factors of Soil Formation: A System of Quantitative Pedology.* McGraw-Hill, New York.

Johnson DW, Todd DE (1990) Nutrient cycling in forests of Walker Branch Watershed, Tennessee: roles of uptake and leaching in causing soil changes. J Environ Qual 19:97–104.

Johnson DW, Todd DE (1998) Harvesting effects on long-term changes in nutrient pools of mixed oak forests. Soil Sci Soc Am J 62:1725–1735.

Jongmans AG, van Breeman N, Lundstrom U, van Hees PAW, Finlay RD, Srinivasan M, Unestam T, Giesler R, Melkerud PA, Olsson M (1997) Rock-eating fungi. Nature 389:682–683.

Joslin JD, Kelly JM, van Miegroet H (1992) Soil chemistry and nutrition of North American spruce–fir stands: evidence for recent change. J Environ Qual 21: 12–30.

Keys JE, Carpenter C, Hooks S, Koenig F, McNab WH, Russell WE, Smith ML (1995) Ecological Units of the Eastern United States—First Approximation. 1:3,500,000. USDA Forest Service, Atlanta, GA.

Knoepp JD, Swank WT (1994) Long-term soil chemistry changes in aggrading forest ecosystems. Soil Sci Soc Am J 58:325–331.

Kuchler AW (1964) *Potential Natural Vegetation of the Conterminous United States.* American Geographic Society, New York.

Likens GE, Driscoll CT, Buso DC, Siccama TG, Johnson CE, Lovett GM, Fahey TJ, Reiners WA, Ryan DF, Martin CW, Bailey SW (1996) The biogeochemistry of calcium at Hubbard Brook. Biogeochem 41:89–173.

Malcolm RL, McCracken RJ (1968) Canopy drip: a source of mobile soil organic matter for mobilzation of iron and aluminum. Soil Sci Soc Am Proc 32: 834–838.

McNab WH, Avers PE (eds) (1994) *Ecological Subregions of the United States: Section Descriptions.* USDA Forest Service, Washington, DC.

Milfred CJ, Olson GW, Hole FD (1967) *Soil Resources and Forest Ecology of Menominee County, Wisconsin.* Soil Series 60. Univ Wis Geol Nat Hist Surv Bull 85. University of Wisconsin Press, Madison, WI.

Norton SA (1980) Geologic factors controlling the sensitivity of aquatic ecosystems to acidic precipitation. In: Shriner DS, Richmond CR, Lindberg SE (eds) *Atmospheric Sulfur Deposition—Environmental Impact and Health Effects.* Ann Arbor Science Publishers, Ann Arbor, MI, 521–531.

Quandt L, Waltman SW (1998) *Dominant Soil Orders for the US.* USDA Natural Resources Conservation Service, National Soil Survey Center, Lincoln, NE.

Ruhe RV (1969) *Quaternary Landscapes in Iowa.* Iowa State University Press, Ames, IA.

Schaetzl RJ, Isard SA (1996) Regional-scale relationships between climate and strength of podzolization in the Great Lakes Region, North America. Catena 28:47–69.

Soil Survey Staff (1996) *Keys to Soil Taxonomy.* 7th ed. USDA Natural Resources Conservation Commission, Washington, DC.

Sverdrup H, Warfinge P (1995) Estimating field weathering rates using laboratory kinetics. In: White AF, Brantley SL (eds) *Chemical Weathering Rates of Silicate Minerals.* Reviews in Mineralogy, Vol. 31. Mineralogical Society of America, Washington, DC, 485–541.

Tiedeman CR, Goode DJ, Hseih PA (1997) *Numerical Simulation of Ground-water Flow Through Glacial Deposits and Crystalline Bedrock in the Mirror Lake Area, Grafton County, New Hampshire.* USGS Prof Paper 1572. Denver, CO.

Ugolini FC, Corti G, Agnelli A, Piccardi F (1996) Mineralogical, physical, and chemical properties of rock fragments in soil. Soil Sci 161:521–542.

Warfinge P, Sverdrup H (1992) Calculating critical loads of acid deposition with Profile—a steady state soil chemistry model. Water Air Soil Poll 63:119–143.

White AF, Brantley SL (eds) (1995) Chemical Weathering Rates of Silicate Minerals Reviews in Mineralogy, Vol. 31. Mineralogical Society of America. Washington, DC.

3. Climate and Atmospheric Deposition Patterns and Trends

Warren E. Heilman, John Hom, and Brian E. Potter

One of the most important factors impacting terrestrial and aquatic ecosystems is the atmospheric environment. Climatic and weather events play a significant role in governing the natural processes that occur in these ecosystems. The current characteristics of the vast number of ecosystems that cover the northeast and north central United States are, in part, the result of climate, weather, disturbance, and atmospheric pollution patterns that exist in the northeast and north central United States. For example, basic ecosystem processes (e.g., heat and moisture exchanges with the atmosphere, photosynthesis, and respiration) along with species diversity and ecosystem health throughout the region all depend, to some degree, on these patterns. Furthermore, future characteristics of ecosystems in the region will depend on future climate, weather, disturbance, and pollution patterns that may develop in response to natural or human-caused changes in our atmospheric environment.

This chapter provides an introduction to a number of climate principles and an overview of some of the current (baseline) patterns of climate, climate variability, extreme weather events, atmospheric pollution, and deposition of atmospheric pollutants that characterize the region. In particular, baseline patterns of temperature, extreme temperature events, spring freezes, midwinter thaw events, precipitation, extreme precipitation events, flood and drought occurrences, short-term precipitation variability, extreme wind events, hurricane occurrence, and fire-weather are presented. Atmospheric pollution and deposition discussions focus on spatial and temporal trends of ozone pollution, and nitrogen, sulfur, and acidic deposition patterns in the region. The list of climate-, weather-, and pollution-related variables relevant to ecosystems in the region and covered in this chapter is not exhaustive. However, the list does include some of the more important climate and pollution variables that play a significant role in influencing basic ecosystem processes, species diversity, and ecosystem health in the region. In addition, an overview of potential climate changes that may occur in the region based on projections from current state-of-the-art climate models is also presented.

Definition of Climate

Weather at a particular location or over a particular region is usually highly variable, and the physical processes inherent in weather events are characterized by spatial and temporal scales that are much smaller than the spatial and temporal scales associated with climatic processes. McIntosh (1972) defines climate at a locality as the synthesis of the day-to-day values of the meteorological variables that affect the locality, where synthesis is more than just simple averaging. In other words, climate can be considered the synthesis of weather. The main meteorological variables used to define climate include precipitation, temperature, humidity, sunshine, wind speed, and wind direction. Associated with these meteorological variables are climate and weather events such as frost, extreme precipitation episodes, storms, hurricanes, tornadoes, drought, extreme maximum and minimum temperature episodes, and short- and long-term precipitation and temperature variability characteristics that also define the overall climate of a region. All of these climate and weather events that can impact a locality or region are influenced by surface characteristics such as land usage, vegetation coverage, land–water variations, lake and sea ice, and topography. The different patterns of climate- and weather-related variables that characterize a particular region are manifestations of both the large-scale state of the atmosphere and the small-scale, complex, atmospheric–surface interactions that occur continuously and everywhere.

Description of Climatic Forcing Factors

Nearly all the exchange of energy between the Earth and space is accomplished through radiative transfer processes (Wallace and Hobbs, 1977). When averaged over long periods of time, the amount of radiation absorbed by the Earth and its atmosphere is nearly equal to the amount of radiation emitted. The energy source for driving the Earth's climate is the Sun (Trenberth et al., 1996), and the Earth–atmosphere system is very nearly in radiative equilibrium with the Sun (Wallace and Hobbs, 1977). The energy received from the Sun is spread mostly across the electro-magnetic spectrum from infrared to ultraviolet wavelengths, with the bulk of the energy found in the visible portion of the spectrum (wavelengths between 0.39 and 0.76 μm). On average, the flux of energy at the top of the Earth's atmosphere is 342 W m^{-2}. About 31% of this incoming energy is reflected back to space due to back-scattering by aerosol particles and reflection by clouds (77 W m^{-2}), and reflection by the Earth's surface (30 W m^{-2}). The reflection of energy back to space leaves about 235 W m^{-2} for warming the Earth–atmosphere system. A complete description of the Earth energy balance can be found in Kiehl and Trenberth (1997).

The energy available for warming the Earth–atmosphere system (about 235 W m^{-2}) is eventually radiated back to space in the form of infrared radiation. Of the 235 W m^{-2} of energy available for warming the Earth–atmosphere system, about 40 W m^{-2} of energy is re-emitted by the Earth's surface and moves relatively unimpeded through the atmosphere into space. The rest of the energy can mainly be attributed to (1) absorbed shortwave solar radiation by the atmosphere that is emitted as longwave radiation to space and (2) infrared radiation emitted from the Earth's surface that is absorbed by the atmosphere and re-emitted to space. However, the atmosphere also emits longwave radiation back to the Earth's surface, thereby acting to warm the Earth's surface and produce average surface temperatures much higher than what would be predicted from blackbody radiation principles alone. Although 99% of the Earth's atmosphere is composed of nitrogen (N) and oxygen (O), these gases do not absorb infrared radiation. However, water vapor, which makes up between 0 and 2% of the Earth's atmosphere, carbon dioxide (CO_2), and other trace gases in the atmosphere are important absorbers and emitters of infrared radiation. These atmospheric constituents are referred to as greenhouse gases because they act as a partial blanket for the thermal radiation emitted from the surface and produce substantially warmer surface conditions than if no greenhouse gases were present (Trenberth et al., 1996).

Atmospheric CO_2 concentrations have increased by more than 25% in the past century due in large part to the combustion of fossil fuels and the removal of forests (Trenberth et al., 1996). According to the Intergov-ernmental Panel on Climate Change (IPCC), projections point to a future rate of increase of CO_2 concentrations in the atmosphere such that concentrations will double from preindustrial levels within the next 50 to 100 years (Houghton et al., 1994). It has been estimated that an increase of this magnitude could lead to an average global warming of the Earth's surface on the order of 2.5°C (Houghton et al., 1990), although the estimates of future warming are still under debate by the scientific community. Furthermore, other greenhouse gas concentrations in the atmosphere besides CO_2 are also observed to be increasing. Human activities such as biomass burning, landfill development, rice-paddy development, agricultural practices, animal husbandry, and industry contribute to enhanced atmospheric concentrations of methane (CH_4), nitrous oxide (N_2O), and tropospheric ozone (O_3). These gases tend to reinforce the changes in radiative forcing from increased CO_2 concentrations. Schimel et al. (1996) provide a summary of the known concentration trends and radiative forcing impacts of these gases.

Human activities have also affected aerosol concentrations in the atmosphere. Aerosol particles are responsible for the scattering of solar radiation back to space, which tends to cool the Earth's surface. Some aerosol particles can also absorb solar radiation, and thus increase local temperatures. Finally, many aerosol particles can act as nuclei for the formation of cloud droplets, thereby influencing cloud formation on a regional basis and the reflection and absorption of solar and infrared radiation. Each year, human activities lead to more than 350 Tg of aerosols in the atmosphere (Schimel et al., 1996). These activities include industrial emissions of dust, soot emissions from fossil fuel burning, soot emissions from biomass combustion, sulfur dioxide (SO_2) emissions (gaseous precursor of sulfate aerosols) from power stations and biomass combustion, and nitrogen oxide (NO_x) emissions (gaseous precursor of nitrate aerosols) from fossil fuel combustion. Because the residence time of aerosols in the atmosphere is short compared with many greenhouse gases, their impact is generally regional in scale (Trenberth et al., 1996; Schimel et al., 1996). The overall cooling effect of aerosols in the atmosphere can mask the warming effect of increased atmospheric CO_2 concentrations, especially on regional scales. However, their presence does not cancel the global-scale warming effects of greenhouse gases that reside in the atmosphere for long periods of time (Trenberth et al., 1996).

The presence of greenhouse gases and aerosols in the atmosphere produce large-scale changes in the radiative forcing over the entire Earth–atmosphere system. These large-scale changes in radiative forcing have the potential for altering atmospheric and oceanic dynamics. On an annual mean basis, the tropics receive more incoming solar radiation than what is emitted as longwave radiation, while the midlatitudes and high latitudes emit more longwave radiation than the shortwave solar radiation received. The net temperature gradients from the equator to the poles that exist because of this global radiation pattern produce large-scale atmospheric and oceanic circulation patterns that redistribute heat from the equator to the poles. The band of atmospheric westerlies that exists over the midlatitudes results from these temperature gradients and the Earth's Coriolis force. Within the band of westerlies, flow instabilities can develop (i.e., baroclinic instability [Holton, 1979]) that ultimately evolve into cyclones and anticyclones and migrate within the band of westerlies. We observe migrating cyclones and anticyclones as part of our day-to-day weather experience. In addition to the large-scale radiative forcing changes and associated temperature changes that occur from enhanced greenhouse gas concentrations in the atmosphere, there exists the potential for induced changes in the patterns of cyclone and anticyclone development from enhanced greenhouse gas concentrations. Potential climate change on regional scales, such as over the northeast and north central United States, will depend to a large extent on the development and behavior of weather systems in response to the large-scale radiative forcing changes caused by elevated greenhouse gas concentrations.

Regional Climate vs. Global Climate

Climatic conditions over any region, including the northeast and north central United States, are a manifestation of both large-scale climate processes (e.g., global atmospheric circulation patterns, oceanic circulation patterns, global radiative forcing in the presence of greenhouse gases) and smaller, regional-scale climate processes that depend more on specific Earth–atmosphere interactions (e.g., topographic impacts on temperature, precipitation, and wind; land–water variations; land usage; urbanization; vegetation variations). The high degree of uncertainty in determining potential climate changes over specific regions of the Earth is due, in part, to the extreme complexity of Earth–atmosphere interactions and an inability to account for critical small-scale surface–atmospheric interactions in climate models due to computational limitations. For this reason, climate model projections such as the typical 2- to 3-°C increase in global mean surface temperatures due to elevated greenhouse gas concentrations should not be interpreted to mean that average surface temperatures over the region will also increase by 2 to 3°C. The specific surface characteristics within the region and the associated complexities of surface–atmosphere interactions and weather system development and evolution in the region affect the region's climate, and will continue to do so in the future.

Spatial and Temporal Temperature Patterns

Species composition and ecosystem structure in the northeast and north central United States and elsewhere are influenced by the atmospheric environment. This environment limits the types of species or organisms that can thrive and the amounts of plant tissues that can be sustained (Melillo et al., 1996). Descriptions of spatial and temporal patterns of near-surface maximum and minimum temperatures are critical in any assessment of the atmospheric environment and its impact on species composition and structure. The spatial and temporal patterns of temperature in the north central and northeastern United States presented in this section are based on kriged National Climate Data Center (NCDC) daily maximum and minimum temperature data from observation sites within the Cooperative Observer Network over the period 1950 to 1993 (EarthInfo, 1995). The spatial interpolation scheme of kriging does not explicitly account for small-scale elevation variations that may exist between observation sites, which can have a major impact on diurnal temperature trends. For this reason, only the broad patterns of temperature variations are considered and discussed in this section. Any analyses and discussions of local temperature variations require the use of interpolation or modeling schemes that account for the small-scale topographic variations that exist between observation sites.

Average Daily Maximum Temperature Patterns

Average daily maximum temperature patterns over the northeast and north central United States reflect not only latitudinal variations over the region, but variations in land–water coverage, elevation, and longitude as well. Fig. 3.1 (color insert) shows the average January, April, July, and October daily maximum temperatures over the region. During the late fall and winter months, the lowest average daily maximum temperatures in the region can typically be found over the northwestern sections of the region (Minnesota and northern Wisconsin) where cold Canadian air masses moving southward have a strong influence on temperatures (Fig. 3.1a). The Great Lakes tend to moderate the southward- and eastward-moving Canadian air masses (Weisberg, 1981) so that daily maximum winter temperatures over the central and eastern Great Lakes states are somewhat higher than over the western Great Lakes states. The highest maximum temperatures during the winter months can be found over the southern tier of states in the region, with the eastern half of Virginia typically having the warmest conditions in the region. The consistently low maximum temperatures indicated over northern New Hampshire during the winter months as well as throughout the year are due to the cold conditions on Mt. Washington, where a temperature reporting station is located. The use of temperature data from this station in the kriging interpolation scheme tends to reduce interpolated maximum and minimum (discussed in the next section) temperature values in the area immediately surrounding Mt. Washington. However, the spatial influence of Mt. Washington's low maximum and minimum temperatures on the interpolated maximum and minimum temperature values in the areas surrounding Mt. Washington is minimized because the number of other reporting stations in Maine, New Hampshire, Vermont, and Massachusetts that were used to develop the kriged pattern of temperatures in the 4-state area was relatively large (50). Furthermore, only broad regional patterns are being considered in this discussion.

During the spring months, temperatures increase over the northern sections of the region, but the predominant north–south temperature gradient over the region persists (Fig. 3.1b). Unlike the winter months, when eastern Virginia typically has the warmest conditions over the entire region, the spring months usher in a more uniform longitudinal maximum temperature distribution over the southern tier of states in the region. Topographic influences on the regional-scale temperature patterns are evident in the spring months, particularly over the Appalachian Mountains in West Virginia. During the spring months, when gradually warmer conditions begin to characterize the northern sections of the region, the relatively cold Great Lakes water temperatures tend to reduce average daily maximum temperatures in the vicinity of the Great Lakes. In Michigan, for example, average daily maximum temperatures over some of the interior sections of the lower peninsula are about 2°C higher than over some areas close to the shore of Lake Michigan. The cold water in Lake Michigan and Lake Superior also reduces the maximum temperatures observed in northeastern Minnesota and the upper peninsula of Michigan.

By July, the highest maximum temperatures in the region are found in Missouri, with parts of Missouri experiencing average daily maximum temperatures near 32°C (Fig. 3.1c). Average maximum temperatures decrease from west to east over the southern tier of states, with a relative minimum in maximum temperatures persisting over eastern West Virginia and western Virginia during the summer months. Average maximum temperatures increase again from the Appalachian Mountains in these states to the Atlantic coast. The coolest sections of the region during the summer months are northeastern Minnesota, northern Wisconsin, northern Michigan, north central Pennsylvania, northeastern New York, northern New Hampshire, and northern and eastern Maine. These areas of relatively low maximum temperatures are the result of Great Lakes cooling effects, high elevations in the Appalachian Plateau and Adirondack Mountains, and normal latitudinal variations.

The fall months are characterized by a return of relatively cold conditions over the northwestern and northeastern sections of the region (Minnesota, Wisconsin, and Maine) and a maximum in temperatures over eastern Virginia (Fig. 3.1d). As air temperatures over the northern sections of the region fall below water temperatures in the Great Lakes, the modifying effects of the Great Lakes on maximum daily temperatures become more pronounced, especially over Michigan and near Lake Erie and Lake Ontario.

Average Daily Minimum Temperature Patterns

Average daily minimum temperature patterns over the northeast and north central United States are very similar to the average daily maximum patterns. Fig. 3.2 (color insert) shows the average daily minimum temperature patterns over the region for the months of January, April, July, and October. As with the maximum temperatures during the late fall and winter months, minimum temperatures are lowest over northern Minnesota and Wisconsin (Fig. 3.2a). Northern Maine also has average daily minimum temperatures comparable to those observed in Minnesota and Wisconsin. The relatively warm water in Lake Michigan, Lake Huron, and Lake Erie during the winter months keeps the observed average minimum temperatures over much of Michigan (especially near the Lake shores) higher than at similar latitudes in the states of Minnesota and Wisconsin. Over the southern half of the region, there is a general increase in average daily minimum temperatures with decreasing latitude. However, the presence of the Appalachian Mountains in West Virginia and western Virginia leads to slightly lower average minimum temperatures in these areas compared with similar latitudes in Kentucky and eastern Virginia.

The spring and summer months of April to September are characterized by smaller overall minimum temperature variations over the entire region. For example, average daily minimum temperatures in April range from $-9°C$ in northern New Hampshire to about $10°C$ in southeastern Missouri (Fig. 3.2b), compared with January average daily minimum temperatures ranging from about $-24°C$ in northwestern Minnesota to about $0°C$ in eastern Virginia (Fig. 3.2a). In July, average minimum temperatures range from about $21°C$ in southeastern Missouri and eastern Virginia to about $6°C$ in northern New Hampshire (Fig. 3.2c). The spring and summer seasons are also characterized by an increase in topographic impacts on minimum temperature gradients near the southern Appalachian Mountains. A significant west-to-east minimum temperature change is clearly evident from central Kentucky through southern and western Virginia into eastern Virginia (Fig. 3.2b,c).

During the fall months of October to December, the average minimum temperature variations over the region increase over what is observed during the late spring and summer months. In October (Fig. 3.2d), average daily minimum temperatures range from about $11°C$ in eastern Virginia to about $-3°C$ in northern New Hampshire. By December, average minimum temperatures range from about $0.5°C$ to about $-20°C$ over the same area. From October to December, the impact of the Great Lakes on average minimum temperatures is pronounced. Across the northern half of the northeast and north central United States, average minimum temperatures are lowest over the northwestern and northeastern sections of the region. The north central portion of the region (including the state of Michigan, northern Ohio, and western New York) typically experiences minimum temperatures that are higher than at similar latitudes in the northwestern and northeastern sections of the region. The relatively warm water in Lake Michigan, Lake Huron, and Lake Erie during the fall months tends to keep nighttime temperatures higher in the north central portion of the region than in the northwestern and northeastern sections. The warm water also tends to modify the temperatures of cold air masses that move southward into the region from Canada during the fall months. The highest minimum temperatures in the region during the fall months are typically found in southeastern Missouri, Kentucky, eastern Virginia, eastern Maryland, and Delaware.

Surface Temperature Trends

The IPCC has reported that annual, globally averaged, surface air temperature anomalies relative to average temperatures measured between 1961 and 1990 generally increased from about $-0.5°C$ in 1890 to about $0.3°C$ in 1990 (Houghton et al., 1966; based on analyses of Jones, 1994). On regional scales, however, annual and seasonal temperature trends are quite variable across the globe. Over North America, the most significant observed temperature changes over the last 40 years have occurred from the north central U.S. (including the western Great Lakes area) through the northwestern sections of Canada into Alaska. Annual average surface temperatures from the period 1955 to 1974, to the period 1975 to 1994 have increased from 0.25 to 1.5°C over this region (Jones, 1994). Over the far northeastern sections of the northeast and north central United States, annual average surface temperatures have increased by about 0.25°C over the same two periods. Other sections of the region have experienced annual average surface temperature increases of less than 0.25°C.

On a seasonal basis, annual average surface temperature differences between the 1955 to 1974 and 1975 to 1994 periods over the region were largest during the winter (December to February) and spring (March to May) seasons (Jones, 1994). During the winter and spring months, the northwestern sections of the region experienced seasonally averaged temperature increases on the order of 0.25 to 0.75°C over the two periods. The far northeastern sections of the region also experienced increases on the order of 0.25 to 0.5°C. The only section of the region that experienced an overall decrease in seasonally averaged surface temperatures during the winter and spring months was the southern portion of the region extending from Illinois and Indiana southward. Over this section of the region, average wintertime surface temperatures decreased on the order of 0 to 0.25°C. During the summer and fall months, average temperature differences from the 1955 to 1974 period to the 1975 to 1994 period were generally smaller both globally and regionally. Surface temperatures averaged about 0.25 to 0.5°C warmer during the summer months (June to August) for the period 1975 to 1994 compared with the 1955 to 1974 period for most of the region. The fall season (September to November) was the only season characterized by lower average surface temperatures during the 1975 to 1994 period compared with the 1955 to 1974 period over the region; average fall temperatures were about 0.25 to 0.5°C lower during the latter period.

Diurnal Temperature Range Trends

Several recent studies indicate that worldwide diurnal temperature ranges have decreased since 1950 (Horton, 1995; Karl et al., 1993b; Houghton et al., 1992). The decrease in diurnal temperature range can be attributed to a worldwide increase in minimum land–surface air temperatures. Karl et al. (1993b) found that minimum land–surface air temperature increases have been about twice the magnitude of maximum temperature increases. Over most of the continental U.S. and most of the northeast and north central United States, overall diurnal temperature ranges from 1981 to 1990 relative to the 1951 to 1980 values have decreased on the order of 0 to 0.5°C (Houghton et al., 1996). The observed decreases in diurnal temperature ranges over this period have been attributed to increases in cloud cover (Plantico et al., 1990; Henderson-Sellers, 1992; Dessens and Bücher, 1995; Jones, 1995). The diurnal temperature range decreases have been observed in both rural and urban areas, and thus, cannot be attributed to urban heat island effects.

Extreme Maximum Temperature Occurrences

Occurrences of extreme maximum temperatures vary considerably over the northeast and north central United States. The far western sections of the region are often influenced by continental air masses that can be very cold or warm, and minimally affected by bodies of water that tend to moderate their temperatures. Sections of the region further east are more often influenced by northward moving maritime air masses from the Gulf of Mexico that are usually warm and moist but produce fewer temperature extremes than continental air masses. The variations in extreme maximum temperature occurrences over the region based on 1950 to 1993 daily maximum temperature observations are shown in Fig. 3.3 (color insert). In this discussion, "extreme" is defined as a daily maximum temperature for a particular month greater than or equal to 11.1°C (20°F) above or below the average daily maximum temperature for that month. Although other definitions of "extreme" are possible and even statistically significant or biologically relevant, the application of the 11.1°C difference threshold across the entire region yields a general characterization of the broad baseline patterns of very high and low daily maximum temperature occurrences in the region. It is one of many ways to assess baseline maximum temperature variability in the region. Portions of northwestern and southwestern Minnesota, central and western Iowa, and eastern Missouri experienced more than 500 occurrences from 1950 to 1993 of daily maximum temperatures greater than or equal to 11.1°C above the average daily maximum temperature for a particular month (Fig. 3.3a). This corresponds to an average of more than 11 extreme-warm events each year. The southern and Atlantic coastal sections of the region along with areas near Lake Superior and Lake Michigan experienced the fewest extreme-warm events over the 44-year period, averaging about 3 to 4 events each year. The remaining areas in the region averaged between 7 and 10 extreme-warm events per year over the 1950 to 1993 period.

It is during the period from October through May that most extremely high daily maximum temperature events occur in the region; such events are infrequent during the summer months. Beginning in October, the likelihood of extremely high maximum temperatures typically increases over western Minnesota and northern Wisconsin, and the upper peninsula of Michigan. About one event each October can be expected in these areas. Less than one event every two years during the month of October can be expected over the southern half of the region. During the months of November and December, more extremely high maximum temperature events begin to occur over portions of Iowa, eastern Missouri, and the northern Ohio River Valley. Each year, these areas typically experience 1 to 2 days during each of these months when the 11.1°C difference threshold is exceeded for daily maximum temperatures. Areas in the vicinity of Lake Superior, Lake Michigan, and the Atlantic coast tend to have fewer episodes of extremely high maximum temperatures during the winter months due to the moderating influence of these bodies of water on warm air masses that move into these regions. In January, most extremely high maximum temperature episodes tend to occur over the extreme northwestern portion of Minnesota, eastern Missouri, and the Ohio River Valley (~2 to 3 events each January). In February and March, the area of relatively frequent extreme-warm events (~2 to 3 events each month) shifts northward somewhat to encompass the region from southern Iowa and northern Missouri eastward to western Pennsylvania. By April, high maximum temperature episodes are rare in the southern-tier states of the region. Western Minnesota, northern Wisconsin, northern Michigan, central Pennsylvania, and western New York tend to experience more extreme events (~2 events) than any other portion of the region during this month. The overall impact of the month-to-month variations in high maximum temperature occurrences over the region over the 44-year period from 1950 to 1993 leads to the total extreme-event pattern shown in Fig. 3.3a.

Fig. 3.3b shows the total number of occurrences of extremely low daily maximum temperatures (≥11.1°C below the average daily maximum temperature for a particular month) over the region from 1950 to 1993. As with the extremely high maximum temperature event, this type of extreme temperature event is more likely to occur over the far western sections of the region where continental air masses play a significant role in daily temperature trends. Between 11 and 15 episodes of extremely low maximum temperatures can be expected each year over western Minnesota, western Iowa, and western Missouri. This contrasts with most of Michigan and the Atlantic coastal region where fewer than 5 episodes typically occur each year. Extremely low maximum daily temperatures are more likely to occur during the fall, winter, and spring seasons than during the summer, although these types of episodes during the summer are more common than extremely high summertime maximum temperature episodes over the region.

Extreme Minimum Temperature Occurrences

The influence of continental air masses on extreme minimum temperature occurrences is similar to their influence on extreme maximum temperature occurrences. Fig. 3.4 (color insert) shows the distribution of extreme minimum temperature occurrences across the northeast and north central United States based on 1950 to 1993 minimum temperature observations and using a similar definition of extreme for minimum temperatures as that used for maximum temperatures (daily minimum temperatures $\geq 11.1°C$ above or below the average daily minimum temperature for a particular month). Most anomalously high and low minimum temperature episodes tend to occur in the northern sections of Minnesota and Wisconsin ($\sim$12 to 16 events each year). Parts of northern New York, Vermont, and New Hampshire also tend to experience more extreme minimum temperatures than other parts of the region. Fig. 3.4a suggests that anomalously high minimum temperatures are relatively infrequent in the Atlantic coastal states from Virginia to southern Maine, and in Michigan ($\sim$1 to 5 events each year). The typical number of yearly high minimum temperature events decreases from northern Minnesota to southern Iowa, northern Missouri, and central Illinois, but increases again over the southwestern and south central sections of the region. Analyses of the monthly patterns of anomalously high minimum temperature occurrences indicate that the maxima in yearly extreme-event occurrences over northern Minnesota, Wisconsin, New York, Vermont, and New Hampshire, as shown in Fig. 3.4a, are wintertime (December to March) events. During the month of April, anomalously high daily minimum temperatures become much less frequent over northern Minnesota; most extreme events of this type during the month of April occur in Michigan ($\sim$1 event each April). Following the summertime period when anomalously high minimum temperature events are rare over the entire region, extreme events tend to become more frequent over the western sections of the region in October ($\sim$1 event each October) and over the southern sections of the region in November ($\sim$1 to 2 events each November).

The pattern of anomalously low minimum daily temperature occurrences over the region is shown in Fig. 3.4b. The spatial variations in occurrences of this type of extreme temperature event are very similar to the variations in the anomalously high minimum temperature events shown in Fig. 3.4a. Northern Minnesota and Wisconsin, and parts of northern Vermont and New Hampshire experienced between 600 and 800 extremely low minimum temperature events over the period 1950 to 1993 (~13 to 18 events each year). Large bodies of water such as Lake Michigan, Lake Huron, Lake Erie, and the Atlantic Ocean reduce the probabilities of extremely cold conditions in western and eastern Michigan, western New York, and along the East Coast from Massachusetts to Virginia. The vast majority of extreme events of this type occur during the months of November to March over the region. For example, in the month of January, parts of northern Minnesota, northern Wisconsin, and northern New York typically experience 3 to 4 days when minimum temperatures exceed 11.1°C below normal. Other sections of the region, excluding the Atlantic coastal states from Virginia northward to Massachusetts, typically experience about two days each January when minimum daily temperatures exceed the defined extreme threshold. The Atlantic coastal section of the region typically experiences one extreme minimum January temperature event each year or every two years.

Late Spring Freeze Occurrences

The annual timing of temperature extremes is as important as the magnitude of the extremes. Freezes that occur late in the spring, after trees have begun to flush, can damage established trees, destroy seed crops, and kill regeneration. Fig. 3.5 shows an example of the effect of such freezes on shoot lengths for oak (*Quercus* spp.) and ash (*Fraxinus* spp.) seedlings in northern Wisconsin. A freeze was observed on calendar day 146 in research plots at the Willow Springs Oak Regeneration Study (Zasada et al., 1994) on the Chequamegon National Forest near Rhinelander, Wisconsin, in 1994 that killed the new growth on ash seedlings but not on oak, while a freeze on day 152 killed oak shoots. In both freezes, temperatures dropped to about −5°C at a height of 0.25 m above the ground. These observations demonstrate the impact of late spring freezes, and the variations in impact between tree species.

Fig. 3.6a is a map depicting average temperatures for late spring freezes following 250 growing degree days (GDDs), base 5°C. Fig. 3.6b is a similar map, for freezes following 300 GDDs (see color insert). Both maps show kriged data derived from NCDC records of daily maximum and minimum temperatures from Cooperative Observer Network stations for the period 1961 through 1990 (EarthInfo, 1995). The effects of proximity to large bodies of water are apparent from the maps. Regions close to

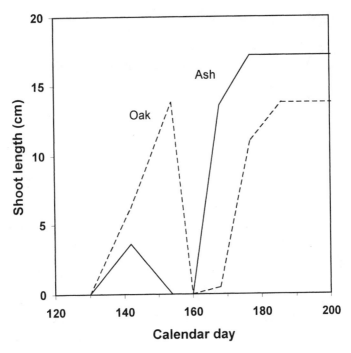

Figure 3.5. Measured oak and ash shoot lengths at the Willow Springs Oak Regeneration Study site on the Chequamegon National Forest near Rhinelander, WI from 30 April 1994 (calendar day 120) to 19 July 1994 (calendar day 200).

water are inclined to have milder freezes. Some coastal regions of New England, and the area around Sheboygan, Wisconsin, have no freezes after reaching 250 GDDs.

This lack of freezes following 250 or 300 GDDs along coastal regions (white areas in Figs. 3.6a and 3.6b) is somewhat deceptive. It does not mean that these areas are especially warm, or that they experience abrupt transitions from below- to above-freezing temperatures. Rather, they are areas that tend to warm slowly in the spring. Inland locations may reach 250 GDDs by large steps on a few days, while coastal locations do so in small steps over many days. This can be seen by comparing Fond du Lac and Sheboygan, Wisconsin. Both stations are at roughly the same latitude, yet Fond Du Lac reaches 250 GDDs on calendar day 125, while Sheboygan doesn't reach it until day 130; this 5-day difference widens to 7 days if one considers the 300-GDDs dates.

Andresen and Harman (1994) examined trends in springtime freezes in the western lower peninsula of Michigan. They found that stations in this area are reaching GDD thresholds earlier in the year than they have in the past, and that there is a trend toward more freezes occurring after a given threshold as a result. Such a trend could lead to decreased regeneration of freeze-sensitive tree and plant species, as they are subjected to more freeze events over the years.

Midwinter Thaw Occurrences

At the opposite end of the spectrum from freezes during warm seasons are warm spells during cold seasons, that is, periods of thawing that occur in the middle of winter. Previous studies have indicated that midwinter thaws play a significant role in red spruce injury in the northeast and north central United States (DeHayes et al., 1990; DeHayes, 1992; Friedland et al., 1984). For the present discussion, a midwinter thaw is a period between December and March when the temperature rises above 0°C for at least one day, followed by the temperature dropping below −16°C at some time during the next three days. The use of these temperature thresholds is based on observed episodes associated with tree dehardening and subsequent tissue damage in the northern Great Lakes region. Fig. 3.7 (color insert) shows the distribution of the number of such thaw–freeze events occurring from December 1961 through March 1990, a total of 29 complete winters, based on NCDC records of daily maximum and minimum temperature observations from Cooperative Observer Network stations (EarthInfo, 1995). Thaw–freeze episodes of this nature are most frequent in high-elevation regions of New England, where they average two or three per year. Areas of Iowa, Minnesota, Missouri, and Wisconsin experience about one event each year, while the remainder of the region experiences less than one event per year.

Spatial and Temporal Precipitation Patterns

As with the spatial and temporal patterns of temperature in the northeast and north central United States, any assessment of the atmospheric environment and its impact on species composition and structure must include descriptions of precipitation patterns. Moisture influences a variety of fundamental ecosystem processes, including C gain through photosynthesis and C loss through respiration (Melillo et al., 1996). Species distributions across the region are, in part, the result of characteristic precipitation patterns that occur. Extreme precipitation events or drought episodes can have significant impacts on ecosystem health and modify landscapes in the region. There are also social and economic implications of the typical and anomalous precipitation patterns that occur in the region (e.g., human responses to severe flooding or drought).

The spatial and temporal precipitation patterns discussed in the following sections are based on kriged daily precipitation data from the NCDC for the years 1950 to 1993 (EarthInfo, 1995). Because local elevation changes can influence precipitation amounts (Daly et al., 1994; Groisman and Easterling, 1994), only the broad precipitation patterns over the region are assessed; local topographic influences on local precipitation amounts at locations between reporting stations are not explicitly accounted for in the kriging interpolation scheme. As shown in the figures in the following sections, the spatial influence of precipitation observations at individual reporting stations (e.g., Mt. Washington in New Hampshire) on the kriged precipitation patterns in the vicinity of the stations is minimized because of the relatively large number of reporting stations in each state. Again, only the broad patterns of precipitation are assessed in this discussion; sporadic and isolated precipitation maxima and minima that appear across the region are not the focus of the discussions in the following sections.

Average Monthly Precipitation Patterns

Precipitation patterns over the northeast and north central United States are highly variable. Fig. 3.8 (color insert) shows the average monthly precipitation amounts over the region for the months of January, April, July, and October. During the winter months, monthly precipitation amounts generally increase from northwest to southeast over the western half of the region (Fig. 3.8a). Average monthly precipitation amounts are generally less than 38 mm over Minnesota, Wisconsin, Iowa, and the northern sections of Illinois and Missouri during January and February, while maximum winter precipitation between 76 to 127 mm occurs over Kentucky, West Virginia, the southern portions of Illinois, Indiana, and Ohio, and along the Atlantic coast. The Great Lakes exert a strong influence on wintertime precipitation over the states of Michigan, northeastern Ohio, western Pennsylvania, and western New York. Relatively warm water during the winter tends to increase atmospheric instability over the Great Lakes, thereby increasing the probability of lake-effect snow events to the south of Lake Superior, east of Lake Michigan, and east of Lake Erie and Lake Ontario (Paulson et al., 1991). The Appalachian Mountains also exert an influence on wintertime precipitation. Precipitation amounts are larger along the western slopes of the Appalachian Mountains in winter, especially in West Virginia and western Pennsylvania. Drier conditions prevail near the eastern slopes of the mountains, especially in central Virginia, Maryland, Pennsylvania, and New York. Average monthly winter precipitation amounts increase from the eastern slopes of the Appalachian Mountains to the Atlantic coast.

During the spring months, precipitation amounts increase over the northwestern sections of the region (Fig. 3.8b), with an accompanying westward and northward shift of the wintertime precipitation maximum over Kentucky to the western sections of Missouri, and the states of Iowa, eastern Minnesota, western Wisconsin, and northern Illinois by June. Precipitation maxima and minima persist over the western and eastern slopes of the Appalachian Mountains, respectively, during the spring months. The increased heating of the Earth's surface during the spring months increases the probability of atmospheric convective activity and thunderstorm development over the western sections of the region. Warmer air masses during the spring months are able to hold more moisture so that precipitation events associated with the passage of frontal boundaries in the region can be more significant than winter precipitation events. The Great Lakes exert an opposite influence on precipitation amounts than that which occurs during the winter months. Great Lakes water temperatures tend to be lower than adjacent land temperatures during the spring and summer months. This inhibits atmospheric convective activity over the Great Lakes and leads to diminished precipitation amounts in Michigan and western New York by late spring. The precipitation maxima along the Atlantic coast also disappear by late spring. By June, the driest conditions in the region typically occur over the northeastern sections of Michigan and most of Virginia.

In July, central Michigan is typically the driest section in the region due to the influence of Lake Michigan (Fig. 3.8c). Relatively dry conditions also tend to prevail over western New York and the eastern slopes of the Appalachian Mountains. The region of maximum precipitation covers a broad area from the western slopes of the Appalachian Mountains (West Virginia, western Pennsylvania, and eastern Kentucky) through the Ohio River Valley to Illinois, northern Missouri, Iowa, Wisconsin, and eastern Minnesota. Maxima also occur in eastern Pennsylvania, New Jersey, eastern Maryland, Delaware, and northern New Hampshire. During the midsummer to late summer months, the driest conditions in the region occur over the northern Ohio River Valley, western Minnesota, and the eastern slopes of the Appalachian Mountains. By September, the largest precipitation amounts occur in northern Wisconsin, western Missouri, northeastern New York, eastern Pennsylvania, and northern New Hampshire.

During the fall months, precipitation maxima return to the southern and Atlantic coast states in the region (Fig. 3.8d). Relatively dry conditions reappear over the northwestern sections of the region (Minnesota, Wisconsin, and Iowa). The fall months also bring a return to enhanced precipitation over western Michigan and parts of western New York in response to the relatively warm water temperatures in Lake Michigan and Lake Erie. The accumulated effect of the average monthly precipitation patterns over the region is shown in Fig. 3.9 (color insert), which depicts the average yearly precipitation amounts over the region. The largest yearly precipitation amounts extend in a band from western Pennsylvania through West Virginia and Kentucky into southeastern Missouri. The year-around influence of the Appalachian Mountains produces a "precipitation shadow" effect along the eastern slopes of the mountains. Yearly precipitation amounts increase as you move eastward from this relatively dry region to the Atlantic coast. On a yearly basis, Minnesota, northwestern Iowa, and eastern Michigan are the driest sections in the region.

Recent Precipitation Trends

There have been numerous studies performed recently that have examined various precipitation trends over large geographical regions. Diaz and Quayle (1980) performed time series analyses of U.S. precipitation data weighted by area and found relatively large and statistically significant increases in contemporary (1955 to 1977) mean autumn precipitation over the 1895 to 1920 mean autumn precipitation for the states of New York, Pennsylvania, West Virginia, Maryland, and Virginia. More recently, Groisman and Easterling (1994) analyzed precipitation data from Canada and the United States over the past 100 years and found that annual precipitation has increased in southern Canada to the south of 55°N by 13% and increased by 4% in the contiguous United States. They determined that these precipitation increases can be attributed mainly to precipitation increases that have occurred in eastern Canada and the adjacent northern regions of the United States, which includes states within the region.

Karl et al. (1993a) found a 2 to 3% per decade increase in annual precipitation over the contiguous United States in the past four decades. They also found that the interannual variability of the ratio of solid to total precipitation increased substantially during the 1980s. In analyzing snow-cover trends over North America, they identified several temperature-sensitive snow-cover regions where area-averaged maximum temperature, snow cover, and snowfall are highly correlated with each other. The western sections of the region, including the states of Indiana, Illinois, Wisconsin, Minnesota, and Iowa fall within a December to March temperature-sensitive snow-cover region, while the New England states fall within the boundaries of an April to May temperature-sensitive snow-cover region. Karl et al. (1993a) found after analyzing 19 years of North American snow-cover climatology that decreases in snow-cover extent occur simultaneously with increases in North American temperatures, and that a global warming of 0.5°C could result in roughly a 10% decrease in the mean annual North American snow-cover extent. The previously noted states within the region would likely experience this potential snow-cover retreat.

Vining and Griffiths (1985) examined the precipitation variability from decade to decade at 10 stations located across the U.S. for the period 1890 to 1979. Three stations in their analysis were within the region: Minneapolis, Ann Arbor, and New York City. Regression lines of decadal precipitation variances indicate a slight increase in precipitation variability over the period of record for most of the United States. The observed changes in variances for the three stations in the region as measured by regression coefficient magnitudes were larger than the U.S. average, although none of the trends was statistically significant.

Nicholls et al. (1996) reported global trends in precipitation over the last half of this century as well as changes over the entire century using two data sets: "Hulme" (Hulme, 1991; Hulme et al., 1994) and the "Global Historical Climate Network" (Vose et al., 1992; Eischeid et al., 1995). Increases in precipitation from the period 1955 to 1974 to the period 1975 to 1994 have occurred over most continents with the notable exceptions of central Africa and west central South America, where decreases as high as 50% have been observed. Precipitation increases at the higher latitudes are clearly evident. Average precipitation increases on the order of 10% have occurred over the north central and northeastern United States during this period. Over the period 1900 to 1994, the north central and northeastern United States experienced an average increase in precipitation on the order of 2 to 5% per decade.

Karl and Knight (1998) examined 20th century trends of precipitation amount, frequency, and intensity across the United States. Their analyses indicate that precipitation has increased by about 10% across the contiguous United States and that the increase can be attributed mainly to increases in heavy and extreme daily precipitation events (upper 10% of all daily precipitation amounts). Within the region, this upward trend is most evident over states in the northwestern part of the region (Minnesota, Wisconsin, Michigan, and Iowa) and over the New England states in the spring, summer, and autumn seasons.

Extreme Precipitation Occurrences

The occurrence of heavy or extreme precipitation events is also an important climatic factor that impacts the natural, social, and economic resources of the northeast and north central United States. Heavy or extreme precipitation events are generally associated with thunderstorm activity or hurricane movement into the northeastern region of the United States. However, prolonged or frequent periods of lesser amounts of precipitation can also adversely impact resources in the region. The extreme flooding in the eastern Dakotas and western Minnesota during the spring of 1997 was the result of an unusually large number of snowstorms and blizzards that cumulatively produced record or near-record snowfall amounts for the winter season. As one indicator of the spatial variability of extreme precipitation over the region, Fig. 3.10 (color insert) shows the number of extreme precipitation occurrences during the months of January, April, July, and October over the period 1950 to 1993, where extreme is defined as daily events that result in at least 5.08 cm of liquid precipitation. Data for Fig. 3.10 were obtained from NCDC daily precipitation observations from stations in the Cooperative Observer Network within the region (EarthInfo, 1995). Although this definition of extreme is one of many that could be utilized and it biases the analysis toward liquid precipitation events, the analysis does provide a regional assessment of one form of extreme precipitation occurrence that influences basic ecosystem processes, particularly during growing seasons.

As expected in this type of analysis, extreme daily precipitation events are generally confined to the southern and Atlantic coastal sections of the region during the winter months (Fig. 3.10a). Over the 44-year period, about 10 to 15 extreme events have occurred during January, February, and March individually over parts of Kentucky and the southern sections of Illinois, Missouri, and Indiana. Connecticut, Massachusetts, and the Mt. Washington region of New Hampshire also experienced relatively large numbers of extreme wintertime precipitation events over the 44-year period. Extreme events in most of the other sections of the region were fairly rare.

During the months of April, May, and June, more extreme precipitation events tend to occur in the western and northwestern sections of the region. In early spring (Fig. 3.10b), the area of increased probability of extreme precipitation expands northward from Kentucky and southern Missouri to encompass much of Illinois and Iowa. By June, more extreme precipitation events tend to occur in northern Missouri, Iowa, and Minnesota, as opposed to the wintertime pattern in which most of the extreme events occurred in Kentucky and southeastern Missouri. The northward and westward propagation of this extreme precipitation maximum is the result of springtime thunderstorm development over the central United States. Much of the northern sections of the northern-tier states (Minnesota, Michigan, New York, Vermont, and Maine) in the region are characterized by infrequent extreme precipitation events during the spring. West Virginia has also experienced relatively few springtime extreme precipitation events over the 44-year period. During the month of June, there is a general decrease in frequency of extreme events as you move eastward from the states of Minnesota, Iowa, and Missouri, to the Ohio River Valley, and then a general increase in the frequency of extreme events as the Atlantic coast is approached. However, the nature of the extreme events in the eastern sections of the region in June is quite spotty in comparison with the large area of maximum extreme events in the western sections of the region.

The summer months of July, August, and September typically bring on more frequent extreme precipitation events along the Atlantic coast from Virginia to Massachusetts. Hurricane occurrence and the potential for hurricane landfall along the eastern U.S. coast are greatest during the late summer and early autumn months (NOAA National Hurricane Center, 1999). Numerous extreme precipitation and flooding events associated with hurricanes tracking along the eastern U.S. coast or making landfall have been recorded during the 20th century (Paulson et al., 1991) and contribute to this overall pattern of extreme precipitation occurrence in the region. It is noteworthy that in the state of Virginia, average summertime precipitation is relatively low in comparison with adjacent states (see Fig. 3.8c) and yet a relative maximum in extreme precipitation events exists over a significant portion of the state. The springtime maximum in extreme events over the western sections of the region is also prevalent in the summer months (Fig. 3.10c), although by September, less frequent extreme events characterize Minnesota and northwestern Iowa. In July, Michigan, New York, northern Vermont, and Maine are characterized by few extreme events in comparison with other states in the region. In August and September, Ohio shows a decline in the number of extreme events over the number occurring in July.

The spatial patterns of extreme precipitation occurrence for the fall months of October, November, and December reveal a transition from the summertime to wintertime patterns. With the onset of colder temperatures in the northern sections of the region, extreme precipitation events become much less frequent. By October (Fig. 3.10d), extreme precipitation maxima in the western sections of the region are confined to western Missouri, while in the eastern part of the region, maxima cover the Atlantic coastal states of Virginia, Delaware, New Jersey, Connecticut, Rhode Island, Massachusetts, and southern Maine. Between these sections of extreme precipitation maxima, few extreme events have occurred, particularly in West Virginia, Ohio, Michigan, Wisconsin, Iowa, and Minnesota. In November and December, there is a return to the wintertime pattern discussed previously, in which most prevalent extreme precipitation occurrences are in Kentucky and the New England coastal sections.

Floods

Floods are the result of weather phenomena that deliver more precipitation to a drainage basin than can be stored or absorbed by the basin (Hirschboeck, 1991). Included in these weather phenomena are convective thunderstorms, hurricanes, and frontal passages. The sources of atmospheric moisture for these weather phenomena are mainly oceans and lakes. Hirschboeck (1991) presented an overview of the primary large-scale moisture delivery pathways over the United States. For atmospheric moisture moving over the conterminous United States, the primary air-mass source regions are the Pacific Ocean, the Atlantic Ocean, the Gulf of Mexico, and the Arctic Ocean. The Great Lakes are also a source of moisture for air masses moving over the north central and northeastern United States. There is a seasonal dependence of atmospheric moisture transport from the source regions to the different sections of the United States, which in turn affects the average precipitation, streamflow, and flooding characteristics across the United States. During the winter months, most moisture delivery pathways over the region originate from air masses situated over the north central and northeastern United States. The southeastern portions of the region are also typically influenced by moisture originating from the Gulf of Mexico. During the spring months, the northward transport of Gulf moisture intensifies, leading to much of the southern and eastern sections of the region being influenced by weather phenomena containing moisture from the Gulf of Mexico. The northern-tier states in the region are typically influenced by the southward transport of moisture from the Arctic Ocean region in early spring, although the cold air masses originating over the Arctic Ocean region have much lower moisture contents than the Gulf air masses. During the summer, the entire eastern half of the United States is dominated by precipitation events in which the moisture has been transported northward from the Gulf of Mexico. The average precipitable water associated with these summertime air masses originating over the Gulf of Mexico is significant. The fall months in the region are also dominated by Gulf-influenced precipitation events. However, moisture originating from the Pacific Ocean may be a factor in some precipitation events over the region, particularly over the western Great Lakes section.

The occurrence of significant precipitation and associated floods depends on appropriate atmospheric uplifting mechanisms that lead to condensation and cloud formation. Convective processes such as thunderstorms, mesoscale convective systems, and convection due to orographic lifting can lead to precipitation events that cause local or widespread flooding. Thunderstorms are primarily responsible for flash flooding in small drainage basins (Hirschboeck, 1991) and their occurrence over the region is seasonally dependent. The average number of days during which thunderstorms develop ranges from less than one over the northern-tier states in the region during the winter months to between 10 and 25 days over the entire region during the summer months (Hirschboeck, 1991).

Unlike typical short-duration thunderstorms that have an areal extent less than $100 \, \text{km}^2$ and the potential for mainly localized flooding, longer-lasting mesoscale convective complexes (MCCs) typically have areal extents up to $200,000 \, \text{km}^2$ and can cause severe flash flooding (Bosart and Sanders, 1981; Maddox, 1983). Mesoscale convective complexes are defined as large, highly organized, multiple-celled, and convectively induced thunderstorm systems lasting longer than 6 hours with cloud shields having an area greater than or equal to $100,000 \, \text{km}^2$, infrared cloud-shield temperatures less than or equal to $-32°C$, and interior cloud region temperatures less than or equal to $-52°C$. Almost 1 of every 4 MCCs results in injuries or fatalities, and they can produce other weather phenomena, such as tornadoes, hail, high winds, and intense electrical storms. Numerous MCCs have been documented over the midwestern

United States, mainly during the spring and summer months. During the spring months, most complexes tend to occur over the south central United States and the southwestern sections of the region. During the summer months, MCCs are prevalent over the Great Plains and the western sections of the region. The summertime pattern of MCCs is consistent with spatial patterns of summertime extreme precipitation occurrences as shown in Fig. 3.10c. The fall months are characterized by many fewer MCCs, although the fall MCCs that have been previously documented tend to occur mainly over the western Great Lakes section of the region.

Another type of convective system that can produce massive amounts of precipitation and significant flooding is the mesoscale convective system (MCS). Chappell (1986) defines an MCS as a multicell storm or group of interacting storms that have organized features, such as a squall line or a cluster of thunderstorms. Slow-moving MCSs have been responsible for some of the most severe localized flooding on record (Hirschboeck, 1991). For example, the July 1977 floods in western Pennsylvania were the result of a quasi-stationary MCS.

The movement of extratropical cyclones through the region also has the potential for causing major flooding in large drainage basins. Unlike many convective thunderstorms that develop within a single air mass, precipitation associated with extratropical cyclones is dependent on increased atmospheric instability due to the convergence of different air masses at frontal boundaries. There are seasonally dependent preferred tracks of extratropical cyclone movement over the conterminous United States, and many of these tracks directly impact the region. In the northeastern United States, significant winter precipitation can result from extratropical cyclones originating over the eastern Rocky Mountain region and western Canada, as well as cyclones that originate along the Atlantic coast and track northeastward along the coast, sometimes creating coastal storm surges. In the spring months, extratropical cyclone paths tend to shift slightly northward. The increased heating during the spring months compared with wintertime heating leads to greater density variations in converging air masses associated with extratropical cyclones. This fact in addition to more precipitable water being available in warmer air masses is the reason springtime extratropical cyclones usually result in much larger precipitation amounts than wintertime extratropical cyclones. Saturated soils, frozen ground surfaces, and concurrent snowmelt during the passage of extratropical cyclones over the region in the spring can increase the flooding potential. During the summer, cyclone paths are mainly confined to the northern states and Canada. Precipitable water contents are large during the summer months, enhancing the potential for flooding in the region. Extratropical cyclone paths in the fall shift slightly southward from the preferred summertime paths, and flooding is somewhat less common because of typically lower precipitable water contents in the cooler fall months.

Precipitation associated with orographic lifting is an important atmospheric process in certain sections of the region. Air masses originating over the Gulf of Mexico or the Atlantic Ocean that are forced up the slopes of the mountainous terrain in the eastern United States can produce significant precipitation and local flash floods. Frequent orographic lifting of air masses during winter in the Appalachian Mountain section of the region can lead to significant accumulations of snow. Widespread flooding during spring snowmelt is possible in this region.

Because many of the states within the region experience significant snowfall during the winter and spring months, flooding can be a direct consequence of the melting of accumulated snow. Flooding can be made more adverse when the ground is frozen or when rain falls on snow, thereby enhancing snowmelt. Over most of the region, winter and spring flooding is most often the result of rain from extratropical cyclones falling on frozen ground covered with snow in conjunction with ongoing snowmelting. Most of Minnesota, northwestern Wisconsin, the western part of the upper peninsula of Michigan, and the northern half of Maine average more than four months of frozen ground per year, while average snow depths in these regions and over much of Michigan, Pennsylvania, part of West Virginia, New York, and the rest of the New England states exceed 101 cm per year (Hirschboeck, 1991).

The northeast and north central U.S. region has experienced numerous floods this century. A compiled listing of some of the major floods that have occurred in each state within the region can be found in Paulson et al. (1991).

Droughts

Many definitions of drought have been proposed, although most definitions refer to abnormal dryness (McNab and Karl, 1991). Mather (1974) defined drought as a phenomenon that occurs when the supply of moisture from precipitation or that stored in the soil is insufficient to fulfill the optimum water needs of plants. Changnon (1987) points out that the definitions of drought are dependent on specific components of the hydrologic cycle, including precipitation, surface runoff, soil-moisture storage, streamflow conditions, and groundwater availability. Precipitation deficits are usually the first indicators of drought occurrence, while streamflow and groundwater levels often respond to these precipitation deficits much later and are usually the last indicators of drought occurrence.

Precipitation deficits and associated droughts that periodically occur in different regions of the country are manifestations of the atmosphere's large-scale general circulation (Namias, 1983). There are specific circulation patterns that can lead to prolonged periods of below-normal precipitation over certain regions of the United States. The atmospheric mechanism responsible for most drought episodes is persistent subsidence of air. This air warms during subsidence, and the relative humidity of the air is low. Over the Great Plains, including the western states in the region (Minnesota, Iowa, and Missouri), summer drought is most often associated with a deep warm anticyclone situated over the central United States. Westerlies along the northern U.S. border often bring additional dry subsiding air into the anticyclone system (Namias, 1983). This atmospheric circulation pattern results in high surface temperatures over the Great Plains and low relative humidity throughout the lower troposphere.

Droughts over the eastern sections of the region are typically caused by different types of atmospheric circulation patterns. Namias (1983) noted that drought in the northeastern United States can occur during a persistent northward displacement of the jet stream, which can lead to sinking motion south of the jet stream over the northeastern United States. Another drought-producing mechanism identified by Namias (1983) in the northeastern United States. is increased cyclonic activity off the northeastern U.S. coast. Temperatures are typically cooler than normal during drought periods associated with this circulation pattern. A third type of circulation pattern that often leads to reduced precipitation over the eastern half of the region is characterized by the westward propagation of the Bermuda high pressure system into the southeastern United States. This westward shift can result in dry conditions over the southeastern and northeastern U.S. because the northward transport of moisture from the Gulf of Mexico by southerly winds to the west of the high pressure system is too far to the west for precipitation to occur in these regions. This particular pattern has also been identified by Heilman (1995) as being conducive to severe wildfire occurrence in the southeastern and northeastern United States.

One particularly useful measure of drought severity is the Palmer Drought Severity Index (PDSI) developed by Palmer (1965). This index takes into account the different degrees of dryness required for drought to occur in regions having different average precipitation amounts, where the criterion for drought is a "deviation from normal experience" (Mather, 1974). Drought severity as measured by the PDSI is based on the numerical values of the index. Positive values of the PDSI indicate an excess of soil moisture, values of 0 to −0.5 are considered near normal, −0.5 to −1 indicate an incipient drought, −1 to −2 indicate a mild drought, −2 to −3 indicate a moderate drought, −3 to −4 indicate a severe drought, and −4 or below indicate an extreme drought. The National Drought Mitigation Center (1997) has provided decade-long analyses of severe or extreme drought occurrence (PDSI ≤ -3) in each climate division across the U.S. for much of the 20th century. In the 1940s, drought occurrence in the northeast and north central United States was relatively rare, with severe or extreme drought only affecting the western and far northeastern sections of the region less than 20% of the time. Western Montana, southern Arizona, and southwestern Kentucky were the driest parts of the country during this decade. The western sections of the region experienced more drought conditions in the 1950s. Climate divisions in southern Iowa and western Missouri experienced severe or extreme drought conditions between 30 and 50% of the time in the 1950s. This drought pattern was part of a larger drought pattern that characterized most of the southern Great Plains and parts of the Rocky Mountain states. In the 1960s, more climate divisions in the northern and northeastern sections of the region experienced drought conditions. Climate divisions in southern Wisconsin, northeastern Illinois, southern Michigan, northwestern Ohio, and most of the New England states experienced drought conditions between 20 and 30% of the time. Drought conditions during the 1970s were not common throughout the region. Only a few climate divisions in the northwestern sections of the region experienced drought conditions more than 10% of the time. The 1980s were also characterized by infrequent drought conditions in the region. Even though 1988 was a year of extreme drought in the Great Plains, the entire decade of the 1980s was marked by only the western sections of the region experiencing drought conditions between 10 and 20% of the time, with two climate divisions in Minnesota and one in Illinois experiencing drought more than 20% of the time. Between 1990 and 1995, severe or extreme drought was rare in the region.

Great Lakes Water Levels

Water levels in the Great Lakes are a useful indicator of the long-term precipitation and evaporation trends for a large portion of the region. Since the late 1960s, water levels in Lake Michigan, Lake Huron, and Lake Erie have been above the long-term averages (Nicholls et al., 1996). Since 1988, water levels in Lake Superior have been slightly below normal. Changnon (1987) examined climate fluctuations and changes in Lake Michigan water levels, and found that the record high water levels observed during 1985 and 1986 were due mainly to above-normal precipitation since 1981. A decrease in evapotranspiration across the Lake Michigan basin resulting from increased cloudiness and a decrease in observed temperatures since about 1940 was also found to have contributed to higher water levels in Lake Michigan. Changes in Great Lakes water levels can have both positive and negative impacts. For example, impacts on shipping, hydropower, and recreational boating can be positive when water levels increase. In contrast, high water levels can have an adverse effect on shorelines and numerous environmental conditions (Changnon, 1987).

Short-Term Precipitation Variability Patterns

Weather patterns responsible for local and regional precipitation events have relatively short time scales. Convective thunderstorms, cyclones,

Color Plate VII

(a)

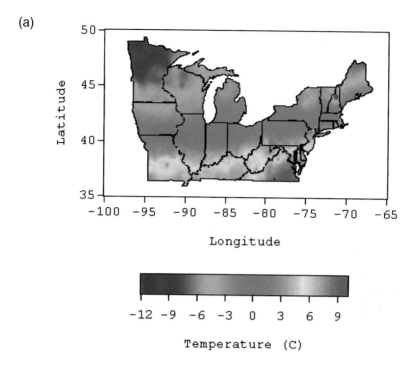

Temperature (C)

(b)

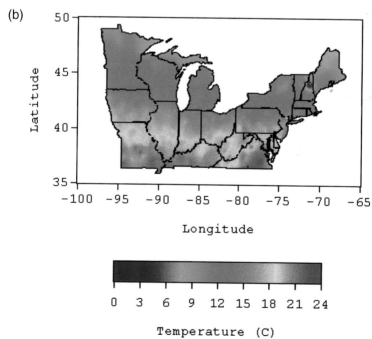

Temperature (C)

Figure 3.1. Average daily maximum temperatures (°C) in the region during the months of (a) January, (b) April, (c) July, and (d) October based on maximum temperature observations from 1950–1993. *(Continued)*

Color Plate VIII

(c)

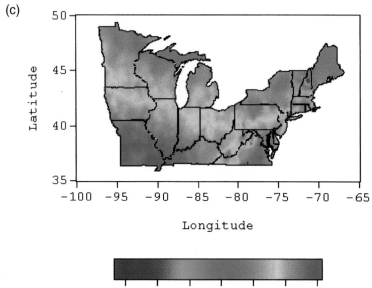

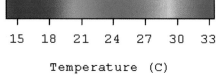

Temperature (C)

(d)

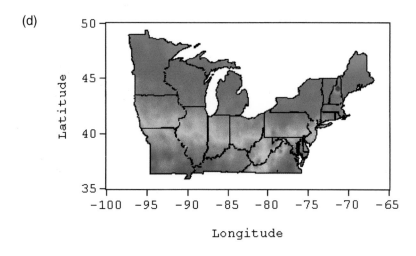

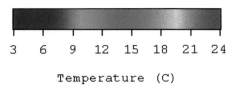

Temperature (C)

Figure 3.1c,d (*Continued*).

Color Plate IX

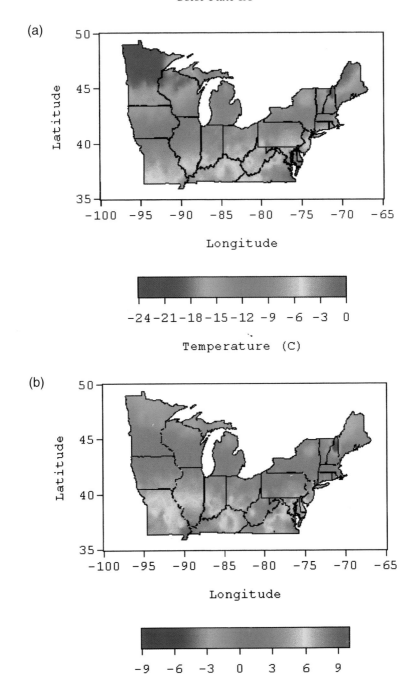

Figure 3.2. Average daily minimum temperatures (°C) in the region during the months of (a) January, (b) April, (c) July, and (d) October based on minimum temperature observations from 1950–1993. (*Continued*)

Color Plate X

(c)

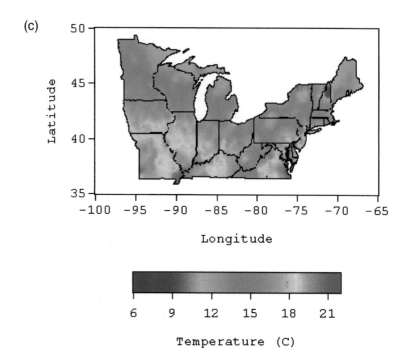

Temperature (C)

(d)

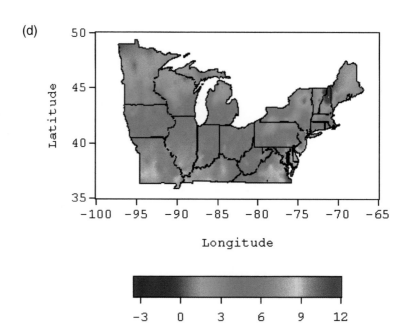

Temperature (C)

Figure 3.2c,d (*Continued*).

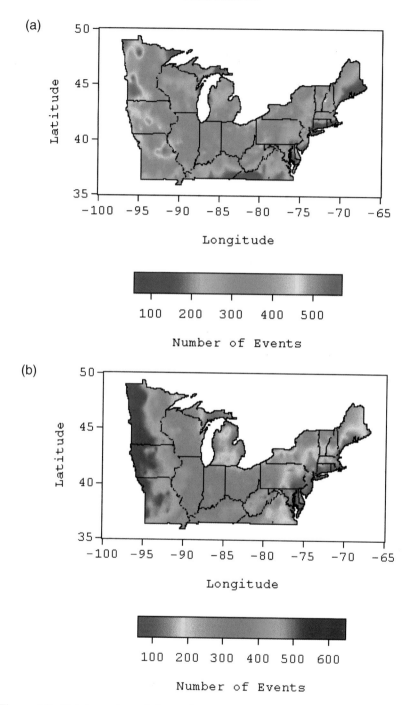

Figure 3.3. Total number of days when daily maximum temperatures exceeded 11.1°C (a) above normal and (b) below normal in the region for the period 1950–1993.

Color Plate XII

(a)

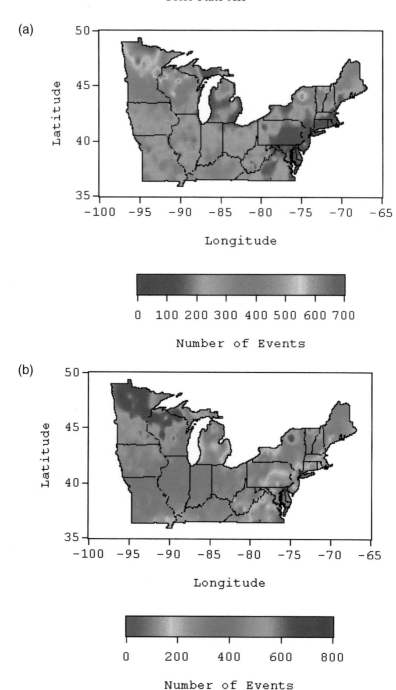

Number of Events

(b)

Number of Events

Figure 3.4. Total number of days when daily minimum temperatures exceeded 11.1°C (a) above normal and (b) below normal in the region for the period 1950–1993.

Color Plate XIII

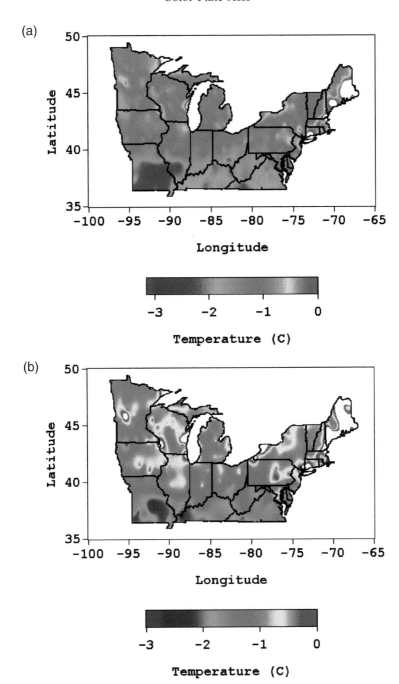

Figure 3.6. Average late-spring freeze temperatures (°C) over the region following (a) 250 growing-degree days and (b) 300 growing-degree days (base 5°C). Areas where no late-spring freezes occurred following 250 or 300 growing-degree days over the period 1961–1990 appear as white areas in the figure.

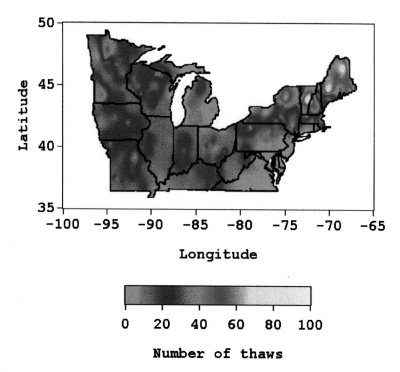

Figure 3.7. Total number of thaw-freeze episodes over the region from December 1961 to March 1990.

Color Plate XV

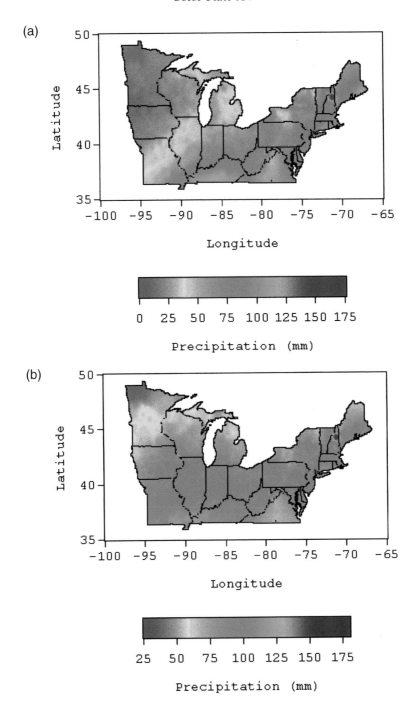

Figure 3.8. Average total precipitation amounts (mm) over the region during the months of (a) January, (b) April, (c) July, and (d) October based on daily precipitation observations from 1950–1993.　　　　　　　　　　　　　　　(*Continued*)

Color Plate XVI

(c)

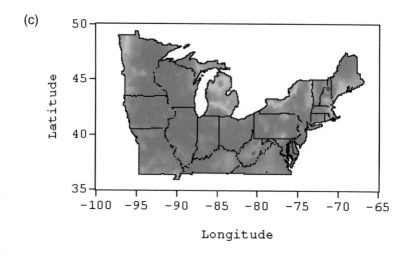

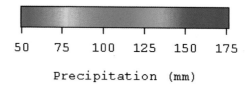

Precipitation (mm)

(d)

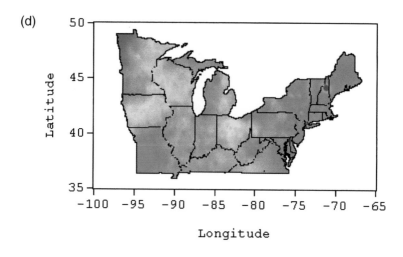

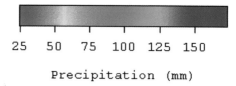

Precipitation (mm)

Figure 3.8c,d (*Continued*).

Color Plate XVII

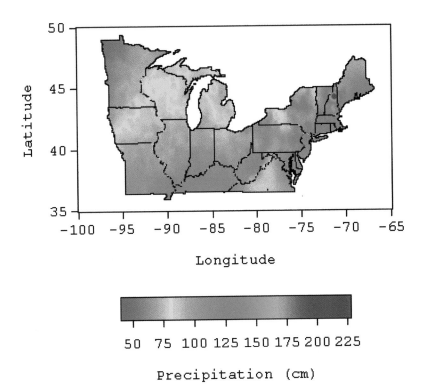

Figure 3.9. Average annual precipitation amounts (cm) over the region based on daily precipitation observations from 1950–1993.

Color Plate XVIII

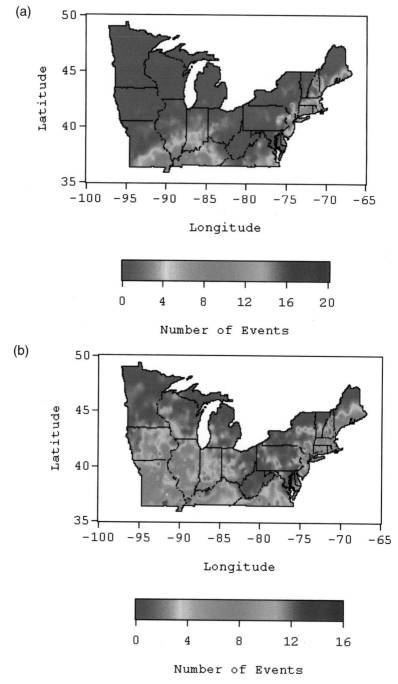

Figure 3.10. Total number of occurrences of daily precipitation exceeding 5.08 cm over the region during the months of (a) January, (b) April, (c) July, and (d) October for the period 1950–1993. (*Continued*)

(c)

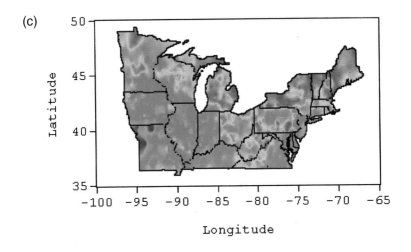

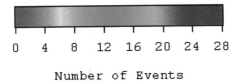

Number of Events

(d)

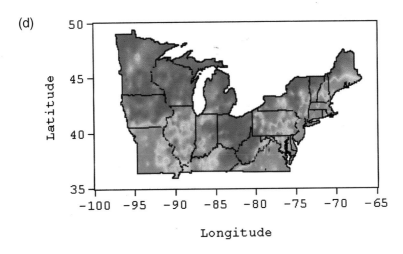

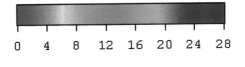

Number of Events

Figure 3.10c,d (*Continued*).

Color Plate XX

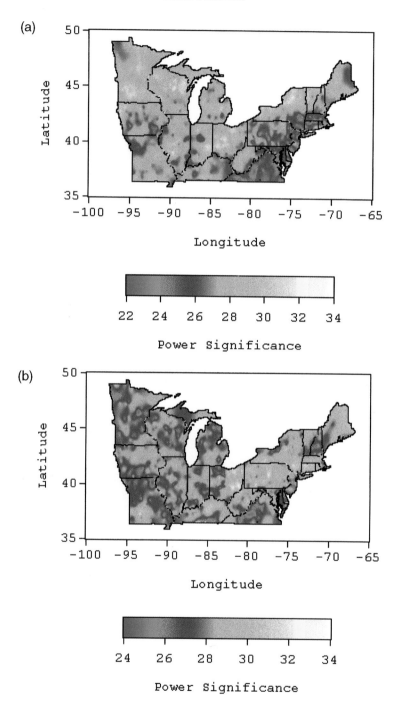

(a)

(b)

Figure 3.11. Normalized power spectrum values that show the relative significance across the region of summer (July–September) precipitation events that occur every (a) 2–4 days, (b) 4–8 days, (c) 8–16 days, and (d) 16–32 days.

(c)

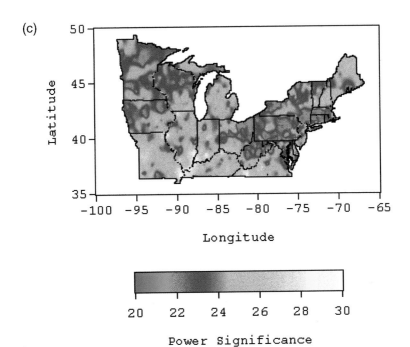

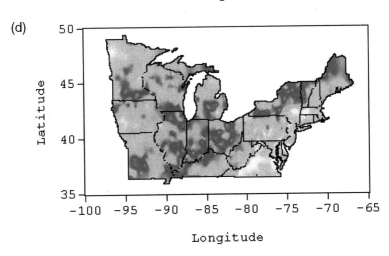

Figure 3.11c,d (*Continued*).

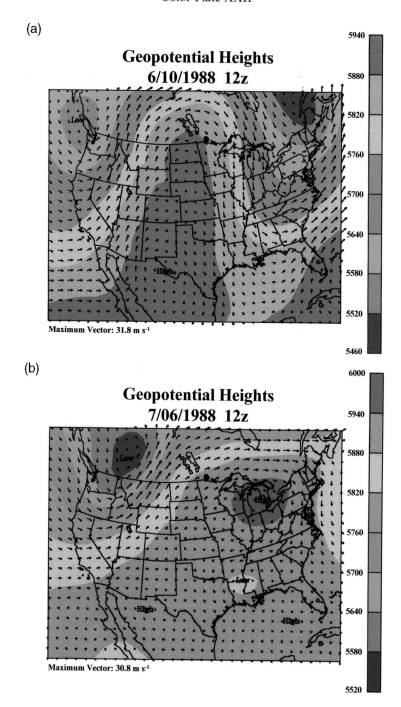

Figure 3.13. Recent examples of three 500 mb geopotential height (contours in meters) and circulation (vectors in m s^{-1}) patterns associated with severe wildfires in the north-central U.S. that occurred on (a) 10 June 1988, (b) 6 July 1988, and (c) 19 April 1989. (*Continued*)

(c)

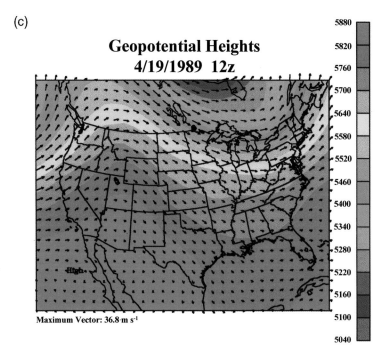

Figure 3.13c (*Continued*).

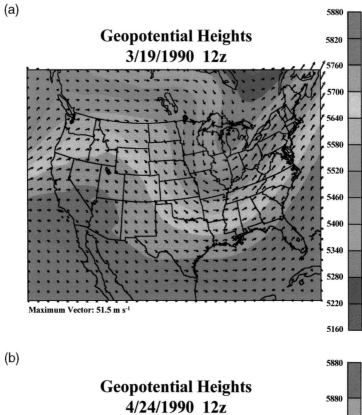

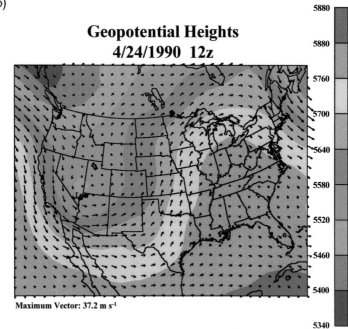

Figure 3.14. Recent examples of two 500 mb geopotential height (contours in meters) and circulation (vectors in m s^{-1}) patterns associated with severe wildfires in the northeastern U.S. that occurred on (a) 19 March 1990 and (b) 24 April 1990.

(a)

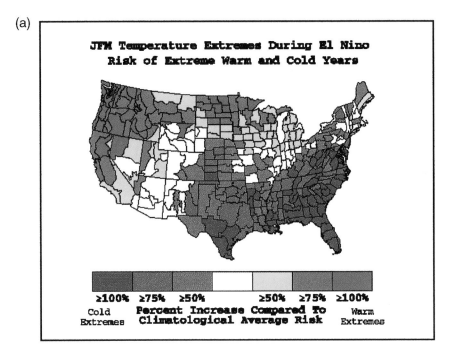

(b)

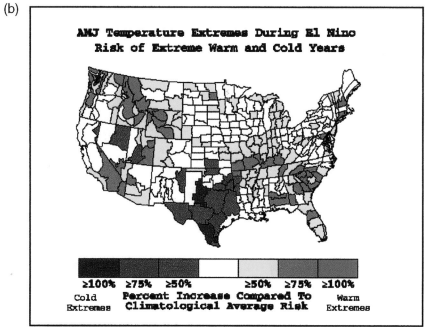

Figure 3.15. Risk of seasonal extreme temperature occurrences in each climate division across the U.S. during ENSO episodes for the periods (a) January–March, (b) April–June, (c) July–September, and (d) October–December (from NOAA-CIRES Climate Diagnostics Center, 1997). (*Continued*)

(c)

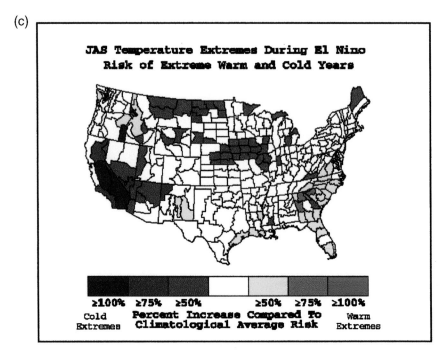

(d)

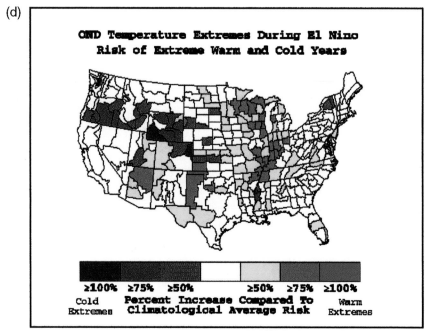

Figure 3.15c,d (*Continued*).

Color Plate XXVII

(a)

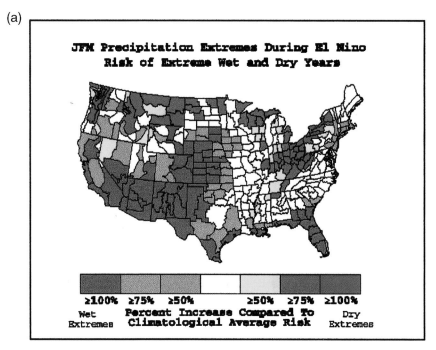

(b)

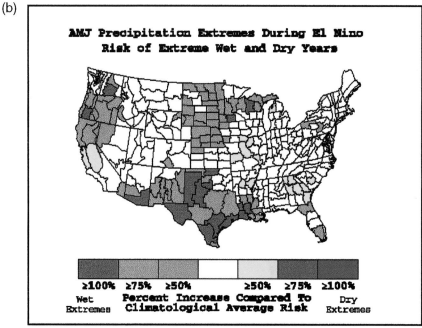

Figure 3.16. Risk of seasonal extreme precipitation occurrences in each climate division across the U.S. during ENSO episodes for the periods (a) January–March, (b) April–June, (c) July–September, and (d) October–December (from NOAA-CIRES Climate Diagnostics Center, 1997). (*Continued*)

(c)

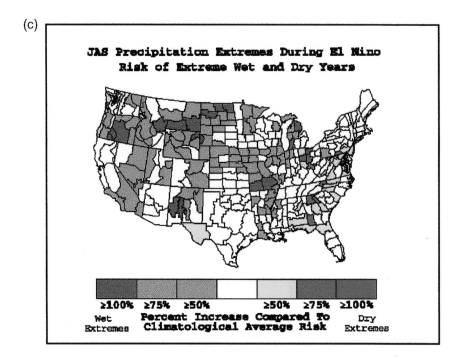

(d)

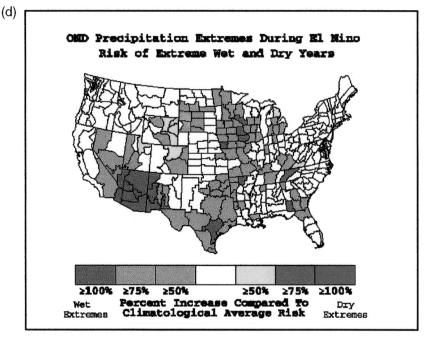

Figure 3.16c,d (*Continued*).

Color Plate XXIX

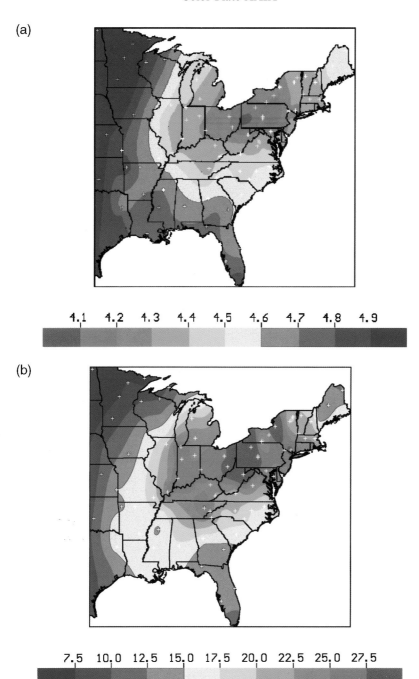

Figure 3.17. (a) Annual trend of pH at NADP/NTN sites (+) in the eastern U.S. from 1983 to 1994. (b) Annual trend of sulfate deposition (kg/ha SO$_4^+$) at NADP/NTN sites (+) in the eastern U.S. from 1983–1994. *(Continued)*

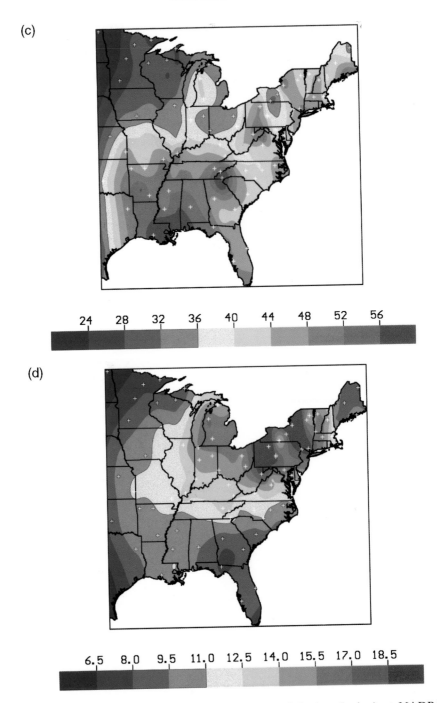

Figure 3.17. (*Continued*) (c) Annual trend of precipitation (inches) at NADP/ NTN sites (+) in the eastern U.S. from 1983–1994. (d) Annual trend of nitrate deposition (kg/ha NO_3^-) at NADP/NTN sites (+) in the eastern U.S. from 1983– 1994.

(e)

(f)

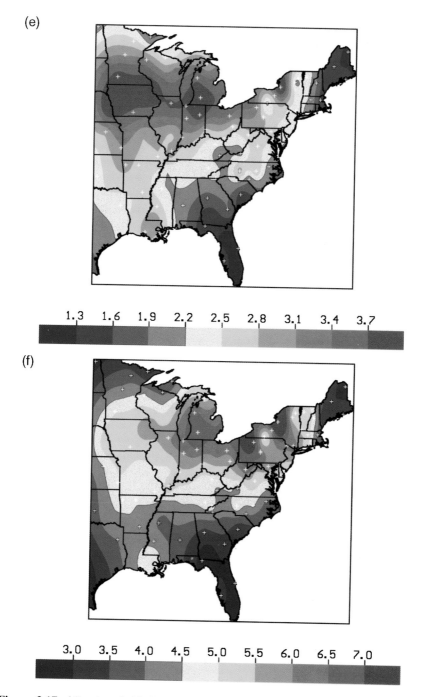

Figure 3.17. (*Continued*) (e) Annual trend of ammonium deposition at NADP/NTN sites (+) in the eastern U.S. from 1983–1994. (f) Annual trend in total N wet deposition ($NO_3^- + NH_4^+$) at NADP/NTN sites (+) in the eastern U.S. from 1983–1994.

Color Plate XXXII

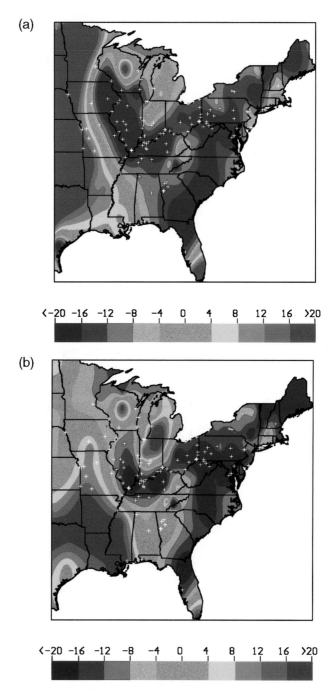

Figure 3.18. Percent departure of 1995 annual (a) H$^+$, (b) sulfate, (c) nitrate, and (d) precipitation data from 1983–1994 trends modeling. Electric power plants affected by Phase I of the CAAA-90, Title IV, are indicated by plus (+) signs.   (*Continued*)

Color Plate XXXIII

(c)

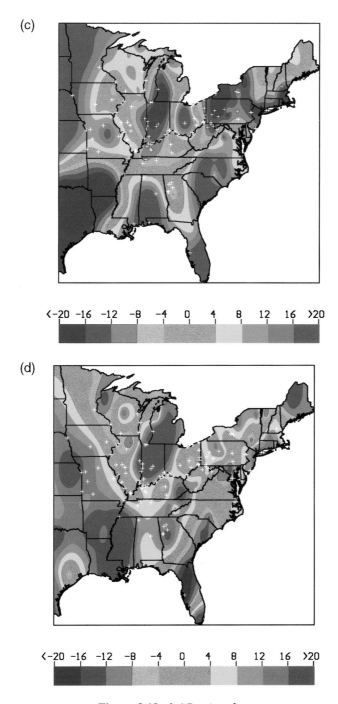

Figure 3.18c,d (*Continued*).

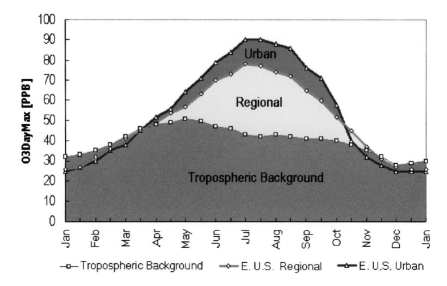

Figure 3.19. Typical monthly average of daily maximum ozone for background, regional, and urban areas.

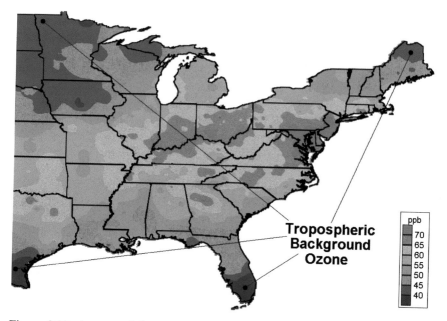

Figure 3.20. Average daily maximum ozone for the eastern (OTAG) region, 1991–1995.

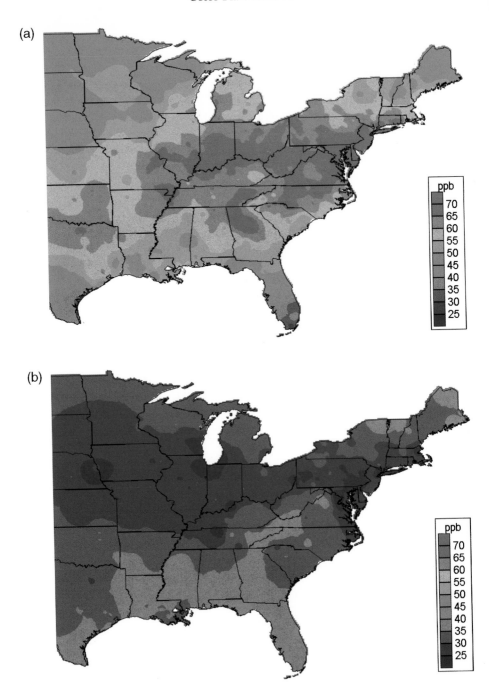

Figure 3.21. Seasonal ozone: maximum daily ozone for (a) summer (June–August), (b) winter (December–February), and (c) seasonal difference (summer–winter) for the eastern U.S. 1991–1995.   (*Continued*)

(c)

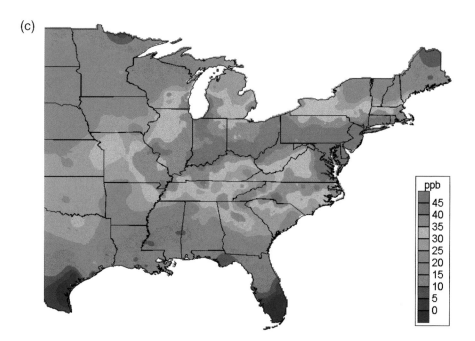

ppb
45
40
35
30
25
20
15
10
5
0

Figure 3.21c (*Continued*).

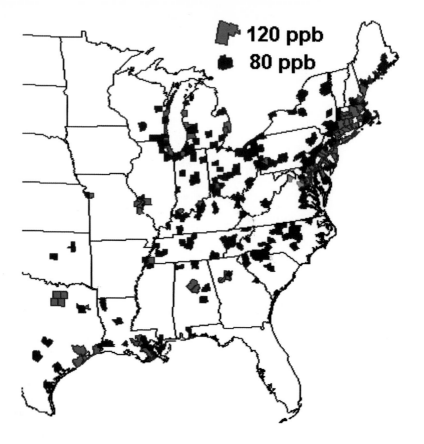

Figure 3.23. Ozone nonattainment counties at 120 and 80 ppb.

Winds and Ozone on High Ozone Days

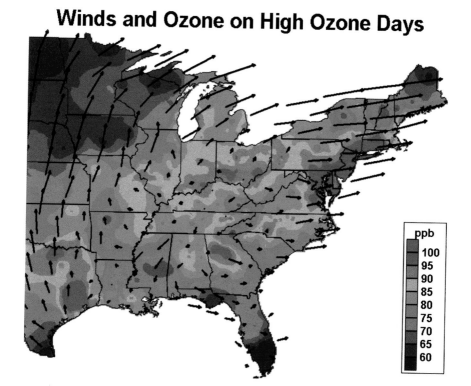

Figure 3.24. Regional transport patterns with high local ozone for the eastern U.S. 1991–1995.

Color Plate XL

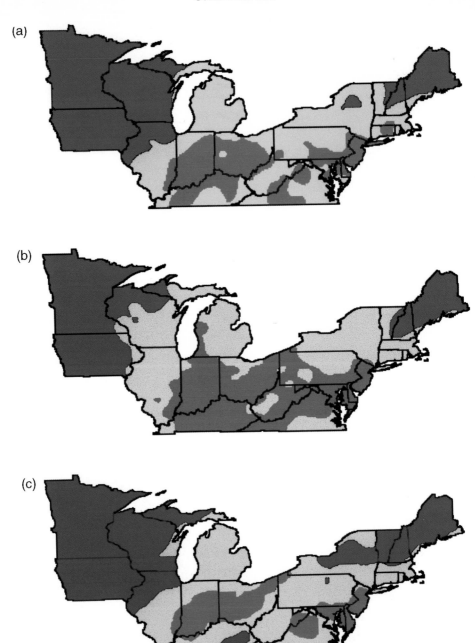

Figure 3.25. Seasonal ozone exposure W126 index for the Northern Region for (a) 1994, (b) 1995, and (c) 1996.

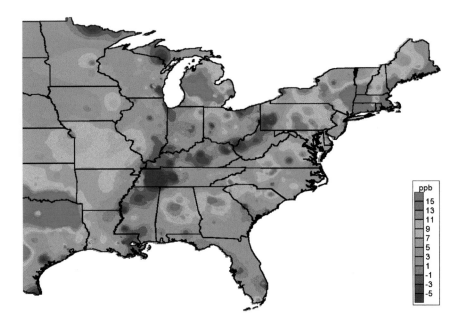

Figure 3.26. Weekly cycle in ozone: difference at the end of the workweek versus the weekend (Friday–Sunday).

anticyclones, air-mass movements, and frontal passages, which all play a role in precipitation occurrences, have time scales on the order of hours to a few weeks. While spatial patterns of monthly averages of precipitation and extreme precipitation events describe one particular facet of precipitation variability across the region, another important variability characteristic is the pattern of precipitation frequency in different seasons across the region on time scales ranging from a couple of days to a month. In order to examine the prevalent short-term precipitation variability modes during each season across the region, power spectrum analyses (Press et al., 1992) were performed on 1950 to 1993 daily precipitation records from the NCDC from 1204 stations across the region (EarthInfo, 1995). Power significance values within specific frequency/period "windows" (i.e., 2 to 4 days, 4 to 8 days, 8 to 16 days, and 16 to 32 days) were determined for each station during the spring, summer, fall, and winter seasons. These power significance values provide an indication of the prevalence or significance of precipitation events at each station within a particular frequency/period "window" compared with precipitation events occurring within the other 3 frequency/period "windows," regardless of how precipitation amounts are distributed across the region in each season.

Power spectrum analyses performed on precipitation data from the winter season, defined as the months of January, February, and March, suggest that precipitation events occurring every 2 to 4 days in the region are generally most prevalent in central Wisconsin, eastern Iowa, and along the Atlantic coast, with the most power significance occurring in eastern Maine. The 2 to 4 days variability mode is least significant in Minnesota, central Missouri, and along the Ohio River Valley. The eastern half of the region is characterized by the prevalence of winter precipitation events occurring on slightly longer time scales of 4 to 8 days. The western sections of the region generally experience fewer significant winter precipitation events on a 4 to 8 day cycle than the eastern sections of the region. Winter precipitation events that occur every 8 to 16 days or longer are least significant along the Atlantic coast states and West Virginia. Central Wisconsin is also characterized by fewer 8- to 16-day precipitation events than in surrounding states. Winter precipitation variability on the longest time scales of 16 to 32 days is most prevalent in Minnesota, Wisconsin, northwestern Iowa, central Missouri, and the upper peninsula of Michigan, while in the Atlantic coastal states, the 16- to 32-day precipitation variability mode is less significant.

In the spring season, defined as the months of April, May, and June, the spatial trends in the significance of 2- to 4-day precipitation events are weaker than in the winter months. Relatively high and low power significance values tend to be randomly scattered throughout the region. However, distinctive spatial patterns exist across the region for the 4- to 8-day, 8- to 16-day, and 16- to 32-day precipitation variability modes.

Springtime precipitation events that occur every 4- to 8-days are most prevalent in the states of Vermont, New Hampshire, eastern New York, New Jersey, and eastern Pennsylvania. They are least prevalent in those states comprising the western half of the region. For the 8- to 16-day springtime precipitation variability mode, Illinois, southern Indiana, southeastern Iowa, and Virginia all tend to have relatively high power significance values. Relatively low power significance values are prevalent over the western (Minnesota, Wisconsin, and most of Iowa) and northeastern (New Hampshire, Vermont, New York, Massachusetts, Connecticut, and Rhode Island) sections of the region. Finally, lower frequency precipitation events that occur every 16 to 32 days are most prevalent in the northwestern sections (western Wisconsin, Minnesota, and western Iowa) of the region as well as Maine during the spring months; they are least prevalent over the rest of the region, especially in Missouri and Ohio.

Fig. 3.11 (color insert) shows the distribution of power significance values across the region during the summer months (July, August, and September) for the 2- to 4-day, 4- to 8-day, 8- to 16-day, and 16- to 32-day precipitation variability modes. Summertime convective precipitation activity leads to relatively large power significance values over most of the region for 2- to 4-day precipitation events (Fig. 3.11a). It is only over the Atlantic coastal region from Massachusetts to Virginia that 2- to 4-day precipitation events are less significant. This is in contrast to the significance of 2- to 4-day precipitation events over the region during the winter months, when precipitation events occurring every 2 to 4 days are more likely to occur along the Atlantic coast and less likely over most other sections of the region (especially in Minnesota and much of the Ohio River Valley). The 4- to 8-day variability mode pattern in the summer is quite variable (Fig. 3.11b); only the area from eastern Ohio eastward through Pennsylvania, southern New York, Massachusetts, and Connecticut tends to be characterized by more precipitation events occurring at intervals of 4 to 8 days than for other areas of the region. Fig. 3.11c suggests that the southern-tier states in the region are more likely to experience precipitation events occurring every 8 to 16 days than the northeastern and northwestern states in the region. At summertime precipitation periods of 16 to 32 days, the Atlantic coastal states from Virginia to Massachusetts along with most of Iowa and northern Minnesota tend to experience a higher proportion of this class of precipitation events than do the Ohio River Valley and northern New England states (Fig. 3.11d). This pattern also contrasts sharply with the 16- to 32-day wintertime variability mode pattern that suggests the Atlantic coastal states experience a *lower* proportion of these precipitation events than most other areas of the region.

In the fall months (October, November, and December), the eastern sections of the region tend to be characterized by a larger proportion of higher frequency precipitation events (2- to 4-day and 4- to 8-day periods) than the western sections of the region. Conversely, more lower frequency (8- to 16-day and 16- to 32-day periods) precipitation events characterize the western sections of the region than the eastern sections during the fall season. During the fall months, the western portions of the region experience more cool and dry air masses moving southward from Canada, thereby decreasing the frequency of precipitation events in the area in comparison with the higher frequency convective precipitation events that occur in the area during the summer months.

Extreme Weather Events

Extreme weather events, such as tornadoes, destructive straight-line winds, and hurricanes, along with fire-weather episodes that produce atmospheric environments conducive for severe wildland fires can all act as direct or indirect agents of landscape change in some or all areas of the northeast and north central United States. Extreme winds associated with thunderstorms, tornadoes, and hurricanes can damage and uproot trees, destroy agricultural crops, damage or destroy property, and can have a lasting impact on local microclimates and basic ecosystem functions and processes (e.g., species regeneration, wildlife habitat, forest sensitivity to insects and diseases, frost occurrence within ecosystems, heat and moisture exchange between the surface and atmosphere). If surface fuel conditions are appropriate, fire-weather episodes can lead to severe wildland fires that also disturb ecosystems and impact basic ecosystem functions and processes in the region. The following sections provide a brief overview of tornado, straight-line wind, hurricane, and fire-weather occurrences in the region.

Tornado and Straight-Line Wind Events

There are three general types of extreme wind events: tornadoes, hurricanes, and straight-line winds. The last category includes winds of varying duration and areal coverage, ranging from microbursts (lasting less than 10 minutes and covering only a few square kilometers) to mesoscale convective complex and squall-line winds (lasting less than an hour or two and covering a few hundred to 1000 square kilometers) up to frontal winds (lasting several hours and covering areas on the order of 10,000 square kilometers). Local topography can weaken or intensify any of these types of extreme winds.

The most common location for tornadoes to form in North America is the region east of the Rocky Mountains and west of the Appalachian Mountains (NOAA Severe Storms Laboratory, 1999). Tornadoes occur most frequently during the late afternoon or early evening periods in the spring and summer months. The typical tornado damage path is about 2 to 3 km long and about 50 m wide. The erratic behavior of tornadoes can result in path lengths that vary from basically a single point to more than 150 km and path widths that vary from less than 10 m to more than 1.5 km. The forward speed of tornadoes can range from nearly stationary to more than 25 m s^{-1}, with the typical speed being in the 5 to 10 m s^{-1} range (NOAA Severe Storms Laboratory, 1999). Tornadoes are categorized according to the Fujita Scale, a 6-category "wind-speed" scale based on the amount of observed damage from a tornado:

F_0: 64–115 km h^{-1} Light damage
F_1: 116–179 km h^{-1} Moderate damage
F_2: 180–251 km h^{-1} Considerable damage
F_3: 252–329 km h^{-1} Severe damage
F_4: 330–417 km h^{-1} Devastating damage
F_5: 418–508 km h^{-1} Incredible damage

Since 1950, only 20 tornadoes have been classified as F_5 tornadoes in the region, and they occurred in the states of Wisconsin, Minnesota, Illinois, Ohio, Indiana, Missouri, and Michigan (NOAA Storm Prediction Center, 1999).

Table 3.1 shows the distribution of reported tornadoes over the period 1950 to 1994 along with the Consumer Price Index–adjusted costs of all tornado damages in each state within the region plus the states of Kentucky and Virginia. Approximately 200 tornadoes are reported each year in the region, mostly between March and July. Most of the reported tornadoes have occurred in the western sections of the region. The states of Iowa, Missouri, and Illinois ranked 6th, 7th, and 9th in the United States, respectively, for the number of reported tornadoes during the period 1950 to 1994. Although the states of Indiana, Minnesota, and Ohio ranked 15th, 18th, and 21st, respectively, for reported tornado occurrence during the same period, the states ranked 2nd, 6th, and 7th, respectively, for the cost of tornado damages incurred. The tornadoes that occurred during the 1950 to 1994 period caused 1545 fatalities and 28,280 injuries in the region plus the states of Kentucky and Virginia. The state of Michigan ranked first in the region and 5th nationally in the number of reported fatalities (237) attributed to tornadoes during the 1950 to 1994 period, while Ohio ranked first in the region and 4th nationally in the number of tornado-related injuries (237) (NOAA Storm Prediction Center, 1999).

The effect of intense winds, be they tornadic, hurricane-based, or straight-line, on trees and forests is usually negative. Any strong wind can break numerous branches, uproot poorly anchored trees, or snap off the crowns of trees. Due to their seasonality, tornadoes have a tendency to cause greater damage to trees that flush earlier, as the leaves increase the drag the tree creates in the wind. Direction can also be a factor of

Table 3.1. Number of Tornadoes, State Rank for Tornado Occurrence, Consumer Price Index (CPI) Adjusted Cost of Tornado Damages, and State Rank for Tornado Costs for Each State in the Region Plus the States of Kentucky and Virginia, Based on 1950 to 1994 Reported Tornado Occurrences (From NOAA Storm Prediction Center, 1999)

State	Number of Tornadoes	State Rank	CPI Adjusted Cost	State Rank
Iowa	1374	6	$7,092,119	10
Missouri	1166	7	$7,393,827	9
Illinois	1137	9	$8,238,192	8
Indiana	886	15	$16,486,543	2
Wisconsin	844	17	$4,107,568	19
Minnesota	832	18	$10,153,546	6
Michigan	712	20	$3,450,385	22
Ohio	648	21	$9,654,648	7
Pennsylvania	451	24	$6,150,330	13
Kentucky	373	28	$2,825,786	23
Virginia	279	29	$1,247,380	28
New York	249	30	$1,840,968	26
Maryland	135	34	$387,377	35
Massachusetts	134	35	$6,177,932	12
New Jersey	112	37	$530,843	33
West Virginia	83	38	$216,385	39
Maine	82	39	$71,046	41
New Hampshire	72	41	$90,713	40
Connecticut	61	42	$3,853,888	20
Delaware	52	44	$56,285	42
Vermont	32	47	$35,124	45
Rhode Island	8	49	$19,796	47

importance when considering wind damage to trees. Moderate, ambient winds that flow from a particular direction tend to cause trees to thicken their trunks along a line parallel to the wind direction, thereby enabling trees to effectively brace themselves against the ambient winds. If a sudden strong gust strikes in the same direction from which the ambient wind usually blows, it may cause less damage than if it comes from a perpendicular heading and strikes the tree broadside.

Hurricanes in the Northeastern United States

Hurricanes originating in the Atlantic Ocean or Gulf of Mexico have the potential for causing significant flooding, storm surges, and wind damage in the northeastern United States. Between 1900 and 1996, a total of 835 hurricanes or tropical storms developed in the Atlantic Ocean or Gulf of Mexico (Unisys Corporation, 1999). The distribution by year of Atlantic or Gulf hurricane or tropical storm occurrence is shown in Fig. 3.12.

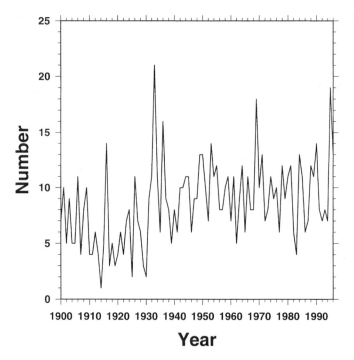

Figure 3.12. Total number of hurricanes or tropical storms that developed in the Atlantic Ocean or Gulf of Mexico each year from 1990 to 1996. (Data from Unisys Corporation, 1999.)

According to Landsea et al. (1996), there has been no significant change in the frequency of tropical storms or hurricanes between 1944 and 1995, but there has been a decrease in the number of intense hurricanes. Between 1991 and 1994, fewer hurricanes occurred in the Atlantic basin than in any other 4-year period since 1994. Peak numbers of hurricanes or tropical storms occurred in 1916, 1933, 1936, 1969, and 1995.

Of the total number of hurricanes or tropical storms that developed in the Atlantic Ocean or Gulf of Mexico between 1900 and 1996, 158 made landfall over the continental United States, with 41 making landfall in Atlantic coastal states within the region. Table 3.2 shows the distribution of hurricane landfall occurrences by state in the region between 1900 and 1996. More hurricanes and tropical storms have made landfall in New York (9) and Connecticut (8) in this period than in any other states within the region. Although hurricanes that make landfall typically produce most damage along coastal areas, extreme flooding and wind damage is possible in further inland areas, especially if convective storms and tornadoes develop in association with inland-moving hurricanes. For example, in 1955, Hurricane Diane made landfall over North Carolina and produced serious flooding in Pennsylvania as well as the coastal states

Table 3.2. Number of Hurricane Direct Hits on States in the Region Between 1900 and 1996 by Saffir/Simpson Categories (Category 1: 119 to 153 km h^{-1}, Category 2: 154 to 177 km h^{-1}, Category 3: 178 to 209 km h^{-1}, Category 4: 210 to 249 km h^{-1}, Category 5: >249 km h^{-1}) (From NOAA National Hurricane Center, 1999)

State	Category 1	Category 2	Category 3	Category 4	Category 5
Virginia	2	1	1	0	0
Maryland	0	1	0	0	0
Delaware	0	0	0	0	0
New Jersey	1	0	0	0	0
New York	3	1	5	0	0
Connecticut	2	3	3	0	0
Rhode Island	0	2	3	0	0
Massachusetts	2	2	2	0	0
New Hampshire	1	1	0	0	0
Maine	5	0	0	0	0

of New Jersey, New York, Delaware, Connecticut, Rhode Island, and Massachusetts.

Flooding can be a major problem, regardless of the strength of the hurricane as measured by its wind speed. Flooding and high winds associated with hurricanes that move up the Atlantic coast or move inland over the northeastern United States can cause severe damage to the region's natural resources, both in the short term and long term. Blowdowns from intense hurricane winds immediately impact forest stands, while trees damaged from high waters and intense winds are susceptible to diseases and insect infestations over time.

Fire-Weather Patterns

Wildland fires typically burn between 150,000 and 600,000 acres each year in the region (USDA Forest Service, 1992). On average, more than 98% of the wildland fires that occur in the region are human-caused. This is in contrast to the western regions of the United States where human-caused fires only account for about 40 to 50% of the total number of reported fires in a typical year. Lightning plays a more significant role in fire occurrence in the western states. Regardless of the cause of wildland fires, the severity of wildland fires in different regions of the United States depends to a large degree on the atmospheric conditions prior to and during fire episodes. There are specific atmospheric circulation, temperature, and moisture patterns that tend to be associated with severe wildland fires in each region of the United States. While these circulation, temperature, and moisture patterns typically do not generate weather that is considered "extreme," they do create conditions that are conducive to wildland fires that can be very destructive and "extreme" in nature.

Heilman (1995) performed empirical-orthogonal-function analyses of the 500-mb heights over the United States at the onset of past severe wildland fires (fires that burned 404.69 ha or more) in six different regions of the United States between 1971 and 1991 in order to determine which large-scale atmospheric circulation patterns and associated temperature and moisture patterns are most conducive to fire occurrence in these regions.

Three middle tropospheric circulation patterns were found most prevalent at the onset of severe wildland fires in the north central United States, including the states of Minnesota and Iowa. Examples of these circulation patterns are shown in Fig. 3.13 when severe wildland fires occurred in the north central United States (see color insert). Fig. 3.13a shows the first pattern consisting of a strong 500-mb ridge centered over the central Great Plains and extending northward into Canada, with the eastern and western regions of the United States dominated by 500-mb troughs. Wind vectors overlaying the 500-mb geopotential height field show southerly and southwesterly flow to the west of the ridge and northerly flow to the east of the ridge. Analyses of the lower tropospheric temperatures during circulation episodes of this type when severe wildland fires occurred in the north central states show positive average departures from monthly means over the entire Great Plains, with maximum departures reaching 6°C over western Minnesota. Lower tropospheric relative humidity departures from monthly mean values under this type of circulation pattern were found to be about 8 to 12% below normal over the western Great Lakes region. The warmer- and drier-than-normal atmospheric conditions over the north central United States under this type of circulation pattern during the spring and fall fire seasons in the north central U.S. can lead to higher probabilities of severe fires occurring in this region.

Fig. 3.13b shows an example of the second middle tropospheric circulation pattern most conducive to severe fire occurrence in the north central United States. This pattern consists of a 500-mb ridge centered over the eastern half of the United States, with the western states dominated by a 500-mb trough. This results in a southwesterly flow over the northern Great Plains, including the states of Minnesota and Iowa. Lower tropospheric temperatures averaged from 6 to 9°C above normal over the western Great Lakes region (Minnesota, Iowa, Wisconsin, and Michigan) when this circulation pattern developed and severe fires occurred in the north central U.S. This pattern also produces relatively dry lower atmospheric conditions over much of the region. Analyses indicate relative humidity departures on the order of 8 to 14% below normal from Wisconsin eastward and southeastward to the Atlantic coast.

The final middle tropospheric circulation pattern associated with severe fires in the north central United States is depicted in Fig. 3.13c. This pattern results in a strong northwesterly flow over the northern Great Plains in response to a strong ridge and trough over the western and eastern

United States, respectively. Unlike the previous circulation patterns, this pattern produces cooler-than-normal lower atmospheric conditions over most of the region. In the western Great Lakes region, lower atmospheric temperatures are typically on the order of 1 to 2°C below the monthly mean. Even though cooler-than-normal conditions prevail over the region under this type of circulation pattern, the lower atmosphere can become very dry. Relative humidity values approached 14 to 16% below normal over the western sections of the region when this circulation pattern appeared during 1971 to 1991 severe wildland fire episodes in the north central states. The extremely dry lower atmosphere is very conducive to fire occurrence in the region if surface fuel conditions are adequate.

When severe wildland fires occur in states east of Minnesota, Iowa, and Missouri, there are typically two types of atmospheric circulation patterns most conducive to their occurrence. The first pattern is characterized by a 500-mb ridge over the western half of the United States and a prominent trough over the eastern half of the United States and southeastern Canada. Refer to the color insert for an example of this pattern as shown in Fig. 3.14a. This circulation pattern transports cool dry air into the northeastern United States. An analysis of average lower atmospheric temperatures when severe wildland fires occurred during circulation patterns of this type between 1971 and 1991 indicates anomalies between −1°C and −5°C over the states northeast of Ohio and West Virginia. The lower atmosphere is typically very dry under this type of circulation pattern. Corresponding relative humidity values are typically 4 to 16% below normal over most of the region in these cases.

An example of the second circulation pattern associated with severe wildland fires in the northeastern United States is shown in Fig. 3.14b. This type of circulation pattern is associated with the westward shift of the Bermuda high pressure system off the southeastern U.S. coast, which results in a blocking of the northward transport of Gulf moisture into the region. Very hot and dry conditions can occur over the eastern U.S. under this circulation pattern, and the probabilities for severe wildland fire occurrence in the eastern half of the United States increase in such cases. The bulk of the moisture from the Gulf region is transported northward in a band to the west of the Mississippi River, where precipitation can be significant. The lower atmospheric temperature and relative humidity patterns associated with this circulation pattern reflect the conduciveness to fire occurrence in the region. Significant positive temperature anomalies occur over most of the eastern half of the United States with this circulation pattern, and negative relative humidity anomalies are prevalent over the eastern half of the region. Average temperature anomalies were generally greater than 5°C over the Great Lakes states during circulation episodes of this type between 1971 and 1991 when severe fires occurred, and average relative humidity anomalies reached as high as −24% over West Virginia and Pennsylvania during the same episodes.

In addition to the middle tropospheric circulation patterns that have been shown to be related to severe wildfire occurrence in the region by Heilman (1995), there are specific surface atmospheric pressure patterns and associated surface circulation patterns that characterize severe fire occurrence in the northeastern United States. Takle et al. (1994) studied surface pressure patterns corresponding to reduced precipitation, high evaporation potential, and enhanced forest-fire danger in the states of West Virginia, Ohio, Pennsylvania, and New York. Their analyses indicate most severe fires in this region occur when (1) an extended surface high pressure system covers most of the eastern United States, (2) a high pressure system is situated just off the Atlantic coast, or (3) a high pressure system is centered in the western Great Lakes region. The near-surface circulations associated with these high-pressure patterns are consistent with the middle tropospheric circulation patterns identified by Heilman (1995) as being conducive to severe fire occurrence in the northeastern United States. Takle et al. (1994) also examined the surface pressure fields generated by the Canadian Climate Centre global circulation model of the present ($1 \times CO_2$) climate and $2 \times CO_2$ climate to determine whether wildfire potential in the eastern U.S. may change under increased atmospheric CO_2 concentrations. The model simulations suggested an increase in frequency of surface circulation patterns in which evaporation generally exceeds precipitation, which is generally conducive to higher wildfire probabilities if surface fuel conditions are adequate.

El Niño–Southern Oscillation Effects

Atmospheric processes span a wide range of temporal and spatial scales. The climate and weather patterns that influence the northeast and north central United States region are the result of these many atmospheric processes at work from the global scale down to the microscale. One particular large-scale atmospheric process that periodically impacts weather and climate across many regions of the Earth, including the region, is the El Niño–Southern Oscillation phenomenon. This phenomenon can have a significant impact on global-scale atmospheric circulation patterns that produce regional temperature and precipitation anomalies across certain areas of the Earth. If the regional temperature and precipitation anomalies associated with an El Niño–Southern Oscillation episode are strong enough in the region, ecosystems in this region can be affected via altered heat and moisture flux regimes, changes in winter snowpack, changes in soil-moisture conditions, fewer or more frequent wildland fire occurrences, altered frequencies of damaging insect infestations and vegetation disease, and other mechanisms. A description of the El Niño–Southern Oscillation phenomenon and its typical seasonal impacts on temperature and precipitation patterns in the region are provided in this section.

What Is El Niño–Southern Oscillation?

Within 30° of latitude from the equator, winds near the surface of the Pacific Ocean generally blow from east to west. These winds are known as the trade winds. The trade winds push surface water away from South America and toward Asia and as they do so, the intense tropical sunlight warms the water. This gradual warming results in a typical sea-surface temperature difference of about 8°C between the western Pacific Ocean and the eastern equatorial Pacific coast (NOAA Pacific Marine Environmental Laboratory, 1997a). The lower sea-surface temperatures off the Pacific coast of South America are due to the upwelling of cold water from deeper levels. Under normal conditions, the upwelling brings nutrient-rich water to the surface and is important for supporting diverse marine ecosystems and major fisheries. Rainfall is normally more abundant over the warmer western Pacific Ocean region.

Sometimes, however, the east-to-west trade winds decrease in intensity, resulting in less upwelling of colder water from lower levels in the ocean and warmer-than-normal sea-surface temperatures off the Pacific coast of South America. The change in the trade winds is part of what is called the Southern Oscillation; the sea-surface temperature anomaly is El Niño. The overall pattern of weakened easterlies and higher sea-surface temperatures is commonly referred to as El Niño–Southern Oscillation (ENSO). Precipitation also tends to increase over the warmer waters of the eastern Pacific Ocean during ENSO episodes, often resulting in flooding in Peru. Drought conditions, on the other hand, often prevail in Indonesia and Australia during ENSO episodes. On average, ENSO episodes tend to occur every 2 to 7 years and last from 12 to 18 months (NOAA Pacific Marine Environmental Laboratory, 1997b).

El Niño–Southern Oscillation Impacts on Temperature and Precipitation

Although the changes in the equatorial sea-surface temperatures in the eastern Pacific Ocean that characterize ENSO episodes occur far away from the continental United States, they can have a profound effect on weather and climate in the United States. The weather and climate we experience in the north central and northeastern United States is controlled to a large extent by the large-scale atmospheric circulations that transport heat and moisture over the entire Earth. The rise in sea-surface temperatures off the equatorial Pacific coast of South America during ENSO and the associated change in the trade winds that characterize ENSO episodes can alter the normal atmospheric circulations over other parts of the Earth, including North America. Analyses of past ENSO episodes indicate that the typical paths of low-pressure and high-pressure systems that control daily weather fluctuations in the midlatitudes are altered during the ENSO events (NOAA Climate Prediction Center, 1999). These changes can cause certain regions of the United States, including the north central and northeastern United States, to experience relatively large temperature and precipitation deviations from normal, and the deviations are seasonally dependent (Green et al., 1997).

Fig. 3.15 (color insert) shows the climate divisions in the continental United States where there has been a greater likelihood of a warmer or colder season than one would expect by chance during an ENSO event based on 100 years of past monthly climate division temperature data and monthly standardized Southern Oscillation Index values (NOAA-CIRES Climate Diagnostics Center, 1997). It has been during the winter and early spring months that ENSO episodes usually have had the most broad-scale impact on temperatures in the region (Fig. 3.15a). During the January to March period of ENSO episodes, the western Great Lakes region and the state of Maine have usually experienced warmer-than-normal conditions, while most of the southern and northeastern portions of the region experienced colder-than-normal conditions. During the midspring to late spring months of past ENSO episodes, there have been fewer climate divisions within the region that typically experienced higher-than-normal temperatures compared with the number of climate divisions during the winter months (Fig. 3.15b). Temperatures during this period have tended to be higher than normal in an area extending from Missouri eastward and northward into Illinois, Indiana, Kentucky, and Michigan. The overall pattern of significant monthly temperature anomalies in the region during the summer months of ENSO episodes has generally been confined to the states of Iowa, western Illinois, and northern Maine (Fig. 3.15c). Evidence from past ENSO episodes suggests that climate divisions in these areas of the region tend to experience colder-than-normal conditions during the summer. Finally, temperature observations during past ENSO episodes suggest the likelihood of higher-than-normal temperatures over the region in the fall months of an ENSO year is greatest over the Ohio River Valley and the western Great Lakes states (Fig. 3.15d).

Fig. 3.16 (color insert) shows the climate divisions in the continental U.S. where there has been a greater likelihood of a wetter or drier season than one would expect by chance during an ENSO event based on 100 years of past monthly climate division precipitation data and monthly standardized Southern Oscillation Index values (NOAA-CIRES Climate Diagnostics Center, 1997). Wintertime monthly precipitation observations during past ENSO episodes suggest that a large area within the region could experience relatively low precipitation totals during an ENSO event (Fig. 3.16a). The precipitation observations indicate that climate divisions in northern and eastern Wisconsin, northern and eastern Michigan, eastern Illinois, Indiana, Ohio, Kentucky, West Virginia, Pennsylvania, and New York have usually experienced anomalously low monthly precipitation amounts during the winter months. Only portions of Iowa, Minnesota, and some climate divisions along the eastern U.S. coast typically have relatively high precipitation amounts during ENSO winters. Data presented in Fig. 3.16b–d suggest that during the spring, summer, and fall seasons, only a few climate divisions scattered across the region typically experience relatively low monthly precipitation amounts. During the months of April, May, and June of past ENSO events, the northern Great Plains plus the northern sections of Wisconsin and Michigan have typically been relatively wet (Fig. 3.16b). The summer months of July, August, and September during past ENSO events have been a period of relatively wet conditions for the entire state of Missouri, much of Iowa, and portions of Minnesota, Illinois, Indiana, Wisconsin, and Michigan (Fig. 3.16c). Finally, precipitation observations during past autumn ENSO episodes indicate the likelihood of relatively wet conditions over much of the western sections of the region, including most climate divisions in the states of Minnesota, Iowa, Missouri, and Wisconsin, and some climate divisions in the states of Michigan, Illinois, Indiana, and Kentucky (Fig. 3.16d).

Overview of General Circulation Model Climate Scenario Simulations

Coupled atmosphere–ocean general circulation models (GCMs) have been and continue to be used to provide scenarios of future climate conditions at the global scale under different assumptions of atmospheric greenhouse gas concentrations and emission scenarios. These models include 3-dimensional representations and interactions of the atmosphere, oceans, and land surface on a global time-dependent basis, along with specifications of the chemical composition of the atmosphere and the vegetation on Earth's surface (Gates et al., 1996). Although GCMs are limited in their ability to simulate cloud and radiative effects, the hydrologic balance over land surfaces, and the heat flux at the ocean surface, these models are the most powerful tools currently available for assessing what future climatic conditions may be like (Gates et al., 1996). There have been numerous scenarios of future climate conditions under increased atmospheric greenhouse gas concentrations that have been generated from the current class of GCMs, as outlined by Kattenberg et al. (1996). This section highlight some of the results of the simulations that are particularly important for ecosystem processes and functions in the region: scenarios of annual and seasonal temperature changes, seasonal precipitation changes, soil-moisture changes, and changes in extreme event occurrences.

Patterns of Annual and Seasonal Mean Temperature Changes

At the global scale, GCM simulation results suggest that global mean temperatures could increase between 1 and 4.5°C relative to the present global mean temperature by the year 2100 due to increased CO_2 concentrations in the atmosphere (Kattenberg et al., 1996). All GCM simulations suggest that the greatest warming will occur over land instead of the oceans because of the diminished impact of evaporative cooling over land in comparison with the oceans. The simulations also suggest that annual mean temperatures will increase most significantly at higher latitudes. For example, simulations performed with the Australian Bureau of Meteorology Research Centre (BMRC) and the Australian Commonwealth Scientific and Industrial Research Organization (CSIRO) GCMs indicate that a 1% increase per year in atmospheric CO_2 concentrations will likely result in an increase in annual mean temperature of 3 to 5°C over the 60 to 90°N latitude range at the time of doubled CO_2 concentrations. These simulations also suggest that the region will experience a corresponding 2 to 3°C increase in annual mean temperature (Kattenberg et al., 1996).

On a seasonal basis, GCM climate scenario simulations indicate that warming on a global scale will be most significant in late autumn and winter, with summertime warming small. Northern hemispheric wintertime (December to February) surface temperature changes from the period 1880 to 1889 to the period 2040 to 2049 have been projected to be on the order of 3 to 5°C over the 60 to 90°N latitude region using the Max-Planck Institute for Meteorology (MPI) GCM (Hasselmann et al., 1995), which takes into account atmospheric aerosol effects that have a cooling impact. Similar temperature changes are projected over the 60 to 90°S latitude region during the southern hemispheric winter (June to August). Average wintertime surface temperatures in years 2040 to 2049 over parts (central and south central) of the region and the north Atlantic have been projected to be 0 to 1°C cooler than the 1880 to 1889 temperatures by the MPI GCM (Kattenberg et al., 1996). Other parts of the region are projected to have surface temperatures about 0 to 1°C higher. During the northern hemispheric summer (June to August), the northwestern sections of the region are projected to be 0 to 2°C cooler on average in years 2040 to 2049 compared with the 1880 to 1889 surface temperatures in the region, while the rest of the region is projected to be 0 to 1°C warmer. The direct forcing by sulfate (SO_4) aerosols is responsible for the net cooling effect projected by the MPI GCM over certain regions of the Earth, although the uncertainty in specifying future S emissions and resulting SO_4 concentrations in the atmosphere must be recognized in the simulation results (Kattenberg et al., 1996).

Patterns of Seasonal Mean Precipitation Changes

An increase in global precipitation has been projected by all GCMs under the various scenarios of increased atmospheric CO_2 concentrations. The GCM simulations suggest that wintertime precipitation will generally increase over the northern latitudes and midlatitudes as a result of his atmospheric water vapor content under overall warmer conditions and ₋ₕₑ transport of more water vapor to the northern high latitudes (Kattenberg et al., 1996; Manabe and Wetherald, 1975). When aerosol effects are included in the GCM simulations, the overall increase in simulated global wintertime precipitation is diminished. Kattenberg et al. (1996) reported on the results of nine different GCM climate scenario simulations. Six of the nine GCMs have projected an increase in wintertime precipitation over central North America when the "current climate" atmospheric CO_2 concentrations are doubled. These increases range from 4 to 18% of the average wintertime "current climate" precipitation over central North America. Colman et al. (1995) reported a projected increase in wintertime precipitation over most of the region using the BMRC GCM, assuming a 1% per year increase in atmospheric CO_2 concentrations over the "current climate" concentration; precipitation increases ranged from 0 to 0.5 mm d^{-1}.

Small changes in summertime precipitation under a doubled atmospheric CO_2 environment have been projected by most GCMs over central North America, although a majority of the nine GCM simulations reported by Kattenberg et al. (1996) indicate a slight decrease (2 to 8%) in average precipitation from "current climate" summertime precipitation amounts. A more significant decrease in precipitation over the eastern United States, including much of the region, has been projected with the BMRC GCM (Colman et al., 1995) under a doubled atmospheric CO_2 environment; precipitation decreases ranged from 0.5 to 1 mm d^{-1}.

Patterns of Seasonal Mean Soil-Moisture Changes

Although patterns of precipitation can provide a useful indication of trends in the Earth's hydrologic cycle, soil-moisture patterns are often more useful indicators because they integrate the combined effects of precipitation, evaporation, and runoff (Kattenberg et al., 1996). The current class of GCMs provide projections of soil-moisture conditions under a "changed climate" due to enhanced atmospheric CO_2 concentrations. However, the current class of GCMs are limited in their ability to simulate land–surface interactions because of the simplicity of their land–surface parameterization schemes. Nevertheless, most GCMs suggest that mean soil moisture will generally increase in the high northern latitudes in winter under a "changed climate." Simulation results from the CSIRO GCM (Gordon and O'Farrell, 1997) suggest an average wintertime soil-moisture increase of 0 to 1 cm along the Atlantic coastal states under a doubled atmospheric CO_2 environment (assuming a 1% per year increase in CO_2 concentrations from the "current climate" conditions). In the western sections of the region, the CSIRO GCM climate simulations indicate a corresponding 0 to 1 cm decrease in average wintertime soil-moisture contents compared with the average "current climate" conditions.

Summertime soil-moisture conditions are projected to be drier in the northern midlatitudes by most GCMs under a doubled atmospheric CO_2 environment because of the enhanced evaporation in summer under higher global temperatures. Projected soil-moisture decreases tend to be more pronounced over geographical regions where summertime precipitation is reduced. However, within the broad northern midlatitude bands, where overall soil moisture has been projected to decrease under a doubled atmospheric CO_2 environment, some sections are projected to experience an increase in soil moisture. For example, the CSIRO GCM (Gordon and O'Farrell, 1997) simulations suggest that the eastern and north central regions of the U.S. will encounter an average summertime soil-moisture increase of 0 to 2 cm at the time of CO_2 doubling compared with "current climate" soil-moisture conditions.

When atmospheric aerosol effects are included in the GCM simulations, wintertime soil-moisture changes from the current mean conditions are diminished, while summertime soil moisture increases markedly over North America. This summertime effect is a manifestation of reduced warming over certain regions when atmospheric aerosols are present, thereby reducing the amount of evaporation from soil surfaces.

Patterns of Changes in Extreme Events

The occurrences of extreme regional or local weather events are dependent on atmospheric dynamic processes that span a wide range of temporal and spatial scales. Many extreme events, such as heavy rain or snow, frost or freeze episodes, and high wind events, are influenced by small-scale atmospheric processes which, in turn, can depend on regional and local topographic features, vegetation characteristics, and land–water variations. The relatively coarse resolution of the current class of GCMs does not permit these models to resolve the smaller-scale atmospheric dynamic processes that can play a major role in the development of extreme regional and local weather events. For this reason, it is very difficult to draw any conclusions about potential changes in extreme weather events under a doubled CO_2 environment from GCM climate scenario simulations. However, the IPCC has made some tentative assessments of the potential occurrence of various types of extreme weather and climate events under a doubled CO_2 environment based on reasoning from physical principles and down-scaling techniques (Kattenberg et al., 1996). These assessments include:

- Significant changes in the frequency of extreme events can result from changes in the mean climate or climate variability.

- An overall warming of the atmosphere tends to lead to an increase in the number of extremely high temperature events and a decrease in the number of extremely low temperature events during the winter.
- Daily temperature variability under a doubled CO_2 environment may decrease in certain regions, while daily precipitation variability may increase over some areas.
- Simulations from several GCMs suggest precipitation intensity may increase under a doubled CO_2 environment, and more frequent or severe drought periods may occur in a warmer climate.

Although these general assessments of the IPCC are not specific to the region, they do provide a sense of the types of extreme event changes that are possible under a doubled CO_2 environment over some regions of the Earth.

Atmospheric Deposition and Ozone Patterns

Climate change, atmospheric and ozone (O_3) deposition exert strong influences on the forest ecosystems in the northeast and north central United States. The principal components that determine the atmospheric deposition patterns are the air pollution concentration gradients resulting from regional emissions sources, the meteorological conditions that are conducive to the deposition of acidic compounds and O_3, the topography of the region, as well as the prevailing air transport patterns. These factors contribute to the wide gradient of atmospheric deposition found in the region, ranging from low, unpolluted background levels in the northern plains, to the highest national levels in the East. This section focuses on the spatial and temporal trends for deposition of sulfur (S), nitrogen (N), and tropospheric O_3 to the region. These substances have been intensively studied in the region to determine the long-term effects of atmospheric deposition and O_3 on forest ecosystem productivity and health (see Chapter 5).

The air quality of the region is influenced by the high density of population centers, industries, power generation plants, and transportation corridors in this region. The high chronic levels of air pollution in the region come from different sources. Stationary sources include factories, power plants, and smelters. Mobil sources include cars, trucks, planes, and trains, and local area sources include natural processes such as wildfires, geologic venting, and biogenic emissions.

In addition to the local and regional sources of pollutants in the region, acid deposition, O_3, and their contributing precursors are transported from source areas in the Midwest, which has the some of the largest producers of primary pollutants in the United States (USEPA, 1998a). The primary pollutants that are directly emitted from these sources are transformed in the atmosphere into secondary pollutant forms, such as nitrate (NO_3^-) and sulfate (SO_4) deposition and O_3. Acidic deposition occurs when emissions of sulfur dioxide (SO_2) and nitric oxides (NO_x) in the atmosphere interact with water, oxygen, and oxidants to form acidic compounds, such as nitric acid (HNO_3) and sulfuric acid (H_2SO_4). The emissions sources, atmospheric conditions that create ozone and acidic deposition, regional transport and precipitation patterns, and topography are major factors that determine the pattern and deposition rate of air pollutants to forest systems.

Emissions

The primary anthropogenic cause of acid deposition is the burning of fossil fuels. In the United States, about 70% of the annual SO_2 and 30% of the NO_x emissions are produced by electricity-generating power plants that burn fossil fuels, of which 97% of the SO_2 emissions comes from coal-burning plants (USEPA, 1998b). The Ohio River Valley, with older power plants that burn high-sulfur coal, leads the United States in regional emissions of SO_2 and NO_x. Consequently, areas receiving the most acid rain are the Northeast and Canada, downwind from these emissions sources.

Ground-level O_3 is formed by the reaction of volatile organic compounds (VOCs) and NO_x in the presence of heat and sunlight. The largest source of VOCs are motor vehicles and other mobile sources with the remainder from power plants and other sources of combustion (USEPA, 1998b). The largest source of naturally produced VOCs in the United States are produced by coniferous tree species, making up 60% of the estimated total natural VOC emissions, and deciduous trees, contributing 30%. The annual U.S. production of natural and anthropogenic VOC emissions are estimated to be nearly equal in mass, but VOCs emitted from natural sources have greater reactivity for potential O_3 production (Allen and Gholz, 1996).

Atmospheric Deposition

As these primary pollutants (NO_x, SO_2, VOCs) are transported by weather patterns over the region, they are transformed by a variety of chemical reactions to secondary pollutants, such as sulfuric and nitric acid aerosols, particulate sulfate and nitrate, and ozone. These pollutants can remain in the atmosphere, affecting visibility and air quality, or can be deposited onto terrestrial surfaces and bodies of water.

Atmospheric deposition occurs via three main pathways: wet deposition, in which material is dissolved in droplets and deposited as rain or snow; dry deposition, involving the direct deposition of gases and particles (aerosols) to surfaces; and cloud-water deposition, involving material dissolved in cloud droplets and intercepted by forest canopies. Atmospheric deposition and air quality data discussed in this section are limited to regional acidic deposition and O_3 patterns.

Acidic deposition is described in this section as the input of wet and dry deposition of inorganic S, N, and H^+ though rain, cloud water, and as aerosols. Sulfur forms include gaseous SO_2 and SO_4 in rain and aerosols. Nitrogen forms are nitrate and ammonium ions in rain and aerosols, nitric acid vapor, and gaseous nitric oxides. Hydrogen ion deposition is through precipitation as rain and snow, and as cloud water.

Monitoring Networks

The National Atmospheric Deposition Program, National Trends Network (NADP/NTN) was initiated in 1978 to monitor the long-term trends in wet acidic deposition. Precipitation is sampled weekly and analyzed for nitrate, ammonium (NH_4^+), sulfate, hydrogen ion (pH), as well as calcium, magnesium, potassium, sodium, chloride, and phosphate (Ca^{2+}, Mg^{2+}, K^+, Na^+, Cl^-, PO_4). The 200 monitoring sites provide national coverage, with stations located mostly in rural locations, away from point sources and large urban centers. The National Oceanic and Atmospheric Administration (NOAA) Atmospheric Integrated Research Monitoring Network (AIRMoN-wet) operates a smaller, more intensive daily sampling network of the NADP.

The Clean Air Status and Trends Network (CASTNet), provides weekly monitoring for dry deposition parameters, including filter-pack measurements of nitric acid and sulfur dioxide, and fine-particle nitrate, ammonium, and sulfate. The network consists of about 50 sites, located primarily on the East and West Coasts. Dry deposition is much more difficult to measure than wet deposition. It is not measured directly, but calculated using measured air concentrations and model estimates of deposition velocity from which dry deposition rates are calculated as the product of deposition velocity and air concentrations. Dry deposition consists of aerosols (gases), small particles, and large particles, and the deposition velocities vary with each component, adding to the difficulty in finding one suitable method. The AIRMoN-dry deposition network operates as a smaller program for developing dry deposition methodologies in CASTNet applications.

Cloud-water deposition was monitored in the Appalachian Mountains from 1986 to 1988 by the Mountain Cloud Chemistry Program (MCCP) at six study sites, and sampling continues at selected sites through the CASTNet Mountain Cloud Deposition Program (MADpro). The purpose of MCCP was to examine the spatial and temporal variation in cloud-water deposition and to determine the importance of cloud-water input to tree canopies relative to wet deposition at high-elevation sites.

The Clean Air Act required each state to establish a network of air monitoring stations called the State and Local Air Monitoring Stations (SLAMS). In order to obtain more timely and in-depth information about air quality at strategic location, The U.S. Environmental Protection Agency (EPA) established the National Air Monitoring Stations (NAMS) as part of the SLAMS network to meet more stringent air monitoring criteria. Approximately a third of these sites are designated rural or remote, but in practice, most are in close proximity to major population centers. The primary focus of this long-term monitoring network is to determine the O_3 concentrations to which large numbers of people are exposed. These monitoring sites are centered in or near urban centers where ambient air concentrations reflect local emissions. In recent years, there has been more interest in the effects of O_3 in rural areas. The CASTNet program also maintains O_3 monitoring at rural sites, where forest and agricultural concerns are the focus.

The EPA manages the Aerometric Information Retrieval System (AIRS) as the primary repository of the nationwide database on the criteria pollutants that must meet the National Ambient Air Quality Standards (NAAQS). Other research groups analyze O_3 production and O_3 transport patterns. The Ozone Transport Assessment Group (OTAG) was established by the Environmental Commissioners of States (ECOS) with the active participation of all 37 states east of the Mississippi (OTAG region). It provides analysis of transport patterns to develop a strategy to deal with long-range transport problems associated with regulating O_3 and its precursors. The Ozone Transport Commission (OTC), consisting of the 12 Northeastern and Mid-Atlantic states, was established by the Clean Air Act Amendment of 1990 to understand and assess the O_3 problems and evaluate control strategies. The North American Research Strategy on Tropospheric Ozone (NARSTO) consists of the Unites States, Canada, and Mexico, and conducts studies on the causes of severe O_3 episodes and the interaction between O_3, NO_x, and VOCs.

Nitrogen and Sulfur Compounds

The N compounds that are of concern for atmospheric deposition and forest interaction are reactive N species in oxidized (NO, NO_2, HNO_3, and NO_3^-) and reduced (NH_3, NH_4^+, and organic nitrogen) forms. About 90% of the nitrogen oxides and about 96% of the sulfur oxides emitted into the atmosphere are anthropogenic by origin (Lovett, 1994; Allen and Gholz, 1996). The primary form of nitrogen oxide emissions is nitric oxide (NO). In the presence of VOCs and sunlight, this gas is rapidly converted to nitrogen dioxide (NO_2), which further reacts to form nitric acid vapor (HNO_3). Nitrogen dioxide is also decomposed by sunlight to produce O_3, therefore both acidic deposition and O_3 formation are intimately tied to NO_x emissions. Nitric acid vapor can be dissolved by rain and cloud-water to form nitrate (NO_3^-) ions and can be deposited as wet and cloud-water deposition. Dry deposition of nitric acid adsorbs rapidly to the canopy due to a high deposition velocity. Most of the ammonia (NH_3) in the atmosphere is thought to originate from volatilization of animal waste, from fertilized farmlands, and alkaline soils. Atmospheric NH_3 can be deposited as dry deposition or dissolved in rain and cloud water, forming ammonium (NH_4^+), and deposited as wet and cloud-water deposition. Organic N deposition is poorly understood and rarely monitored. It is composed primarily of particulate material from soils, vegetation (pollen, VOCs), animal waste, and reactions of NO_x with organic compounds.

The primary forms of S deposition that are important to forests systems are SO_2 and SO_4. These oxides of sulfur are mainly emitted from coal-burning electricity-generating facilities, predominantly in the form of SO_2 gas. The electricity-generating facilities are often located in rural areas and have a large impact on rural airsheds. Sodium dioxide is oxidized in the gaseous and aqueous phase to ultimately form H_2SO_4 vapor and H_2SO_4 aqueous acid solution. It can react with other aerosols to form particulate SO_4. About 20% of the SO_2 emitted in the United States is converted to SO_4 during its atmospheric lifetime. The main removal pathway for SO_2 is dry deposition to surfaces, and SO_4 can be deposited as wet, cloud-water, and dry deposition.

Acidic Deposition: Regional Trends

The northeastern and north central subregions of the region are distinguished by having the nation's highest levels of acidic, sulfur, and nitrogen deposition (Fig. 3.17, color insert). Maps based on NADP/NTN wet deposition data for the eastern U.S. were produced using surface estimation algorithms for H^+, SO_4^{2-}, NO_3^-, NH_4^+, total inorganic N, and precipitation averaged from 1983 to 1994 to characterize wet-deposition spatial trends (Lynch et al., 1996a). Mean annual H^+ concentration from 1983 to 1994 ranged from pH >5.3 at pristine sites in northwestern Minnesota to acidity levels a full pH unit lower (pH < 4.1) for much of the Ohio River Valley and along the length of the Appalachian Mountains and Appalachian Plateau region (Fig. 3.17a; NADP, 1996, 1998).

The regional pattern of deposition reflects the emissions source areas in the Midwest, the Ohio River Valley, and western Pennsylvania, and the prevailing wind patterns, which transport the elevated levels of acidic deposition to the eastern United States. Sulfate and nitrate deposition patterns over the region were similar to H^+ deposition (pH), with the lowest deposition in the northwestern corner of the region, gradually increasing eastward along a longitudinal gradient. The deposition gradient shifts its axis to the northeast along the Ohio River Valley, reaching the highest values in a region bordered by eastern Ohio, northern West Virginia, western Pennsylvania, and western New York. At the northern

Appalachian Plateau region and the Allegheny Mountains, deposition loads decrease slightly on the eastern side, corresponding to lower precipitation totals in the central Pennsylvania region. Higher deposition loads return with higher precipitation off Lake Ontario, reaching from the Adirondack Mountains to the Catskills, then decreases along a longitudinal gradient, with high values for the Green Mountains and White Mountains, but reaching low deposition levels in northern Maine. Ammonium shows a distinctly different spatial pattern, with higher deposition in the Midwest and eastern Great Plains states (Fig. 3.17e), stemming from greater NH_3 volatilization from fertilizers and animal wastes in the agricultural regions.

Dry deposition contributes a significant portion of the total N and S acid deposition, and can account for the majority of the N deposition in some subregions (Tables 3.3 and 3.4). CASTNet data show that dry deposition could account for up to 52% of total (wet + dry) N deposition in the Northeast and up to 45% of total S deposition (USEPA, 1998c; Allen and Gholz, 1996). Dry deposition is more variable spatially than wet deposition, as it is depleted with greater distance from the source. Dry deposition is a more significant contributor in and near major source regions, and wet deposition is more significant in areas with heavy precipitation, such as mountainous regions. In general, calculated dry

Table 3.3. Regional Averages of Total Nitrogen Deposition by Year and Percentage of Dry Deposition (INS = Insufficient Data)

Region	Total Deposition (kg ha^{-1})							
	1989	1990	1991	1992	1993	1994	1995	Average
Northeast	7.6	7.7	6.9	6.7	7.6	7.5	7.1	7.3
Upper NE	3.9	3.1	3.3	3.1	3.3	3.3	3.3	3.3
Midwest	6.3	6.5	5.9	5.8	6.9	6.2	6.5	6.3
Upper MW	4.8	5.0	INS	5.4	4.8	4.4	4.5	4.8
South Central	6.6	5.6	5.3	5.2	5.7	5.6	6.6	5.8
Southern Periferal	4.0	6.1	3.8	3.6	3.8	4.4	4.2	4.3
West	INS	1.7	1.7	1.5	1.4	1.7	1.2	1.5
East	6.1	6.1	5.6	5.5	6.0	5.7	5.9	5.8

Region	Percent Dry Deposition							
	1989	1990	1991	1992	1993	1994	1995	Average
Northeast	43.1	41.1	45.8	42.3	42.4	47.3	51.8	44.8
Upper NE	20.0	20.3	18.8	18.2	21.0	21.6	21.7	20.2
Midwest	41.2	42.2	45.9	42.8	41.3	50.7	51.3	45.1
Upper MW	38.4	34.7	INS	34.6	34.0	37.0	40.9	36.6
South Central	49.9	53.7	49.6	51.1	52.1	51.1	59.5	52.4
Southern Periferal	36.7	41.0	36.6	40.0	41.5	37.9	48.5	40.3
West	INS	49.3	50.3	54.5	47.1	57.4	38.2	49.5
East	42.1	43.9	44.9	43.3	43.3	46.7	50.5	45.0

Table 3.4. Regional Averages of Total Sulfur Deposition by Year and Percentage of Dry Deposition (INS = Insufficient Data)

Region	Total Deposition (kg ha^{-1})							
	1989	1990	1991	1992	1993	1994	1995	Average
Northeast	17.1	18.1	15.3	14.9	16.3	15.3	10.9	15.4
Upper NE	6.8	6.3	6.4	6.1	6.0	6.1	5.2	6.1
Midwest	16.3	16.3	13.7	13.3	15.0	13.1	11.3	14.1
Upper MW	6.9	8.7	INS	8.2	7.5	6.4	5.0	7.1
South Central	14.5	11.9	10.7	10.5	11.5	11.3	9.9	11.5
Southern Periferal	7.4	7.3	6.7	6.3	6.4	7.1	5.7	6.7
West	INS	1.8	1.8	1.5	1.7	1.7	1.1	1.6
East	13.4	13.7	12.1	11.4	12.2	11.3	9.1	11.9

Region	Percent Dry Deposition							
	1989	1990	1991	1992	1993	1994	1995	Average
Northeast	41.6	40.4	44.7	41.1	41.2	43.0	45.4	42.5
Upper NE	17.2	17.7	13.1	15.0	16.1	15.9	14.7	15.7
Midwest	39.7	39.1	43.5	38.9	39.0	44.9	42.6	41.1
Upper MW	32.5	26.8	INS	23.0	26.4	28.4	29.0	27.7
South Central	36.1	36.5	33.4	36.3	38.0	35.5	40.8	36.7
Southern Periferal	24.4	26.2	23.0	27.3	27.5	23.4	27.8	25.7
West	INS	34.9	33.6	33.6	37.1	46.8	21.8	34.6
East	37.2	37.2	39.4	36.5	37.4	38.4	39.8	38.0

deposition fluxes for the CASTNet sites (USEPA, 1998c) showed that SO_2 and HNO_3 were the dominant forms of S and N dry deposition, with SO_2 accounting for about 70% of the dry S deposition at eastern sites and HNO_3 accounting for approximately 65% of the dry N deposition. Higher dry deposition percentage occurred along the Ohio River Valley into New York State and also into the Mid-Atlantic sites, as well as drier regions in the south. A low ratio of dry/wet deposition may reflect higher precipitation patterns for those regions.

The ratio of cloud deposition to precipitation deposition varies from 1:1 in the northern sites to 2:1 and 4:1 in the southern sites, providing a substantial fraction of the total chemical deposition to high-elevation forests. Cloud water with a pH of 3.0 was routinely sampled at the MCCP sites. In high-elevation spruce–fir forests, cloud-water pH was approximately 1 pH unit lower than precipitation pH (pH 2.8 to 3.8 vs. pH 3.8 to 4.9) at the same site (NAPAP, 1991). Concentrations of the major ions, H^+, NO_3^-, SO_4^{2-}, and NH_4^+ were substantially higher in cloud water than in precipitation, ranging from a factor of 5× to 20×, depending on location (Mohnen, 1992). Cloud frequency (% hours), in which clouds cover these high-elevation sites may reach 20 to 40% of the time during the growing season, depending on elevation. Total deposition of S and N above cloud base is nearly twice as great as below cloud base.

The MCCP sites experienced the highest concentrations with cloud water originating from trajectories that passed over the Ohio River Valley. The data support the idea that forests exposed to cloud immersion are exposed to higher atmospheric loading despite their distance from major sources. Factors that increase the efficiency of atmosphere-to-surface exchange of pollutants in mountains include cloud immersion, the high surface area of mountain conifers, and generally higher wind speeds.

Total Deposition

Total N and S depositions for forest and rural areas have been calculated for the CASTNet monitoring sites in the region by combining the wet, dry, and cloud-water deposition. Annual average total N deposition from 1989 to 1994 ranged from 2.5 kg ha^{-1} in northern Maine to 8.5 kg ha^{-1} in central Pennsylvania. Annual average total S deposition for the 6-year period ranged from 5.2 kg ha^{-1} for Maine to 18.7 kg ha^{-1} in southern Indiana. Estimates of total N and total S depositions reached as high as 28 kg N ha^{-1} and 36 kg S ha^{-1} in regions of the United States that are subject to cloud-water and high regional loading (Johnson et al., 1991).

The total deposition is most likely underestimated because of limitations in directly measuring dry deposition and cloud-water deposition to forest canopies, and by accounting for only the major inorganic forms of N and S. The contribution that organic N deposition (both wet and dry) adds to the total deposition is poorly understood. At local scales, topography (elevation, slope, aspect) and vegetation canopy cover can cause significant variation in deposition rates, such as at mountainous forest sites. At higher elevations, the orographic uplifting of air masses causes increased precipitation. Wet deposition can increase from direct cloud-water inputs and dry deposition, both of which are thought to increase with elevation.

At larger regional scales, patterns of wet and dry deposition depend of the regional emissions sources, transport patterns, area and local emissions sources, weather and circulation patterns, topography, and land use.

Temporal Trends in Acidic Deposition

An analysis by Lynch et al. (1996a) of the ionic concentration of precipitation at NADP/NTN sites from 1983 to 1994 found that the annual mean SO_2 concentrations decreased east of the Mississippi, with major drops in the Ohio River Valley and the Mid-Atlantic states. Sulfate concentrations at 92% of the selected NADP/NTN monitoring sites in the United States have decreased since 1983. The trends were significant ($p < 0.05$) at 37% of the sites. Hydrogen ion concentration decreased at 82% of the monitoring sites for this period in the Northeast, and these trends were significant at 35% of the sites. No sites in the Northeast saw a significant increase in SO_4 or H^+ during this period. Nationally, NO_3^- concentrations remained relatively unchanged, with 56% of the sites with decreasing trends, countered by 44% of the sites with increasing trends. However, 14% of the sites had significant increasing trends, while only 1% showed significant decreases. Ammonium concentration increases were larger than those for NO_3^-, with 80% of the sites showing increasing NH_4^+ concentrations since 1983 and 22% of the sites having significantly higher concentrations (Lynch et al., 1996a). The largest increases in NO_3^- and NH_4^+ were in the western states. Unlike the long-term decreasing trend for S concentrations, N concentrations in precipitation have increased in the United States since 1983.

Emissions Reductions and Deposition Rates

In Phase I of Title IV of the 1990 Clean Air Act Amendments (CAAA), Congress set emissions reduction goals for SO_2 beginning on January 1, 1995, at selected high-emitting electricity-generating facilities located primarily in the eastern United States. The first years of compliance in the SO_2 and NO_x emissions programs were 1995 and 1996, respectively. In order to evaluate the effect of the reductions program on acidic deposition, NADP/NTN deposition monitoring data from 1983 to 1994 were analyzed and a seasonal trend model was developed to estimate the expected precipitation chemistry for 1995. The observed 1995 precipitation chemistry was compared with the predicted 1995 estimates based on the pre–Title IV 1983 to 1994 trend model to evaluate the percentage of departure that Title IV compliance for 1995 had on the spatial and temporal patterns for NO_3^-, SO_4, and H^+ concentrations (Lynch et al., 1996a).

In the first year of compliance, SO_2 emissions dropped dramatically and were 39% below the allowable level under Title IV. This resulted in lower SO_4 concentrations in precipitation in the eastern United States, (Fig. 3.18b), particularly along the Ohio River Valley and states downwind in the Mid-Atlantic region. Mean annual SO_4 and H^+ concentrations for the eastern states observed in 1995 were below predicted values at 89 and 79% of the selected monitoring sites, respectively, using the 1983 to 1994 trend model. Measured SO_4 concentrations were about 11% less than estimated for the Northeast, and as much as 25% less in the Mid-Atlantic and Ohio River Valley regions (Fig. 3.18a). Concurrent with these SO_4 reductions have been similar levels of reductions in H^+ concentrations, which decreased 12%. The spatial pattern was identical to the decrease in SO_4 (Figs. 3.18a,b, color insert).

Maximum reductions in SO_4 and H^+ concentrations occurred in the Ohio River basin and the Mid-Atlantic regions immediately downwind of the major stationary sources targeted by Phase I of the CAAA-90, Title IV, as indicated by the plus (+) signs in Fig. 3.18. Precipitation deviations from the long-term (1983 to 1994) average (Fig. 3.18d) does not explain the observed decreases in SO_4 and H^+ concentrations in 1995 (Fig. 3.18a,b). Lower precipitation volumes are associated with higher concentrations, and most of the eastern states had below-average precipitation volumes in 1995. Precipitation volumes would not selectively reduce only SO_4 and H^+. Nitrate and other ions would be similarly affected, but were not. The lower precipitation volumes resulted in higher 1995 concentrations in virtually all ions except for SO_4 and H^+. The two ions declined independent of precipitation volume. These results clearly support the conclusion that Phase I of the CAAA-90, Title IV, has reduced acid deposition in the eastern United States (Lynch et al., 1996a).

Nitrate concentrations in 1995 were higher than predicted from the reference trend model (Fig. 3.18c, color insert) corresponding to the lower precipitation for that year. Nitrogen oxide reductions for Title IV did not begin until 1996, and had substantially less impact in reducing acid deposition than the 1995 reductions in SO_2 from the targeted stationary sources. In contrast, stationary sources contribute only 30% of the total NO_x emissions for the United States. Because of the wide variety of sources for N emissions, NO_x are more diffcult to regulate and reduce.

Spatial Pattern of Ozone

The northern hemispheric global average O_3 concentration in the lower troposphere is about 40 ppb. This represents the background conditions for O_3 values for the northern United States. The average daily midday O_3 concentration over the entire eastern United States is about 60 ppb. Concentrations over urban industrial areas average 80 ppb, about twice the lower tropospheric background (Fig. 3.19, color insert). Lowest O_3 levels are in northwestern Minnesota and northern Maine, with the highest regional levels occurring in the Washington–New York corridor, where the average daily maximum ozone concentration exceeds 70 ppb. A second area of high O_3 is centered over the Ohio River Valley, including the Kentucky–Indiana–Ohio borders (Fig. 3.20, color insert).

Temporal Patterns of Ozone

Ozone concentrations vary greatly across time and space. The northern U.S. exhibits a strong degree of seasonality for O_3 production. This pattern is predictable as seasonal O_3 tracks air temperature closely. Highest rates develop during the summer months and drop to background levels during the cold winter months. High temperatures can increase the emission rates of natural and anthropogenic VOCs and high temperatures correlate with the meteorological conditions that enhance the photochemical formation of O_3: clear skies, increased solar radiation, low wind speeds, and transport patterns from the south and west.

During the summer months, high ground-level O_3 concentrations are observed within and downwind of many large urban areas in the eastern United States. Ozone patterns for 1991 to 1995 (Husar, 1997; OTAG, 1998) show highest summer (June to August) average daily maximum O_3 levels in the Ohio River Valley and the Mid-Atlantic region, extending up the northeast corridor (Fig. 3.21, color insert). Highest average daily maximum O_3 levels correspond with the large anthropogenic NO_x and VOCs densities near urban centers, such as like the Washington–New York corridor and along the industrial Ohio River Valley.

National trends from 1987 to 1996 show that the ambient average daily maximum 1-hour O_3 concentrations decreased 15%, and the estimated number of 120 ppb 1-hour exceedances declined by 73% over the past decade (USEPA, 1997a). The 1996 mean O_3 concentration dropped 6% lower than in 1995. This large decrease in 1996 was attributed to noticeable changes in the meteorological conditions for the eastern region. The summer of 1995 was hot and dry throughout most of the central and eastern United States, making it ideal for peak O_3 formation, while 1996 was dramatically wetter and cooler, with an unseasonably low number of hot days in the Northeast. Days with temperatures equal or greater than 90°F in the Ohio River Valley for the summer of 1995 reached 20 to 33 days across this region, compared with 5 to 10 days above 90°F in 1996. Along the eastern corridor from Philadelphia to Richmond, the temperature climbed to at least 90°F for 40 to 44 days compared with 3 to 15 days in 1996 (NOAA, 1996). The year 1988 was unusually hot and dry for the eastern United States and produced an anomalous O_3 season, producing the highest national composite mean O_3 level (USEPA, 1997a) and the greatest number of exceedances of the 1-hour 120 ppb standard in the past decade (Baumgardner and Edgerton, 1998).

Ozone production varies daily and seasonally, as the atmospheric chemical reactions depend on sunlight and higher temperatures. As sunlight intensity decreases, O_3 levels drop rapidly, and continue to be depleted during the night. This daily course of O_3 production is less evident in rural and mountainous areas where concentrations can remain elevated into the night, especially at higher elevations under atmospheric inversion conditions (Baumgardner and Edgerton, 1998). Typical daily O_3 peaks for remote forested areas occur from 2 PM to 5 PM with nocturnal predawn lows at 50 to 60% of daytime maximum O_3 levels as for Ashland, ME (Fig. 3.22). Daily cycles for Beltsville, MD show greater amplitude of daily peaks and shorter duration of daily peak O_3, with lower nocturnal minimum from nitric oxide titration in more urban areas (Fig. 3.22). High elevation forested sites show little or no dirunal pattern, with the curve nearly flat for Big Meadows, VA (Fig. 3.22). High-elevation sites in the East could therefore have consistently high O_3 exposure but not high peak concentrations. Topography can explain some of the different patterns between rural sites. Mountaintops are usually above the nocturnal

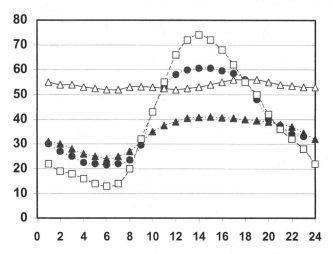

Figure 3.22. Diurnal hourly average ozone concentrations for sites in the Mid-Atlantic region showing the influence of terrain and urbanization in (a) Ashland, ME (▲), (b) Parson, WV (●), (c) Big Meadows, VA (△), and (d) Beltsville, MD (□).

inversion layer, and radiational cooling may circulate O_3-laden air from aloft to the site at night. Lower elevation sites, such as Parson, WV, has complex terrain that is under the nocturnal boundary layer for longer periods, and subject to O_3 loss through surface deposition losses.

New Primary and Secondary Ozone Standards

Recent proposed revisions by EPA on the National Ambient Air Quality Standards, lowering the primary (public health) and secondary (forest and crops) standard from 120 ppb to 80 ppb, and changing the time base average from a 1-hour peak period to an 8-hour average (USEPA, 1997b). Counties that exceeded the current 1-hour primary standard more than once per year (averaged over three consecutive years) were deemed non-attainment areas. Counties must eventually comply with a maximum 8-hour average for each year not to exceed 80 ppb (3-year average of the fourth-highest daily maximum 8-hour average), to be considered in attainment. Therefore, EPA will not designate areas as non-attainment for the new 8-hour O_3 standard until the year 2000.

The spatial pattern of counties in the NGCRP region with 1-hour exceedances is most evident over large urban metropolitan areas along the northeast corridor between Washington and Boston, and between Chicago and St. Louis. On average, the 8-hour daily maximum O_3 is about 85% of 1-hour daily maximum values. However, projections are that more counties will be in non-attainment of the 80 ppb air quality standard than the 120 ppb standard (Fig. 3.23, color insert) as the industrial states north of the Ohio River, from Illinois to Pennsylvania, will have more exceedances based on the 8-hour standard.

Ozone Transport

Non-attainment states must make reductions to meet the new O_3 standard. However, these states do not have control of pollutants entering their airsheds through regional transport. OTAG has analyzed typical transport patterns of high-O_3 incidents (Husar and Renard, 1997; OTAG, 1998). Regional episodes during the summer can develop when accumulated O_3 covers several states in the eastern region. Ozone concentrations rise as temperature and concentrations of precursors rise in the region. As the winds increase, concentrations in the Midwest decrease, and high regional O_3 can be transported eastward, from one state to another, contributing to exceedance of the O_3 standard.

On high-O_3 days, transport winds are slow with clockwise circulation around the south central and eastern states. On low-O_3 days, faster transport winds originate from outside the domain (Fig. 3.24, color insert). Low-O_3 air comes from outside, while high-O_3 originates from inside this domain. High winds disperse local O_3 but contribute to higher regional O_3 through long-range transport. Stagnation over the multistate areas followed by transport of O_3 downwind results in regional O_3 episodes. There is an increasing O_3 trend from west to east in the midsection of the eastern United States from St. Louis to Baltimore (Husar, 1997).

Significant effort for O_3 reductions at their boundaries will be necessary to reach attainment guidelines locally. New attention has been focused on NO_x as precursor of tropospheric O_3 production. NO_x emissions are highest near cities, and correspond with power generation and fossil fuel usage in the industrial Midwest and Ohio River Valley. Although the role of NO_x in O_3 formation was known, the early focus for controlling O_3 was directed at controlling VOCs. To control O_3 near the ground, it is necessary to link emissions reductions of both NO_x and VOC precursors. The reductions in NO_x were expected to have broad regional O_3 benefits, whereas VOC controls would generate greater O_3 improvements on a local scale.

Ozone Statistics

Ozone is potentially the most damaging of the major air pollutants in terms of negative impacts to forest growth and species composition (Barnard et al., 1990). In the eastern seaboard states, 20% of the forested area and 27% of the croplands are in counties that exceed the NAAQS for O_3 in 1989 (Lefohn et al., 1990). This may be a conservative estimate of area affected, as O_3 damage in plants is known to occur at levels as low as 50 to 60 ppb for sensitive plants (Heck et al., 1982). The use of NAAQS maximum hourly means correlated well with long-term averages in urban areas, but not for rural forested sites because of different O_3 concentration patterns in sites distant from emissions sources. Ozone and dry HNO_3 deposition at rural forest sites show a positive correlation using a weighted O_3 concentration averaged and dry nitric acid deposition. Therefore, forest that may see high average O_3 concentrations may also have the higher loadings of N and S dry deposition in these forested landscapes (Taylor et al., 1992).

Many O_3 exposure indices have been developed to compare daily, monthly, seasonal, and annual periods to describe the distribution of O_3 to the biological systems across the United States (Allen and Gholz, 1996; Lefohn et al., 1992). In agricultural and forested areas, it has been argued that cumulative O_3 indices are more biologically relevant than the time-based average 1- and 8-hour standards. The cumulative indices, such as SUM06 and W126, weight more strongly the higher values of O_3 measured over the 12-hour periods.

The SUM06 value is calculated as the sum of all O_3 concentrations greater than or equal to 60 ppb for each day (8 AM to 8:00 PM). The daily values are summed over the O_3 season (March through October). The maximum 3-month sum for the site is used an index value for that year. The W126 index (Lefohn et al., 1988) weights high hourly exposures more heavily (>80 ppb) and de-emphasizes lower, less biologically effective levels (< 30 ppb) using a sigmoidally weighted function.

W126 index maps for 1994, 1995, and 1996 were developed by Lefohn for the Forest Health Monitoring Program (Fig. 3.25, color insert) to show the year-to-year variability for the north central and northeastern regions of the United States. Spatial patterns shows higher W126 index values, expressed as parts per million hours (ppm-h), along the Mid-Atlantic region, down the eastern seaboard, and bounded by the Appalachian Mountains to the west. High W126 extend through lower Michigan and through the Ohio River basin. The lowest seasonal W126 values found in the NGCRP region were in the northern extremes of Minnesota and Maine. Ozone production varies greatly year to year, due to meteorological factors. The years 1995 and 1996 represent contrasting very high and very low ozone years based on hot–dry vs. cool–wet summer climate conditions (NOAA, 1996). Conditions that produce local and regional O_3 are much more variable than wet N or S deposition. Although long-term averages of O_3 concentrations may be necessary to identify O_3 trends to the region, the high O_3 years may be more important biologically, due to potential damage from peak O_3 episodes.

Regional Mapping of Atmospheric Deposition

The interpolated maps of long-term wet distribution chemistry and of O_3 concentration have been particularly useful in describing the spatial and temporal patterns of atmospheric deposition for regional analysis and for providing coverages where data are lacking (Allen and Gholz, 1996; Lefohn et al., 1992; Mohnen, 1992; Smith and Shadwick, 1992; SAMAB, 1996). There is much greater difficulty in modeling and predicting dry deposition and cloud deposition over a landscape or regional basis due to the enormous number of biological, chemical, and physical factors that regulate these deposition processes. The Regional Acid Deposition Model (RADM) (Dennis et al., 1991), a large complex meteorological transport model, has been used to predict wet and dry acidic deposition and oxidants at a 20 to 80 km grid scale. The model considers the emissions of the precursors to acidic deposition, the meteorological processes that transport and mix the atmospheric chemicals over time and space, the physical and chemical reactions that occur, and the meteorological factors and surface properties of the terrain that lead to the deposition of acidic substances. The model is used for episodic runs for areas east of the Great Plains to forecast changes in deposition and air quality resulting from changes in primary emissions, and predicts the amount of acidic deposition received in the assessment area, such as the Chesapeake Bay watershed basin.

Spatial mapping and analysis using climate and atmospheric deposition trends can be used in process models to evaluate the effects of single or multiple stresses on forested systems over an extended geographical area. They require estimates of the deposition of chemicals and acidity of precipitation at much finer geographical scales to resolve important topographic features so the chemical inputs can be accurately matched with site and vegetation conditions. Ollinger et al. (1993) used a regression approach with a digital elevation map (DEM) to produce interpolated maps of deposition for the New England states at 1 km^2 resolution. They found that the gradients for wet and dry deposition rates followed longitude and latitude for the northeastern region, indicating that wet deposition material originates from sources in the Midwest (longitudinal gradient) while dry deposition originates from urban sources along the eastern seaboard (latitudinal gradient).

Lynch et al. (1996b) have produced enhanced regional wet deposition estimates based on modeled precipitation inputs. The model also incorporates topographic features to estimate wet deposition of major ions for the region. The coordinates, elevation, and monthly precipitation from a larger network of NOAA monitoring sites (about 1500) provide an extensive record of precipitation volume, and the wet deposition concentrations from the NADP/NTN network comprise the data set used for the model. The current model is a moving neighborhood, distance-weighted linear least squares regression of precipitation monitoring observations on latitudinal and longitudinal coordinates, elevation, and a set of variates to represent both slope and aspect. The modeled enhancement of wet deposition by this method can be produced at 0.3 to 1 km^2 resolution, but is limited by the accuracy of the site location for the precipitation station, typically characterized at 1 minute resolution for latitude and longitude.

Human-Induced Climate Cycles with Atmospheric Deposition

Direct human influence on regional climate by air pollution has been hypothesized as the cause of weekly cycles identified in climate and pollution data sets. Urban centers have a prominent weekly pollution cycle, characterized by high late-week pollution levels (Friday) as opposed to low early-week levels (Sunday), as the result of weekly patterns of anthropogenic emissions. Cervaney and Balling (1998) argue that this shows direct anthropogenic forcing of regional climate as the 7-day cycle accumulates higher industrial pollutants levels throughout the workweek, thereby supplying a heavier load of condensation nuclei during the weekend. This fosters the growth of heavier cloud cover, which is estimated to supply 20% more rain during the weekends in the Atlantic seaboard region.

Weekly cycles of O_3 levels corresponding to the workweek have been shown with OTAG daily O_3 data analysis (OTAG, 1998). Highest maximum daily O_3 (90th percentile) shows up to a 15 ppb difference in the Northeast (Fig. 3.26, color insert) from Friday to Sunday, coinciding with human activities of the summer work week.

Conclusions

Overall, the NADP/NTN monitoring network shows acid deposition declining. Sulfate declined at 92% of the study sites between 1983 and 1994, with 38% of the sites showing statistically significant decreases. Hydrogen ion concentration decreased at 82% of the monitoring sites for this period in the Northeast, and this trend was significant at 35% of the sites. Nitrate and ammonium exhibited greater regional variability, showing slight increases since 1983. With the implementation of Title IV of the CAAA of 1990, SO_2 emissions dropped dramatically and were 39% below the allowable level. There was a dramatic 10 to 25 % drop in the 1995 wet deposition SO_4 concentration and acidity pH compared with the 1983 to 1994 reference period (see Fig. 3.18). Regional reductions were at some of the highest acid rain receptor regions in the Midwest, Northeast, and the Mid-Atlantic region. Increases in SO_4 concentrations were found in southern Michigan and the southwest region of the map (see Fig. 3.18b,d) and were attributed to lower rainfall than normal in 1995. Mean annual SO_4 and H^+ concentrations for the eastern states observed in 1995 were below predicted values at 89% and 79% of the selected monitoring sites, respectively, using the 1983 to 1994 trend model. Measured SO_4 concentrations for the Northeast were about 11% less than estimated and as much as 25% less in the Mid-Atlantic region and the Ohio River Valley (see Fig. 3.18a). Concurrent with these SO_4 reductions, have been similar levels of reductions in acidity. However, NO_3^- deposition has increased nationally, primarily in the West.

Regional transport of pollutants plays an important role in the Northern Region. Acid deposition, O_3, and their contributing precursors are transported from source areas in the Midwest, which has the some of the largest producers of primary pollutants in the United States. Studies estimate that 25 to 35% of the N from atmospheric deposition into Chesapeake Bay is from the surrounding airshed, outside the Chesapeake watershed boundaries.

Ozone and dry nitric acid deposition at rural forest sites show a positive correlation using a weighted O_3 concentration averages, SUM60, and dry nitric acid deposition. Therefore, forest that may see high average O_3 concentrations may also have higher loadings of N and S dry deposition in these forested landscapes, producing conditions of multiple air pollution stress for forested systems over extended geographical areas (Taylor et al., 1992).

In higher elevations, the patterns of acid deposition and O_3 exposure may often differ from rural areas at lower elevations. In the case of acidic deposition, it may underestimate the total loading due to orographically enhanced deposition. Ozone at high-elevation sites may see different exposure patterns, and the maximum O_3 exposure may shift to late evening or early morning in higher elevations than in lower elevations, which can be biologically important.

Acid deposition has produced chronic loading of N, S, and H^+ to the region's forests and forest soils. The amount of wet deposition that falls is highly correlated with local and regional sources and the total amount of precipitation. Ozone deposition properties differ from acidic deposition, as O_3 levels show greater spatial and temporal variations. Ozone does not accumulate and directly affect biogeochemical cycling as does chronic deposition of acidity, N, and S. Plant response time to O_3 deposition has a short timeframe, minutes to hours, and direct plant damage may be more likely to occur from both chronic levels and peak O_3 events, depending on the plant's sensitivity.

Atmospheric deposition maps and indices give us a regional view on which areas may be suspectible to stress from O_3 and acid deposition. However, the uptake response of plants to O_3 is directly influenced by climate and environmental factors (light, temperature, relative humidity, nutrient and plant water status), which control stomatal opening, and genetic factors, which control plant sensitivity. Drought conditions during the active growing season would restrict photosynthesis and growth, but would also limit plant uptake of O_3. During hot, dry summer O_3 episodes, stomatal closure may actually reduce the risk of O_3 to sensitive plants (SAMAB, 1996). This illustrates the complex relationship between regional atmospheric deposition, climate, plant uptake, and plant ecosystem response, which will be explored in following chapters.

This discussion has centered on the spatial and temporal trends of atmospheric deposition in the Northern Region. The yearly variation in meteorology is a major control on the formation of these air pollutants, on the the total annual N, S, and H^+ loading through wet, dry or cloud deposition, and on the transport and distribution of these deposition compounds throughout the region. Annual precipitation amounts, sunlight and summer temperature conditions, and wind patterns influence the total annual wet deposition to a region, production of ozone, and the transport these compounds to the northern forested regions.

References

Allen ER, Gholz HL (1996) Air quality and atmospheric deposition in Southern U.S. forests. In: Fox S, Mickler RA (eds) *Impact of Air Pollutants on Southern Pine Forests*. Ecological Studies 118. Springer-Verlag, New York, pp 83–170.

Andresen JA, Harman JR (1994) *Springtime Freezes in Western Lower Michigan: Climatology and Trends*. Michigan Agric Exp Sta Res Rep 536. Michigan State Agricultural Experiment Station, East Lansing, MI.

Barnard JE, Lucier AA, Brooks RT et al. (1990) Changes in forest health and productivity in the United States and Canada. In: *National Acid Precipitation Assessment Program. Acidic Deposition State of Science and Technology*. Vol. 3, Report 16. National Acid Precipitation Assessment Program, Washington, DC.

Baumgardner BE, Edgerton ES (1998) Rural ozone across the eastern United States: Analysis of CASTNet data, 1988–1995. J Air Waste Manage Assoc 48:674–688.

Bosart LR, Sanders F (1981) The Johnstown flood of July 1977: a long-lived convective storm. J Atmos Sci 38:1616–1642.

Cerveny RS, Balling RC (1998) Weekly cycles of air pollutants, precipitation and tropical cyclones in the coastal NW Atlantic region. Nature 394:561–563.

Changnon SA (1987) *Detecting Drought Conditions in Illinois*. Illinois State Water Survey Circular 164–87.

Chappell CF (1986) Quasi-stationary convective events. In: Ray PS (ed) *Mesoscale Meteorology and Forecasting*. American Meteorological Society, Boston, Massachusetts, 289–310.

Colman RA, Power SB, McAvaney BJ, Dahni RR (1995) A non-flux-corrected transient CO_2 experiment using the BMRC coupled atmosphere/ocean GCM. Geophys Res Lett 22:3047–3050.

Daly C, Neilson RP, Phillips DL (1994) A statistical-topographic model for mapping climatological precipitation over mountainous terrain. J Appl Meteor 33:140–158.

DeHayes DH (1992) Winter injury and developmental cold tolerance of red spruce. In: Eager C, Adams MB (eds) *The Ecology and Decline of Red Spruce in the Eastern United States*. Springer-Verlag, New York, pp 295–337.

DeHayes DH, Waite CE, Ingle MA, Williams MW (1990) Winter injury susceptibility and cold tolerance of current and year-old needles of red spruce trees from several provenances. Forest Sci 36:982–994.

Dennis RL, Barchet WR, Clark TL, Seilkop SK (1991) Evaluation of regional acidic deposition models. In: *National Acid Precipitation Assessment Program*.

Acidic Deposition State of Science and Technology. Vol. 3, Report 16. National Acid Precipitation Assessment Program, Washington, DC.

Dessens J, Bücher A (1995) Changes in minimum and maximum temperatures at the Pic du Midi in relation with humidity and cloudiness, 1882–1984. Atmos Res 37:147–162.

Diaz HF, Quayle RG (1980) The climate of the United States since 1895: spatial and temporal changes. Mon Wea Rev 108:249–266.

EarthInfo (1995) *Database guide for EarthInfo CD^2 NCDC Summary of the day.* EarthInfo, Inc., Boulder, Colorado.

Eischeid JK, Baker CB, Karl TR, Diaz HF (1995) The quality control of long-term climatological data using objective data analysis. J Appl Meteor 34:2787–2795.

Friedland AJ, Gregory RA, Karenlamp L, Johnson AH (1984) Winter damage to foliage as a factor in red spruce decline. Can J For Res 14:963–965.

Gates WL et al. (1996) Climate models—evaluation. In: Houghton JT, Meira Filho LG, Callander BA, Harris N, Kattenberg A, Maskell K (eds) IPCC (Intergovernmental Panel on Climate Change) *Climate Change 1995: The Science of Climate Change.* Cambridge University Press, Cambridge, United Kingdom, pp 229–284.

Gordon HB, O'Farrell SP (1997) Transient climate change in the CSIRO coupled model with dynamic sea-ice. Mon Wea Rev 125:875–907.

Green PM, Legler DM, Miranda CJ, O'Brien JJ (1997) *The North American Climate Patterns Associated with the El Niño–Southern Oscillation.* COAPS Project Report Series 97-1. Center for Ocean-Atmospheric Prediction Studies, Florida State University, Tallahassee, FL. < http://www.coaps.fsu.edu/lib/booklet/ >

Groisman PY, Easterling DR (1994) Variability and trends of total precipitation and snowfall over the United States and Canada. J Climate 7:184–205.

Hasselmann K et al. (1995) Detection of anthropogenic climate change using a fingerprint method. In: Ditlevsen P (ed) *Proceedings of "Modern Dynamical Meteorology," Symposium in Honor of Aksel Wiin-Nielsen, 1995.* European Centre for Medium-Range Weather Forecasts Press. Reading, England, pp 203–221.

Heck WW, Taylor OC, Adams RM (1982) Assessment of crop loss with ozone. J Air Poll Contr Assoc 32:353–361.

Heilman WE (1995) Synoptic circulation and temperature patterns during severe wildland fires. In: *Proceedings of the Ninth Conference on Applied Climatology.* American Meteorological Society, Dallas, TX, pp 346–351.

Henderson-Sellers A (1992) Continental cloudiness changes this century. GeoJournal 27(3):255–262.

Hirschboeck KK (1991) Climate and floods. In: Paulson RW, Chase EB, Roberts RS, Moody DW (eds) *National Water Summary 1988–1989: Hydrologic Events and Floods and Droughts.* Water Supply Paper 2375. United States Geological Survey (USGS), Denver, CO, pp 67–88.

Holton JR (1979) *An Introduction to Dynamic Meteorology.* Academic Press, New York.

Horton EB (1995) Geographical distribution of changes in maximum and minimum temperatures. Atmos Res 37:102–117.

Houghton JT, Callander BA, Varney SK (eds) IPCC (Intergovernmental Panel on Climate Change) (1992) *Climate Change 1992: The Supplementary Report to the IPCC Scientific Assessment.* Cambridge University Press, Cambridge, United Kingdom.

Houghton JT et al. (eds) IPCC (Intergovernmental Panel on Climate Change) (1994) *Climate Change 1994: Radiative Forcing of Climate Change and an*

Evaluation of the IPCC IS92 Emission Scenarios. Cambridge University Press, Cambridge, United Kingdom.

Houghton JT, Jenkins GJ, Ephraums JJ (eds) IPCC (Intergovernmental Panel on Climate Change) (1990) *Climate Change: The IPCC Scientific Assessment.* Cambridge University Press, Cambridge, United Kingdom.

Houghton JT, Meira Filho LG, Callander BA, Harris N, Kattenberg A, Maskell K (eds) IPCC (Intergovernmental Panel on Climate Change) (1996) *Climate Change 1995: The Science of Climate Change.* Cambridge University Press, Cambridge, United Kingdom.

Hulme M (1991) An intercomparison of model and observed global precipitation climatologies. Geophys Res Lett 18:1715–1718.

Hulme M, Zhao Z-C, Jiang T (1994) Recent and future climate change in East Asia. Int J Climatol 14:637–658.

Husar RB, Renard WP (1997) *Ozone as a Function of Local Wind Speed and Direction: Evidence of Local and Regional Transport.* Center for Air Pollution Impact and Trend Analysis (CAPITA), Washington University, St Louis, MO. < http://capita.wustl.edu/OTAG/reports/otagwind/OTAGWIN4.html>

Husar RB (1997) *Seasonal Pattern of Ozone over the OTAG Region.* Center for Air Pollution Impact and Trend Analysis (CAPITA), Washington University, St. Louis, MO. < http://capita.wustl.edu/otag/reports/otagseas/otagseas.html >

Johnson DW, Van Miegroet H, Lindberg SE, Harrison RB, Todd DE (1991) Nutrient cycling in red spruce forests of the Great Smoky Mountains. Can J For Res 21:769–787.

Jones PD (1994) Hemispheric surface air temperature variations: A reanalysis and an update to 1993. J Climate 7:1794–1802.

Jones PD (1995) Maximum and minimum temperature trends in Ireland, Italy, Thailand, Turkey and Bangladesh. Atmos Res 37:67–78.

Karl TR, Groisman PY, Knight RW, Helm RR Jr. (1993a) Recent variations of snow cover and snowfall in North America and their relation to precipitation and temperature variations. J Climate 6:1327–1344.

Karl TR et al. (1993b) A new perspective on recent global warming: asymmetric trends of daily maximum and minimum temperature. Bull Am Met Soc 74:1007–1023.

Karl TR, Knight RW (1998) Secular trends of precipitation amount, frequency, and intensity in the United States. Bull Am Met Soc 79:231–241.

Kattenberg A et al. (1996) Climate models—projections of future climate. In: Houghton JT, Meira Filho LG, Callander BA, Harris N, Kattenberg A, Maskell K (eds) IPCC (Intergovernmental Panel on Climate Change) *Climate Change 1995: The Science of Climate Change.* Cambridge University Press, Cambridge, United Kingdom, pp 285–357.

Kiehl JT, Trenberth KE (1997) Earth's annual global mean energy budget. Bull Am Met Soc 78:197–208.

Landsea CW, Nicholls N, Gray WM, Avila LA (1996) Downward trends in the frequency of intense Atlantic hurricanes during the past five decades. Geophys Res Lett 23:1697–1700.

Lefohn AS, Benkovitz CM, Tanner M, Shadwick DS, Smith LA (1990) Air Quality Measurements and Characterization for Terrestrial Effects Research. NAPAP State of Science and Technology Report No. 7. National Acid Precipitation Assesment Program, Washington, DC.

Lefohn AS, Laurence JA, Kohut RJ (1988) A comparison of indices that describe the relationship between exposure to ozone and reductions in the yield of agricultural crops. Atmos Environ 22:1229–1240.

Lefohn AS, Knudsen HP, Shadwick DS, Hurmann K (1992) Surface ozone exposures in the eastern United States (1985–1989). In: Flagler RB (ed) *The Response of Southern Commercial Forest to Air Pollution*. TR-21. Air and Waste Management Association, Pittsburgh, PA, pp 81–93.

Lovett, GM (1994) Atmospheric deposition of nutrients and pollutants in North America: an ecological perspective. Ecol Appl 4:629–650.

Lynch JA, Bowersox VC, Grimm JW (1996a) *Trends in Precipitation Chemistry in the United States, 1983–1994: An Analysis of the effects of 1995 of Phase I of the Clean Air Act Amendments of 1990, Title IV*. USGS Open-File Rep 09-0346. US Geological Survey (USGS), Washington, DC.

Lynch JA, Grimm JW, Corbett ES (1996b) Enhancement of regional wet deposition estimates based on modeled precipitation inputs. In: Hom J, Birdsey R, O'Brian K (eds) *Proceedings, 1995 Meeting of the Northern Global Changes Program*. Gen Tech Rep NE-214. USDA Forest Service, NE Forest Experiment Station, Radnor, PA.

Maddox RA (1983) Large-scale meteorological conditions associated with midlatitude, mesoscale convective complexes. Mon Wea Rev 11:1475–1493.

Manabe S, Wetherald RT (1975) The effects of doubling the CO_2 concentration on the climate of a general circulation model. J Atmos Sci 32:3–15.

Mather JR (1974) *Climatology: Fundamentals and Applications*. McGraw-Hill, New York.

McIntosh DH (1972) *Meteorological Glossary*. Chemical Publishing, New York.

McNab AL, Karl TR (1991) Climate and droughts. In: Paulson RW, Chase EB, Roberts RS, Moody DW (eds) *National Water Summary 1988–1989: Hydrologic Events and Floods and Droughts*. Water Supply Paper 2375. United States Geological Survey (USGS), Denver, Colorado, pp 89–98.

Melillo JM, Prentice IC, Farquhar GD, Schulze ED, Sala OE (1996) Terrestrial biotic responses to environmental change and feedbacks to climate. In: Houghton JT, Meira Filho LG, Callander BA, Harris N, Kattenberg A, Maskell K (eds) IPCC (Intergovernmental Panel on Climate Change) *Climate Change 1995: The Science of Climate Change*. Cambridge University Press, Cambridge, United Kingdom, pp 445–481.

Mohnen VA (1992) Atmospheric deposition and pollutant exposure of eastern U.S. forest. In: Eagar C, Adams MB (eds) *Ecology and Decline of Red Spruce in the Eastern United States*. Ecological Studies 96. Springer-Verlag, New York, pp 64–124.

Namias J (1983) Some causes of United States drought. J Climate Appl Meteor 22:30–39.

National Drought Mitigation Center (1997) An Historical look at the Palmer Drought Severity Index. < http://enso.unl.edu/ndmc/climate/palmer/pdsihist.htm >

[NAPAP] National Acid Precipitation Assessment Program (1991) *1990 Integrated Assessment Report*. U.S. National Acid Precipitation Assessment Program, Washington, DC.

[NADP] National Atmospheric Deposition Program (1996) Precipitation chemistry in the United States, 1994. In: *NADP/NTN Annual Data Sumary*. Natural Resource Ecology Laboratory, Colorado State University, Ft Collins, Colorado, pp 25–34.

[NADP] National Atmospheric Deposition Program (1998) *1997 Wet Deposition*. Illinois State Water Survey, Champaign, Illinois.

[NAPAP] National Acid Precipitation Assessment Program (1998) *NAPAP Biennial Report to Congress: An Integrated Assessment*. National Science and Technology Council, Committee on Environment and Natural Resources. Washington, DC.

Nicholls N, Gruza GV, Jouzel J, Karl TR, Ogallo LA, Parker DE (1996) Observed climate variability and change. In: Houghton JT, Meira Filho LG, Callander BA, Harris N, Kattenberg A, Maskell K (eds) IPCC (Intergovernmental Panel on Climate Change) *Climate Change 1995: The Science of Climate Change.* Cambridge University Press, Cambridge, United Kingdom, pp 133–192.

[NOAA] National Oceanic and Atmospheric Administration (1996) *Special Climate Summary 96/4.* Summer 1996 National Weather Service. National Centers for Environmental Prediction. Climate Prediction Center. Climate Operations Branch and Analysis Branch. < http://nic.fb4.noaa.gov/products/special_summaries/96_4/ >

[NOAA-CIRES Climate Diagnostics Center] National Oceanic and Atmospheric Administration (1997) Risk of seasonal climate extremes in the US related to ENSO. < http://www.cdc.noaa.gov/~cas/atlas.html >

[NOAA Climate Prediction Center] National Oceanic and Atmospheric Administration (1999) *Warm Episode Seasonal 250-hPa Wind: Climatological, Actual, and Departure from Normal.* < http://www.cpc.ncep.noaa.gov/products/analysis_monitoring/ensostuff/hist/histcirc.html >

[NOAA National Hurricane Center] National Oceanic and Atmospheric Administration (1999) *U.S. Mainland Hurricane Strikes by State, 1900–1996.* < http://www.nhc.noaa.gov/paststate.html >

[NOAA Pacific Marine Environmental Laboratory] National Oceanic and Atmospheric Administration (1997a) *What is an El Niño?* < http://www.pmel.noaa.gov/toga-tao/el-nino-story.html >

[NOAA Pacific Marine Environmental Laboratory] National Oceanic and Atmospheric Administration (1997b) *Definitions of El Niño, La Niña, and ENSO.* < http://www.pmel.noaa.gov/toga-tao/ensodefs.html >

[NOAA Storm Prediction Center] National Oceanic and Atmospheric Administration (1999) *Historical Tornado Archive.* < http://www.spc.noaa.gov/archive/tornadoes/index.html >

Ollinger SV, Aber JD, Lovett GM, Millham SE, Lathrop RG, Ellis JM (1993) A spatial model of atmospheric deposition for the northeastern US. Ecol Appl 3:459–472.

[OTAG] Ozone Transport Assessment Group (1998) *OTAG Technical Supporting Document: Final Report.* Ozone Transport Assessment Group. Environmental Council of States. < http://www.epa.gov/ttn/otag/finalrpt/ >

Palmer WC (1965) *Meteorological Drought.* Research Paper 45. US Weather Bureau, US Department of Commerce, Washington, DC.

Paulson RW, Chase EB, Roberts RS, Moody DW (eds), (1991) *National Water Summary 1988–89: Hydrologic Events and Floods and Droughts.* Water Supply Paper 2375. US Geological Survey, Denver, CO.

Plantico MS, Karl TR, Kukla G, Gavin J (1990) Is recent climate change across the United States related to rising levels of anthropogenic greenhouse gases? J Geophys Res 95:16617–16637.

Press WH, Teukolsky SA, Vetterling WT, Flannery BP (1992) *Numerical Recipes in FORTRAN: The Art of Scientific Computing.* Cambridge University Press, New York.

Schimel D et al. (1996) Radiative forcing of climate change. In: Houghton JT, Meira Filho LG, Callander BA, Harris N, Kattenberg A, Maskell K (eds) IPCC (Intergovernmental Panel on Climate Control) *Climate Change 1995: The Science of Climate Change.* Cambridge University Press, Cambridge, United Kingdom, pp 65–131.

Takle ES, Bramer DG, Heilman WE, Thompson MR (1994) A synoptic climatology for forest fires in the NE US and future implications from GCM simulations. Int J Wildland Fire 4:217–224.

Taylor GE, Ross-Todd BM, Allen E P, Conklin P (1992) *Patterns of tropospheric ozone in forested landscapes of the Integrated Forest Study.* In: Johnson DW, Lindberg SE (eds) Atmospheric Deposition Forest Nutrient Cycling: A Synthesis of the Integrated Forest Study. Ecological Studies 91. Springer-Verlag, New York, pp 50–71.

Trenberth KE, Houghton JT, Meira Filho LG (1996) The climate system: an overview. In: Houghton JT, Meira Filho LG, Callander BA, Harris N, Kattenberg A, Maskell K (eds) IPCC (Intergovernmental Panel on Climate Change) *Climate Change 1995: The Science of Climate Change.* Cambridge University Press, Cambridge, United Kingdom, pp 51–64.

Unisys Corporation (1999) *Atlantic Tropical Storm Tracking by Year.* < http:// weather.unisys.com/hurricane/atlantic/index.html >, 1 pp.

[USDA Forest Service] United States Department of Agriculture, Forest Service (1992) *1984–1990 Forest Fire Statistics.* USDA Forest Service, Washington, DC.

[USEPA] United States Environmental Protection Agency (1997a) *National Air Quality and Emissions Trends Report, 1996.* EPA 454/R.97-013. USEPA, Office of Air Quality Planning and Standards, Research Triangle Park, North Carolina.

[USEPA] United States Environmental Protection Agency (1997b) *Final Revisions to the Ozone and Particulate Matter Air Quality Standards.* EPA 456/F-97-004. USEPA, Office of Air and Radiation, Washington, DC.

[USEPA] United States Environmental Protection Agency (1997a) *Acid Rain Program Emissions Scorecard 1996.* EPA 430/R 97-031. USEPA, Office of Air and Radiation, Acid Rain Division, Washington, DC.

[USEPA] United States Environmental Protection Agency (1998b) *National Air Quality and Emissions Trends Report, 1997.* EPA 454/R.98-016. USEPA, Office of Air Quality Planning and Standards, Research Triangle Park, North Carolina.

[USEPA] United States Environmental Protection Agency (1998c) *Clean Air Status and Trends Network (CASTNet) Deposition Summary Report (1987–1995).* EPA 600/R-98-027. USEPA, Office of Research and Development, Washington, DC.

Vining KC, Griffiths JF (1985) Climatic variability at ten stations across the United States. J Climat Appl Meteor 24:363–370.

Vose RS, Schmoyer RL, Steurer PM, Peterson TC, Heim R, Karl TR, Eischeid J (1992) *The Global Historical Climatology Network: Long-Term Monthly Temperature, Precipitation, Sea Level Pressure, and Station Pressure Data.* Report ORNL/CDIAC-53, NDP-041, Oak Ridge National Laboratory, Oak Ridge, TN.

Wallace JM, Hobbs PV (1977) *Atmospheric Science: An Introductory Survey.* Academic Press, New York.

Weisberg JS (1981) *Meteorology: The Earth and Its Weather.* Houghton Mifflin, Boston, MA.

Zasada JC, Teclaw RM, Isebrands JG, Buckley DS, Heilman WE (1994) *Effects of Canopy Cover and Site Preparation on Plant Succession and Microclimate in a Hardwood Forest in Northern Wisconsin.* Proceedings of the 1994 Society of American Foresters/Canadian Institute of Forestry Convention, 18–22 September 1994, Anchorage, Alaska, Society of American Foresters, Bethesda, MD, pp 479–480.

4. Forest Declines in Response to Environmental Change

Philip M. Wargo and Allan N.D. Auclair

Decline diseases are intimately linked to stress and environmental change. There is strong evidence that, as a category, decline diseases have increased significantly in response to the climate, air chemistry, and other changes documented in the northeastern United States over the past century, and particularly the last two decades. No other forest response to environmental change and stress is expected to be as dramatic. Decline diseases occur in response to multiple, often overlapping and interacting, stressors that typify the ongoing and future environmental changes expected in the Northeast.

Forest declines have a special significance in the Northeast since, more than in any other region, they form the conceptual and experimental basis on which the theories and models of decline diseases have evolved. New significant scientific advances from recent research in the Northeast hold considerable promise for better understanding and effective treatment of these complicated diseases. Several studies are in progress on the development of an early warning of the risk to dieback and on innovative experiments on managing and adapting to future declines (Auclair, 1997).

Decline Disease: A Unique and Complex Response
to Environmental Stress

Definition of Decline Disease

Decline denotes a deterioration of tree health, often leading to mortality. This may be a normal and inevitable phase of a tree's life before death (Castello et al., 1995a). The term decline also has been used to characterize reductions in radial growth observed on tree species (Hornbeck and Smith, 1985; LeBlanc, 1990; Van Deusen et al., 1991). However, decline disease is not always associated with decreases in growth (Van Deusen et al., 1991; Wargo et al., 1993). In this chapter, decline refers to the disease condition, while growth reduction is considered part of the syndrome of the decline disease (Wargo et al., 1993).

In forest pathology, decline is a major category of tree diseases characterized by premature and progressive deterioration in tree health related to stress and secondary-organism attack (Manion, 1987). The interaction of stressors with "secondary organisms" (Houston, 1981) results in dieback, decline, and mortality of trees. Forest trees typically "decline" in response to a plethora of stressors.

Stressors are any adverse environmental factors that induce damage or injurious strain to the living organism (Levitt, 1972). Stressors may be biotic or abiotic, may occur separately or in concert, and affect plants individually, as populations, or as communities. Stress may occur chronically over a significant period or it may be of short duration and acute. The strain caused by these stressors may be physical or chemical, or both, and sometimes is irreversible, that is, a portion or all of a plant is killed (Levitt, 1972).

Declines have occurred in the past and will continue to occur in the future. Because of the multiple factors involved in their development, declines are especially difficult to diagnose and to attribute cause and effect, and thus, to reproduce experimentally. The term decline was commonly assigned to diseases whose primary cause was unknown or "unexplained." Sinclair (1967) was one of the first to recognize that these diseases have a unique etiology and proposed the concept of complex causality.

Symptoms of Decline Diseases

Symptoms of decline diseases are general and similar among the various species and regions experiencing decline diseases. Manion (1991) lists 10 of the most common symptoms of decline, including reduced growth rate, dieback, and impaired energy relationships. The rate at which symptoms appear and their magnitude reflect primarily the condition of the host population, including its age (Auclair et al., 1997) and vigor and vitality (Wargo, 1978). Other factors include the genetic uniformity or heterogeneity of the tree populations, the intensity and frequency of stress, and the numbers and aggressiveness of secondary organisms attacking the weakened trees. Subsequent mortality can occur on scattered individual trees, in small clusters or groups of trees, or over extensive areas depending on the interaction of these factors (Houston, 1981; Wargo, 1981b).

Crown, Stem, and Root Characteristics

Crown. Anomalous leaf condition and dieback of the crown are among the first and most conspicuous features of trees that are stressed and undergoing decline. Leaf anomalies include reduced leaf size, scorch of leaf margins, veinal necrosis, seasonally premature leaf coloration, and leafdrop. Some pathologists have argued that these leaf anomalies can result from specific events, such as spring frost, severe drought, or heat stress, and do not necessarily lead to progressive dieback or decline. For this reason, Auclair et al. (1996, 1997) omitted leaf anomalies in their reconstruction of dieback episodes in northern hardwoods.

Progressive crown dieback over one or several consecutive years is a common feature of decline disease. Typically, the crowns of deciduous trees dieback from the tips inward toward the trunk. Clusters of epicormic shoots may distinguish the crown in later stages. In conifers, the older needles often succumb first, giving the appearance of clusters of young, live foliage at the outer ends of the branches. The rate at which these features develop may be related to the level and persistence of evapotranspiration stress on the crown. Hence, rapid dieback could occur in affected trees during periods of exceptional drought and/or heat stress. Conversely, cool wet weather could result in renewed vigor and health, as was dramatically evidenced on species of birch (*Betula* spp.) in the early 1950s (Auclair et al., 1996).

In their studies on sugar maple (*Acer saccharum* Marsh.) blight, Houston and Kuntz (1964) were the first to relate bud mortality and dieback to defoliation. Defoliation of sugar maple by a number of insects was associated with dieback and decline of sugar maple (Giese et al., 1964; Giese and Benjamin, 1964). Wargo (1981b) showed similar relationships in studies on oak (*Quercus* spp.) decline and mortality and defoliation by the gypsy moth (*Lymantria dispar* L.). In detailed studies on dieback in defoliated sugar maple, Gregory and Wargo (1986) showed that dieback was related to survival of terminal and axillary buds on defoliated branches. After early season defoliations, the developing buds in the axils of the removed leaves abscised, but axillary and terminal buds on the refoliated terminal shoots survived through winter. In late season defoliation, most buds of refoliated shoots did not survive and the next year's growth depended on axillary buds formed prior to defoliation. Thus, when progressing from early to late defoliations, the next year's shoot growth depended decreasingly on the last-formed and increasingly on the first-formed portions of the previous year's shoot. Bud survival, and hence terminal bud and twig dieback, was related to the scale and leaf primordia production in buds formed after defoliation. Insufficient scale primordia increased bud susceptibility to winter injury and dieback the following spring.

Stem. A common feature of trees showing dieback is a reduction in radial and terminal growth. Reductions in radial growth have been most commonly reported and sometimes precede crown symptoms by several to many years (Staley, 1965—oak decline; Bauce and Allen, 1991; Kolb and McCormick, 1993—sugar maple decline; Wargo et al., 1993—red spruce decline). Reductions in terminal elongation on declining red spruce (*Picea rubens* Sarg.) trees also have been observed (Tobi et al., 1995; Wargo et al., 1993).

Another feature observed on trees showing dieback is the inability to rapidly transmit water to the crown. Greenidge (1951) used dye treatments to demonstrate air blockages or embolisms in the stem cross sections of yellow birch (*Betula alleghaniensis* Britton). The degree of blockage was roughly proportional to the severity of dieback in the crown. In trees with severe dieback, the dye dumped at the base of the tree, barely reaching the bottom of the crown.

Sperry et al. (1988) ascribed the development of embolized branches in sugar maple to winter insolation at the time of extreme cold, and to drought in the late summer. Tree stresses most likely to cause irreversible embolisms include freezing and drought stresses (Tyree and Sperry, 1988, 1989). There is great individuality in the susceptibility within and between tree species to emboli formation; differences noted do not necessarily or systematically relate to vessel and tracheid structure (Zimmerman, 1983). Contrary to early concepts on embolism formation, conifers, as with deciduous species, can embolize and show as severe water transport dysfunction.

Deciduous trees are particularly susceptible to embolisms due to freezing stress (Cochard and Tyree, 1990). Auclair et al. (1996) noted that freezing (prolonged winter thaw followed by sudden freezing and/or root freeze and root kill) was correlated to the onset of major dieback episodes in northern hardwoods. Once putatively injured, the affected trees seem to have been especially sensitive to drought which, when it followed a freezing event, determined the rate and severity of dieback in the crown. Auclair et al. (1996) demonstrated that each major episode of dieback during the century (1910 to 1995) coincided with occurrences of potential winter freezing events followed by drought during the growing season. This suggests a direct link between changing levels of climatic stresses in the region and the extent of decline disease.

Root. Reduction in carbon reserves, especially obvious in the roots, is a major symptom of declining trees. Below normal levels of starch have been observed in declining oak (Staley, 1965; Wargo, 1981c), sugar maple (Wargo et al., 1972), and red spruce (Wargo et al., 1993). Carbohydrate dynamics seems to integrate the interaction of stressors, tree vitality, and subsequent decline (Wargo, 1999).

Another common feature in declining trees is a high level of rootlet mortality. Greenidge (1951) excavated 74 declining yellow birch trees and quantified the levels of rootlet mortality. Levels of fine root loss ranged from 10 to 40% in slightly affected trees, and up to 40 to 90% in severely declining trees. Similar relationships of fine root loss were also observed in declining red spruce (Wargo et al., 1993) and oak species (Staley, 1965).

This observation has particular significance to forests in the Northeast undergoing climatic change. Root tissues are by far the least frost hardened of any tree tissue (Larcher and Bauer, 1981; Sakai and Larcher, 1987), are relatively shallow, and usually are covered by snowpack from December through early March. "Open winters" (i.e., low snowpack accumulation or prolonged winter thaws resulting in the complete subsidence of snowpack) create risk for root kill in the event of cold air temperatures (Pomerleau, 1991; Robitaille et al., 1995). Auclair et al. (1996) quantified the incidence and magnitude of potential root freezing stress in southern Quebec (Lennoxville) and in northern Vermont (Burlington). Throughout the last century, at intervals of 8 to 12 years, episodes of conditions that could induce severe root freezing occurred at both stations. Since the mid-1970s (especially in southern Quebec) there has been a strong trend for winters without snowpack. The incidence of dieback has followed these trends.

Mortality, Recovery, and Regeneration Patterns

Mortality. The common outcome in the event of mounting stress is progressive deterioration in tree health. In the event of persistent stress and, in most instances, attack by secondary organisms, mortality is likely. Tree death can occur within one growing season but has been known to occur as long as 6 to 10 years after the first symptoms are apparent. This extended dying period is significant ecologically because the declining tree occupies the site for an extended period and reduces the levels of resources (light, water, and nutrients) otherwise available for growth of associated healthy individuals.

A common observation has been that decline diseases, including mortality, most affect older or larger trees in the population. Possible reasons for this include change in root/shoot ratio, the sapwood/canopy area ratio, and physiological or metabolic senescence (Mueller-Dombois, 1992). Another possible explanation is the much greater risk of emboli formation in larger and older trees (Zimmerman, 1983).

There are marked differences among species of northern hardwoods in susceptibility to mortality. Using periodic plot inventories of forest growth and mortality in Vermont from 1960 to 1995, Auclair (1997) estimated that decline accounted for 73% of the total volume loss (reduced growth plus mortality) in white (*Betula papyrifera* Marsh.) and yellow birch compared with only 26% in sugar maple. White ash (*Fraxinus americana* L.) (53%) and red spruce (58%) were intermediate. This is consistent with the high levels of tree death (70 to 100%) in episodes of birch dieback in the Northeast in the 1940s and 1950s. Although the overall observed tree death rates in sugar maple (2 to 7%) have been notably less in the 1980s and 1990s, mortality can be much higher in this species in the event of persistent multiple or interacting stresses (see Case Study 1 later in this chapter).

Recovery. Decline diseases in the Northeast have been highly episodic. Onset of a major episode can be surprisingly rapid, as can the subsidence of the episode and resumption of partial or full health of the affected trees. Trees showing decline have been known to recover, especially if favorable growth conditions resume at an early point in the decline spiral and secondary-organism attack is absent or minimal. Sugar maples that had as much as 40% branch mortality survived and had good crowns after three years of recovery (Gross, 1991). Oak species defoliated by the gypsy moth continued to deteriorate for five years after the last defoliation and then recovered to their original healthy condition within five years (Campbell and Sloan, 1977).

Regeneration. Seedlings and saplings are often unaffected and remain healthy, even at the time of severe dieback and decline of the canopy trees. Vigorous regeneration often can be of the same species as the declining trees, though exceptions do occur. Dieback and decline is widespread in eucalyptus (*Eucalyptus* spp.) forests across Australia and many, if not most, declining stands are not regenerating (Heatwole and Lowman, 1986). Case Study 1, presented later in this chapter, shows that regeneration in declining sugar maple forests in northwestern Pennsylvania is inadequate to sustain closed forest.

Regeneration during and after a tree decline is one of the least studied aspects of decline disease. For example, relatively little is known about the pattern of regeneration and succession that followed the massive mortality of birch throughout the 1940s and 1950s. Regeneration in declining red spruce forests in the Northeast requires better data on replacement species before a full understanding of the changes effected by decline is possible.

Two Theories on Causal Mechanisms of Decline Diseases

It is generally accepted that decline diseases are characterized by the widespread premature senescence and mortality of canopy trees over a relatively short period, and by the absence of evidence of a single causal factor (Ciesla and Donaubauer, 1994). Also, it is generally accepted that environmental stressors, both abiotic and biotic, play an important role in the etiology of declines (Sinclair, 1967; Sinclair and Hudler, 1988).

The feature that distinguishes declines from other forest diseases is a chain of stress–response events that results in progressive deterioration in tree health and mortality in at least part of the population. The sequence begins with the increased vulnerability of the host (for reasons not always clear), followed by inciting or triggering stress (usually first evident in the form of one or more obvious decline symptoms), and by the additional stress of attacking disease organisms and insects.

There are several different theories or "models" of decline disease. A general point of agreement in all models is that stress has a major and unique role in the etiology of decline diseases. Another point of agreement is that declines are by nature complicated. Typically, multiple abiotic and biotic stresses act individually or in concert to amplify their effect. It is this multiplicity that marks forests as especially vulnerable to decline disease because of the plethora of simultaneous environmental changes now occurring in eastern North America.

Sinclair–Manion–Houston Models

Sinclair (1967) proposed three categories of "causal factors" [sic]: (1) predisposing factors that weaken trees and reduce their ability to tolerate adverse conditions, (2) inciting factors that trigger the decline event, and (3) contributing factors that intensify and perpetuate the disorder. In a recent version of their theory, Sinclair and Hudler (1988) listed four factors in decline causality: (1) perennial or continual irritation by one factor, (2) drastic injury plus secondary stress, (3) interchangeable predisposing and contributing factors, and (4) synchronous cohort senescence. They regarded the fourth as a variation of their third causality mechanism.

Manion (1991) proposed that the three factors in Sinclair's (1967) model interact. The result is a spiraling decrease in tree health, and frequently culmination in tree death. Houston (1967, 1981, 1992) proposed a stress-altered tree/secondary pathogen model for decline diseases. Stressors effect physical, physiological, and/or biochemical changes in healthy tissues and predispose trees to attack by secondary organisms (Houston, 1992). These changes result directly in disease or dysfunction if severe enough, or render tissues susceptible to pathogenic organisms (insects and microbes) that usually are resisted by the tree (Crist and Schoeneweiss, 1975; Schoeneweiss, 1978, 1981a,b; Wargo, 1972, 1975, 1988, 1996; Wargo and Houston, 1974). Mortality usually can be ascribed to attacks by one or more of these naturally occurring opportunistic organisms (Houston, 1992; Wargo, 1977, 1980, 1981a).

Houston's (1992) "predisposing factors" are equivalent to Sinclair's (1967) and Manion's (1991) "inciting factors" and many of their suggested "predisposing factors." Houston's (1992) "predisposing factors" are related to a narrower interpretation of the definition of predisposition, which means "to bring about susceptibility to infection." The "predisposing factors" in the Sinclair–Manion model influence the response of trees to the "inciting factor," but they do not necessarily predispose trees to infection.

Are the levels of abiotic and biotic stresses increasing in the Northeast? Do these changes indicate a greater or lessor risk of decline than historically, and into the future? There is the strong impression and considerable documentation on increasing levels of extreme climatic stress, lower air quality (especially oxides of nitrogen [NO_x] and ozone [O_3]), and pest outbreaks (including explosive deer population) in parts of the Northeast. These models imply that the increased levels of stress are interacting and adversely impacting the forests in the region. A higher level of decline disease is a likely outcome.

Mueller-Dombois Model

Mueller-Dombois proposed a natural dieback phenomenon due to synchronous cohort senescence as an alternative to the decline disease theory (Mueller-Dombois, 1983a,b, 1992; Mueller-Dombois et al., 1983). This theory focused on decline as a response to aging of cohort populations and to changes in stand demographics. The four additional components of the Mueller-Dombois (1992) theory are: (1) simplified forest structure, (2) edaphically extreme sites, (3) periodically recurring perturbations, and (4) biotic agents.

The synchronous cohort senescence model is compatible with the Sinclair–Manion–Houston "disease models" of decline. In the Sinclair and Manion models, synchronous cohort senescence would be a predisposing factor. In the Houston model, cohort aging is considered part of the forest relationships that influence the vulnerability of trees to stress-triggering events.

Many, if not most, forests in the Northeast are at or nearing biological maturity. Effective fire suppression, pest management and control measures, forest conservation, and a limited rate of harvesting since about 1930 have favored survival and increasing forest age. Auclair et al. (1997) noted that the timing of the first major successive episodes of dieback on ash, birch, sugar maple, and red spruce in the Northeast during this century corresponded closely with the species' age to maturity. Species populations that were young and vigorous early in the century did not show dieback despite high levels of freezing and drought stress (Auclair, 1997; Auclair et al., 1996). This observation is consistent with the concept of cohort senescence. It implies that as an increasing number of tree species in the region reach biological maturity simultaneously, dieback will become extensive. The pattern of significant forest loss to dieback in the Northeast since the mid 1970s is believed to be due to multiple species having reached maturity (Powell et al., 1993) at a time of unusually high climate stress in the region (Auclair et al., 1996, 1997; Auclair, 1997). In addition, there have been expanded invasions by exotic insect pests as well as recent increases in epizootics of native insect pests. The increase in forest susceptibility to both exotic and native pests could reflect the aging of the tree species along with changes in species composition throughout much of the Northeast.

The cohort senescence theory implies that the high level of risk is likely to persist or even increase as forests age. Increasing levels of climatic stress, lower air quality (especially NO_x and O_3), and pest outbreaks in the region do not bode well for sustaining forest health of aging tree populations in the absence of strong countermeasures. A possible management strategy is to selectively increase the harvesting of mature stands and ensure their replacement with young forests of stress-resistant tree species.

History and Distribution of Decline Diseases in the Northeast

Decline diseases have occurred in most major tree species in the United States. Some examples include ash, beech (*Fagus grandifolia* Ehrh.), birch, maple, oak, pine (*Pinus* spp.), red spruce, and sweet gum (*Liquidambar styraciflua* L.) in the eastern United States (Houston, 1987). Episodes have been reported as early as the 1870s and probably occurred much earlier than that. The history of many of these decline disease episodes is reviewed in Houston (1987), Millers et al. (1989), Walker et al. (1990), and Auclair et al. (1997).

Relationship to Previous Disturbances

The development of decline diseases also may reflect the effect of previous disturbance. Widespread forest harvesting for timber, chemical wood, and firewood in the Northeast began about 1860 and continued until about 1930. In the 1860s, extensive areas in marginal cropland were abandoned to regrowth. At the same time, the use of railways was expanded, including the use of narrow-gauge rail lines to harvest hardwoods in terrain not easily accessed by rivers or lakes used for transport of timber (Bormann and Likens, 1979). Extensive disturbance in the 1860s appears to have sharply "synchronized" forest establishment, and hence the timing of maturation and decline of different tree species (Auclair et al., 1997).

In addition, the effects of previous disturbance history can be a significant factor affecting disease development. The extent and magnitude of decline and mortality on oak (*Quercus* spp.) often concurs with areas formerly occupied by American chestnut (*Castanea dentata* [Marsh.] Borkh.), which was essentially removed from the forests by the chestnut blight fungus (*Cryphonectria parasitica* [Murrill] Barr). Both the dominance of oak and its occurrence on sites where it was excluded by chestnut increase the susceptibility of oak stands to persistent gypsy moth defoliation and decline (Healy et al., 1997). The majority of sugar maple stands experiencing decline in northwestern Pennsylvania may occupy "off-sites" or occur in much higher densities than before on sites colonized aggressively by maple after extensive cutting and burning early in the century (see Case Study 1 later in this chapter).

Timing and Episodicity

The onset and subsidence of decline diseases is typically rapid. Major episodes over large regions can erupt within 1 to 2 years and disappear with equal rapidity. The incidence of successive episodes of dieback and decline in the Appalachian region illustrates this pattern (Fig. 4.1a). Several characteristic features are evident. First, early episodes involved one or several species in each episode. Since the mid-1970s, many hardwood species have been affected simultaneously. Second, the onset date corresponds within 5 years to the estimated age to maturity of each species. Third, the severity and extent of dieback episodes has not changed over the century (Fig. 4.1a); in the more recent episodes, the number of tree species in the vegetation affected simultaneously has increased, giving the appearance of an increase in decline (Fig. 4.1b).

Spatial Incidence and Heterogeneity

Decline diseases typically show a high level of spatial variability. The fact that different age classes and different species are affected in different degrees results in a complex or heterogeneous appearance of decline within a forest stand. Even within the same population cohort, trees rooted to different depths or that occupy different soils may show differences in the severity and rate of disease development. Superimposed on this are subtle interactions among numerous stressors, genetic variations, opportunistic organisms, and predisposing factors.

Regional differences between decline episodes have not been well studied. Auclair et al. (1997) noted a tendency for episodes of hardwood dieback to be distinctly more severe and abrupt in the more northern areas. This possibly reflects the magnitude of the inciting/predisposing stress, the simplified composition and age structure of the northern populations, and differences in a wealth of site legacies, such as soil nutrient status, structure, and drainage, soil microbial populations, and disturbance history (Vogt et al., 1996).

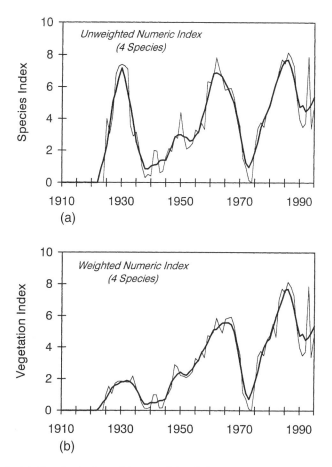

Figure 4.1. (a) Species index of forest dieback in Vermont from 1910 to 1995. This index is the sum of the numerical indices (Auclair et al., 1997) of four species (ash spp., birch spp., sugar maple, and red spruce) divided by the number of species showing dieback in any given year (i.e., unweighted by the number of species). (b) Vegetation index of forest dieback in Vermont from 1910 to 1995. This index is the sum of the numerical indices of the four species divided by four (i.e., weighted by the total number of species affected, 1910 to 1995).

Environmental Stress and Change in the Northeast

Primary Stressors

Lists of environmental factors that can act as primary stressors are found in Houston (1987), Manion (1991), and Millers et al. (1989). Drought and defoliation are the most common stressors associated with decline disease in the Northeast. Stress from sucking insects or defoliation from late spring frosts or fungal leaf pathogens also have been associated with decline disease (Millers et al., 1989; Wargo, 1996). Defoliation by exotic insects has played a major role in oak decline (Wargo, 1981b, 1996). Defoliation by native defoliators, sometimes in combination with anthracnose fungi, has been related to decline episodes in sugar maple (Hall, 1995; Millers et al., 1989). Similar abiotic and biotic stressors have been associated with decline diseases in Europe (Ciesla and Donaubauer, 1994; Siwecki and Liese, 1991).

Primary Abiotic Stressors

The Northeast has experienced more change in climate, air chemistry, land use, site alterations, and other human impacts than any other region in the United States. The most significant of these include: (1) region-wide notable increases in mean annual temperature and precipitation over the century and a trend toward unseasonable and intense rainfall events since 1960 (Karl et al., 1996); (2) increased frequency and magnitude of winter thaw–freeze and root freeze events over the century (Auclair et al., 1996); (3) significant increases in tropospheric ozone (O_3), acidic deposition, and heavy metals from air pollution as well as increases in atmospheric CO_2 and NO_x concentrations; (4) widespread historic human disturbances, including forest harvesting, burning, cropland abandonment, rail and road construction, and suburbanization; extensive areas are off site following the reestablishment of aggressive tree species on sites not usually occupied by them under competition and gap replacement; (5) current extensive site and stand alteration through forest land use and management (including changes in species composition, such as the increased dominance by hardwoods, and changes in forest age structure), alteration of drainage patterns (e.g., by road construction), forest thinning, and forest harvest. Trees occupying borders of city and suburban streets and croplands and the introduction of exotic ornamentals represent a special but common variation in the Northeast.

Primary Biotic Stressors

Some insects and diseases can kill trees directly, but many cause damage that weakens trees and predisposes them to other insects or pathogenic organisms that ultimately kill the tree. Thus, their primary role is as a stressor of forest trees. Defoliating insects are the major members of this group, often causing spectacular, sudden damage over extensive forest areas in a diverse array of tree species, especially deciduous species (Millers et al., 1989; Houston, 1987). Sucking insects, such as adelgids, aphids, and scales, also have acted as primary biotic stressors and have triggered episodes of decline disease (Houston, 1987; Millers et al., 1989). Some foliar pathogens also can cause defoliation and predispose trees and trigger decline diseases (Hall, 1995; Wargo, 1996). These organisms rarely kill trees by themselves.

Interpreting the ecological role of these primary biotic stressors is complicated. Many of these organisms are nonindigenous and exotic to the ecosystem they are disturbing. These biotic stressors may be responding to abnormal abundance of certain tree species within a landscape or to species that are growing off site. Both situations can provide the trigger for pest outbreaks. Population dynamics of some of these organisms also are related to abiotic stressors, especially water deficits. Whether drought weakens trees and enables population growth of these organisms and subsequent attack, or whether drought enables population growth of these organisms that weaken the trees, is not clearly understood.

Secondary Organisms

Most of the pathogenic organisms (i.e., insect pests and traditional disease organisms) involved in decline disease syndromes are considered incapable of attacking and causing disease in healthy trees. Houston (1992) referred to these as "secondary-action organisms" to indicate their role as secondary in the time sequence of decline disease but not in importance to the syndrome. This group of organisms includes a variety of fungi and insects, and probably bacteria and viruses (not intensively studied) (Castello et al., 1995b) that can kill fine roots, buds, twigs, inner bark, cambium, and/or xylem on the major branches and boles of mature forest trees. Most of these organisms have narrow host ranges but some can infect a range of hosts.

The organisms that have been investigated and associated with important decline diseases of forest trees (1) are ubiquitous inhabitants of natural forest ecosystems whose evolved roles in the absence of major external stress events are ecologically beneficial, (2) are unable to succeed in living tissues not previously predisposed by stress, and (3) affect stressed trees principally by invading and killing meristematic regions of roots, stems, twigs, or buds (Houston, 1992). Secondary-action organisms act as ecosystem "roguers," killing individuals or groups of trees that have been stressed and are no longer fully functional members of the ecosystem (Wargo, 1995). Some secondary-action organisms also act in a dual role as scavengers. These decay the woody substrate they have killed, releasing additional resources for use by healthier existing or replacement trees (Wargo, 1995). The *Armillaria* root disease fungus is an excellent example of a roguer–scavenger organism (Wargo, 1980, 1995).

Have the abundance and/or aggressiveness of secondary-action organisms changed in the Northeast? The population levels of these organisms typically respond in proportion to the amount of tree mortality. Losses in forest volume to reduced growth and mortality due to recent dieback (1976 to 1995) increased 2.6-fold compared with the long-term historical (1910 to 1975) average (Fig. 4.2a). We can safely assume that the presence of secondary-action organisms has followed these upward trends. This trend of volume loss was close to the annual pest-related tree mortality

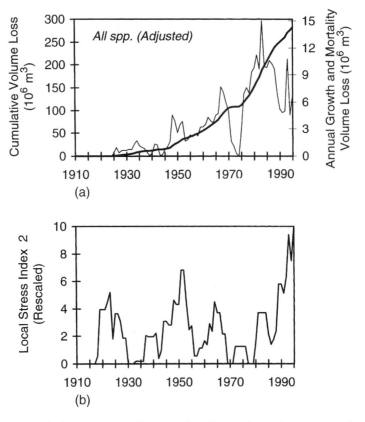

Figure 4.2. (a) Century trends in annual and cumulative loss to mortality and reduced growth due to dieback in the Appalachian region (Vermont, New York, Pennsylvania, West Virginia, Maryland) of the northeastern United States. (b) The local stress index based on extreme freezing and drought stresses at Burlington, VT, 1910 to 1995.

rates inventoried in northeastern hardwoods. The overall increase in tree mortality measured in permanent forest inventory plots from 1977 to 1991 was 2.2-fold compared with the previous 14-year period (1962 to 1976) (Powell et al., 1993).

Four Case Studies of Decline Diseases

Case Study 1. Biotic and Land Use Factors: Decline of Sugar Maple in Northwestern Pennsylvania

Multiple tree species are being affected by decline disease on the Allegheny National Forest in northwestern Pennsylvania. Sugar maple and red maple (*Acer rubrum* L.) are in varying stages of decline, beech stands are declining in response to beech bark disease, butternut (*Juglans cinerea* L.) has declined sharply since the 1960s due to butternut canker, oaks are declining from persistent defoliation by the gypsy moth, and nearly pure stands of black cherry (*Prunus serotina* Ehrh.) are being defoliated by the cherry scallop shell moth (*Calocalpe undulata* L.), possibly setting the stage for a defoliation-induced decline of that species. Decline diseases reported in the past have been discrete episodes of "tree declines" or species declines involving one or several tree species (Millers et al., 1989). The current declines are exceptional in the large number of tree species being affected simultaneously.

In response to this mortality, except for beech and striped maple (*Acer pennsylvanicus* L.), stands are failing to regenerate back to forests because of heavy browsing by excessive populations of white-tail deer and competition from heavy forest-floor cover of ferns and grasses, also a result of excessive deer browsing. Closed-canopy forests are shifting to savanna-like vegetation with scattered trees and shrubs growing among ferns and grasses. This may be a unique situation or it may indicate what may happen in the future when stressors, exacerbated by global change, interact with biotic agents.

The current maple forests are the result primarily of production of charcoal and alcohol from massive clearcutting for chemical wood from about 1890 to 1930. The abundance of sugar maple is 3 to 5 times greater than its predisturbance composition of about 5%, and stands are of relatively similar age, that is, 70 to 90 years old (Whitney, 1999). These forests are experiencing significant decline disease in response to a variety of factors (Kolb and McCormick, 1993). Recently, stands have been defoliated by a number of native insects, including the elm spanworm (*Ennomos subsignarius* Hbn.) and the forest tent caterpillar (*Malacosoma disstria* Hubner), and one exotic insect, the pear thrips (*Taeniothrips inconsequens* Uzl.) (Long et al., 1997). Some stands have experienced up to four successive years of defoliation. Also, trees are being affected by the maple borer (*Glycobius speciosus* Say.), attacked very aggressively by several *Armillaria* spp. (Marcais and Wargo, 1998), and colonized by *Steganosporium*, a twig and branch canker fungus. In addition, the forest recently experienced significant drought in 1988, 1991, and 1995. Mortality can be as high as 65% in some stands, especially on dry and nutrient-poor sites.

Sugar maple decline is more severe in stands growing in unglaciated soils (>500,000 years ago) than in stands growing in recently glaciated soils (~12,000 years ago) (Horsley et al., 1999). Not only does the disease seem less severe on glaciated sites but also reproduction is greater and more vigorous. This has led to speculation that acidic deposition also might be a factor on unglaciated soils (Drohan and Sharpe, 1997; Sharpe and Sunderland, 1995).

Some stands are not regenerating back to sugar maple. This is primarily the result of excessive deer browsing but also from extreme competition for light from ferns and grasses in the understory. Millers et al. (1989) list a total of 65 different species of trees that have been affected by decline diseases in the eastern United States this century. In these cases, however, the affected forests, have characteristically and quickly regenerated back to forest tree species.

Case Study 2. Winter Injury and Acidic Deposition: Decline of Red Spruce in the Northeast

Midwinter thaws play a significant role in winter injury on red spruce (DeHayes et al., 1990; Friedland et al., 1984). Injury is related to needle freezing followed by drying rather than strictly to "acute frost desiccation" or "winter desiccation" (DeHayes, 1992; Hadley et al., 1991; Herrick and Friedland, 1990; Johnson, 1992; Johnson et al., 1992; Perkins et al., 1991; Vann et al., 1992).

DeHayes (1992) found that current-year foliage of red spruce exposed to experimentally induced winter thaws for five days dehardened by 5 to 14°C and did not attain full cold tolerance for at least five days. Cold tolerance of current-year foliage of red spruce was reduced by 3 to 14°C during a natural thaw in January (Strimbeck et al., 1995). Under a regime of subfreezing temperatures, pre-thaw cold tolerance levels were reestablished within 10 to 20 days.

Rapid drops in winter temperature after midwinter thaws (vs. extreme low temperatures) also have been implicated from field observations (Manion and Castello, 1993) of "winter reddening" of red spruce foliage. Perkins and Adams (1995) confirmed these observations experimentally. Curry and Church (1952) and more recently Tobi et al. (1995) reported that radial growth was reduced after winter injury. This reduction occurred in the growing season immediately after winter injury. The reduction of internode growth occurred in the third growing season after injury. This suggests that a series of winter injuries in succession or in close years can account for the reduced radial growth noted on red spruce since about 1960 (Cook and Zedaker, 1992). A lower resistance to other stresses also has been inferred (LeBlanc and Raynal, 1990).

The most recent research indicates that acidic deposition may enhance the probability of winter injury on red spruce (DeHayes, 1992; DeHayes et al., 1991; Fowler et al., 1989; Peart et al., 1991; Sheppard, 1994; Vann et al., 1992). DeHayes et al. (1991) demonstrated that cold tolerance was reduced from 3 to 5°C in seedlings treated with acidic mist. The role of acidic deposition in decline diseases is not completely understood. Aluminum-induced calcium depletion has been hypothesized as a major mechanism through which acid deposition causes decline in red spruce. This hypothesis is discussed in detail in Chapter 6.

Case Study 3. Extreme Climatic Fluctuations and Forest Age: Decline of Northern Hardwoods

Recent evidence indicates that forest maturity plays a pivotal role in the onset of decline diseases in northern hardwoods. Under high levels of climatic stress, it is apparent that tree populations will not die back unless they have reached or are nearing biological maturity or large size (Auclair et al., 1997). Once tree populations near maturity, they are vulnerable to extreme climate and other stressors. Auclair et al. presented convincing evidence that extreme climatic fluctuations have triggered major episodes of dieback in mature population cohorts of northern hardwood tree species as well as in other forest types and geographical regions (Auclair, 1987, 1993a,b, 1997; Auclair et al., 1992, 1996, 1997).

Forest Maturity

A comparative study of the actual onset dates of nine principal episodes of dieback across subregions of the northern hardwoods in the U.S. and Canada indicated a close relationship of maturity to the expected date of onset (i.e., within five years). The latter was based on the age to maturity (Millers et al., 1989), adjusted by a "lag interval" based on the tree's growth rate. This adjustment was necessary since the six subregions represented widely different temperature and precipitation regimes that affected growth and hence the rate at which the tree reached maturity (Auclair et al., 1997).

The implications of this finding are significant with respect to the question of decline diseases vs. environmental change in the Northeast. Prior to about 1975, trees did not decline in response to climatic and other environmental stressors until the populations of successive individual species had approached maturity. High levels of freezing and drought stresses early in this century (pre = 1925) failed to trigger major dieback events. Conversely, dieback and decline is now (post = 1975) widespread in response to both an aged forest structure and high levels of climatic stress (Auclair et al., 1996, 1997). The estimated levels of mortality since 1975 have been unprecedented; approximately 58% of all climate-related dieback this century has occurred since 1976 (see Fig. 4.2a). There is some question whether defoliator-induced diebacks have followed this pre = 1975/post = 1975 pattern (Millers et al., 1989).

Extreme Freezing and Drought Stress

Sudden freezing, such as prolonged winter thaw followed by a rapid return to subzero (°C) temperatures, and/or severe summer drought are likely to incite air emboli in sapwood and other irreversible damages to stem and root tissue (Tyree and Sperry, 1989; Cochard and Tyree, 1990). Root freezing can result in similar damages as demonstrated in field experiments in which snow was removed or suspended above the soil surface over the winter months (Pomerleau, 1991; Robitaille et al., 1995). Root freezing as a result of deep soil frost triggered dieback symptoms on the test plots the following summer. The marked absence of snowpack over the 1930s, late 1940s to mid-1950s, the 1960s, and since the late 1970s (Fig. 4.3) concurs closely with the timing of severe and extensive dieback episodes in eastern Canada and the northeastern U.S.

Auclair (1997) developed an index of freezing stress (thaw freeze and root freeze) and one of drought stress (soil water deficit and heat stress) for the Northeast that extends over the century (1910 to 1995). The indices, based on extreme events evident in daily meteorological records at Burlington, Vermont, were combined to produce an overall annual estimate of the level of extreme climatic stress on forests of the region (see Fig. 4.2b). The combined "local stress index" indicated that climate stress during the 1990's has been significantly higher than historical levels, and that the onset of each principal episode of dieback occurred at a time of high climatic stress. Conversely, subsidence of dieback or "recovery" occurred when the levels of stress had decreased about one-tenth of the index maximum.

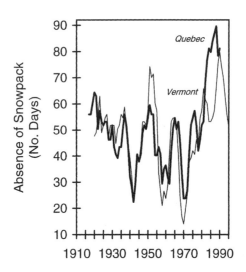

Figure 4.3. Five-year running mean of the number of days over the winter period (December to mid-March) with snowpack of 5 cm or less. The absence of snowpack in southern Quebec (Lennoxville) increased 57% and in northern Vermont (Burlington) 24% in the 1976 to 1995 period compared with pre-1976 levels. Peak levels (i.e., open winters) were intervals marked by significant dieback episodes in northern hardwoods.

Case Study 4. Global Climate Change: Simultaneous Declines in the Northeastern United States and Central Europe

In the mid-1970s, a general decline ("Waldsterben" and later "neuartige Waldschaden") occurred throughout central Europe, including Austria, Belgium, France, Germany, and Italy. Both conifers, such as Norway spruce (*Picea abies* [L.] Karsten), white fir (*Abies alba* Mill.), and Scots pine (*Pinus sylvestris* L.), and angiosperms, such as beech (*Fagus sylvatica* L.) and oaks (primarily *Quercus robur* L. and *Quercus petraea* Liebl.) were among the species reported to be affected (Schutt and Cowling, 1985).

In the eastern United States, red spruce was reported to be declining and dying throughout its range in the Appalachian Mountains in the late 1950s, and again in the mid-1970s (Bruck, 1984; Siccama et al., 1982; Vogelmann et al., 1985; Weiss et al., 1985). Surveys throughout the Appalachian range indicated that mortality was greater in high-elevation stands (>800 m) (Craig and Friedland, 1991; Johnson and Siccama, 1983; Miller-Weeks and Smoronk, 1993; Scott et al., 1984; Weiss et al., 1985).

The simultaneous events in Europe and North America in and after the mid-1970s led to much speculation about their cause. At the time, it was believed that an unprecedented decline of forests that occurred in the late 1970s and early 1980s was attributable to air pollution effects (Skelly and Innes, 1994). Although there was lack of clear evidence for traditional stressors and opportunistic pathogens to be involved in these declines (Carey et al., 1984; Wargo et al., 1987, 1993; Bruck, 1989; Weidensaul et al., 1989), there was also lack of clear evidence that air pollutants, particularly acid deposition, was involved (Skelly and Innes, 1994; Kandler, 1990, 1992a,b, 1993).

Working with global climate phenomena, including anomalies in the global mean annual air temperatures and El Niño–Southern Oscillation Index, Auclair proposed that global climate change was responsible for the simultaneous forest declines in geographically distant regions (Auclair, 1987; Auclair et al., 1992, 1996). Fig. 4.4 identifies the onset of major diebacks in northern hardwoods (northeastern United States, southeastern Canada) and in central Europe in relation to changes in global mean annual temperature. The onset of successive dieback episodes coincided with rapid temperature increases; recovery occurred as global temperatures relaxed or dropped from previous highs. It was assumed that periods of rapid change in global temperature were marked by intervals of exceptional climatic stress at the regional level, both in North America and in Europe.

In Fig. 4.5a, anomalies in global temperature (from the 1910 to 1995 mean annual temperature) and in the Southern Oscillation Index (i.e., 1000 mb pressure difference between Tahiti and Darwin, Australia) were combined to give an integrated estimate of the risk to dieback from

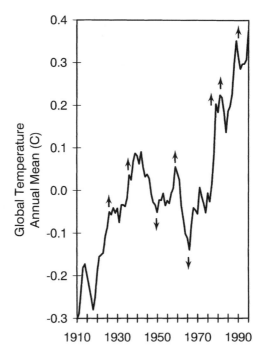

Figure 4.4. Relation of global warming (i.e., five-year running mean of the global mean annual temperature) to the onset (↑) and recovery (↓) of forest dieback episodes in eastern North America and central Europe.

climatic stress. Especially evident is the high level of current climatic stress; from 1976 to 1995, inclusive, the index was 6-fold higher than historically (i.e., from 1910 to 1975). Fig. 4.5b shows the close relationship between the global stress index and the incidence of local freezing events in the northeastern U.S. (Burlington, VT). As in the northeastern U.S., extreme fluctuations in winter temperatures have been implicated in triggering the extensive forest declines in Europe from the mid-1970s onward (Auclair, 1993a; Hartmann and Blank, 1993; Hartmann et al., 1991).

Future Research Needs

It remains important that forest agencies and managers in the Northeast have access to a rich array of accurate, timely information on forest condition and health. Potential benefits are improved capability to anticipate and manage effectively decline at a time of rapid change. Three areas in which research can facilitate this goal are:

1. Development of Geographical Information Systems (GIS) Regional Maps and Early Warning System (EWS) of Risk to Forest Health.

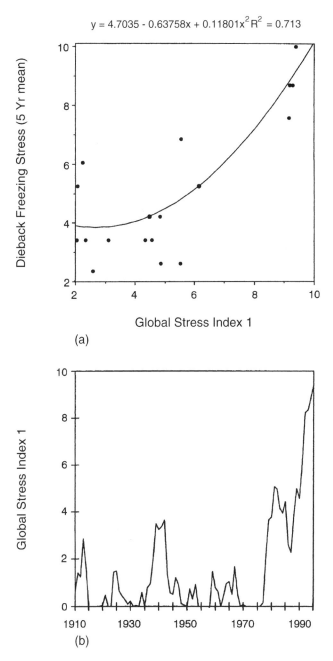

$y = 4.7035 - 0.63758x + 0.11801x^2 R^2 = 0.713$

(a)

(b)

Figure 4.5. (a) Century trends in the global climatic stress index based on the average of the anomalies in global mean annual temperature and the Southern Oscillation Index. (b) The relation of the index of dieback freezing stresses (at Burlington, VT) to the global stress index. A strong relationship is evident, especially at values >4.0. Values are five-year running means of each index.

A GIS/EWS for risk of dieback and decline could identify "hot spots" in the Northeast that historically have been at risk to dieback injuries, the "real-time" situation now, and how the risk of dieback and decline has changed and is changing as global warming and other stresses continue. The EWS would alert managers and the public to expected events and impacts on health of the forests. Anticipatory planning is a decided advantage in adjusting both expectations and management actions to minimize the impacts.

2. Development of a suite of indicators of forest susceptibility to stress and vulnerability to disturbance from stressor–host–pathogen interactions. Several biochemical, physiological, dendrochemical, and microbial methods show promise as early warning indicators. However, further evaluation is needed before they can be developed into useful tools for land managers.

3. Evaluate the factors (site, species, stress, pathogen, etc.) that affect the spatial and temporal development of decline disease within forests and across landscapes. Tree–host genetic variability in relationship to susceptibility and vulnerability to stress and pathogens is poorly understood. Legacy factors (Vogt et al., 1996) such as previous disturbance from insects, fire, pathogens, forest management, and land use changes, and how they have influenced tree species abundance and distribution on the landscape, must be evaluated.

4. Development of management prescriptions to reduce damage and growth/volume loss to decline diseases. One option is to "weatherproof" and "pestproof" the forest by selectively identifying areas/locals at high risk and treating them in such a way as to minimize injury and mortality. For example, actions may include

- Increased monitoring of areas identified to be at high risk to ensure rapid response by managers and forest agencies. Links with the Forest Service's Forest Inventory and Analysis Program and the Forest Health Monitoring Program are needed.
- Specialized observations on the etiology of disease, dieback, and decline. This could include successive measurements of the inciting injury, moisture stress, infection by disease organisms, insect outbreaks, growth loss and mortality, and tree regrowth/recovery processes.
- Salvage harvesting in advance of or early in the dieback episode to ensure high quality and economic value of wood products (i.e., free of pests, fungi, decay).
- Harvesting of "stands at risk" and the replacement of these with trees that are younger, more vigorous, and more resistant to freezing, drought, and other stresses.
- Management practices that discriminate against tree species off site and favor those adapted to site and timber stand.

- Thinning or improvement cuts to reduce competition for moisture on drought-prone sites.

5. Process-level climate experiments (in the laboratory and in the field) designed to better identify the action and mechanism of stressors (e.g., role of freezing stresses in the development of irreversible emboli).

Conclusions

Forest scientists have made significant progress in understanding the role of natural and some anthropogenic stressors in decline disease. The major challenge will be to sort out declines due to natural factors from those due to anthropogenic factors, as well as those resulting from the interactions of both types of factors. Natural stressors, such as persistent defoliation, drought, and frost, have triggered episodes of decline. As global atmospheric chemistry and climate change, the potential for natural and anthropogenic stressors interacting to affect forest health is likely to increase. The extent and severity of forest declines has been high since the mid-1970s. During this period, much new information and insight into decline processes has opened the possibilities for predicting the incidence of decline and for better managing and adapting forests to future conditions.

References

Auclair AND (1987) Climate theory of forest decline. In: *Woody Plant Growth in a Changing Chemical and Physical Environment.* Lavender DP (ed) *Proceedings of the International Union of Forestry Research Organizations Working Party on Shoot Growth Physiology (S2.01-11), 27–31 July 1987,* University of British Columbia, Vancouver, British Columbia, Canada. pp 1–30.

Auclair AND (1993a) Extreme winter temperature fluctuation, root and sapwood injury, and oak decline in central Europe. In: Luisi N, Lerario P, Vannini A (eds) *Recent Advances in Studies on Oak Decline. Proceedings of an International Congress, 13–18 September 1992, Brindisi, Italy.* Tipolitografia Radio, Bari, Italy, pp 139–148.

Auclair AND (1993b) Extreme climatic fluctuations as a cause of forest dieback in the Pacific Rim. Water Air Soil Pollut 66:207–229.

Auclair AND (1997) *Retrospective Analysis of Forest Dieback, Extreme Climatic Fluctuations, and Global Climate Changes from 1910 to 1995.* Interim Task Report to the Office of Global Programs. National Oceanic and Atmospheric Administration, Washington, DC.

Auclair AND, Eglington PD, Minnemeyer SL (1997) Principal forest dieback episodes in northern hardwoods: development of numeric indices of area extent and severity. Water Air Soil Pollut 93:175–198.

Auclair AND, Lill JT, Revenga C (1996) The role of climatic variability and global warming in the dieback of northern hardwoods. Water Air Soil Pollut 91:163–186.

Auclair AND, Worrest RC, Lachance D, Martin HC (1992) Climatic perturbation as a general mechanism of forest dieback. In: Manion PD, Lachance D (eds)

Forest Decline Concepts. American Phytopathological Society, St. Paul, MN, pp 38–58.

Bauce E, Allen DC (1991) Etiology of a sugar maple decline. Can J For Res 21:686–693.

Bormann FH, Likens GE (1979) *Pattern and Process in a Forested Ecosystem*. Springer-Verlag, New York.

Bruck RI (1984) Decline of montane boreal ecosystems in central Europe and the southern Appalachian Mountains. Technical Association of the Pulp and Paper Industry Proc., 159–163.

Bruck RI (1989) Survey of diseases and insects of Fraser fir and red spruce in the southern Appalachian Mountains. Eur J For Pathol 19:389–398.

Campbell RW, Sloan RJ (1977) *Forest Stand Responses to Defoliation by the Gypsy Moth*. Forest Science Monograph 19. Society of American Foresters, Washington, DC.

Carey AC, Miller EA, Geballe GT et al. (1984) *Armillaria mellea* and decline of red spruce. Plant Dis 68(9):794–795.

Castello JD, Leopold DJ, Smallidge PJ (1995a) Pathogens, patterns, and processes in forest ecosystems. Bioscience 45(1):16–24.

Castello JD, Wargo PM, Jacobi V, Bachand GD, Tobi DR, Rogers MAM (1995b) Tomato mosaic virus infection of red spruce on Whiteface Mountain, New York: prevalence and potential impact. Can J For Res 25:1340–1345.

Ciesla WM, Donaubauer E (1994) *Decline and Dieback of Trees and Forests*. United Nations Food and Agriculture Organization (FAO) Forestry Paper 120. FAO, Rome, Italy.

Cochard H, Tyree MT (1990) Xylem dysfunction in Quercus: vessel tyloses, cavitation and seasonal changes in embolism. Tree Physiol 6:393–407.

Cook ER, Zedaker SM (1992) The dendrochronology of red spruce decline. In: Eagar C, Adams MB (eds) *Ecology and Decline of Red Spruce in the Eastern United States*. Springer-Verlag, New York, pp 192–231.

Craig BW, Friedland AJ (1991) Spatial patterns in forest composition and standing dead red spruce in montane forests of the Adirondacks and northern Appalachians. Environ Monit Assess 18:129–143.

Crist CR, Schoeneweiss DF (1975) The influence of controlled stresses on susceptibility of European white birch stems to attack by *Botryosphaeria dothidea*. Phytopathol 65:369–373.

Curry J, Church T (1952) Observations of winter drying of conifers in the Adirondacks. J For 50:114–116.

DeHayes DH (1992) Winter injury and developmental cold tolerance of red spruce. In: Eagar C, Adams MB (eds) *The Ecology and Decline of Red Spruce in the Eastern United States*. Springer-Verlag, New York, pp 295–337.

DeHayes DH, Thornton FC, Waite CE, Ingle MA (1991) Ambient cloud deposition reduces cold tolerance of red spruce seedlings. Can J For Res 21:1292–1295.

DeHayes DH, Waite CE, Ingle MA, Williams MW (1990) Winter injury susceptibility and cold tolerance of current and year-old needles of red spruce trees from several provenances. For Sci 36:982–994.

Drohan JR, Sharpe WE (1997) Long-term changes in forest soil acidity in Pennsylvania, U.S.A. Water Air Soil Pollut 95:299–311.

Fowler DJ et al. (1989) Effects of acid mist on the frost hardiness of red spruce seedlings. New Phytol 113:321–355.

Friedland AJ, Gregory RA, Karenlamp L, Johnson AH (1984) Winter damage to foliage as a factor in red spruce decline. Can J For Res 14:963–965.

Giese RL, Benjamin DM (1964) *Studies of Maple Blight. Part II. The Insect Complex Associated with Maple Blight.* Res Bull 250. University of Wisconsin, Madison, WI, pp 20–57.

Giese RL, Houston DR, Benjamin DM, Kuntz JE (1964) *Studies of Maple Blight. Part I. A New Condition of Sugar Maple.* Res Bull 250. University of Wisconsin, Madison, WI, pp 1–19.

Greenidge KNH (1951) *Dieback: a disease of yellow birch* (Betula lutea *Michx.*) *in eastern Canada.* Ph.D. Thesis, Department of Biology, Harvard University, Boston, MA.

Gregory RA, Wargo PM (1986) Timing of defoliation and its effect on bud development, starch reserves, and sap sugar concentration in sugar maple. Can J For Res 16:10–17.

Gross HL (1991) Dieback and growth loss of sugar maple associated with defoliation by the forest tent caterpillar. For Chron 67:33–42.

Hadley JL, Friedland AJ, Herrick GT, Amundson RG (1991) Winter desiccation and solar radiation in relation to red spruce decline in the northern Appalachians. Can J For Res 21:269–272.

Hall TJ (1995) Effect of forest tent caterpillar and *Discula campestris* on sugar maple in Pennsylvania. Phytopathol 85(10):1129.

Hartmann G, Blank R (1993) Etiology of oak decline in northern Germany: history, symptoms, biotic and climatic predisposition, pathology. In: Luisi N, Lerario P, Vannini A (eds) *Recent Advances in Studies on Oak Decline. Proceedings of an International Congress, 13–18 September 1992, Brindisi, Italy.* Tipolitografia Radio, Bari, Italy, pp 277–284.

Hartmann G, Blank R, Lewark S (1991) Oak decline in northern Germany: distribution, symptoms, probable causes. In: Siwecki R, Liese W (eds) *Oak Decline in Europe: Proceedings of an International Symposium, 15–18 May 1990, Kornik, Poland.* Polish Academy of Sciences, Kornik, Poland, pp 69–74.

Healy W, Gottschalk K, Long R, Wargo P (1997) Changes in eastern forests: chestnut is gone, are the oaks far behind? In: Transactions of the 62nd North American Wildlife and Natural Resources Conference. March 14–18, Washington, DC. Wildlife Management Institute, Washington, DC, pp 249–263.

Heatwole H, Lowman M (1986) *Dieback: Death of an Australian landscape.* Reed Books, French Forest, New South Wales, Australia.

Herrick GT, Friedland AJ (1990) Winter desiccation and injury of subalpine red spruce. Tree Physiol 8:23–36.

Hornbeck JW, Smith RB (1985) Documentation of red spruce decline. Can J For Res 15:1199–1201.

Horsley SB, Long RP, Bailey SW, Hallett RA, Hall TJ (1999) Factors contributing to sugar maple decline along topographic gradients on the glaciated and unglaciated Allegheny Plateau. In: Horsley SB, Long RP (eds) *Sugar Maple Ecology and Health. Proceedings of an International Symposium, 2–4 June 1998, Warren, PA.* Gen Tech Rep NE-261. United States Department of Agriculture (USDA) Forest Service, Northeastern Research Station, Radnor, PA, pp 60–63.

Houston DR (1967) The dieback and decline of northeastern hardwoods. Trees 28:12–14.

Houston DR (1981) Stress triggered tree diseases—the diebacks and declines. NE-INF-41-81. United States Department of Agriculture (USDA) Forest Service, Northeastern Forest Experiment Station, Bromall, PA.

Houston DR (1987) Forest tree declines of past and present: current understanding. Can J Plant Pathol 17:349–360.

Houston DR (1992) A host–stress–pathogen model for forest dieback–decline diseases. In: Manion PD, Lachance D (eds) *Forest Decline Concepts.* American Phytopathological Society Press, St. Paul, MN, pp 3–25.

Houston DR, Kuntz JE (1964) *Studies of Maple Blight. Part III. Pathogens Associated with Maple Blight.* Res Bull 250. University of Wisconsin, Madison, WI, pp 59–78.

Johnson AH (1992) The role of abiotic stress in the decline of red spruce in high elevation forests of the eastern United States. Annu Rev Phytopathol 30: 349–367.

Johnson AH et al. (1992) Synthesis and conclusions from epidemiological and mechanistic studies of red spruce decline. In: Eagar C, Adams MB (eds) *The Ecology and Decline of Red Spruce in the Eastern United States.* Springer-Verlag, New York, pp 385–411.

Johnson AH, Siccama TG (1983) Acid deposition and forest decline. Environ Sci Tech 17:294–306.

Kandler O (1990) Epidemiological evaluation of the development of Waldsterben in Germany. Plant Dis 74:4–12.

Kandler O (1992a) Historical declines and diebacks of central European forests and present conditions. Environ Tox Chem 11:1077–1093.

Kandler O (1992b) Development of the recent episode of Tannensterben (fir decline) in eastern Bavaria and the Bavarian Alps. In: Huttl RF, Mueller-Dombois D (eds) *Forest Decline in the Atlantic and Pacific Region.* Springer-Verlag, New York, pp 216–226.

Kandler O (1993) The air pollution/forest decline connection: the "Waldsterben" theory refuted. Unasylva 174(44):39–49.

Karl TR, Knight RW, Easterling DR, Quayle RG (1996) Indices of climate change for the United States. Bull Am Meteorol Soc 77:279–292.

Kolb TE, McCormick LH (1993) Etiology of sugar maple decline in four Pennsylvania stands. Can J For Res 23:2395–2402.

Larcher W, Bauer H (1981) Ecological significance of resistance to low temperature. In: Lange OL, Nobel PS, Osmond CB, Ziegler H (eds) *Physiological Plant Ecology.* Vol. 1. Springer-Verlag, New York, pp 403–437.

LeBlanc DC (1990) Red spruce decline on Whiteface Mountain, New York. I. Relationships with elevation, tree age, and competition. Can J For Res 20:1408–1414.

LeBlanc DC, Raynal DJ (1990). Red spruce decline on Whiteface Mountain, New York. II. Relationships between apical and radial growth decline. Can J For Res 20:1415–1421.

Levitt J (1972) *Responses of Plants to Environmental Stresses.* Academic Press, New York.

Long RP, Horsley SB, Lilja PR (1997) Impact of forest liming on growth and crown vigor of sugar maple and associated hardwoods. Can J For Res 27: 1560–1573.

Manion PD (1987) Decline as a phenomenon in forests: pathological and ecological considerations. In: Hutchinson TC, Meema KM (eds) *The Effects of Atmospheric Pollutants on Forests, Wetlands and Agricultural Ecosystems. Proceedings of the North Atlantic Treaty Organization (NATO) Advanced Workshop.* Springer-Verlag, Berlin, Germany, pp 267–275.

Manion PD (1991) *Tree Disease Concepts. 2nd ed.* Prentice-Hall, Englewood Cliffs.

Manion PD, Castello JD (1993) Snow depth identifies late winter as the "window" for freezing injury of red spruce. Phytopathol 83:1351.

Marcais B, Wargo PM (1998) Influence of liming on the abundance and vigor of Armillaria rhizomorphs. In: *Proceedings of the 9th International Union of Forestry Research Organizations Conference on Root and Butt Rots, 31 August to 8 September 1997, Carcans, France.* Institut National de la Recherche, Paris, France.

Millers I, Shriner DS, Rizzo D (1989) History of hardwood decline in the eastern United States. Gen Tech Rep NE-126. United States Department of Agriculture (USDA) Forest Service, Northeastern Forest Experiment Station, Broomall, PA.

Miller-Weeks M, Smoronk D (1993) *Aerial Assessment of Red Spruce and Balsam Fir Condition in the Adirondacks Region of New York, the Green Mountains of Vermont, the White Mountains of New Hampshire, and the Mountains of Western Maine, 1985–1986.* NA-TP-16-93. United States Department of Agriculture (USDA) Forest Service, Northeastern Area, State and Private Forestry, Radnor, PA.

Mueller-Dombois D (1983a) Canopy dieback and successional processes in Pacific forests. Pacific Sci 37:317–325.

Mueller-Dombois D (1983b) Population death in Hawaiian plant communities: a causal theory and its successional significance. Tuexenia 3:117–130.

Mueller-Dombois D (1992) A natural dieback theory, cohort senescence as an alternative to the decline disease theory. In: Manion PD, Lachance D (eds) *Forest Decline Concepts.* American Phytopathological Society, St. Paul Minnesota, pp 26–37.

Mueller-Dombois D, Canfield JE, Halt RA, Buelow GP (1983) Tree-group death in North America and Hawaiian forest: a pathological problem or a new problem for vegetation ecology? Phytoenologia 11:117–137.

Peart DR, Jones MB, Palmlotto PA (1991) Winter injury to red spruce at Mount Moosilauke, New Hampshire. Can J For Res 21:1380–1389.

Perkins TD, Adams GT (1995) Rapid freezing induces winter injury symptomatology in red spruce foliage. Tree Physiol 15:259–266.

Perkins TD, Adams GT, Klein RM (1991) Desiccation or freezing? Mechanisms of winter injury to red spruce. Am J Bot 78:1207–1217.

Pomerleau R (1991) *Experiments on the Causal Mechanisms of Dieback on Deciduous Forests in Quebec.* Info Rep LAU-X-96. Forestry Canada, Sainte Foy, PQ, Canada.

Powell DS, Faulkner JL, Darr DR, Zhu Z, MacCleery DW (1993) Forest statistics of the United States, 1992. US Department of Agriculture, Washington, DC.

Robitaille GR, Boutin R, Lachance D (1995) Effects of soil freezing stress on sap flow and sugar content of mature sugar maples (*Acer saccharum*). Can J For Res 25:577–587.

Sakai A, Larcher W (1987) *Frost Survival of Plants.* Ecological Studies. Vol. 62. Springer-Verlag, Berlin, Germany.

Schoeneweiss DF (1978) The influence of stress on diseases of nursery and landscape plants. J Arboricult 4:217–225.

Schoeneweiss DF (1981a) Infectious diseases of trees associated with water and freezing stress. J Arboricult 7:13–18.

Schoeneweiss DF (1981b). The role of environmental stress in diseases of woody plants. Plant Dis 65:308–314.

Schutt P, Cowling EB (1985) Waldsterben, a general decline of forests in central Europe: symptoms, development, and possible causes. Plant Dis 69:548–558.

Scott JT, Siccama TG, Johnson AH, Breisch AR (1984) Decline of red spruce in the Adirondacks, New York. Bull Torrey Bot Club 111:438–444.

Sharpe WE, Sunderland TL (1995). Acid-base status of upper rooting zone soil in declining and non-declining sugar maple (*Acer saccharum* Marsh.) stands in Pennsylvania. In: Gottschalk KW, Fosbroke SL (eds) *Proceedings, 10th Central Hardwood Forest Conference, 5–8 March 1995, Morgantown, West Virginia*. Gen Tech Rep NE-197. United States Department of Agriculture (USDA) Forest Service, Northeastern Forest Experiment Station, Radnor, PA, pp 172–178.

Sheppard LJ (1994) Causal mechanisms by which sulfate, nitrate and acidity influence frost hardiness in red spruce: review and hypothesis. New Phytol 127(1):69–82.

Siccama TG, Bliss M, Vogelmann HW (1982) Decline of red spruce in the Green Mountains of Vermont. Bull Torrey Bot Club 109:163–168.

Sinclair WA 1967. Decline of hardwoods: possible causes. Proc Int Shade Tree Conf 42:17–32.

Sinclair WA, Hudler GW (1988) Tree declines: four concepts of causality. J Arboricult 14:29–35.

Siwecki R, Liese W (eds) (1991) *Oak Decline in Europe. Proceedings of an International Symposium, 15–18 May 1990, Kornik, Poland*. Polish Academy of Sciences, Kornik, Poland.

Skelly JM, Innes JL (1994) Waldsterben in the forests of central Europe and eastern North America: fantasy or reality? Plant Dis 78:1021–1032.

Sperry JS, Donnelly JR, Tyree MT (1988) Seasonal occurrence of xylem embolism in sugar maple (*Acer saccharum*). Am J Bot 75:1212–1218.

Staley JM (1965) Decline and mortality of red and scarlet oaks. For Sci 11:2–17.

Strimbeck GR, Schaberg PG, DeHayes DH, Shane JB, Hawley GJ (1995) Midwinter dehardening of red spruce during a natural thaw. Can J For Res 25:2040–2044.

Tobi DR, Wargo PM, Bergdahl DR (1995). Growth response of red spruce after known periods of winter injury. Can J For Res 25:669–681.

Tyree MT, Sperry JS (1988) Do woody plants operate near the point of catastrophic xylem dysfunction caused by dynamic water stress? Plant Physiol 88:574–580.

Tyree MT, Sperry JS (1989) Vulnerability of xylem to cavitation and embolism. Annu Plant Physiol Molec Biol 40:19.

Van Deusen PC, Reams GA, Cook ER (1991) Possible red spruce decline. Contributions of tree-ring analysis. J For 89:20–24.

Vann DR, Strimbeck GR, Johnson AH (1992) Effects of ambient levels of airborne chemicals on the freezing resistance of red spruce foliage. For Ecol Manage 51:69–79.

Vogelmann HW, Badger GJ, Bliss M, Klein RM (1985) Forest decline on Camels Hump, Vermont. Bull Torrey Bot Club 112:274–287.

Vogt KA et al. (1996) *Ecosystems: Balancing Science with Management*. Springer-Verlag, New York.

Walker SW, Auclair AND, Martin HC (1990) *History of Crown Dieback and Deterioration Symptoms of Hardwoods in Eastern Canada*. Federal LRTAP Liaison Office, Atmospheric Environment Service, Environment Canada, Downsview, Ontario, Canada.

Wargo PM (1972) Defoliation-induced chemical changes in sugar maple roots stimulate growth of *Armillaria mellea*. Phytopathol 62:1278–1283.

Wargo PM (1975) Lysis of the cell wall of *Armillaria mellea* by enzymes from forest trees. Physiol Plant Pathol 5:99–105.

Wargo PM (1977) *Armillaria mellea* and *Agrilus bilineatus* and mortality of defoliated oak trees. For Sci 23:485–492.

Wargo PM (1978) Defoliation by the gypsy moth—how it hurts your tree. USDA Home Garden Bull 223. US Department of Agriculture, Washington, DC.

Wargo PM (1980) *Armillaria mellea*: an opportunist. J Arboricult 6:276–278.

Wargo PM (1981a) Defoliation and secondary-action organism attack: with emphasis on *Armillaria mellea*. J Arboricult 7:64–69.

Wargo PM (1981b) Defoliation, dieback and mortality. In: Doane CC, McManus ML (eds) *The Gypsy Moth*: *Research Toward Integrated Pest Management*. Tech Bull 1584. United States Department of Agriculture (USDA) Forest Service, Washington, DC, pp 240–248.

Wargo PM (1981c) Measuring response of trees to defoliation stress. In: Doane CC, McManus ML (eds) *The Gypsy Moth*: *Research Toward Integrated Pest Management*. Tech Bull 1584. United States Department of Agriculture (USDA) Forest Service, Washington, DC, pp 248–267.

Wargo PM (1988) Amino nitrogen and phenolic constituents of bark of American beech, *Fagus grandifolia*, and infestation by beech scale, *Cryptococcus fagisuga*. Eur J For Pathol 18:279–290.

Wargo PM (1995) Disturbance in forest ecosystems caused by pathogens and insects. In: *Forest Health through Silviculture. Proceedings of the 1995 National Silviculture Workshop, 8–11 May 1995, Mescalero, NM*. Gen Tech Rep RM-GTR-267. United States Department of Agriculture (USDA) Forest Service, Rocky Mountain Research Station, Ft Collins, CO.

Wargo PM (1996) Consequences of environmental stress on oak: predisposition to pathogens. Ann Sci For 53:359–368.

Wargo PM (1999) Integrating the role of stressors through carbohydrate dynamics. In: Horsley SB, Long RP (eds) *Sugar Maple Ecology and Health*: *Proceedings of an International Symposium, 2–4 June 1998, Warren, PA*. Gen Tech Rep NE-261. United States Department of Agriculture (USDA) Forest Service, Northeastern Research Station, Radnor, PA, pp 107–113.

Wargo PM, Bergdahl DR, Tobi DR, Olson CW (1993) *Root Vitality and Decline of Red Spruce*. Contributiones Biologiae Arborum. Vol. 4. Ecomed Publishers, Landsberg am Lech, Germany.

Wargo PM, Carey AC, Geballe GT, Smith WH (1987) Effects of lead and trace metals on growth of three root pathogens of spruce and fir. Phytopathol 77:123.

Wargo PM, Houston DR (1974) Infection of defoliated sugar maple trees by *Armillaria mellea*. Phytopathol 64:817–822.

Wargo PM, Parker J, Houston DR (1972) Starch content in roots of defoliated sugar maple. For Sci 18:203–204.

Weidensaul TC, Fleck AM, Hartzler DM, Capek CL (1989) *Quantifying Spruce Decline and Related Forest Characteristics at Whiteface Mountain, New York. Summary report.* Ohio Agric Res Devel Ctr, Ohio State University, Wooster, OH.

Weiss MJ, McCreery L, Miller I, O'Brien JT, Miller-Weeks M (1985) *Red Spruce and Balsam Fir Decline and Mortality*. NA-TP-11. United States Department of Agriculture (USDA) Forest Service, Northeastern Area, State and Private Forestry, Broomall, PA.

Whitney GG (1999) Sugar maple: Abundance and site relationships in the pre- and postsettlement forest. In: Horsley SB, Long RP (eds) *Sugar Maple Ecology and Health. Proceedings of an International Symposium, 2–4 June 1998, Warren, PA*. Gen Tech Rep NE-261. United States Department of Agriculture (USDA) Forest Service, Northeastern Research Station, Radnor, PA, pp 14–19.

Zimmermann MH (1983). *Xylem Structure and the Ascent of Sap*. Springer-Verlag, New York.

2. Global Change Impacts on Tree Physiology

5. Interacting Effects of Multiple Stresses on Growth and Physiological Processes in Northern Forest Trees

Judson G. Isebrands, Richard E. Dickson, Joanne Rebbeck, and David F. Karnosky

Global climate chagnge is a complex and controversial subject, both technically and politically. Recently, the Intergovernmental Panel on Climate Change (IPCC) of the United Nations concluded that "the balance of evidence suggests a discernible human influence on global climate," and that "further accumulation of greenhouse gases will commit the earth irreversibly to global climate change with its consequent ecological, economic, and social disruption" (Houghton et al., 1996; Brown et al., 1997; Kerr, 1997). One of the concerns is that changing climate will have major effects on future forest composition, productivity, sustainability, and biological as well as genetic diversity (Houghton et al., 1996).

   Two pollutants that are generally considered to have the greatest impacts on plant growth and are indisputably increasing concomitantly in the atmosphere as a consequence of human activity are carbon dioxide (CO_2) (Keeling et al., 1995) and tropospheric ozone (O_3) (Taylor et al., 1994; Chameides et al., 1997). Both pollutants are increasing concurrently, and are expected in rural agricultural and forested areas to have the greatest impacts on plant growth. Increasing CO_2 typically enhances plant growth (Kimball et al., 1990; Koch and Mooney, 1996), while O_3 has a deleterious affect on plant growth (Adams et al., 1989; Yunus and Iqbal, 1996). Less is known about the responses of plant growth to the interaction of concomitant increasing CO_2 and O_3 (Allen, 1990; Krupa and Kickert, 1993). Similarly, little is known about the interacting effects of elevated CO_2 and O_3 on forest trees and forest ecosystems, but overall responses in trees to elevated CO_2 (Ceulemans and Mousseau, 1994) and O_3 (Raineer et al., 1993; Taylor et al., 1994) are similar to that of other plants. As with crop plants, it is known that inherent differences exist among and within forest tree species in their response to certain atmospheric stressors (Kozlowski and Constantinidou, 1986a,b; Taylor, 1994; Karnosky et al., 1996; Hogsett et al., 1997).

The literature is exhaustive on the subject of single factor effects of CO_2, O_3, and other stressors on plants. Thus, it is beyond the scope of this chapter to review all these contributions. Anyone seeking more information on CO_2 effects on plants should consult references by Strain (1987), Kimball et al. (1990), Koch and Mooney (1996), and Yunus and Iqbal (1996); and for effects on woody plants, Ceulemans and Mosseau (1994), Curtis (1996), Wullschleger et al. (1997) and Mickler and Fox (1998). For information on O_3 effects on plants, the reader should refer to Adams et al. (1989), Mooney et al. (1991), Krupa and Kickert (1993), and Yunus and Iqbal (1996); and for effects on woody plants and forest ecosystems, Taylor et al. (1994), Taylor (1994), Fox and Mickler (1996), and Hogsett et al. (1997).

Multiple Stress Effects

Plant responses to CO_2 and O_3 become even more complex when other known stressors such as nitrogen availability, temperature and water extremes, and pests, are combined with the effects of CO_2 and O_3 (Allen, 1990). Traditionally, research on the effects of air pollutant stresses on agricultural crops and forest trees examined one or at most two stresses at the same time with a single plant species. This approach is not surprising given the logistic problems and expense of multiple-factor experiments. However, these single and occasionally multiple-factor experiments have not been very helpful for understanding the long-term response of a single species under natural conditions, much less long-term community or ecosystem responses. Given the strong genetic component of response (Taylor, 1994; Karnosky et al., 1997), and that essentially every biotic and abiotic factor measurable in the environment modifies the responses to air pollutants, and that these responses change with time as trees acclimate to these variable stresses, it is impossible to experimentally determine future stands or community dynamics. However, single- or multiple-factor experiments with as many other environmental factors as possible held constant or increased in response to plant demand, are still necessary to provide biological and mechanistic input into process-based models (Taylor et al., 1994; Lee and Jarvis, 1995; Lloyd and Farquhar, 1996; Thornley and Cannell, 1996). In this chapter we focus our discussion on the responses of aspen (*Populus tremuloides* Michx.), yellow poplar (*Liriodendron tulipifera* L.), and white pine (*Pinus strobus* L.) to ozone (O_3), carbon dioxide (CO_2), their interactions ($O_3 \times CO_2$), and the interactions with water and nitrogen (N) availability, and with pests. Responses of other tree species may also be included when information is limited.

In general, increasing atmospheric CO_2 concentrations will increase photosynthetic rates, leaf production, height growth, and dry weight production. In contrast, increasing atmospheric concentrations of O_3 will decrease photosynthetic rates, increase leaf senescence, and decrease dry weight production. The amount of change, however, depends on many internal plant factors (e.g., plant and tissue age, plant growth strategy, genotypic response, ability to adapt to changing environmental conditions, etc.) and external environmental factors (e.g., light, nutrients, water, temperature, magnitude and duration of exposure, etc.) (Miller et al., 1997). Because elevated CO_2 concentrations impact many metabolic processes and usually increase growth, higher CO_2 concentrations may also compensate for other environmental stresses (Allen, 1990; Ceulemans and Mousseau, 1994). Compensation may work primarily through an increase in the efficient use of other limiting resources. Such increased efficiency is commonly found in CO_2–water use interactions (Eamus and Jarvis, 1989; Eamus, 1991; Bowes, 1993; Lee and Jarvis, 1995; Anderson and Tomlinson, 1998; Tomlinson and Anderson, 1998). An increase in CO_2 concentration commonly increases photosynthetic rate and decreases stomatal conductance and transpiration rates. Increased carbon fixation and decreased water use leads to an increase in water use efficiency. This decrease in water use may be beneficial during short periods of water stress. However, high CO_2 concentrations also often increase total leaf area such that improved water use per unit area is offset by increased leaf area. Thus, total water use during drought may be greater, increasing total water stress (Kerstiens et al., 1995; Beerling et al., 1996).

The interactions of CO_2 and nutrients have important implications for plant growth. The lack of consideration or lack of adequate control of nutrient supply is probably a major factor in much of the conflicting experimental results of growth responses to increased CO_2 (Curtis, 1996; Pettersson et al., 1993; Wullschleger et al., 1997). Even when several levels of nutrients are supplied initially, plant growth constantly changes the ratio of supply to internal demand. Ideally, nutrients should increase exponentially to maintain a constant relative growth rate (Ingestad and Ågren, 1995; Coleman et al., 1998). Most studies show that CO_2 enrichment increases growth even though light and/or nutrients are limiting growth (Conroy and Hocking, 1993). A common assumption is that response to CO_2 will be less if light or nutrients are limiting (law of the minimum). This response is often true if absolute growth increases are considered (low N plants plus CO_2—low N plants). However, the proportional or percent growth increase in response to CO_2 in N-limited plants is often equal to or greater than that found in nonlimited plants (Bowes, 1993; Lloyd and Farquhar, 1996). However, the opposite response is also frequently found (Curtis et al., 1995; Gebauer et al., 1996). Response seems to be strongly controlled by plant life history and the relative allocation of carbon (C) and N to leaf and root growth (Bazzaz and Miao, 1993; Laurence et al., 1994; Lloyd and Farquhar, 1996). A common response to CO_2 enrichment is a greater increase in C fixation rate than N uptake rate. If much of this C is used in leaf growth rather than root growth and N uptake, leaf N concentration will decrease (based on area or dry weight). The increase in total leaf area, however, even if N uptake rate decreases, results in greater total plant N content (Idso et al., 1996; Tissue et al., 1997).

Trees growing in the field are seldom exposed to a single environmental stress. With increasing atmospheric pollutants (e.g., CO_2, O_3, N deposition), trees must respond to these new stresses in addition to more common stresses, such as drought, low light, and nutrient deficiencies. Carbon dioxide enrichment may partially or totally ameliorate growth decreases in response to these common stresses. However, the addition of O_3 stress is often additive to growth impacts of these other common stresses (Greitner et al., 1994), and CO_2 may or may not compensate (Volin et al., 1998). Ozone damage is often greater in fast-growing plants that are watered and fertilized (Winner, 1994; Dickson et al., 1998), or the converse, slow-growing, stressed plants are less sensitive to O_3 damage, particularly if water-stressed (Tingey and Hogsett, 1985). Published results are contradictory. For example, N fertilization had no effect on O_3-induced decreases in growth and leaf senescence of hybrid poplar (Gunthardt-Goerg et al., 1996); however, N fertilization increased O_3 impact on growth of Norway spruce (*Picea abies* [L.] Karsten) (Lippert et al., 1996) and radish (*Raphanus sativus* L.) (Pell et al., 1990); but decreased growth in aspen (Karnosky et al., 1992a,b) and birch (*Betula* spp.) (Pääkkonen and Holopainen, 1995). These differences in response reflect differences in experimental protocols, plant growth strategy, and C allocation patterns. Large differences in response are also found between species (Tjoelker et al., 1993) and among clones or genotypes within species (Taylor, 1994; Karnosky et al., 1996; Dickson et al., 1998).

Knowledge about the combined impacts of chronicly elevated CO_2 and O_3 on ecosystems remains limited and largely speculative. It has been shown that litter decomposition exposed to elevated CO_2 and/or O_3 is significantly decreased, suggesting nutrient cycling within ecosystems may be impacted (Boerner and Rebbeck, 1995; Scherzer et al., 1998). Insect herbivory and other impacts on plants are also significant factors in ecosystem responses to environmental stresses. There is sufficient preliminary data to predict with confidence that both CO_2 and O_3 will significantly affect fundamental plant processes, which will translate into altered tree susceptibility to all major guilds of plant-feeding insects (i.e., folivores, phloem and xylem sappers, phloeo- and xylophages, and rhizophages). Under some circumstances, this may precipitate rapid, substantive changes in plant competitive abilities and thereby drastically alter the normal compositional and successional trajectories of plant communities (Maron, 1998). Elevated CO_2 decreases leaf N levels and increases starch content, fiber, leaf temperatures, and concentrations of phenolics and tannins (Mooney et al., 1991; Bazzaz and Fajer, 1992; Lincoln et al., 1993; Trier et al., 1996). Responses for most insect folivores are typically increased leaf consumption, coupled with decreased growth, survival, and fecundity (Lincoln et al., 1993). However, this research applies primarily to external folivores. No one has yet investigated the responses of leaf miners or those of stem and root borers. Miners typically selectively feed on mesophyll and parenchyma tissues and may be able to avoid serious CO_2-induced nutrient dilutions and increases in raw fiber (Trier and Mattson, 1997). No one knows how the biochemical micro-environment of stem borers is likely to change, and how borers may respond. Carbon dioxide-induced increases in cambial growth may enhance rapid callus formation and thus more certain containment and death of stem-invading larvae. Theory also predicts that the standard hypersensitive or rapid induced (secondary chemical based) resistance (RIR) in stem tissues to foreign invasions may be enhanced given the fact that RIR and plant growth are usually positively linked (Herms and Mattson, 1992).

Under elevated ozone, decreasing whole plant growth and module longevity are typical responses (Pye, 1988). This decrease is due to the phytotoxic effects of elevated O_3, which typically decreases Rubisco (Brendley and Pell, 1998), lowers rates of photosynthesis and decreases leaf surface area due to premature leaf abscission (especially in sensitive species). Ozone enrichment also generally diminishes branch growth and longevity of short shoots (Matyssek et al., 1993), leading to weakened radial growth and diminished growth and maintenance of fine roots (Coleman et al., 1996). In their seedling and sapling stages, many indeterminately growing tree species produce large populations of leaves, which are strong sinks throughout the growing season. As older source leaves continually abscise with O_3 stress, total C fixation decreases, resulting in new leaf growth at the expense of allocation to fine roots and storage. Shoot growth in the next season may subsequently be affected (Andersen and Rygiewicz, 1991). Ozone fumigation of plants has resulted in increased plant susceptibility to many species of herbivores (Herms et al., 1996). However, there have been no long-term investigations comparing many different guilds of insects. Herms et al. (1996) found that growth of four species of leaf feeders was enhanced by O_3 treatment of trembling aspen plants. It is likely that other guilds, such as leaf miners and stem borers, will also be enhanced. This enhancement is likely because high levels of O_3 may interfere with the various local and system signal transduction pathways that plants have evolved to produce both generalized and specific defenses against pathogens and insects (e.g., widespread cell membrane damage, stressed C budgets, etc.).

This chapter focuses on the effects of interacting multiple stresses on growth and physiological processes of northern forest trees with emphasis on the investigations conducted as part of the United States Department of Agriculture (USDA) Forest Service's Northern Global Change Research Program. In our portion of that program, parallel studies of the effects of interacting CO_2 and O_3 on forest trees were conducted on plants in pots in controlled environments and in open top chambers, and in the ground in open top chambers, in a cooperative study at Delaware, OH (USDA Forest Service, Northeastern Research Station), Rhinelander, WI (USDA Forest Service, North Central Research Station), and Alberta, Michigan (Michigan Technological University). Emphasis was on trembling aspen at Rhinelander and Alberta, yellow poplar at Delaware, and on common seed sources of white pine at Delaware and Alberta. Because different responses were observed among species and locations, results are presented here on a species basis.

Trembling Aspen

Trembling aspen is the most widely distributed tree species in North America and is a significant ecological and commercial species for the eastern deciduous, boreal, and Rocky Mountain forest biomes (Barrett, 1980; Powell et al., 1992; Hackett and Piva, 1994). It is highly responsive to most stresses and has a high degree of natural genetic variability. Aspen is highly responsive to CO_2 (Brown, 1991; Sharkey et al., 1991), O_3 (Karnosky, 1976; Karnosky et al., 1996; Wang et al., 1986; Berrang et al., 1986, 1989, 1991; Karnosky et al., 1992a,b, 1996), nitrogen (Coleman et al., 1998), water stress (Griffin et al., 1991), and herbivores (Lindroth et al., 1993). The ease of cloning aspen is also an important aspect of this species because various clones can be propagated for use in physiological and molecular studies in which replication is often limited and natural genetic variability of seedling experimental material may mask responses. Karnosky et al. (1996) showed a decrease of 30 to 40% in coefficients of variation for O_3 responses of clones vs. seedlings. This finding was especially striking as the clonal comparisons had only about half as many plants as the seedling studies. By selecting clones with a range of stress tolerances (Table 5.1), the range of natural variability found in seedlings can be mimicked (Karnosky et al., 1996) and the ease of detecting treatment effects is greatly enhanced.

Table 5.1. Origin and Background Ozone-Sensitivity Information of the *Populus Tremuloides* Plants in This Chapter (Karnosky et al., 1996)

Plants	Orgin (County)	Foliar Ozone Sensitivity	Growth Ozone Sensitivity
Clone 216	Wisconsin (Bayfield)	Tolerant	Tolerant
Clone 253	Michigan (Leelanau)	Sensitive	Sensitive
Clone 259	Indiana (Porter)	Sensitive	Sensitive
Clone 271	Indiana (Porter)	Intermediate	Intermediate
Seedlings	Michigan (Houghton)	Untested	Untested

Tropospheric Ozone

It is well known that O_3 can affect aspen growth and that this impact varies considerably with genotype (see Fig. 5.1 color insert). Wang et al. (1986) showed an 18 to 20% decrease in aspen growth in nonfiltered vs. filtered air in Dutchess County, New York. In contrast, Karnosky et al. (1992a) found decreases of stem biomass following single season exposures of 80 ppb O_3 varying from 0 to 74%, depending on clone. For the highest seasonal doses (similar to those in the lower Great Lakes region), decreases of 43, 21, and 33%, respectively, were found in leaf, stem, and root biomass across all clones tested. Single season responses may compound with further exposures, particularly in the more sensitive clones such as 259 (Fig. 5.2). Ozone also affected crown architecture by influencing the ratio of long and short shoots retained and by affecting

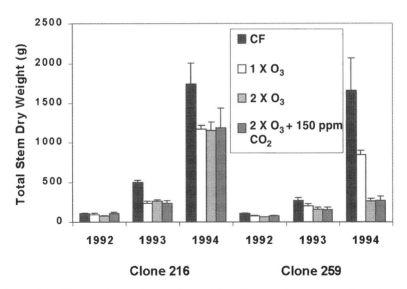

Figure 5.2. Changes in total stem dry weight of aspen clones exposed to elevated O_3 and $O_3 + CO_2$ for 3 years in open-top chambers.

stem and branch diameter and length/weight ratios. Carbon allocation was also impacted (Coleman et al., 1995b). Because the lower leaves of aspen that supply photosynthates to roots senesce prematurely with exposure to O_3, root growth appears to be particularly sensitive to O_3 (Coleman et al., 1996).

Much of the O_3 response in aspen appears to be controlled by the photosynthetic apparatus. Clonal tolerances to O_3 as determined by previous studies were highly correlated to photosynthetic responses (Coleman et al., 1995a). The seasonal decrease in biomass attributable to O_3 in aspen can be largely explained by decreases in whole-tree photosynthesis estimated for different leaf categories in combination with total leaf area per each category (Coleman et al., 1995a). Premature leaf senescence (Fig. 5.3) plays an important part in the decrease in whole-tree photosynthesis by eliminating photosynthesizing lower leaves (Coleman et al., 1996). Although some photosynthetic compensation occurs in the upper leaves under O_3 exposure, it is not enough to make up for the lost leaf surface area.

The differences in the tolerance of the clones examined in our studies (Karnosky et al., 1992a,b, 1996, 1997) cannot simply be explained by differences in photosynthesis (Ps) rates or stomatal conductance (Coleman et al., 1995a) that would lead to differing internal O_3 doses as predicted by Reich (1987). Sheng et al. (1997) found increased superoxide dismutase (SOD) activity in our O_3-tolerant aspen clones as compared with O_3-sensitive clones during long-term O_3 exposures (Fig. 5.4). In particular, the manganese (Mn) SOD and the copper/zinc (Cu/Zn) SOD were simultaneously elevated in an O_3-tolerant aspen clone (271) following both long-term and short-term exposures. Because plants have evolved with numerous oxidative stress tolerance mechanisms, it seems likely that multiple mechanisms are involved in determining differences in O_3 tolerance. However, the molecular and biochemical mechanisms controlling the physiological differences in O_3 sensitivity of various trembling aspen genotypes remain largely unknown.

Most O_3 exposure research with aspen used open-top chambers (OTCs) as first described by Heagle et al. (1973). While these chambers are closer to outdoor environmental conditions than indoor chambers, there is still a large chamber effect on aspen growth (Hendrey and Kimball, 1994; Karnosky et al., 1996). For example, following three years of growth in the ground, our nonchambered open-plot aspen trees were some 40% less in height and stem biomass than the 1× ambient O_3 chamber trees and 65% less than the charcoal-filtered trees. Characterizing O_3 risks to forests based on OTC studies (see Hogsett et al., 1997) remains speculative because of the large chamber effects found with trees. It is still unclear as to whether these chamber effects actually alter response to O_3. Future experiments must address this question (McLeod and Long, 1999). Additional approaches are needed to examine the impact of O_3 on forests

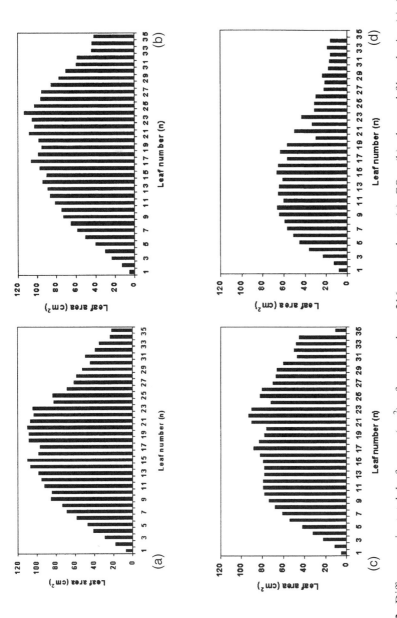

Figure 5.3. Differences in total leaf area (cm^2) of aspen clone 216 exposed to (a) CO_2, (b) charcoal-filtered air, (c) O_3, and (d) $CO_2 + O_3$ for one growing season. Leaves are numbered from the top of the plant.

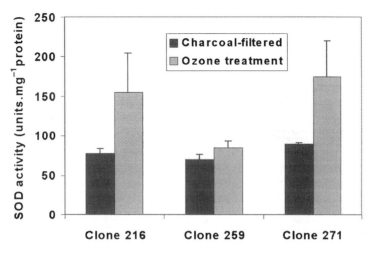

Figure 5.4. Superoxide dismutase (SOD) activity for three aspen clones (216, 259, 271) grown in charcoal-filtered air and O_3 treatment (From Sheng et al., 1997).

and to examine larger-scale forest community and ecosystem-level responses that are impossible to determine in OTCs. For example, to determine the relative growth rates of selected O_3-sensitive and O_3-tolerant clones, trees are grown in pots along a gradient from low to high O_3 in southern Wisconsin. With low O_3 levels, the sensitive and tolerant clones generally grow at about the same rate. However, under elevated O_3, as in southeastern Wisconsin, growth is severely restricted for sensitive clones compared with tolerant clones.

In another study, clones varying in O_3 sensitivity were planted in the field in similarly arranged plantations at areas of high, medium and low O_3 in the Great Lakes region. Ozone bioindication plots, competitive interaction plots, and growth and yield plots were established at each of the sites with identical sets of aspen clones. Preliminary results have shown large clonal growth differences related to O_3 sensitivity.

Carbon Dioxide

Increases in photosynthesis, individual leaf area, whole-crown leaf area, leaf area duration, and LAI have been observed in aspen trees grown in elevated CO_2 (Ceulemans and Mousseau, 1994; Ceulemans et al., 1994; Curtis et al., 1995). Trembling aspen response to elevated CO_2 is similar to that found for other aspen species. For example, researchers have found significant increases in whole-leaf photosynthesis in trembling aspen trees grown under twice ambient CO_2, particularly in the lower canopy. Leaves in the lower canopy transport C to the roots. Therefore, substantial increases in relative below-ground C allocation were found in elevated CO_2 (Zak et al., 1993).

Aspen grown under elevated CO_2 typically have greater tissue C/N ratios than aspen trees grown under ambient CO_2 (Lincoln et al., 1993). In addition, concentrations of phenolic compounds, including simple phenolics, condensed tannins, and hydrolyzable tannins, increase in trees exposed to CO_2 (Roth and Lindroth, 1994; Lindroth et al., 1995). Thus, elevated CO_2 may have a major impact on the herbivory of aspen leaves (Herms et al., 1996).

Ozone and Carbon Dioxide

We know little about the response of forest trees to the interaction of CO_2 and O_3. There are suggestions that increasing CO_2 may ameliorate O_3 damage (Allen, 1990; Taylor et al., 1994). Amelioration is probably valid for some species (Mortensen, 1995). However, our results with 3 years of study with trembling aspen growing in OTCs suggests that CO_2 at the level of 150 ppm over ambient does not compensate for decreases in growth and biomass caused by elevated O_3 (see Fig. 5.2). In fact, elevated CO_2 plus O_3 decreased photosynthetic rates and carboxylation efficiencies in older leaves in some otherwise O_3-tolerant aspen clones (Kull et al., 1996) and decreased overall leaf size compared to CO_2 alone (see Fig. 5.3). The causes of the lack of compensation for O_3 by elevated CO_2 for aspen have not yet been explained. Some authors have shown decreased antioxidant activity in the presence of elevated CO_2 that might counteract increases in antioxidants found in tolerant clones (Sheng et al., 1997), but antioxidant levels were similar for both O_3-treated and $O_3 + CO_2$-treated aspen plants in one study (Karnosky et al., 1997).

Nitrogen Interaction

Our work with the interactions of CO_2, O_3, and N fertilization on hybrid poplars and aspen clones has shown that CO_2 enrichment may compensate for O_3 impacts on growth, but the degree of compensation has both environmental and genetic components. To examine some of these interactions we tested 5 hybrid poplar clones selected for a range of growth rates in large pots with adequate water and fertilizer in OTCs, and treated with $CO_2 + O_3$ (Dickson et al., 1998). The decrease in dry weight in response with O_3 compared with controls was greater in the more productive clones (NM-6, 50% and 31.3 g vs. DN-70, 41% and 16.5 g), while the increase in dry weight in response to increased CO_2 (Fig. 5.5) was essentially the same in all clones (NM-6, 36% vs. DN-70,

160 J.G. Isebrands et al.

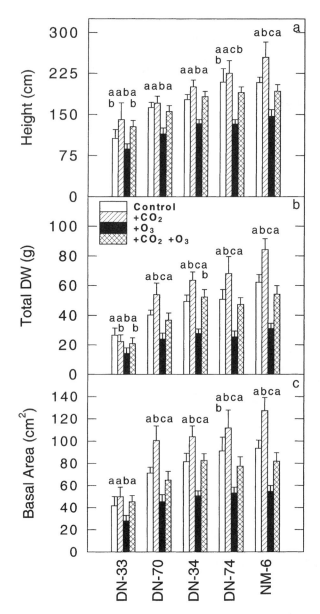

Figure 5.5. (a) Height, (b) total dry weight (DW), and (c) basal area of 5 hybrid poplar clones exposed to CO_2, O_3, and $CO_2 + O_3$ for one growing season (From Dickson et al., 1998).

34%). In addition, CO_2 exposure alleviated the detrimental response to elevated O_3 (see Fig. 5.5). However, the O_3 exposure also negated the increase in growth from CO_2.

In OTCs, we also tested 3 aspen clones that vary in sensitivity to O_3. The water and fertilizer regimes were the same as for the poplar hybrids, but the aspen were grown in large pots for unlimited root growth and access to nutrients. Carbon dioxide exposure increased average dry weights 37% and O_3 exposure decreased average dry weight 28%. However, there were large clonal differences in response to CO_2 and O_3 in combination. Carbon dioxide exposure of clone 271 increased total dry weight by 71% compared with controls, while dry weight production from the O_3-and $CO_2 + O_3$-exposed plants did not differ. Ozone exposure, however, completely negated the CO_2 response. In contrast, with clone 216 (O_3-intermediate) and clone 259 (O_3-sensitive), CO_2 exposure increased total dry weight only 20% while O_3 exposure decreased dry weight 38% (216) and 50% (259), and the addition of CO_2 to the O_3 exposure did not ameliorate yield losses from O_3.

In another study of the interactions of CO_2, O_3, and N availability, the clones 216 and 259 were grown in growth chambers, in large pots and with different N fertilization regimes. In the controlled-access regime, plants were fertilized daily with a complete nutrient solution based on a 3% relative addition rate (RAR) (Ingestad and Lund, 1986). The 3% RAR is designed to maintain these aspen clones at about 2% total plant nitrogen content but limit maximum potential growth rate (Coleman et al., 1998). In the unlimited access regime, plants were fertilized daily with a complete fertilizer solution containing N at a concentration (16 mmol) considerably in excess of maximum growth requirements. When N was limiting growth, there was no response to increased CO_2 by either clone; O_3 decreased total dry weight (216, 10% and 259, 35%), and CO_2 added to the O_3 exposure did not ameliorate the O_3 response (Fig. 5.6). In contrast, when N was

Liquid Fertilizer

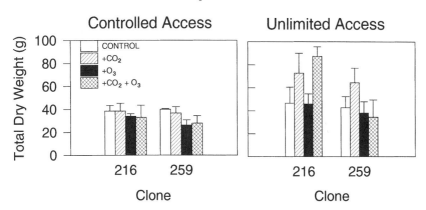

Figure 5.6. Total dry weight of two aspen clones exposed to CO_2, O_3, and $CO_2 + O_3$ growing with limiting and excess nitrogen fertilizer (From Coleman et al., 1998).

not limiting growth, CO_2 increased dry weight production in both clones (216, 60% and 259, 49%); O_3 exposure had little effect on either clone (259 decreased 12%), and CO_2 added to the O_3 exposure completely counteracted the O_3 response in clone 216 (actually increased dry weight production 85% over controls) but did not counteract the O_3 response of clone 259 (see Fig. 5.6). These results indicate that it will be very difficult to predict average species response to increasing concentrations of CO_2 and O_3, other environmental stresses, and their interactions, because genotype has such a large and variable effect on plant responses. Ecological risk assessment based on a limited number of studies of seedling populations (Hogsett et al., 1997) cannot account for the potentially large impact on sensitive genotypes. Risk assessment could be considerably strengthened, however, if information were available on the potential range of genotypic response expected within a species (Taylor, 1994).

Multiple Stresses and Insects

Studies of aspen foliage from plants treated in exposure chambers have shown that increased CO_2 and O_3 alter the chemical composition of the foliage, which in turn influences both its resistance to insect attack and its nutritional value for insect growth. Using plant material from growth chamber experiments, we found that elevated O_3 generally increased insect growth, while elevated CO_2 generally decreased insect growth. However, the increase in foliage volume expected under increased CO_2 may be partially offset by increased insect feeding, thus decreasing the potential increase in tree growth from CO_2 (Herms et al., 1996). Likewise, a decrease in foliage volume from O_3 damage may not decrease plant growth as much if insects consume less foliage from the O_3-damaged trees. Lindroth et al. (1993) demonstrated that light environment affects the dynamics of O_3 interactions with insects. Preliminary results from field studies of aspen show that foliar insect populations increase under O_3 and CO_2 (Mattson et al., unpublished).

Yellow Poplar

Yellow poplar is an ecologically and economically important hardwood species with a wide geographical range (Fowells, 1965). It grows throughout most of the eastern United States from southern New England west through Michigan and southern Ontario. At the northern end of its range, yellow poplar is usually found in stream bottoms and valleys at elevations below 300 m (Beck, 1990). It is a fast-growing shade-intolerant species with an indeterminate growth habit. It thrives on many soil types but generally grows best on moderately moist, well-drained, and loose-textured soils. It does not grow well in very dry or very wet situations.

Tropospheric Ozone

Because of yellow poplar's wide geographical range, it is potentially exposed to elevated levels of tropospheric ozone (O_3), the most widespread and phytotoxic of the atmospheric pollutants on forest ecosystems (Lefohn and Pinkerton, 1988; Linzon and Chevone, 1988; Simini et al., 1992). Yellow poplar has been rated as O_3-sensitive based on foliar symptoms and induced leaf abscission and has been used as a bioindicator (Davis and Skelly, 1992a,b). Although considered O_3-sensitive, Simini et al. (1992) reported foliar stipple injury and premature leaf abscission without any significant negative growth effects for field-planted yellow poplar seedlings exposed to ambient O_3 in OTCs in Pennsylvania. Consistent growth responses to O_3 have not been reported with this species because different studies have shown both decreases and increases in growth (Kress and Skelly, 1982; Mahoney et al., 1984; Jensen, 1985; Chappelka et al., 1988; Jensen and Patton, 1990; Tjoelker and Luxmoore, 1991; Cannon et al., 1993). Most of the O_3 growth and physiological response studies of seedling yellow poplar have used 1-year-old bare root stock in environment-controlled systems, such as greenhouses, growth chambers, or continuously stirred tank reactors (Kress and Skelly, 1982; Chappelka et al., 1988; Roberts, 1990; Jensen and Patton, 1990; Cannon et al., 1993; Cannon and Roberts, 1995). Typical results of such studies are those of Chappelka et al. (1988) who reported decreases in net photosynthesis (P_n) and stomatal conductance (g_s) in seedlings exposed to elevated ozone (100 to 150 ppb O_3) without growth or biomass effects, while Jensen (1985) reported relative growth rate was decreased 35% in seedlings exposed to 100 ppb O_3 for 20 weeks compared with control seedlings.

Results from OTC exposures of potted yellow poplar seedlings have been inconsistent. Tjoelker and Luxmoore (1991) exposed potted yellow poplar seedlings to O_3 ranging from 32 to 108 ppb (7-hour seasonal mean) in OTCs for 18 weeks and reported no significant effect on P_n, water use efficiency or, final whole-plant biomass but did report increased leaf abscission. They hypothesized that the indeterminate growth habit of yellow poplar permitted compensatory leaf growth which could have ameliorated O_3 effects on biomass production. Rebbeck (1996a) reported a stimulation in growth and plant biomass after one season of exposure to 107 ppm h^{-1} O_3 in OTCs, but after two seasons of exposure, root/shoot ratios and leaf area declined with increasing exposure to O_3 (Fig. 5.7). Net photosynthesis of yellow poplar leaves decreased 21 to 42% exposed to 1.7 times ambient O_3 compared with control seedlings, while g_s was generally unaffected (Rebbeck and Loats, 1997) (Figs. 5.8 and 5.9). There may be a

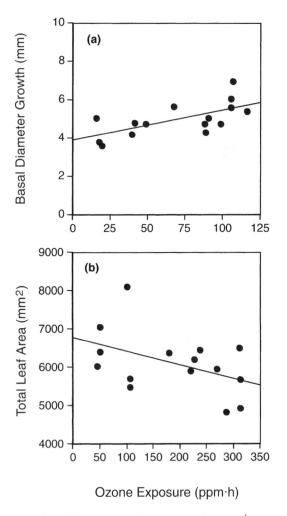

Figure 5.7. The relationship between O_3 exposure (ppm h^{-1}) and yellow-poplar growth: (a) basal diameter growth (mm) in 1990 and (b) total leaf area (mm^2) in 1991 (From Rebbeck, 1996a).

significant lag in the expression of growth effects following decreases in C fixation after exposure to elevated O_3. To date, there are no published reports on the response of saplings or mature yellow poplar to gaseous pollutants. Studies are underway to investigate the effects of elevated O_3 plus elevated CO_2 on plantation-grown yellow poplar seedlings over five growing seasons to determine how this species responds as it ages and increases in size. The goal is to extrapolate seedling pollutant response data to saplings and older trees grown under more realistic growing conditions.

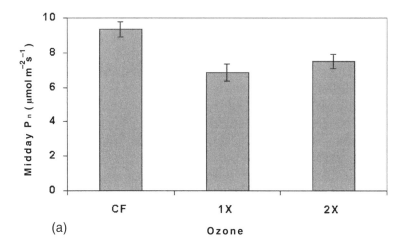

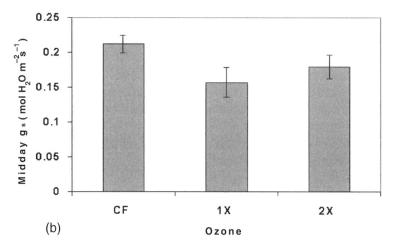

Figure 5.8. Seasonal midday (a) net photosynthesis (P_n) and (b) stomatal conductance (g_s) of yellow-poplar foliage exposed to CF, 1×, and 2× ambient O_3 for two growing seasons (From Rebbeck and Loats, 1997).

Carbon Dioxide

Enhanced growth of seedling yellow poplar exposed to increased concentrations of CO_2 was found by Norby and coworkers in the southeastern United States (O'Neill et al., 1987; Norby and O'Neill, 1991; Norby et al., 1992; Wullschleger et al., 1992; Gunderson et al., 1993). O'Neill et al. (1987) found that newly germinated yellow poplar seedlings exposed to

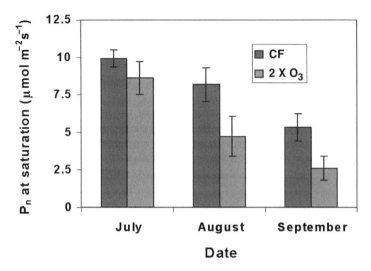

Figure 5.9. Net photosynthesis (P_n) at light saturation calculated from light response curves for leaves of yellow-poplar seedlings exposed to either charcoal-filtered air (CF) or twice ambient ozone ($2\times O_3$) during the second season of exposure (From Rebbeck and Loats, 1997).

692 ppm CO_2 for 24 weeks in growth chambers had significantly greater root (99%), leaf (69%), stem diameter (20%), and total dry weights (73%) while specific leaf area was significantly less (−21%) than for seedlings grown in ambient CO_2 (367 ppm). Norby and O'Neill (1991) exposed fertilized and unfertilized seedlings to ambient, +150 ppm, or +300 ppm CO_2 for 24 weeks and reported the only increase in dry weight associated with enriched CO_2 occurred in roots (25 to 40%). Leaf area was slightly decreased in elevated CO_2. Some of the commonly reported physiological and growth responses associated with elevated CO_2, such as decreased stomatal conductance (g_s) and photosynthetic down-regulation/acclimation were not consistently observed in these studies. In fact, stomatal conductance and P_n increased with increasing CO_2 concentration throughout the study. Nutrient deficiency did not impede growth enhancement in enriched CO_2. In a subsequent 3-year OTC study of field-planted, unfertilized and unirrigated yellow poplar seedlings exposed to elevated CO_2, Gunderson et al. (1993) found no photosynthetic down regulation (12 to 144% enhancement) and limited effects on g_s. These responses were consistent across leaf age and canopy position. Foliar total chlorophyll content was decreased 27% in seedlings exposed to +300 μl l^{-1} CO_2 for 24 weeks. Significant decreases in respiration of foliage exposed to elevated CO_2 were also observed (Wullschleger et al., 1992). Despite observed enhancement in leaf-level P_n and lower rates of leaf respiration, whole-plant biomass production did not increase (Norby et al., 1992). They suggested

that the lack of above-ground response resulted from changes in C allocation patterns that decreased leaf production and increased fine root production.

Ozone and Carbon Dioxide

To predict how yellow poplar might respond to future climate changes, a 5-year OTC study to investigate the response to enriched CO_2 in the presence of elevated O_3 was conducted in Delaware, Ohio. It was hypothesized that negative O_3 effects would be ameliorated by exposure to elevated CO_2, and that the response of older trees to these atmospheric gases would be similar to seedlings. Throughout the study, most of the typical enhancement responses observed were associated with exposure to enriched CO_2, with few or no effects associated with exposure to elevated O_3 (Rebbeck, 1996b). After the first season of exposure (20 weeks), no impacts on stem height or basal diameter were observed in seedlings grown in twice ambient O_3 ($2{\times}O_3$, cumulative exposure of 136 ppm h^{-1}) (Table 5.2). Basal stem diameter of seedlings exposed to twice ambient O_3 + twice ambient CO_2 ($2{\times}O_3$ + $2{\times}CO_2$) increased (13 to 21%) compared with seedlings grown in charcoal-filtered, one times ambient ($1{\times}O_3$), or $2{\times}O_3$-air (Rebbeck, 1993; Rebbeck et al., 1993). Stimulated height and diameter growth of yellow poplar exposed to $2{\times}O_3$ + $2{\times}CO_2$-air continued for three growing seasons (Rebbeck, 1996b). After two years of exposure, $2{\times}O_3$ + $2{\times}CO_2$ grown yellow poplar tended to have greater leaf (6%), stem (25%), branch (14%), and root (20%) biomass, and total leaf area (12%) compared with all other treatments. Photosynthetic enhancements (24 to

Table 5.2. Total Stem Height and Diameter of Yellow Poplar Seedlings Exposed to Ambient and Elevated O_3 + Elevated CO_2 in Delaware, Ohio, from 1992 through 1994

Treatment	Total Stem Height (cm)		
	1992	1993	1994
CF	94.85 ± 3.14a	274.97 ± 10.36a	365.9 ± 20.8a
$1{\times}O_3$	90.84 ± 4.44a	267.35 ± 8.66a	343.9 ± 20.2a
$2{\times}O_3$	96.89 ± 4.08a	285.44 ± 8.43a	382.2 ± 24.3a
$2{\times}O_3$ + CO_2	102.88 ± 5.04b	304.70 ± 10.47b	460.0 ± 11.1b

Treatment	Basal Diameter (mm)		
	1992	1993	1994
CF	9.85 ± 0.5a	13.99 ± 0.78a	45.17 ± 2.73a
$1{\times}O_3$	8.93 ± 0.5a	29.24 ± 1.39a	39.46 ± 3.37a
$2{\times}O_3$	9.47 ± 0.5a	29.71 ± 1.17a	43.85 ± 2.98a
$2{\times}O_3$ + CO_2	11.33 ± 0.5b	34.41 ± 1.61b	53.94 ± 3.18b

* Each value is a mean of 36 trees ± 1 standard error. Means followed by different letters are significantly different at $P < 0.05$ (Rebbeck, 1996b).

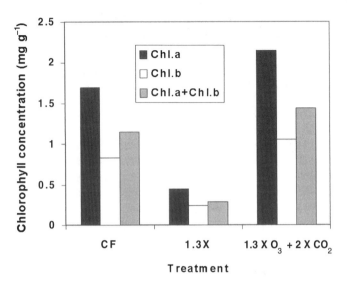

Figure 5.10. Response of mean foliar chlorophyll concentrations of yellow poplar foliage exposed to elevated O_3 (1.3 × ambient ozone) alone or in combination with twice ambient CO_2 (1.3×O_3 + 2×CO_2) for one growing season (From Carter et al., 1995).

48%) were observed for seedlings exposed to 2×O_3 + 2×CO_2-air for one and two seasons (Rebbeck et al., 1995; Rebbeck, 1996b), while foliar chlorophyll (Fig. 5.10) and N content decreased (Carter et al., 1995; Rebbeck et al., 1995; Scherzer and Rebbeck, 1995). After four seasons of treatment, O_3 alone had no effect on foliar N concentrations; however, 2×O_3 + 2×CO_2-air decreased N concentrations by 18 to 40% (Scherzer et al., 1998). Preliminary analyses indicate, after 5 seasons of exposure, continued enhancement of growth and photosynthesis in 2×O_3 + 2×CO_2-air. Because there were limited O_3 effects on the growth or physiology of these seedlings, it was difficult to ascertain the ameliorative effects of enriched-CO_2. Most of our enriched-CO_2 responses observed in Ohio were similar to those reported in the southeastern United States (Norby and O'Neill, 1991; Norby et al., 1992; Wullschleger et al., 1992; Gunderson et al., 1993). These findings suggest that field-planted yellow poplar, when exposed to enriched CO_2 and grown with limited nutrients, limited soil moisture, and ambient or elevated O_3 will display enhanced growth and photosynthetic assimilation.

Nitrogen and Water

Little direct information on the growth and physiology of yellow poplar is available in response to CO_2 and O_3 combined with water stress and nitrogen stress (limited or excess). However, the influence of water stress on modulating the response of yellow poplar to O_3 has been studied (Roberts, 1990; Cannon et al., 1993; Cannon and Roberts, 1995). In growth chamber studies, water stress significantly decreased P_n, g_s, and transpiration (T_s) (by more than 70%) but O_3 treatments generally had little effect on physiological or growth parameters. The combination of water stress and 70 ppb O_3 significantly decreased root biomass and root/shoot ratio (Cannon et al., 1993). Roberts (1990) reported a significant additive effect of drought and O_3 on xylem water potential. These studies indicate that significant interactions involving low levels of O_3 and short periods of water stress may have considerable effects on stomatal physiology. Tjoelker and Luxmoore (1991) reported that yellow poplar whole-plant dry weight did not respond to either O_3 or N (levels ranging from 58 to 172 $\mu g\, g^{-1}$) but allocation to fine roots did increase in low-N soils.

Eastern White Pine

Eastern white pine is a widely distributed component of the eastern US forest and is the largest of the northeastern conifers (Fowells, 1965). White pine has played an important role throughout our history as the "peace tree" of the Native Americans and as a valuable source of lumber for construction of eastern and midwestern cities. It also is important ecologically because it rapidly invades old fields and is intimately associated with oaks (Stine and Baughman, 1992). White pine has long been known to be sensitive to multiple environmental stresses (Kozlowski, 1979), although there are fewer published works on the interacting stresses on white pine than on aspen and yellow poplar. An exception is the pioneering work of Dochinger et al. in the USDA Forest Service, Northeastern Forest Experiment Station in the 1960s who studied the interactions of sulfur dioxide and tropospheric O_3 as influenced by other environmental factors (Dochinger et al., 1970; Houston, 1974).

Tropospheric Ozone

White pine's sensitivity to O_3 has been known for many years (Kozlowski and Constantinidou, 1986a). When white pine experiences acute O_3 episodes, its current-year needles develop silver flecks and tip dieback ("tipburn") (Woodman, 1987). Chronic symptoms include chlorosis and premature needle abscission of older needles. High O_3 levels throughout the southeastern US have been shown to cause extensive damage to white pine on a watershed scale, including premature needle abscission and decreased basal area increment (McLaughlin, 1985; Swank and Vose, 1990). Moreover, McLaughlin et al. (1982) showed that chronic O_3 stress caused declined vigor, increased respiration, and altered C allocation patterns in field-grown white pine. However, Bennett et al. (1994) in a review of air pollution surveys of white pine found that hypersensitive white pine individuals are no longer present in the forest because they have succumbed to past O_3 exposures. More recent tree-ring studies, in Acadia National Park, Maine, suggest white pine tree-ring growth is affected negatively by O_3 level and duration (Bartholomay et al., 1997).

As in aspen, white pine has a high degree of genetic variability in response to O_3 (Houston and Stairs, 1973). Karnosky (1981) found a higher mortality rate in O_3-sensitive genotypes of eastern white pine stands in southern Wisconsin. That study indicated that natural selection may have already altered many stands in higher O_3 regions of the Northeast. To examine possible genetic differences between white pine from O_3 selected (sensitive genotypes lost) and nonselected regions, a large cooperative study was initiated by a consortium of scientists from Michigan Technological University; USDA Forest Service, Northeastern Forest Experiment Station, and the USDA Forest Service, North Central Research Station to compare the responses of seedlings from Ohio (where O_3 levels have been historically high) and from northern Michigan (where O_3 levels have been historically quite low). Seedlings from these two locations were exposed to ambient O_3 ($1\times O_3$),   twice ambient O_3 ($2\times O_3$) and $2\times O_3$ + elevated CO_2 in OTCs at 2 locations: a clean-air site in the upper peninsula of Michigan and a relatively high O_3 site near Delaware, Ohio.

At the Alberta, Michigan site, we found no significant differences in height, stem, root, or current year needle biomass in response to the O_3 treatments, but there were significant O_3 effects on diameter growth and mature needle retention for the Michigan source but not for the Ohio source. At the Ohio site, O_3 had a small stimulating effect on height growth of white pine in the first year. However, after 3 years, there was no significant effect on height or diameter growth from $2\times O_3$ (Rebbeck, 1996b).

Carbon Dioxide

Little is known about the effects of elevated CO_2 on white pine. In three recent reviews of the CO_2 literature (Ceulemans and Monsseau, 1994; Curtis, 1996; Wullschleger et al., 1997), only one reference was cited. Bazzaz et al. (1990) reported a 20% increase in biomass growth, a 14% decrease in leaf area, and an 11% increase in root/shoot ratio with elevated CO_2 in a glasshouse experiment with white pine seedlings. Although there is essentially no information on the response of white pine to elevated CO_2, white pine can be expected to respond like other conifers and C_3 plants with increases in photosynthesis, biomass, and water use efficiency and decreases in N content (Conroy and Hocking, 1993; Ineichen et al., 1995; Wullschleger et al., 1997).

Ozone and Carbon Dioxide

Very little research has been done on the interacting effects of O_3 and CO_2 on white pine. The only available information on the combined effects of O_3 and CO_2 on white pine is from the cooperative OTC experiment in Michigan and Ohio. At the Michigan site, in the first two years after planting, O_3 decreased needle length compared with the charcoal-filtered (CF) treatment, while the needle length in the $O_3 + CO_2$ treatment did not differ from the CF treatment. This result indicates that CO_2 may offset some detrimental effects of O_3. After four years of growth in the OTCs, O_3 increased the loss of older needles and decreased total plant dry weight. Carbon dioxide in the $O_3 + CO_2$ treatment partially counteracted needle senescence and increased stem dry weight and total plant dry weight compared with both O_3 and the CF treatment.

At the Ohio site, there were no significant growth differences attributable to $O_3 + CO_2$ in the first 4 growing seasons. There was a slight stimulatory effect of $O_3 + CO_2$ in height growth in the first year, but in the second year height was less than the control. In the second year, $O_3 + CO_2$ stimulated photosynthesis compared with the O_3 and control treatments. Chlorophyll content of the older needles was decreased by $O_3 + CO_2$ by 40 to 50% compared with the controls. Despite these physiological differences, no significant growth effects were detected with $O_3 + CO_2$ compared with other treatments (Rebbeck, 1996b).

Nitrogen and Other Multiple-Stress Interactions

There have been few studies of the effect of N on white pine's response to interacting multiple stressors. We would expect low nitrogen availability to limit the effect of CO_2 on white pine as in other plants (Conroy and Hocking, 1993). Moreover, we would expect low nitrogen availability to increase the detrimental effects of O_3 on white pine growth. Eberhardt et al. (1988) found no effect of N fertilizer on the O_3 response in white pine and Reich et al. (1988) found that increased N from acid rain in combination with O_3 had a deleterious effect on white pine growth depending upon soil conditions in a microcosm experiment. In a related study, Stroo et al. (1988) found that acid rain (i.e., lower pH) and O_3 exposure decreased mycorrhizae infection and therefore nutrition in white pine. McLaughlin (1985) reported that the interactions of O_3 and SO_2 atmospheric pollutants were primary causal agents in white pine decline in the eastern US, while genetic variation accounted for a 5-fold difference in the response to the interacting stressors. No reports of the effects of interacting stressors on pest and disease incidence in white pine are currently available. Based upon the research to date, we would expect significant multiple-stress pest interactions in white pine as found in our other species.

Summary and Future Direction

The interacting effects of multiple stresses on growth and physiological processes in northern forest trees are complex, and the mechanisms of sensitivity to the combinations of multiple stressors remain unknown. There are major differences among species and genotypes within species in their responses to various interacting stressors. Aspen is particularly sensitive to elevated O_3, CO_2, and their interactions, while hybrid poplars, yellow poplar, and white pine vary in sensitivities. Nutrition has an important effect on the response to multiple stressors, with poorer nutrition usually predisposing plants to the effects of other stressors. Multiple stressors also have an important effect on ecological factors such as insect feeding that deserves more attention.

To date, most studies of multiple-stress interactions in trees have been conducted on small trees growing in pots and/or in controlled environments and OTCs. Such experiments cannot incorporate the host of biotic and abiotic interactions that occur in a forest ecosystem (Koch and Mooney, 1996). Large-scale field experiments are needed in the future to minimize chamber effects and better depict "real world" conditions (Hendrey and Kimball, 1994; McLeod and Long, 1999). To accomplish the goal of studying the interacting effects of multiple stressors on physiological processes and growth of northern temperate forest ecosystems, we have initiated a free air carbon dioxide enrichment (FACE) experiment in Rhinelander, Wisconsin, to examine the interacting effects of elevated CO_2 and O_3 on physiological processes (C and N allocation) growth, survival, and competition of pure and mixed stands of aspen clones, sugar maple (*Acer saccharum* Marsh.), and paper birch (*Betula papyifera* Marsh.) (Karnosky et al., 1999). This study design consists of twelve 30 m FACE rings—a factorial combination of 4 treatments (ambient control; 560 ppm CO_2; 80 to 100 ppb O_3; and $CO_2 + O_3$) with three replications. The FACE technology features a vertical vent pipe delivery system equipped with baffles as well as a centralized O_3 delivery system. This experiment offers the opportunity for an interdisciplinary team of scientists to study the mechanisms of interacting multiple-stress effects on a larger scale. This approach is expected to lead to a better understanding in the future of forest ecosystem responses to the ever-changing complexities of the environment.

Figure 5.1. Two-year-old aspen trees of clone 259, an ozone-sensitive genotype, grown in open-top chambers at Alberta, MI, charcoal-filtered (right) vs. Milwaukee, WI, O_3 profile (left).

References

Adams RM, Glyer JD, Johnson SL, McCart BA (1989) A reassessment of the economic effects of ozone on U.S. agriculture. J Air Pollut Control Assoc 39:960–968.

Allen LH Jr. (1990) Plant responses to rising carbon dioxide and potential interactions with air pollutants. J Environ Qual 19:15–34.

Anderson CP, Rygiewicz PT (1991) Stress interactions and mycorrhizal response: understanding carbon allocation priorities. Environ Pollut 73:217–244.

Anderson PD, Tomlinson PT (1998) Ontogeny affects response of northern red oak seedlings to elevated CO_2 and water stress. I. Carbon assimilation and biomass production. New Phytol 140:477–491.

Barrett JW (1980) Regional Silviculture of the United States. John Wiley, New York.

Bartholomay GA, Eckert RT, Smith KT (1997) Reduction in tree-ring widths of white pine following ozone exposure at Acadia National Park, Maine. Can J For Res 27:361–368.

Bazzaz FA, Fajer ED (1992) Plant life in a CO_2-rich world. Sci Am 266:68–74.

Bazzaz FA, Miao SL (1993) Successional status, seed size, and responses of tree seedlings to CO_2, light, and nutrients. Ecology 74(1):104–112.

Bazzaz FA, Coleman JS, Morse SR (1990) Growth responses of seven major co-occurring tree species of the northeastern United States to elevated CO_2. Can J For Res 20:1479–1484.

Beck DE (1990) *Liriodendron tulipifera* L. Yellow-Poplar. In: Burns RM, Honkala BH (eds) *Silvics of North America. Vol. 2. Hardwoods.* Agric Handbook 654. United States Department of Agriculture (USDA) Forest Service, Washington, DC, 406–416.

Beerling DJ, Heath J, Woodward FI, Mansfield TA (1996) Drought–CO_2 interaction in trees: observations and mechanisms. New Phytol 134:235–242.

Bennett JP, Anderson RL, Mielke ML, Ebersole JJ (1994) Foliar injury air pollution surveys of eastern white pine (*Pinus strobus* L.): a review. Environ Monitor Assess 30:247–274.

Berrang PC, Karnosky DF, Bennett JP (1989) Natural selection for ozone tolerance in *Populus tremuloides*. II. Field verification. Can J For Res 19:519–522.

Berrang PC, Karnosky DF, Bennett JP (1991) Natural selection for ozone tolerance in *Populus tremuloides*: an evaluation of nationwide trends. Can J For Res 21:1091–1097.

Berrang PC, Karnosky DF, Mickler RA, Bennett JP (1986) Natural selection for ozone tolerance in *Populus tremuloides*. Can J For Res 16:1214–1216.

Boerner REG, Rebbeck J (1995) Decomposition and nitrogen release from leaves of three hardwood species grown under elevated O_3 and/or CO_2. In: Collins HP, Robertson GP, Klug MJ (eds) *The Significance and Regulation of Soil Biodiversity.* Kluwer Academic, The Hague, Netherlands, pp 169–177.

Bowes G (1993) Facing the inevitable: plants and increasing atmospheric CO_2. Ann Rev Plant Physiol Plant Mol Biol 44:309–332.

Brendley BW, Pell EJ (1998) Ozone-induced changes in biosynthesis of Rubisco and associated compensation to stress in foliage of hybrid poplar. Tree Physiol 18:81–90.

Brown KR (1991) Carbon dioxide enrichment accelerates the decline in nutrient status and relative growth rate of *Populus tremuloides* Michx. seedlings. Tree Physiol 8:161–173.

Brown LR, Renner M, Flavin C (1997) *Vital Signs* 1997. W.W. Norton, New York.

Cannon WN, Roberts BR (1995) Stomatal resistance and the ratio of intercellular to ambient carbon dioxide in container-grown yellow-poplar seedlings exposed to chronic ozone fumigation and water stress. Environ Exper Bot 35:161–165.

Cannon WN, Roberts BR, Bargar JH (1993) Growth and physiological response of water-stressed yellow-poplar seedlings exposed to chronic ozone fumigation and ethylenediurea. For Ecol Manage 61:61–73.

Carter GA, Rebbeck J, Percy KE (1995) Leaf optical properties in *Liriodendron tulipifera* and *Pinus strobus* as influenced by increased atmospheric ozone and carbon dioxide. Can J For Res 25:407–412.

Ceulemans R, Mousseau M (1994) Effects of elevated atmospheric CO_2 on woody plants. New Phytol 127:425–446.

Ceulemans R, Jiang XN, Shao BY (1994) Growth and physiology of one-year-old poplar (*Populus*) under elevated atmospheric CO_2 levels. Ann Bot 75:609–617.

Chameides WL, Saylor RD, Cowling EB (1997) Ozone pollution in the rural United States and the New NAAQS. Science 276:916.

Chappelka AH, Chevone BI, Seiler J (1988) Growth and physiological responses of yellow-poplar seedlings exposed to ozone and simulated acidic rain. Environ Pollut 49:1–18.

Coleman MD, Dickson RE, Isebrands JG (1998) Growth and physiology of aspen supplied with different fertilizer addition rates. Physiol Plantarum 103:513–526.

Coleman MD, Isebrands JG, Dickson RE, Karnosky DF (1995a) Photosynthetic productivity of aspen clones varying in sensitivity to tropospheric ozone. Tree Physiol 15:585–592.

Coleman MD, Dickson RE, Isebrands JG, Karnosky DF (1995b) Carbon allocation and partitioning in aspen clones varying in sensitivity to tropospheric ozone. Tree Physiol 15:593–604.

Coleman MD, Dickson RE, Isebrands JG, Karnosky DF (1996) Root growth and physiology of potted and field-grown trembling aspen exposed to tropospheric ozone. Tree Physiol 16:145–152.

Conroy J, Hocking P (1993) Nitrogen nutrition of C_3 plants at elevated atmospheric CO_2 concentrations. Physiol Plantarum 89:570–576.

Curtis PS (1996) A meta-analysis of leaf gas exchange and nitrogen in trees grown under elevated carbon dioxide. Plant Cell Environ 19:127–137.

Curtis PS, Vogel CS, Pregitzer KS, Zak DR, Teeri JA (1995) Interacting effects of soil fertility and atmospheric CO_2 on leaf area growth and carbon gain physiology in *Populus x euramericana*. New Phytol 129:253–263.

Davis DD, Skelly JM (1992a) Growth response of four species of eastern hardwood tree seedlings exposed to ozone, acidic precipitation, and sulfur dioxide. J Am Pollut Contr Assoc 42:309–311.

Davis DD, Skelly JM (1992b) Foliar sensitivity of eight eastern hardwood tree species to ozone. Water Air Soil Pollut 62:269–277.

Dickson RE, Coleman MD, Riemenschneider DE, Isebrands JG, Hogan GD, Karnosky DF (1998) Growth of five hybrid poplar genotypes exposed to interacting elevated CO_2 and O_3. Can J For Res 28:1706–1716.

Dochinger LS, Bender FW, Fox FL, Heck WW (1970) Chlorotic dwarf of eastern white pine caused by an ozone and sulfur dioxide interaction. Nature 225:476.

Eamus D (1991) The interaction of rising CO_2 and temperatures with water use efficiency. Plant Cell Environ 14:843–852.

Eamus D, Jarvis PG (1989) The direct effects of increase in the global atmospheric CO_2 concentration on natural and commercial temperate trees and forests. Adv Ecol Res 19:1–55.

Eberhardt JC, Brennan E, Kuser J (1988) The effect of fertilizer treatment on ozone response and growth of eastern white pine. J Arborcult 14:153–155.

Fowells HH (1965) *Silvics of Forest Trees of the United States.* Agric Handbook 271. United States Department of Argiculture (USDA) Forest Service, Washington, DC.

Fox S, Mickler RA (eds) (1996) *Impacts of Air Pollutants on Southern Pine Forests.* Ecological Studies 118. Springer-Verlag, New York.

Fuentes JD, Dann TF (1994) Ground-level ozone in eastern Canada: seasonal variations, trends, and occurrences of high concentrations. J Air Waste Manage Assoc 44:1019–1026.

Gebauer RLE, Reynolds JF, Strain BR (1996) Allometric relations and growth in *Pinus taeda*: the effect of elevated CO_2 and changing N availability. New Phytol 134:85–93.

Greitner CS, Pell EJ, Winner WE (1994) Analysis of aspen foliage exposed to multiple stresses: ozone, nitrogen deficiency and drought. New Phytol 127:579–589.

Griffin DH, Schaedle M, DeVit MJ, Manion PD (1991) Clonal variation of *Populus tremuloides* response to diurnal drought stress. Tree Physiol 8:297–304.

Gunderson CA, Norby RJ, Wullschleger SD (1993) Foliar gas exchange responses of two deciduous hardwoods during 3 threes of growth in elevated CO_2: no loss of photosynthetic enhancement. Plant Cell Environ 16:797–807.

Gunthardt-Goerg MS, Schmutz P, Matyssek R, Bucher JB (1996) Leaf and stem structure of poplar (*Populus x euramericana*) as influenced by O_3, NO_2, their combination, and different soil N supplies. Can J For Res 26:649–657.

Hackett RL, Piva RJ (1994) *Pulpwood Production in the North Central Region, 1992.* Res Bull NC-111. United States Department of Agriculture (USDA) Forest Service, North Central Forest Experiment Station, St. Paul, MN.

Heagle AS, Body DE, Heck WW (1973) An open-top field chamber to assess the impact of air pollution on plants. J Environ Qual 2:365–368.

Hendrey GR, Kimball BA (1994) The FACE Program. Agric For Meteorol 70: 3–14.

Herms DA, Mattson WJ (1992) The dilemma of plants: to grow or defend. Quart Rev Biol 67(3):283–335.

Herms DA, Mattson WJ, Karowe DN, Coleman MD, Trier TM, Birr BA, Isebrands JG (1996) Variable performance of outbreak defoliators on aspen clones exposed to elevated CO_2 and O_3. In: Hom J, Birdsey R, O'Brian K (eds) Proceedings 1995 Meeting of the Northern Global Change Program, 14–16 March, United States Department of Argiculture, Gen Tech Rep NE-214. Radnor, PA, pp 43–55.

Hogsett WE, Weber JE, Tingey D, Herstrom A, Lee EH, Laurence JA (1997) Environmental auditing: an approach for characterizing tropospheric ozone risk to forests. Environ Manage 21:105–120.

Houghton JT, Meira Filho LG, Callander BA, Harris N, Kattenberg A, Maskell K (eds) IPCC (Intergovernmental Panel on Climate Change) (1996) *Climate Change 1995: The Science of Climate Change. Contribution of Working Group I to the Second Assessment Report of the Intergovernmental Panel on Climate Change.* Cambridge University Press, Cambridge, UK.

Houston D (1974) Response of selected *Pinus strobus* L. clones to fumigation with sulfur dioxide and ozone. Can J For Res 4:65–68.

Houston D, Stairs GR (1973) Genetic control of sulfur dioxide and ozone tolerance in eastern white pine. Forest Sci 19:267–271.

Idso SB, Kimball BA, Hendrix DL (1996) Effects of atmospheric CO_2 enrichment on chlorophyll and nitrogen concentrations of sour orange leaves. Environ Exp Bot 36:323–331.

Ineichen K, Wieneken V, Wieken A (1995) Shoots, roots, and ectomycorrhiza formation of pine seedlings at elevated atmospheric carbon dioxide. Plant Cell Environ 18:703–707.

Ingestad T, Ågren GI (1995) Plant nutrition and growth: basic principles. Plant Soil 168–169:15–20.

Ingestad T, Lund AB (1986) Theory and techniques for steady state mineral nutrition and growth of plants. Scand J For Res 1:433–453.

Jensen KF (1985) Response of yellow poplar seedlings to intermittent fumigation. Environ Pollut 38:183–191.

Jensen KF, Patton RL (1990) Response of yellow-poplar (*Liriodendron tulipifera* L.) seedlings to simulated acid rain and ozone. 1. Growth modifications. Environ Exper Bot 30:59–66.

Karnosky DF (1976) Threshold levels for foliar injury to *Populus tremuloides* Michx. by sulfur dioxide and ozone. Can J For Res 6:166–169.

Karnosky DF (1981) Changes in eastern white pine stands related to air pollution stress. Mitt Forst Bundes Wien 137:41–45.

Karnosky DF, Gagnon ZE, Reed DD, Witter JA (1992a) Effects of genotype on the response of *Populus tremuloides* Michx. to ozone and nitrogen deposition. Water Air Soil Pollut 62:189–199.

Karnosky DF, Gagnon ZE, Reed DD, Witter JA (1992b) Growth and biomass allocation of symptomatic and asymptomatic *Populus tremuloides* clones in response to seasonal ozone exposures. Can J For Res 22:1785–1788.

Karnosky DF, Gagnon ZE, , Dickson RE, Coleman MD, Lee EH, Isebrands JG (1996) Changes in growth, leaf abscission, and biomass associated with seasonal tropospheric ozone exposures of *Populus tremuloides* clones and seedlings. Can J For Res 26:23–37.

Karnosky DF, Podila GK, Gagnon Z, Pechter P, Akkapeddi A, Sheng Y, Riemenschneider DE, Coleman MD, Dickson RE, Isebrands JG (1997) Genetic control of responses to interacting tropospheric ozone and CO_2 in *Populus tremuloides*. Chemosphere 36:807–812.

Karnosky DF, Mankovska B, Percy K, Dickson RE, Podila GK, Sober J, Noormets A, Hendrey G, Coleman MD, Kubiske M, Pregitzer KS, Isebrands JG (1999) Effects of tropospheric O_3 on trembling aspen and interaction with CO_2: results from an O_3-gradient and a FACE experiment. Water Air Soil Pollut 116:1–2.

Keeling CD, Whort TP, Wahlen M, VanderPlicht J (1995) Interannual extremes in the rate of rise of atmospheric carbon dioxide since 1980. Nature 375:666–670.

Kerr RA (1997) Greenhouse forecasting still cloudy. Science 276:1040–1042.

Kerstiens G, Townend J, Heath J, Mansfield TA (1995) Effects of water and nutrient availability on physiological responses of woody species to elevated CO_2. Forestry 68:303–315.

Kimball BA (1990) *Impact of Carbon Dioxide, Trace Gases, and Climate Change on Global Agriculture*. American Society of Agronomy. Special Pupl. #53. American Soc Agron, Madison, WI.

Koch GW, Mooney HA (1996) *Carbon Dioxide and Terrestrial Ecosystems*. Academic Press, San Diego, California.

Kozlowski TT (1979) *Tree Growth and Environmental Stresses*. University of Washington Press, Seattle, Washington, DC.

Kozlowski TT, Constantinidou HA (1986a) Responses of woody plants to environmental pollution. Part I. Sources and types of pollutants and plant response. For Abst 47:5–51.

Kozlowski TT, Constantinidou HA (1986b) Responses of woody plants to environmental pollution. Part II. Factors affecting responses to pollution and alleviation of pollution effects. For Abst 47:105–132.

Kress LW, Skelly JM (1982) Response of several eastern forest tree species to chronic doses of ozone and nitrogen dioxide. Plant Dis 66:1149–1152.

Krupa SV, Kickert RN (1993) The greenhouse effect: the impacts of carbon dioxide (CO_2), ultraviolet-B (UV-B) radiation and ozone (O_3) on vegetation (crops). Vegetation 104/105:223–238.

Kull O, Sober A, Coleman MD, Dickson RE, Isebrands JG, Gagnon Z, Karnosky DF (1996) Photosynthetic responses of aspen clones to simultaneous exposures of ozone and CO_2. Can J For Res 26:639–648.

Laurence JA, Amundson RG, Friend AL, Pell EJ, Temple PJ (1994) Allocation of carbon in plants under stress: an analysis of the ROPIS experiments. J Environ Qual 23:412–417.

Lee HSI, Jarvis PG (1995) Trees differ from crops and from each other in their responses to increases in CO_2 concentration. J Biogeo 22:323–330.

Lefohn AS, Pinkerton JE (1988) High resolution characterization of ozone data for sites located in forested areas of the United States. J Am Pollut Contr Assoc 38:1504–1511.

Lincoln DE, Fajer ED, Johnson RH (1993) Plant-insect herbivore interactions in elevated CO_2 environments. Tree 8:64–68.

Lindroth RL, Arteel GE, Kinney KK (1995) Responses of three saturniid species to paper birch grown under enriched CO_2 atmospheres. Funct Ecol 9:306–311.

Lindroth RL, Reich PB, Tjoelker MG, Volin JC, Oleksyn J (1993) Light environment alters response to ozone stress in seedlings of *Acer saccharum* Marsh. and hybrid *Populus* L. New Phytol 124:647–561.

Linzon SN, Chevone BI (1988) Tree decline in North America. Environ Pollut 40:87–99.

Lippert M, Haberle K-H, Steiner K, Payer H-D, Rehfuess K-E (1996) Interactive effects of elevated CO_2 and O_3 on photosynthesis and biomass production of clonal 5-year-old Norway spruce (*Picea abies* [L.] Karst.) under different nitrogen nutrition and irrigation treatments. Trees 10:382–392.

Lloyd J, Farquhar GD (1996) The CO_2 dependence of photosynthesis, plant growth responses to elevated atmospheric CO_2 concentrations and their interaction with soil nutrient status. I. General principles and forest ecosystems. Funct Ecol 10:4–32.

Mahoney MJ, Skelly JM, Chevone BI, Moore LD (1984) Response of yellow poplar (*Liriodendron tulipifera* L.) seedling shoot growth to low concentrations of O_3, SO_2, and NO_2. Can J For Res 14:150–153.

Maron JL (1998) Insect herbivory above- and belowground: individual and joint effects on plant fitness. Ecology 79:1281–1293.

Matyssek R, Keller T, Koike T (1993) Branch growth and leaf gas exchange of *Populus tremula* exposed to low ozone concentrations throughout two growing seasons. Environ Pollut 79:1–7.

McLaughlin SB (1985) Effects of air pollution on forests: a critical review. J Air Pollut Contr Assoc 35:512–534.

McLaughlin SB, McConathy RK, Duvick D, Mann LK (1982) Effects of chronic air pollution stress on photosynthesis, carbon allocation and growth of white pine trees. For Sci 28:60–70.

McLeod AR, Long SP (1999) Free-air carbon dioxide enrichment (FACE) in global change research: a review. Advan Ecol Res 28:1–56.

Mickler RA, Fox S (eds) (1988) *The Productivity and Sustainability of Southern Forest Ecosystems in a Changing Environment*. Ecological Studies 128. Springer-Verlag, New York.

Miller PR, Arbaugh MJ, Temple PJ (1997) Ozone and its known and potential effects on forests in the western United States. In: Sanderman H (ed) *Forest Decline and Ozone*. Springer-Verlag, Berlin, Germany, pp 39–67.

Mooney HA, Drake BG, Luxmoore RJ, Oechel WC, Pitelka LF (1991) Predicting ecosystem responses to elevated CO_2 concentrations. BioScience 41:96–104.

Mortensen LM (1995) Effects of carbon dioxide concentration on biomass production and partitioning in *Betula pubescens* Ehrh. Seedlings at different ozone and temperature regimes. Environ Pollut 87:337–343.

Norby RJ, Gunderson CA, Wullschleger SD, O'Neill EG, McCracken MK (1992) Productivity and compensatory responses of yellow-poplar trees in elevated CO_2. Nature 357:322–324.

Norby RJ, O'Neill EG (1991) Leaf area compensation and nutrient interactions in CO_2-enriched seedlings of yellow-poplar (*Liriodendron tulipifera* L.). New Phytol 117:515–528.

O'Neill EG, Luxmoore RJ, Norby RJ (1987) Elevated atmospheric CO_2 effects on seedlings growth, nutrient uptake, and rhizosphere bacterial populations of *Liriodendron tulipifera* L. Plant Soil 104:3–11.

Pääkkonen E, Holopainen T (1995) Influence of nitrogen supply on the response of clones of birch (*Betula pendula* Roth.) to ozone. New Phytol 129:595–603.

Pell EJ, Winner WE, Vinten-Johansen C, Mooney HA (1990) Response of radish to multiple stresses. I. Physiological and growth responses to changes in ozone and nitrogen. New Phytol 115:439–446.

Pettersson R, McDonald AJS, Stadenberg I (1993) Response of small birch plants (*Betula pendula* Roth.) to elevated CO_2 and nitrogen supply. Plant Cell Environ 16:1115–1121.

Powell DS, Faulkner JL, Darr DR, Zhu Z, MacCleery DW (1992) *Forest Resources of the U.S.* Gen Tech Rep RM-234. United States Department of Argiculture (USDA) Forest Service, Rocky Mountain Forest and Range Experiment Station, Ft Collins, Colorado.

Pye JM (1988) Impact of ozone on the growth and yield of trees: a review. J Environ Qual 17:347–360.

Rainerr M, Gunthardt-Goerg MS, Landolt W, Keller T (1993) Whole-plant growth and leaf formation in ozonated hybrid poplar (*Populus x euramericana*). Environ Pollut 81:207–212.

Rebbeck J (1993) Investigation of long-term effects of ozone and elevated carbon dioxide on eastern forest species: first-year response. In: Flagler R (ed) *Proceedings of the 86th Annual meeting of the Air & Waste Management Association, 13–18 June 1993, Denver, CO*. 93-TA-43.02. pp 1–10.

Rebbeck J (1996a) Chronic ozone effects on three northeastern hardwood species: growth and biomass. Can J For Res 26:1788–1798.

Rebbeck J (1996b) The chronic response of yellow-poplar and eastern white pine to ozone and elevated carbon dioxide: three-year summary. In: Hom J, Birdsey R, O'Brian K (eds) *Proceedings, 1995 Meeting of the Northern Global Change Research Program, 14–16 March 1995, Pittsburgh, PA*. Gen Tech Rep NE-214. United States Department of Argiculture (USDA) Forest Service, Northeastern Forest Experiment Station, Radnor, PA, pp 23–30.

Rebbeck J, Loats KV (1997) Ozone effects on seedling sugar maple (*Acer saccharum* Marsh.) and yellow-poplar (*Liriodendron tulipifera* L.): gas exchange. Can J For Res 27:1595–1605.

Rebbeck J, Scherzer AJ, Loats KV (1993) First season effects of elevated CO_2 and O_3 on yellow-poplar and white pine seedlings. Bull Ecol Soc Am 74(2) (Suppl):403–404.

Rebbeck J, Scherzer AJ, Loats KV (1995) Effects of two years of exposure to ozone and elevated carbon dioxide on the physiological response of white pine and yellow poplar. Ohio J Sci 95:34–35.

Reich PB (1987) Quantifying plant response to ozone: a unifying theory. Tree Physiol 3:63–91.

Reich PB, Schoettle AW, Stroo HF, Amundson RG (1988) Effects of ozone and acid rain on white pine (*Pinus strobus*) seedlings grown in five soils. III. Nutrient relations. Can J Bot 66:1517–1531.

Roberts BR (1990) Physiological response of yellow-poplar seedlings to simulated acid rain, ozone fumigation, and drought. For Ecol Manage 31:215–224.

Roth SK, Lindroth RL (1994) Effects of CO_2-mediated changes in paper birch and white pine chemistry on gypsy moth performance. Oecologia 98:133–138.

Scherzer AJ, Rebbeck J (1995) Effects of two years of exposure to ozone and elevated carbon dioxide on foliar N and P dynamics of yellow-poplar. Ohio J Sci 95:35.

Scherzer AJ, Rebbeck J, Boerner REJ (1998) Foliar nitrogen dynamics and decomposition of yellow-poplar and eastern white pine during four seasons of exposure to elevated ozone and carbon dioxide. For Ecol Manage 109: 355–366.

Sharkey TD, Loreto F, Delwiche CF (1991) High carbon dioxide and sun-shade effects on isoprene emission from oak and aspen tree leaves. Plant Cell Environ 14:333–338.

Sheng Y, Podila GK, Karnosky DF (1997) Differences in O_3-induced superoxide dismutase and glutathione antioxidant expression in O_3 tolerance and sensitive trembling aspen (*Populus tremuloides* Michx.) clones. For Genetics 4:31–41.

Simini M, Skelly JM, Davis DD, Savage JE (1992) Sensitivity of four hardwood species to ambient ozone in north central Pennsylvania. Can J For Res 22:1789–1799.

Stine RA, Baughman MJ (1992) *White Pine Symposium Proceedings: History, Ecology, Policy and Management.* Publ NR-BU-6044-S. University of Minnesota, St Paul, MN.

Strain BR (1987) Direct effects of increasing atmospheric CO_2 on plants and ecosystems. Trends Ecol Evol 2:18–21.

Stroo HF, Reich PB, Schoettle AW, Amundson RG (1988) Effects of ozone and acid rain on white pine (*Pinus strobus*) seedlings grown in five soils. II. Mycorrhizal infection. Can J Bot 66:1510–1516.

Swank WT, Vose JM (1990–91) Watershed-scale responses to ozone events in a *Pinus strobus* L. plantation. Water Air Soil Poll 54:119–133.

Taylor GE Jr. (1994) Role of genotype in the response of loblolly pine to tropospheric ozone: effects at the whole-tree, stand, and regional level. J Environ Qual 23:63–82.

Taylor GE Jr., Johnson DW, Andersen CP (1994) Air pollution and forest ecosystems: a regional to global perspective. Ecol Appl 4:662–689.

Thornley JHM, Cannell MGR (1996) Temperate forest responses to carbon dioxide, temperature and nitrogen: a model analysis. Plant Cell Environ 19:1331–1348.

Tingey DT, Hogsett WE (1985) Water stress reduces ozone injury via a stomatal mechanism. Plant Physiol 77:944–947.

Tissue DT, Thomas RB, Strain BR (1997) Atmospheric CO_2 enrichment increases growth and photosynthesis of *Pinus taeda*: a 4 year experiment in the field. Plant Cell Environ 20:1123–1134.

Tjoelker MG, Luxmoore RJ (1991) Soil nitrogen and chronic ozone stress influence physiology, growth and nutrient status of *Pinus taeda* L. and *Liriodendron tulipifera* L. seedlings. New Phytol 119:69–81.

Tjoelker MG, Volin JC, Oleksyn J, Reich PB (1993) Light environment alters response to ozone stress in seedlings of *Acer saccharum* Marsh. and hybrid *Populus* L. I. *In situ* net photosynthesis, dark respiration and growth. New Phytol 124:627–636.

Tomlinson PT, Anderson PD (1998) Ontogeny affects response of northern red oak seedlings to elevated CO_2 and water stress. II. Recent photosynthate distribution and growth. New Phytol 140:493–504.

Trier TM, Mattson WJ (1997) Needle mining by the spruce budworm provides sustenance in the midst of privation. OIKOS 79:241–246.

Volin JC, Reich PB, Givnish TJ (1998) Elevated carbon dioxide ameliorates the effects of ozone on photosynthesis and growth: species respond similarly regardless of photosynthetic pathway or plant functional group. New Phytol 138:315–325.

Wang D, Karnosky DF, Bormann FH (1986) Effects of ambient ozone on the productivity of *Populus tremuloides* Michx. grown under field conditions. Can J For Res 16:47–55.

Winner WE (1994) Mechanistic analysis of plant responses to air pollution. Ecol Appl 4(4):651–661.

Woodman JN (1987) Pollution-induced injury in North American forests: facts and suspicions. Tree Physiology 3:1–15.

Wullschleger SD, Norby RJ, Gunderson CA (1997) Forest trees and their response to atmospheric carbon dioxide enrichment: a compilation of results. In: Allen LH, Kurkham MB, Olszyk DM, Whitman CE (eds) *Advances in Carbon Dioxide Effects Research*. ASA special pub 61. American Society of Agronomists, Madison, Wisconsin, pp 79–100.

Wullschleger SD, Norby RJ, Hendrix DL (1992) Carbon exchange rates, chlorophyll content, and carbohydrate status of two forest tree species exposed to carbon dioxide. Tree Physiol 10:21–31.

Yunus M, Iqbal M (1996) *Plant Response to Air Pollution*. John Wiley, Chichester, England.

Zak DR, Pregitzer KS, Curtis PS, Teeri JA, Fogel R, Randlett DL (1993) Elevated atmospheric CO_2 and feedback between the carbon and nitrogen cycles. Plant Soil 151:105–117.

6. Physiological and Environmental Causes of Freezing Injury in Red Spruce

Paul G. Schaberg and Donald H. DeHayes

For many, concerns about the implications of "environmental change" conjure up scenarios of forest responses to global warming, enrichment of greenhouse gases, such as carbon dioxide and methane, and the northward migration of maladapted forests. From that perspective, the primary focus of this chapter, that is, causes of freezing injury to red spruce (*Picea rubens* Sarg.), may seem somewhat counterintuitive and inconsistent with the overall theme of the book. However, the dramatically increased incidence of freezing injury to northern montane red spruce forests over the past four decades is, in fact, largely a function of human-induced environmental change. "Environmental change" in the context of this chapter includes both changing climatic patterns and chemical changes in the atmospheric, forest canopy, and/or soil environment that may directly or indirectly result from atmospheric wet (precipitation or cloud water) or dry (direct deposition of gases or aerosols) deposition.

Winter injury to red spruce is now recognized as a significant factor in the decline of montane red spruce forests in northeastern North America. It is well documented that red spruce has been subject to repeated, severe, regionwide winter injury over at least the past 50 years, with this injury most commonly observed in montane forests (Friedland et al., 1984; Johnson et al., 1986; Peart et al., 1991), although it also occurs at lower elevations (Morgenstern, 1969; Peart et al., 1991; DeHayes et al., 1990). In recent years, the frequency of major regionwide winter injury events has averaged about four years with more moderate, localized injury occurring in intervening years. It has been demonstrated that red spruce winter injury is caused by subfreezing temperatures rather than foliar desiccation (DeHayes, 1992) and that the species exhibits only modest levels of midwinter cold tolerance barely sufficient to protect foliage from minimum temperatures commonly encountered in northern montane habitats. As a result, any "environmental change" that might impair the cold tolerance of red spruce foliage by just a few degrees would be expected to result in an increased frequency of freezing injury.

Well-documented environmental changes in the northeastern United States include increased pollutant deposition and exposure, especially in high elevation spruce–fir forests exposed to high concentrations of ozone and cloud water heavily laden with pollutant ions (Mohnen, 1992). There is also evidence that climatic perturbations, such as the frequency of winter thaws, have also increased in at least some locations in the northeastern U.S. (Strimbeck et al., 1995; Schaberg et al., 1996). Because red spruce grows in regions where air temperatures usually remain below 0°C throughout the winter, extended thaws that allow melting and mobilization of water in the plant may stimulate precocious dehardening. Freezing injury would be expected if a prolonged thaw is followed by extreme cold or another form of cold stress resulting from temperature fluctuations, such as rapid freezing or repeated freeze–thaw cycles (Perkins and Adams, 1995; Lund and Livingston, 1998). Such environmental changes could conceivably lead to physiological impairment that may alter the phenological development and/or the full extent of cold tolerance achieved by northern montane red spruce trees, thereby increasing susceptibility to freezing injury. The primary focus of this chapter is to describe the implications of environmental change on red spruce freezing injury, with particular emphasis on environmentally predisposing factors and physiological mechanism(s) responsible for pollution- and/or climate-induced alterations in cold tolerance.

Measurement of Red Spruce Cold Tolerance

Because of the numerous experimental and logistical challenges of assessing cold tolerance of trees *in situ*, most of our understanding of the causes of freezing injury and the physiology of cold tolerance development in red spruce has been derived from laboratory examinations of foliar freezing tolerance. Such assessments are pertinent to both trees under ambient conditions in natural forests and seedlings or trees exposed to altered environmental conditions applied as experimental treatments. However, because different cold tolerance laboratory protocols are often

used by differing investigators, the potential exists for different methodologies to influence assessments of red spruce cold tolerance and
interpretations of potential causal or predisposing factors that may
influence freezing injury.

Laboratory Cold Tolerance Measurements

Laboratory cold tolerance experiments usually involve exposing plant
tissue (e.g., current-year needles) to a series of decreasing subfreezing
temperatures in a systematic manner that includes controlling both the
rate of freezing and subsequent thawing of the tissue. The extent of
cellular disruption or injury is then assayed for tissue exposed to each test
temperature and some protocol is used to describe a temperature or range
of temperatures associated with freezing injury to the tissue under the
conditions of the experiment. Numerous assays for freezing injury exist
and several have been utilized in red spruce research, including visible
tissue discoloration, electrolyte leakage, chlorophyll fluorescence, chlorophyll degradation, and photosynthetic gas exchange (DeHayes and
Williams, 1989; Adams and Perkins, 1993; Hadley et al., 1993; Manter
and Livingston, 1996). Although all of these procedures have been shown
to effectively detect freezing injury, each appears to be measuring some
different aspect or element of cell injury. For instance, post-freezing
electrolyte leakage is an assessment of the loss of plasma membrane
integrity, which may include alterations of membrane permeability or
membrane rupture. In contrast, chlorophyll fluorescence is a reflection of
low temperature-induced reduction in photochemical efficiency and is
expected to be indicative of a disruption within chloroplasts or, more
specifically, the thylakoid membrane. Photosynthetic gas exchange and
chlorophyll content measurements likely reflect yet another set of
temperature–induced physiological perturbations.

Another important procedural element that may influence cold tolerance interpretations is the handling of post-freezing injury data and
designation of a meaningful temperature that is reliably associated with
actual cold tolerance and freezing injury. Data handling and reporting
procedures vary widely. In many cases, tissue injury estimates from
control temperatures are compared with those from subfreezing test
temperatures in a series of statistical tests (effectively an array of t-tests) to
estimate a temperature range associated with injury (Jacobson et al., 1992;
Manter and Livingston, 1996). Alternatively, the degree of tissue injury
across trees or treatments is compared at an arbitrarily chosen point of
comparison, such as an LT_{20} or LT_{50} (Sheppard et al., 1989; Adams and
Perkins, 1993). We have used an analysis of variance approach of
normalized electrolyte leakage data to compute a critical temperature (T_c)
of tissue injury, which is defined as the highest temperature at which tissue
freezing injury can be detected (DeHayes and Williams, 1989). In addition,

we have also developed a curve-fitting procedure that fits a function to each tree or treatment and allows computation of the temperature (T_m) at the midpoint of a sigmoid curve fit to electrolyte leakage data (Strimbeck, 1997). Although T_c and T_m have slightly different applications, the two procedures produce similar results in both an absolute and relative sense. A comparative study conducted in our laboratory revealed an extremely high correlation ($r^2 = 0.84$, $P < 0.01$) between cold tolerance estimates derived using T_c and T_m calculations.

Relevance of Cold Tolerance Measurements to Freezing Injury

Numerous tissue viability and data handling protocols appear to provide useful cold tolerance information and no one method appears to be "best." The selection of procedure may vary with the objectives of the assessment and the expertise and equipment of the research team involved. An additional important consideration, however, should be the relevance of the laboratory cold tolerance estimates to the actual or relative cold tolerance of trees under field conditions.

We have examined the relationship between laboratory-generated cold tolerance estimates (T_c) for red spruce derived from electrolyte leakage assessments and actual visible freezing injury measured under both field (Fig. 6.1a) and laboratory conditions (Fig. 6.1b). We have routinely demonstrated strong correlations between T_c and field freezing injury even though correlations are skewed downward because of variation in T_c among trees sufficiently cold tolerant to escape injury under field conditions (note the range in T_c among trees with 0% injury in Fig. 6.1a). The latter also points out the value of laboratory cold tolerance assessments in providing critical cold tolerance information in the absence of injury-producing conditions in the field. It is important to emphasize that the strong correlations between T_c and field freezing injury indicate a strong relative relationship, and a good predictive value of cold tolerance estimates generated using this laboratory procedure. That is, laboratory cold tolerance assessments using this procedure effectively distinguish trees or treatments with greater or less cold tolerance than others and accurately discern increases or decreases in cold tolerance over time or in response to environmental disturbance or manipulations. However, the extent to which cold tolerance estimates generated in the laboratory using any protocol represent the actual temperature of freezing injury under field conditions is considerably less clear.

In a separate study, we have also demonstrated the efficacy of electrolyte leakage as an assay for freezing-induced tissue injury. Using image analysis as an objective measure of tissue injury detected through foliar discoloration, Strimbeck (1997) found a strong correspondence between increased electrolyte leakage and the initiation of foliar reddening associated with red spruce freezing injury (see Fig. 6.1b). As expected,

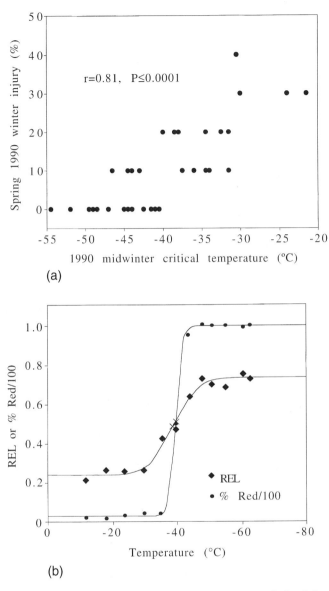

Figure 6.1. (a) Relationship between critical temperatures derived from laboratory freezing experiments with 36 red spruce trees and quantitative estimates of
foliar winter injury of the same trees recorded in a provenance test of red spruce
near Colebrook, NH. (b) Comparison of relative electrolyte leakage (REL) and
image analysis of needle reddening (Red/100).

electrolyte leakage detected injury at temperatures a few degrees higher
(i.e., warmer) than was evident through computer image analysis of foliar
discoloration. We believe that this pre-visual injury represents real low

temperature–induced physiological perturbation at the cell membrane level even though it does not translate into visual symptoms of injury until temperatures are a few degrees lower. Some evidence suggests that this initial injury, evident by electrolyte leakage, but not foliar discoloration, may represent a level of injury that is sublethal and repairable (e.g., Palta et al., 1977; Hadley et al., 1993). Although certainly a possibility, the evidence for this in red spruce or other north temperate conifers is not abundant. Even if such membrane disturbance is repairable, it is likely that there would be an energy cost to the plant. It is more likely that this initial injury detectable by electrolyte leakage is too subtle to be detected visually, as is indicated by the unusually steep visual injury response curve (appearing as a threshold response in Fig. 6.1b) that suggests even image analysis is not sensitive to the very early phases of freezing injury. Microscopic anatomical analyses of injured red spruce needles are consistent with cell-to-cell variation and a gradation in freezing injury at the cellular level (Adams et al., 1991).

Relevance of Cold Tolerance Estimates to Actual Injury Temperatures

A remaining critical question is the extent to which laboratory estimates of cold tolerance reflect the actual temperature of injury under ambient field conditions. At first pass, one might suspect that laboratory estimates might slightly underestimate actual cold tolerance because tissue viability assays often focus on the initiation of freezing injury because it is difficult to determine when tissue is dead. In contrast, however, recent evidence has suggested the opposite is more likely. That is, laboratory estimates more likely overestimate cold tolerance because the systematic and controlled nature of laboratory freezing conditions (e.g., slow and consistent rate of freezing and thawing) potentially represents a less stressful set of conditions than the natural environment, which includes a diverse and interacting set of environmental conditions and inconsistent fluctuations in ambient temperatures.

Data from Lund and Livingston (1998), for instance, suggest that multiple freeze–thaw events, which are common under ambient conditions, may enhance the freezing injury susceptibility of red spruce. Thus, T_c would be expected to be higher under ambient conditions in which temperatures fluctuated above and below freezing. Furthermore, Manter and Livingston (1996) have reported that rapid post-freezing thaw rates can also exacerbate tissue injury that originally results from freezing, but would not be detected in controlled studies that typically utilize a slow post-freezing rate of thaw. Similarly, Hadley et al. (1996) noted that post-freezing environmental conditions dramatically influenced the extent of freezing injury detected by both electrolyte leakage and foliar discoloration. Prefrozen shoots subsequently exposed to the ambient atmosphere with or without direct sunlight had much greater freezing injury symptoms than shoots placed in a constant 3°C refrigerator. In these experiments, red spruce shoots exposed to ambient atmospheric conditions after freezing to lethal temperatures exhibited LT_{10}s (based on visible injury) that were 5 to 15°C higher than those for shoots maintained in a controlled environment.

These data collectively indicate that the actual midwinter freezing tolerance of red spruce is likely less than that reported in controlled freezing tests. Although laboratory freezing studies provide conservative estimates of red spruce cold tolerance, they are strongly correlated with freezing injury under field conditions. Therefore, for consistency throughout this chapter, we will report laboratory-generated estimates of T_c or T_m as the measures of cold tolerance.

Midwinter Cold Tolerance of Red Spruce

Numerous cold tolerance assessments of red spruce have been conducted in several laboratories over the past decade. Despite varying objectives and the use of different procedures and populations in these assessments (all factors that would be expected to produce dramatically different results), a remarkably consistent pattern of midwinter maximum cold tolerance for red spruce has emerged. The average temperature in which freezing injury occurs to current-year needles of red spruce in northern New England and New York laboratory studies during winter is in the −40 to −45°C range (Table 6.1). The most comprehensive of these assessments, which examined the January cold tolerance of 60 native trees across an elevational range on Mt. Mansfield, Vermont, revealed an

Table 6.1. Estimates of the Cold Tolerance of Red Spruce Current-year Foliage Using Various Assessment Techniques

Citation	Assessment Technique	Date	Number of Trees	Estimated Cold Tolerance (°C)
Sheppard et al., 1989	Visual injury	Jan 1987	20	$LT_{10} \cong -37$[a] $LT_{50} \cong -42$[a]
DeHayes, 1992	Electrolyte leakage	Jan 1988	9	$T_c \cong -40$
Hadley and Amundson, 1992	Electrolyte leakage	Jan 1990	10	$T_c \cong -41$[a]
Perkins et al., 1993	Visual injury	Jan 1992	5	$LT_{20} \cong -46$[a]
Adams and Perkins, 1993	Chlorophyll fluorescence	Jan 1992	5	F_0 inflection $\cong -50$[a,b]
Strimbeck et al., 1995	Electrolyte leakage	Jan 1995	10	$T_c \cong -47$
Schaberg et al., 2000b	Electrolyte leakage	Jan 1996	60	$T_m \cong -42$

[a] Estimated from figure.
[b] Occurs after 20% visible injury.

average injury temperature of −41.7°C with a range of −30 to −54°C (Schaberg et al., 2000b). Importantly, 36% of the sampled trees exhibited maximum cold tolerance between −30 and −40°C. Minimum winter temperatures on the mountain fall within this temperature range during 91% of winters. Red spruce trees from southern New England are even more susceptible to freezing injury (DeHayes et al., 1990). Midwinter critical temperatures for 20 trees from a southern New Hampshire and a Massachusetts population averaged −38.7°C with a range from −21 to −49°C. In this case, 45% of the sampled trees exhibited maximum cold tolerance less than −40°C. Furthermore, trees from these provenances consistently suffered the highest incidence of winter injury in a rangewide provenance test located in northern New Hampshire (DeHayes et al., 1990).

Importantly, the maximum depth of midwinter red spruce cold tolerance is considerably less than balsam fir (*Abies balsamea* [L.] Miller), a sympatric associate that does not suffer freezing injury, and is only barely sufficient to avoid freezing injury under normal winter conditions. For example, cold tolerance assessments in our laboratory consistently demonstrate midwinter cold tolerance estimates for balsam fir below −60 to −65°C with some trees escaping injury at temperatures at or below −90°C. Historical temperature records from both Whiteface Mountain, NY, and Mt. Mansfield, VT, make it clear that red spruce, rather than balsam fir, appears to represent the outlier species with respect to cold tolerance. These records indicate that ambient montane temperatures approaching the maximum cold tolerance of at least some red spruce trees occur most winters, again illustrating the unique freezing injury susceptibility of red spruce. Also, freezing injury in red spruce is confined to current-year needles, which are about 10°C less cold tolerant than year-old needles. The latter indicates that red spruce trees have the physiological capacity to develop greater cold tolerance, but current-year needles lag behind older needles in cold tolerance development. In fact, if current-year needles were equal in cold tolerance to year-old needles of red spruce, it is expected that the species would escape freezing injury most years.

Given that laboratory-generated cold tolerance estimates are likely conservative and accurately reflect at least relative cold tolerance, freezing injury to the current-year foliage of some red spruce trees would be expected in many winters in northern montane environments. Field observations of red spruce winter injury support this contention (Morganstern, 1969; Friedland et al., 1984; Johnson et al., 1986, 1988; DeHayes et al., 1990; DeHayes, 1992). If specific pollution events (e.g., high acidic deposition inputs) or unusual climatic conditions (winter thaws) further reduce red spruce cold tolerance during some winters, as laboratory and field experiments have demonstrated (Fowler et al., 1989; DeHayes et al., 1991; DeHayes, 1992; Vann et al., 1992; Strimbeck et al.,

1995; Schaberg et al., 1996), the frequency and extent of red spruce freezing injury would be expected to dramatically increase in years such impacts are prevalent.

Potential Cold Tolerance Perturbations

Gaseous Pollutants

Considerable research has evaluated the impact of natural and anthropogenic factors on foliar cold tolerance. Among the factors assessed, changes in the concentrations of two gaseous agents (ozone and carbon dioxide) have been reported to alter the cold tolerance of spruce.

Although some work has suggested that ozone (O_3) exposure may increase freezing injury for red spruce during autumn (Fincher et al., 1989), the preponderance of evidence indicates that O_3 does not reduce red spruce foliar cold tolerance or increase freezing injury susceptibility. For example, we found no differences in the cold acclimation, depth of winter hardiness, or deacclimation of seedlings exposed to ambient (± 54 nl l^{-1}) versus reduced O_3 (± 24 nl l^{-1}) (DeHayes et al., 1991). In addition, in a study that included 3 independent experiments, we detected no reductions in autumn, winter, or spring cold tolerance for seedlings exposed to O_3 concentrations as high as 4 times ambient levels (Waite et al., 1994). In fact, in 2 of the experiments, seedlings exposed to elevated O_3 were actually more cold tolerant in January than seedlings that received low O_3 (Fig. 6.2).

Controlled exposures of other spruce species to elevated carbon dioxide (CO_2) levels provide an uncertain view of the potential impacts of CO_2 enrichment on red spruce cold tolerance. Margolis and Venzina (1990) evaluated the impact of either a continuous 10-week, or a series of 2-week, treatments of 1000 μl l^{-1} CO_2 on bud development and autumn cold tolerance of containerized black spruce seedlings (*Picea mariana* [Mill.] B.S.P.). They found that all CO_2 enrichments resulted in reduced cold tolerance (Margolis and Venzina, 1990). Because late growing season CO_2 enrichment delayed bud development and reduced autumn cold tolerance for black spruce (Margolis and Venzina, 1990), it seemed possible that cold tolerance perturbations resulted from a phenological disruption of cold acclimation. However, CO_2 enrichment has also been shown to decrease foliar nitrogen (N) concentrations in black spruce (Campagna and Margolis, 1989). And, considering the reported positive influence of N on foliar cold tolerance (DeHayes et al., 1989; Klein et al., 1989), it is also plausible that CO_2 enrichment reduced cold hardiness indirectly via a reduction of N (Margolis and Venzina, 1990). A preliminary evaluation of the combined influence of CO_2 enrichment and N nutrition on the cold tolerance of spruce seedlings was undertaken by Dalen et al. (1997). They evaluated the impact of CO_2 exposure and N fertilization on the autumn

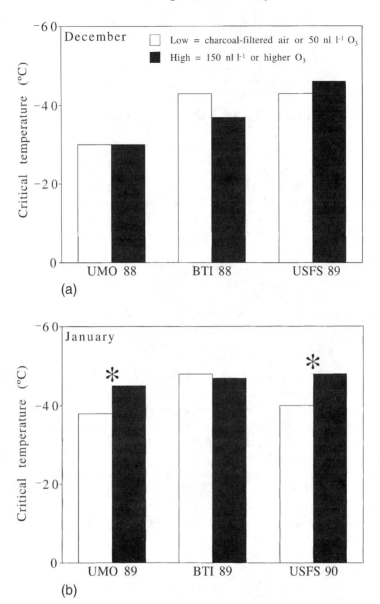

Figure 6.2. Critical temperatures during (a, b) winter (1988 and 1989) and (c) early spring (1989 and 1990) of current-year needles from red spruce exposed to charcoal-filtered air or ozone at the University of Maine (UMO), Boyce Thompson Institute (BTI), and the U.S. Forest Service Research Laboratory in Delaware, Ohio (USFS). An asterisk indicates significant treatment differences ($P \leq 0.05$).

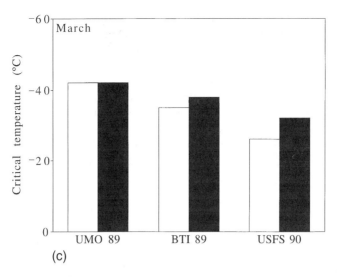

Figure 6.2. (*Continued*)

bud set and winter cold tolerance of Norway spruce (*Picea abies* [L.] Karst.) seedlings. Although CO_2 enrichment and/or N fertilization had no impact on the timing of bud set, elevated CO_2 increased the cold tolerance of high N-treated plants (Dalen et al., 1997). These results suggest that CO_2 enrichment and N nutrition may interact to influence winter cold tolerance through mechanisms separate from those controlling bud development.

Solar Warming and Rapid Freezing

Solar desiccation has been ruled out as the mechanism of winter injury in red spruce (Peart et al., 1991; Perkins et al., 1991; DeHayes, 1992). However, in part because winter injury can be greatest on the sun-exposed, south side of trees (Hadley et al., 1991; Perkins et al., 1991; Boyce, 1995), solar warming has been implicated as a contributor to foliar freezing injury (Hadley et al., 1991; Perkins et al., 1991; Hadley and Amundson, 1992; Strimbeck et al., 1993; Perkins and Adams, 1995). Field measurements have documented that the temperature of sun-exposed foliage can exceed that of the ambient air by over 20°C and that, similar to reports of the directionality of injury, solar warming is greatest for the southern aspect of crowns (Strimbeck et al., 1993).

It has been hypothesized that radiational warming could result in freezing injury if low temperatures follow sun-induced foliar dehardening (Hadley et al., 1991; Hadley and Amundson, 1992) or if rapid freezing occurs when solar illumination ends (Hadley et al., 1991; Perkins et al., 1991; Strimbeck et al., 1991; Perkins et al., 1993; Strimbeck et al., 1993).

Although radiational warming from protracted artificial illumination can result in foliar dehardening (Hadley and Amundson, 1992), it is less certain that radiational dehardening would occur in the field because solar warming there is often "transitory" and does not always raise needle temperatures above the freezing point (Strimbeck et al., 1993). In contrast, the rapid freezing of foliage following solar warming has been documented in the field (Perkins et al., 1991; Strimbeck et al., 1993), and results of experimental simulations indicate that rapid freezing can induce needle discoloration similar to natural winter injury (Perkins and Adams, 1995). However, the span of foliar temperature drop required to induce rapid freezing injury ($\geq 15°C$) has not been recorded under field conditions (Strimbeck, 1997). The current absence of field verification for the temperature drops that induce rapid freezing may indicate that these conditions rarely occur in nature (Strimbeck, 1997), although it is also possible that current sampling has been insufficient to fully characterize field conditions. The apparent ability of rapid freezing to induce injury on year-old foliage is also inconsistent with reports from the field. Freezing injury occurs almost exclusively on the current-year foliage of red spruce (DeHayes et al., 1990; DeHayes, 1992). Yet, rapid freezing can result in injury to year-old foliage when experimental conditions cause severe injury in current-year foliage (Strimbeck, 1997).

To fully elucidate the significance of solar warming and rapid freezing to red spruce winter injury, more comprehensive monitoring of needle microclimate during winter will be required. Still, even with the current level of scientific uncertainty, Strimbeck (1997) concluded that rapid freezing stress may help explain localized injury concentrated on sun-exposed branches. However, it seems improbable that rapid freezing following solar warming could account for reports of injury on shaded foliage (DeHayes et al., 1990; Hadley et al., 1991; Peart et al., 1991) or explain regionwide injury events, which have occurred in the northeastern U.S. at a rate of two to three times per decade (Johnson et al., 1988; DeHayes, 1992).

Precocious Dehardening and Winter Thaws

Although brief daytime thaws during winter appear to have little impact on red spruce cold tolerance (Perkins et al., 1993), considerable evidence indicates that significant precocious dehardening can occur in response to longer thaws. Several studies have documented reductions in cold tolerance of up to 14°C for red spruce seedlings exposed to simulated thaws (5 to 10°C) lasting four to five days (DeHayes, 1992; Schaberg et al., 1996). And, Strimbeck et al. (1995) recently documented that the current-year foliage of mature montane red spruce dehardened an average of 9°C after only three days of exposure to above-freezing (0 to 10°C) temperatures (Fig. 6.3). Individual trees dehardened as much as 14°C (Strimbeck

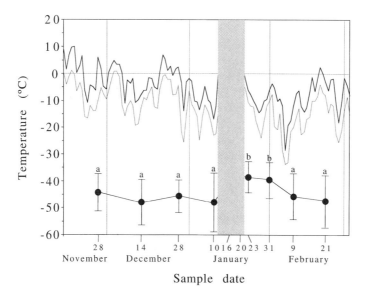

Figure 6.3. Cold tolerance of current-year foliage from mature red spruce trees and ambient air temperatures for Mt. Mansfield, VT. Solid and dotted lines are daily maximum and minimum temperatures, respectively, at the elevation of the sample stand (1000 m). Points and bars are mean, maximum, and minimum critical temperatures (T_c) of current-year foliage from 10 trees. Shaded area indicates extent of thaw weather. Mean T_c values with the same letter are not significantly different at $P \leq 0.01$, using the Duncan's multiple range test.

et al., 1995). More importantly, trees remained partially dehardened and more vulnerable to freezing injury for up to 19 days after consistent subfreezing ambient temperatures resumed (Strimbeck et al., 1995).

Data also indicate that the dehardening response to winter thaws may be unique to red spruce. For example, although red spruce experienced substantial and persistent reductions in cold tolerance during the January 1995 thaw, companion balsam fir trees showed no evidence of dehardening (Strimbeck et al., 1995). Results of a replicated experiment we conducted show the same response for seedlings exposed to simulated thaw treatments (DeHayes, 1992). Red spruce and balsam fir seedlings were exposed to either 5°C for five days, 10°C for five days, these treatments followed by five days of ambient (subfreezing) conditions, or continuous ambient conditions (the control). Red spruce seedlings experienced significant reductions in cold tolerance in response to thaw, whereas no reductions in cold tolerance were detected for balsam fir (Table 6.2). Similar to trees in the field, thaw-induced dehardening often persisted for the red spruce seedlings despite reexposure to subfreezing temperatures.

Table 6.2. Reduction in Cold Tolerance of Current-year Needles from Red Spruce and Balsam Fir Seedlings Exposed to 4 Temperature Treatments Compared with Seedlings Exposed to Ambient Subfreezing Conditions during Winter 1989

Temperature, Treatments	Reduction in Cold Tolerance (°C) on:					
	1/23/89		2/17/89		3/27/89	
	Spruce	Fir	Spruce	Fir	Spruce	Fir
10°C, 5 days	12.0[a]	5.7	13.5[a]	3.1	11.9[a]	1.3
5°C, 5 days	11.1[a]	0.3	8.3[a]	2.1	5.9[a]	–
10°C, 5 days followed by 5 days at ambient	4.3	1.9	7.9[a]	0.5	7.3[a]	–
5°C, 5 days followed by 5 days at ambient	5.8[a]	1.3	4.7	1.9	1.3	−1.1
Ambient	−45.2	−62.9	−41.1	−63.7	−40.5	−49.5

[a] Significant reduction in cold tolerance from seedlings exposed to ambient subfreezing conditions.

Because red spruce reach midwinter (pre-thaw) hardiness levels that are dangerously near ambient temperature lows, protracted reductions in cold tolerance following thaws greatly increase the risk of freezing injury if thaws are followed by extreme cold. In fact, the findings of Strimbeck et al. (1995) may exemplify a consequence of this elevated risk: the least cold tolerant tree following the January 1995 thaw exhibited the greatest amount of freezing injury after minimum temperatures at the site reached −34°C about 16 days after the thaw ended (Strimbeck et al., 1995). Even with thaw-induced dehardening, however, overall freezing injury among study trees was not severe (Strimbeck et al., 1995). From a broader perspective, examination of climatic records shows that midwinter thaws were associated with only two of the last four episodes of regionwide freezing injury (Tobi et al., 1995). Thus, although possibly an important contributor, midwinter thaws alone do not account for all recent episodes of red spruce freezing injury. More likely, combinations of predisposing factors including thaw-induced dehardening, air pollution stress, rapid freezing, and even repeating cycles of freezing and thawing (Lund and Livingston, 1998) may differentially contribute to injury development over time (Strimbeck et al., 1995).

Plant and Soil Nutrition

Nitrogen

Numerous studies have shown that short-term nitrogen (N) additions either have no impact on red spruce freezing tolerance or may even improve hardiness levels. For example, we fertilized red spruce seedlings with one of four concentrations of soil-applied ammonium nitrate

(NH_4NO_3) (0, 300, 1500 or 3000 kg N ha^{-1}) during early-, mid-, or late-summer and observed that seedlings generally acclimated more rapidly, achieved a greater depth of cold tolerance in winter, and deacclimated more slowly with increasing N treatment (DeHayes et al., 1989). The timing of N application also had an impact: N applications in early summer resulted in a slight increase in midwinter cold tolerance, whereas mid- and late-summer additions resulted in considerable increases in hardiness (DeHayes et al., 1989). Klein et al. (1989) reported that misting nutrient-deficient and fertilized red spruce seedlings with simulated cloud water containing either ammonium (NH_4), nitrate (NO_3), or both, had no effect on foliar cold tolerance. However, improving the nutrient status of seedlings that were initially N-deficient reduced their sensitivity to freezing injury (Klein et al., 1989). In addition, L'Hirondelle et al. (1992) reported that N fertilization increased the late autumn cold tolerance of red spruce seedlings, but that the low doses of N applied as part of a simulated acid mist had no impact.

Data also suggest that, at least during the first years of application, N additions have only a limited impact on the cold tolerance of mature trees in the field. White (1996) evaluated the cold tolerance of red spruce trees within two experimental watersheds in Maine: an untreated reference watershed and an adjacent watershed that received granular ammonium sulfate [($NH_4)_2SO_4$] at a dose that supplied approximately 32 kg ammonium (NH_4) ha^{-1} year^{-1} and 86 kg sulfate (SO_4) ha^{-1} year^{-1}. Following four years of treatment, the midwinter cold tolerance of 8 dominant or codominant red spruce trees from each watershed was assessed using electrolyte leakage. No difference between watersheds was detected regarding the temperature at which freezing damage occurred (White, 1996). However, there was some indication that ($NH_4)_2SO_4$ treatment may have increased cold tolerance slightly: relative damage at $-45°C$ was lower in trees from the treated watershed (White, 1996).

Although it is clear that short-term additions of N do not reduce red spruce cold tolerance, this might not hold true for prolonged N additions. Recent studies have determined that chronic N fertilization can result in growth reductions and increased mortality for red spruce trees (McNulty et al., 1996). Alterations in foliar cation nutrition and carbon relations have been associated with and may contribute to N-induced decline (Schaberg et al., 1997). Still, some data suggest that increases in freezing injury may also occur in response to long-term N addition (Perkins et al., 2000), although the early appearance of injury during this study made it impossible to fully accredit injury to any one stressor (Perkins et al., 2000).

Sulfur

Sulfur (S) additions appear to have a variable impact on hardiness levels during autumn, but no impact on winter cold tolerance. For instance, although they did not measure winter cold tolerance, Cape et al. (1991) found that application of sulfate-containing mists reduced the cold tolerance of red spruce seedlings on one of four autumn dates assessed. Using data from this study as a foundation, Sheppard (1994) then proposed a mechanism whereby foliar assimilation of sulfate (SO_4^{2-}) could result in reduced foliar cold tolerance. However, other findings indicate that foliar assimilation of SO_4^{2-} is very limited (Lindberg and Lovett, 1992; McLaughlin et al., 1996). In a separate study, Jacobson et al. (1992) reported that the anionic (SO_4^{2-} and NO_3^-) composition of mist treatments significantly impacted red spruce cold tolerance in the fall, but that the specific impact differed between the two dates assessed: in early October nitrate NO_3^--treated plants had lower cold tolerance than SO_4^{2-}-treated ones, while in mid-October the opposite was true. During the winter, differences in cold tolerance were associated with treatment pH and not the anionic composition of treatments (Jacobson et al., 1992). In a separate acid mist experiment, L'Hirondelle et al. (1992) found that SO_4^{2-}-treated seedlings had greater frost hardiness at the start of cold acclimation, but that the anionic composition of mists had no impact on hardiness levels thereafter. In addition, recent studies have documented significant reductions in cold tolerance for red spruce saplings exposed to acid mists that had either equalized SO_4^{2-} concentrations (Schaberg et al., 2000a) or contained no additional SO_4^{2-} (DeHayes et al., 1999). Data from mature trees fertilized with $(NH_4)_2SO_4$ for 4 years also showed no changes or slight increases in winter cold tolerance following treatment (White, 1996).

Aluminum

High concentrations of aluminum (Al) within soil solutions have been implicated as a contributing factor in the decline of red spruce in the northeastern U.S. (Shortle and Smith, 1988; Lawrence et al., 1995). Considering the preeminent importance of freezing injury to red spruce decline in this region (Johnson et al., 1988; Wilkinson, 1990; DeHayes, 1992; Tobi et al., 1995), it seems likely that if Al influences decline here, this influence could be through some perturbation in cold tolerance.

Schaberg et al. (2000a) recently tested the influence of soil solution Al at a level close to the maximum concentration reported for native red spruce forest soils (e.g., about 210 μM Al, Joslin and Wolfe, 1992) on the mineral nutrition, gas exchange, growth, and foliar cold tolerance of red spruce saplings. Saplings were treated with one of four Ca (0, 25, 75, or 225 μM) and two Al (0 and 200 μM) soil watering treatments in a factorial arrangement and were split between two aerial mist treatments (pH 3 or 5). Al treatment altered the growth, gas exchange, and foliar chemistry of saplings (Table 6.3), but no differences in autumn or winter cold tolerance were found (Table 6.4). Because Al may contribute to the decline of red

Table 6.3. Statistical Significance of Soil Solution Al (0 or 200 µM Al) and Ca (0, 25, 75 or 225 µM Ca) Treatments on Foliar Cation Concentrations, Diameter, Height and Shoot Length, and Gas Exchange of Red Spruce Seedlings

Foliar Elemental Concentration (mg kg^{-1})

	Al	B	Ca	Mg	Mn	P	Zn
Soil Al	*	NS	*	*	*	**	*
Soil Ca	NS	NS	*	NS	NS	NS	*

Growth (mm)

	Stem Diameter	Total Height	Shoot Length
Soil Al	*	*	*
Soil Ca	NS	NS	NS

Gas Exchange (µmol m^{-2} s^{-1})

	Jul 1994		Sep 1994	
	Photo	Resp	Photo	Resp
Soil Al	**	*	*	NS
Soil Ca	NS	*	NS	NS

NS = not significant. * = $P \le 0.05$. ** = $P \le 0.01$.

spruce through a disruption in calcium (Ca) nutrition (Shortle and Smith, 1988; Lawrence et al., 1995), it is important to note that Al treatment dramatically reduced total foliar Ca concentrations (see Table 6.4). Interestingly, however, Al treatment did not decrease Ca concentrations specifically associated with the plasma membranes of mesophyll cells (Table 6.4). This membrane-associated Ca (mCa) is of particular importance to the physiology of conifer needles (DeHayes et al., 1997). Although of possible importance to other aspects of physiology, realistically high soil solution Al concentrations neither reduced critical pools of mCa nor increased the risk of foliar freezing injury.

Calcium

Many factors, including the leaching loss of Ca due to acid precipitation (Likens et al., 1996) and N deposition (Aber et al., 1995), Ca removal via intensive harvesting (Federer et al., 1989), and recent reductions in Ca deposition (Hedin et al., 1994) may be contributing to a depletion of Ca from forest ecosystems within the eastern U.S. Although the long-term impact of Ca depletion has yet to be determined, the decline of red spruce at certain locations has already been linked to a disruption of Ca nutrition (McLaughlin and Kohut, 1992; McNulty et al., 1996; Schaberg et al., 1997). Red spruce in the southern Appalachian Mountains provide

Table 6.4. Influence of Soil Al, Soil Ca, and Acidic Mist Treatments on Current-year Foliar Ca Leaching and Concentration, Membrane-associated Ca (mCa), and Cold Tolerance in Red Spruce

Treatments	Foliar Ca Leaching (µg/l)			Total Foliar Ca (mg kg^{-1})				mCa Fluorescence			Cold Tolerance (°C)		
	Jul 94	Aug 94	Sep 94	Jul 94	Sep 94	Nov 94	Feb 95	Sep 94	Nov 94	Jan 95	Nov 94	Jan 94	Feb 95
Soil Al													
0 µmol l^{-1}	247	162	113	2417[a]	2803[a]	2562[a]	2931[a]	0.21	0.18	0.14	−46.5	−54.0	−47.8
200 µmol l^{-1}	328	133	96	1523	2040	1926	2091	0.23	0.18	0.15	−45.9	−54.6	−46.6
Soil Ca													
0 µmol l^{-1}	249	172	84	1560[a]	2042	1733[a]	1815[a]	0.21	0.18	0.15	−45.4	−53.8	−45.6
225 µmol l^{-1}	373	150	106	2335	2514	2743	3142	0.23	0.17	0.15	−47.2	−55.8	−46.7
Acidic Mist													
pH 3	526[a]	233[a]	148[a]	1969	2289	2262	2488	0.21	0.16[b]	0.12[b]	−44.5[a]	−51.5[b]	−44.9[b]
pH 5	66	64	60	1971	2553	2246	2534	0.22	0.19	0.18	−48.0	−57.0	−49.4

[a] $P \leq 0.01$.
[b] $P \leq 0.05$.

a prime example of this: considerable research has indicated that acid deposition-induced alterations in Ca nutrition and carbon relations have contributed to spruce decline in this region (McLaughlin and Kohut, 1992; McLaughlin et al., 1991, 1993).

Freezing injury of northern red spruce could also be related to a disruption in Ca nutrition. Calcium is important to freezing injury avoidance in herbaceous plants (Pomeroy and Andrews, 1985; Monroy et al., 1993; Crotty and Poole, 1995), and, a possible link between Ca nutrition and the cold tolerance of red spruce is supported by some data. Perkins and Adams (1995) reported that the concentration of several cations, including Ca, was related to freezing injury susceptibility: trees sensitive to injury had significantly lower ($P < 0.10$) Ca, magnesium (Mg) and manganese (Mn) concentrations than resistant trees. Results from cloud water exclusion studies on Whitetop Mountain, VA, also suggest an association between cation nutrition and cold tolerance. Native red spruce seedlings were exposed to or excluded from ambient acid cloud water and then physiologically and chemically assessed (Thornton et al., 1990; DeHayes et al., 1991). Although no treatment differences were found in growth or photosynthetic characteristics, seedlings exposed to acid cloud water had significantly lower concentrations of Ca and Mg in current-year foliage (which is susceptible to freezing injury) but not in year-old needles (which are not typically injured) (Thornton et al., 1990). More importantly, current-year needles of acid-exposed seedlings were also significantly less cold tolerant than comparable foliage from protected plants (DeHayes et al., 1991).

A more specific link between the Ca content and hardiness of red spruce foliage was discovered as part of a preliminary experiment conducted in our laboratory. We conducted laboratory cold tolerance and foliar Ca measurements on 19 mature trees that were part of cold tolerance studies in previous winters. The Ca content of current-year foliage was significantly correlated with the temperature of freezing injury on that sample date ($r = -0.64$; $P = 0.003$; Fig. 6.4) as well as the degree of freezing injury experienced the previous winter ($r = 0.44$; $P = 0.10$). Thus, trees with relatively low foliar Ca were generally less cold tolerant and suffered greater freezing injury the past winter. Interestingly, the correlation between Ca content and cold tolerance appeared to be driven primarily by trees with low rather than high Ca (Fig. 6.4). Trees with low Ca had relatively mild critical temperatures, whereas above some apparent Ca threshold there was no linear relationship between cold tolerance and Ca.

Despite evidence supporting a Ca–cold tolerance relationship, recent data by Schaberg et al. (2000a) indicate that total foliar Ca and cold tolerance levels are not always related. In this study, red spruce saplings received controlled additions of Ca and/or Al via the soil solution and received either pH 3.0 or 5.0 aerial mist treatments (see previous section on

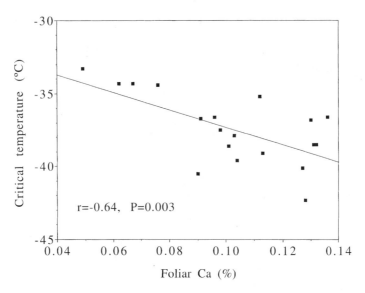

Figure 6.4. Relationship between foliar critical temperature (°C) and foliar calcium concentrations (%) of 31-year-old red spruce trees growing in north-western NH. Observations represent means of three samples from each of 19 trees.

Aluminum for details). Calcium treatment significantly increased foliar Ca incorporation and resulted in higher respiration rates while shoots were elongating (see Table 6.3). Similar to the Al treatments in this study, and perhaps because Ca treatment did not alter physiologically important concentrations of mCa (Table 6.4), Ca additions had no discernible impact on cold tolerance.

Acidic Cloud Deposition

Strong evidence now indicates that exposure to acid cloud water increases the risk of foliar freezing injury for red spruce (Table 6.5). Numerous studies have shown that treatment of red spruce with simulated acid deposition reduces the cold tolerance of current-year needles anywhere from 5 to 12°C (Fowler et al., 1989; Jacobson et al., 1992; Sheppard et al., 1993; Waite et al., 1994; DeHayes et al., 1999; Schaberg et al., 2000a) (see Table 6.5). In addition, studies of seedlings (DeHayes et al., 1991) or branches (Vann et al., 1992) exposed to ambient cloud water show that reductions in cold tolerance comparable to those documented for controlled studies also occur in the field. For example, in a mist exclusion study conducted on Whitetop Mountain, VA, native red spruce seedlings exposed to ambient cloud water were 5°C less cold tolerant in winter than seedlings that had cloud water excluded (DeHayes et al., 1991). Similarly, in a study on Whiteface Mountain, NY, airborne chemicals were excluded

Table 6.5. Studies of Mist pH Effects on the Cold Tolerance of Red Spruce Current Year Foliage

Citation	Subject	Treatments	Treatment ions	Season	Mean Δ cold tolerance
Fowler et al. 1989	seedlings	pH 2.5 to 5.0	H^+, NH_4^+, NO_3^-, SO_4^{2-}	autumn	12°C
DeHayes et al. 1991	seedlings	ambient cloud vs. no cloud	ambient mix	late autumn to winter	3 to 5°C
Vann et al. 1992	mature tree branches	ambient cloud vs. no cloud	ambient mix	winter	10°C
Jacobson et al. 1992	seedlings	pH 2.5 to 4.5	H^+, NO_3^-, SO_4^{2-}	autumn	6 to 7.5°C[a]
Sheppard et al. 1993	seedlings	pH 2.5 vs. 5.0	H^+, NH_4^+, NO_3^-, SO_4^{2-}	autumn	6°C
Sheppard et al. 1993	seedlings	pH 2.7 vs. 5.1	H^+, NH_4^+, NO_3^-, SO_4^{2-}	autumn	8°C
Waite et al. 1994	seedlings	pH 3.0 vs. 4.2	H^+, NO_3^-, SO_4^{2-}	winter	6°C
DeHayes et al. 1999	seedlings	pH 3.0 vs. 5.0	H^+, Cl^-	winter	10°C
Schaberg et al. 2000a	saplings	pH 3.0 vs 5.0	H^+, equalized SO_4^{2-}	winter	8°C

[a] Estimated from figure.

from select branches of 75-year-old red spruce trees using exclusion chambers (Vann et al., 1992). Current-year foliage on branches exposed to ambient chemicals was approximately 10°C less cold tolerant during winter and experienced more freezing injury than comparable foliage protected from ambient inputs for three months during the previous growing season (Vann et al., 1992). Because maximum winter cold tolerance levels of red spruce current-year needles are barely sufficient to protect foliage from the minimum temperatures encountered (DeHayes, 1992), disturbances that decrease cold tolerance even a few degrees are important.

Although the chemical composition of cloud water in the eastern U.S. is dominated by 4 ionic constituents, NO_3^-, SO_4^{2-}, NH_4^+, and hydrogen ions (H^+) (Mohnen, 1992), evidence suggests H^+ ions are responsible for reducing the foliar cold tolerance of red spruce during winter. Studies have shown that misting with solutions containing NO_3^- (L'Hirondelle et al., 1992) or NH_4^+ and NO_3^- (Cape et al., 1991) do not decrease the foliar cold tolerance of treated seedlings. Cape et al. (1991) reported that misting with solutions containing SO_4^{2-} reduced foliar cold tolerance on one out of four autumn dates assessed (see previous section on Sulfur), whereas L'Hirondelle et al. (1992) found that misting with SO_4^{2-} actually increased the frost hardiness of seedlings at the start of cold acclimation. However, neither of these studies examined the impacts of SO_4^{2-} on winter hardiness levels (Cape et al., 1991; L'Hirondelle et al., 1992). Jacobson et al. (1992) showed that, although SO_4^{2-} and/or NO_3^- additions reduced cold tolerance somewhat in autumn, these treatments had no influence on winter hardiness levels. Reductions in winter cold tolerance reported for this study were specifically associated with mist acidity and not the anionic composition of treatments (Jacobson et al., 1992). Similar findings were reported by L'Hirondelle et al. (1992). Furthermore, recent work has shown that misting with an acid treatment that contained no NO_3^-, SO_4^{2-}, or NH_4^+ significantly reduced the autumn and winter cold tolerance of red spruce saplings (DeHayes et al., 1999). In this study, 12 saplings were evenly distributed between two mist treatments: base nutrient solutions acidified with hydrochloric acid (HCl) to pH 3.0 or 5.0 applied for a total of 118 hours from September through November, 1993. Average reductions in cold tolerance (7.2°C in November and 10.4°C in February) were comparable to those reported for studies in which mist treatments included NO_3^-, SO_4^{2-}, NH_4^+, and H^+ (see Table 6.5). Another study also highlighted the importance of H^+ rather than SO_4^{2-} to treatment-induced reductions in cold tolerance (Schaberg et al., 2000a). Here, pH 3.0 mist application resulted in an 8°C reduction in cold tolerance relative to pH 5.0 treatment even though mist solutions contained equal SO_4^{2-} concentrations. The similarity in response across different anionic solutions coupled with reduced H^+, but not SO_4^{2-} (Joslin et al., 1988; McLaughlin et al., 1996), concentrations in throughfall suggest that acid mist–induced

Ca leaching and cold tolerance reductions are likely the result of cation exchange driven by differential H^+ exposure.

Interactions with Acid Deposition

Although it is well established that exposure to acid deposition reduces the cold tolerance of red spruce (DeHayes, 1992), other environmental factors could interact with acid deposition and/or each other to ameliorate, intensify, or in some other way alter its impact on hardiness. Because of this, knowledge of the interactions among stressors is essential to fully assess red spruce's risk of freezing injury during this period of rapid environmental change. Unfortunately, little is currently known about the influence of integrated stress on tree physiology in general, let alone specific influences on the cold tolerance of red spruce. Still, there is some evidence that other environmental perturbations can interact with acid deposition to modify plant physiology, including, at times, freezing tolerance.

A case in point: the interactive impact of acid mist and soil Al on red spruce mineral nutrition and growth. It is well established that soil pH and Al can interact to influence the availability and uptake of Al and base cations (Johnson and Fernandez, 1992). Consequently, pH and Al levels influence potentials for Al toxicity and/or cation deficiencies within plants (Marschner, 1986). Schaberg et al. (2000a) evaluated the interactive impact of soil solution Al and acid mist treatments on red spruce saplings exposed to soil Ca, soil Al and aerial mist treatments (see previous section on Aluminum). In addition to treatment main effects (see Tables 6.3 and 6.4), in a few specific instances the relative influence of Al differed between the two pH mist treatments. For instance, Al-induced reductions in Ca, Mg, zinc (Zn) and phosphorus (P) concentrations were comparatively greater for pH 5–treated saplings than for pH 3–treated plants (Schaberg et al., 2000a). Despite documented influences of Al × pH interactions on foliar cation concentrations, and even though cation chemistry (especially Ca) may have links to cold tolerance physiology (Pomeroy and Andrews, 1985; Monroy et al., 1993; Perkins and Adams, 1995), no interactive impacts of acid mist and Al treatment on cold tolerance were detected.

In contrast to results for acid mist and soil Al treatments, some data indicate that acid mist exposure and N fertilization may interact and influence foliar cold tolerance (L'Hirondelle et al., 1992). As independent factors, both N fertilization and acid mist application influence cold tolerance: N fertilization can enhance cold hardiness (DeHayes et al., 1989) whereas acid mist reduces freezing tolerance (DeHayes, 1992). L'Hirondelle et al. (1992) examined the possible combined effects of N and acid mist treatments by fertilizing pH 3–misted seedlings with either 100 (high-N) or 20 (low-N) $mg\,Nl^{-1}$ and evaluating the influence of treatment on autumn cold acclimation. They found that high-N treatment partially mitigated the impact of acid exposure during the later stages of acclimation (L'Hirondelle et al., 1992). Although N treatment had no influence on cold tolerance during September and October, high-N seedlings were slightly but significantly more cold tolerant than low-N seedlings by November (L'Hirondelle et al., 1992). Because no assessment of cold hardiness was made after November 7 (L'Hirondelle et al., 1992), it is unknown if N treatment would have improved freezing tolerance during winter. Even if N additions do temper the influence of acid mist on cold tolerance, other data suggest that N additions do not fully compensate for acid-induced reductions in hardiness. Several studies have reported that applications of acid mists that include N reduce foliar cold tolerance levels (Fowler et al., 1989; Waite et al., 1994). However, differences in N concentrations among mist treatments may have been insufficient to influence foliar cold tolerance.

Of all the possible combinations of factors that influence red spruce cold tolerance, interactions between acid mist and thaw have the greatest potential to increase the risk of freezing injury during winter. Mean reductions in cold tolerance of 12°C (Fowler et al., 1989) can occur in response to acid mist treatment, and average drops in hardiness up to 14°C have been reported for red spruce exposed to simulated winter thaws (Schaberg et al., 1996). If the combination of acid mist and thaw produced a cold tolerance response that was additive of the individual impacts of these stressors, resulting reductions in hardiness would dramatically increase the risk of freezing injury for red spruce current-year foliage. Possible interactions between acid mist and thaw are of practical importance because atmospheric additions of acid-producing compounds are likely to continue, while midwinter thaws are predicted to occur with greater frequency and intensity in the decades ahead (MacCracken et al., 1991). Although of great potential consequence to the health and survival of red spruce in the northeastern U.S., it is unknown if acid mist and thaw interact to heighten the risk of freezing injury for red spruce.

Potential Explanations of Freezing Injury in Red Spruce

Physiological Responses to Winter Warming

Phenology of Bud Development and Foliar Cold Tolerance

Although numerous climate models predict that mean surface temperatures could increase 1 to 3°C over the next 50 years, it seems unlikely that these changes will be spatially and seasonally uniform. Temperature changes are likely to be greatest at higher latitudes (Ramanathan, 1988; Lorius et al., 1990) where autumn, winter, and spring temperatures could be disproportionately altered (Kettunen et al., 1987; MacCracken et al., 1991).

Potential increases in temperature during the dormant period have raised questions regarding the likelihood of precocious budbreak, and several researchers have used computer models to simulate the impact of climatic warming on bud phenology and growth initiation for trees in northern Europe (Hanninen, 1991; Kellomaki et al., 1995). These models predict that climatic warming could speed ontogenetic development following the chilling period and induce budbreak as early as midwinter (Hanninen, 1991; Kellomaki et al., 1995). Because temperatures below 0°C are likely to occur in winter and early spring despite overall warming, precocious budbreak would likely increase the risk of frost injury (Hanninen, 1991; Kellomaki et al., 1995). Although the assumptions used in these models require further evaluation, the predictions of Hanninen (1991) and Kellomaki et al. (1995) raise concerns about the possible influence of climate change on bud development for all northern tree species, including red spruce. However, it seems unlikely that red spruce would be uniquely impacted. In fact, because red spruce typically begins growth later in the spring than most sympatric species (Burns and Honkala, 1990), it might actually be less likely to experience frost injury as a result of precocious budbreak.

The development and maintenance of foliar cold tolerance is also tied to environmental cues: short days and relatively mild subfreezing temperatures in the fall which initiate the cold acclimation process and low subfreezing temperatures which induce and perpetuate the attainment of maximum cold tolerance levels during winter (Sakai and Larcher, 1987). Warming during the dormant season (autumn through spring) could disrupt the timing and rate of foliar cold acclimation in the autumn, the depth of cold tolerance attained during winter, or the timing and rate of dehardening in the spring. Although disruptions of any of these processes seem possible, evidence suggests that, compared with other northern conifers, red spruce may be most susceptible to perturbations of winter cold tolerance (DeHayes, 1992). For example, levels of cold tolerance for red spruce are comparable with sympatric species such as balsam fir during the autumn and spring, but fall below the hardiness levels of other species during winter (DeHayes, 1992). The fact that red spruce deharden during thaws whereas sympatric balsam fir do not (Strimbeck et al., 1995) may highlight the unique potential for increased freezing injury to red spruce due to warming winter climates.

Midwinter Thaws

In northern latitudes, the greatest disruption of temperature is predicted to occur during winter (Kettunen et al., 1987; MacCracken et al., 1991). Importantly, it is predicted that temperature extremes for winter will increase as thaws become more common while contemporary temperature minima persist (MacCracken et al., 1991). Increases in the number (Schaberg et al., 1996) and severity (Strimbeck et al., 1995) of winter thaws have already been reported for portions of Vermont. The trend in winter thawing degree days (TDD, the summation of daily temperature means for all days from 21 December through 21 March with average temperatures >0°C) for Mt. Mansfield, VT, provides an example of this (Fig. 6.5). The slope of the linear relationship between TDD and year (m = 0.6185) is significantly greater than zero ($P \leq 0.002$) for the period following measurement initiation in 1954 (see Fig. 6.5), indicating that TDD have generally increased over this time. Considering the documented sensitivity of red spruce to freezing injury during winter (DeHayes, 1992) and its propensity for precocious dehardening during thaws (Strimbeck et al., 1995), perturbations of winter temperatures could be of particular concern for this species. However, red spruce do not just deharden during thaws. Significant changes in carbon metabolism also occur.

Northern red spruce typically have relatively low rates of net photosynthesis in the fall and spring, and rates close to zero for much of the winter (Schaberg et al., 1995, 1998). Yet, photosynthetic activity can increase substantially during extended thaws (Schaberg et al., 1995, 1996, 1998). For example, in their evaluation of field photosynthesis for 46 plantation-grown trees in northwestern VT, Schaberg et al. (1995) found that photosynthetic rates more than tripled during two protracted thaws (Fig. 6.6). In addition, although average rates of photosynthesis reached only 14% of the mean for red spruce during the growing season, rates for

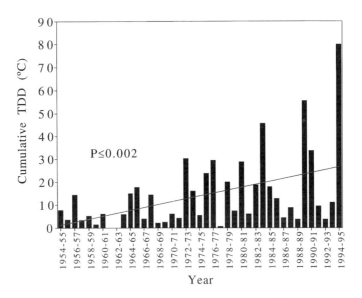

Figure 6.5. Cumulative thawing degree days (TDD) for 21 December through 21 March recorded on Mt. Mansfield, Stowe, VT, from 1954 through 1995 (NOAA records). Cumulative TDD are the summation of mean daily temperatures for days when average temperatures exceeded 0°C. Slope of the linear trend of TDD over time is positive and significantly different from zero ($P \leq 0.002$).

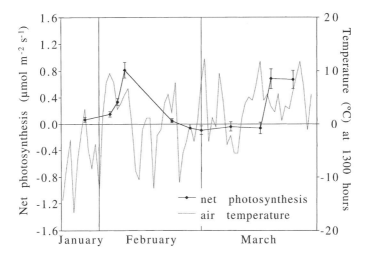

Figure 6.6. Average midday CO_2 exchange rates for red spruce from 3 Vermont seed sources plotted with air temperatures at 1300 hours for the sampling period (January through March 1991). Error bars are ± 1 SE from means for all seed sources combined.

some trees approximated growing season levels (Schaberg et al., 1995). Increases in *in situ* photosynthesis can occur within four days of thaw inception (Schaberg et al., 1996), whereas increases in photosynthetic capacity (maximum photosynthetic rates measured on rehydrated shoots under near-optimal growing season temperature and light conditions) can occur within 48 hours (natural thaw) or even 3 hours (simulated thaw) (Schaberg et al., 1998). The delayed rise in field photosynthesis relative to photosynthetic capacity suggests that other environmental limitations (possibly low water availability) retard the photosynthetic response to thaw in the field (Schaberg et al., 1998). Thaw-induced increases in respiration have also been noted for red spruce seedlings exposed to simulated thaw (Schaberg et al., 1996). Initial increases in respiration were followed by rises in carbon capture so that plants exhibited net carbon uptake by the fourth day of thaw treatment (Schaberg et al., 1996).

Despite likely thaw-induced increases in respiration and potential delays in photosynthetic gain, increases in total carbohydrate and foliar sugar concentrations have been reported for red spruce seedlings in the field following a winter thaw (Snyder, 1990). The physiological importance of possible carbon gain during winter is still unknown. However, winter photosynthesis could benefit leaves by augmenting energy stores when more distal carbon reserves are less available due to cold-induced reductions in phloem transport (Grusak and Minchin, 1989). Increases in foliar carbon reserves during winter may also help offset the metabolic costs of maintaining evergreen foliage.

The temporal association of thaw-induced changes in photosynthesis and cold hardiness raises interesting questions regarding survival and selection. Because both potentially positive (increased carbon capture) and negative (decreased cold hardiness) changes in physiology can occur, red spruce's response to thaw could be characterized as a "tradeoff," with the adaptive consequences of thaw-induced changes in physiology dependent upon long-term probabilities of "costs" vs. "benefits."

This tradeoff could be particularly pertinent within the context of potential climate change. Pollution-induced climatic warming could have broad impacts on the health and distribution of forest trees (Krauchi, 1993), with montane and alpine trees and species with limited genetic diversity among the groups most likely to be affected (Krauchi, 1993). The distribution (Burns and Honkala, 1990) and unusually low genetic variability of red spruce (DeHayes and Hawley, 1992; Hawley and DeHayes, 1994) may place this species at particular risk. Data on the range of maximum cold tolerance levels achieved by red spruce, balsam, and Fraser fir in rangewide provenance tests support the possibility that this limited genetic variability extends to cold tolerance (DeHayes, 1992). In midwinter, the fir provenances differed considerably in cold tolerance whereas pure red spruce provenances had similar average cold tolerances (DeHayes, 1992). However, the specific impacts of climate change would likely depend on the mix of environmental changes experienced. For example, if the number and/or duration of winter thaws increased and winter low temperatures were moderated, then red spruce might benefit from enhanced carbon capture without an increased risk of freezing injury. A more likely possibility, however, is that pollution-induced climate change could cause winter thaws to become increasingly common while existing low temperature extremes persist (MacCracken et al., 1991). Under this scenario, the increased risk of freezing injury following thaw could offset any physiological benefit of winter carbon capture. In fact, if thaw-induced reductions in cold tolerance resulted in foliar injury and loss, these losses would decrease potentials for current and future carbon capture and remove access to carbon reserves within lost foliage. If this occurred, thaw-associated alterations in physiology could actually be detrimental to the carbon reserves of red spruce.

The adaptive consequences of climate change on the health and distribution of red spruce are particularly difficult to predict due to the potential impacts of other pollution-induced stresses. Even if northern climates become uniformly warmer during winter, acid mist–induced reductions in hardiness (Fowler et al., 1989; DeHayes et al., 1991; Vann et al., 1992) could cause red spruce to be vulnerable to freezing injury at temperatures currently thought to be safe. And, if red spruce respond as black spruce seedlings have in a controlled test (Margolis and Venzina, 1990), atmospheric CO_2 enrichment could also reduce cold tolerance and increase the risk of freezing injury, although pollutant additions of N could partially offset acid- and/or CO_2-induced reductions in cold tolerance. In addition, acid deposition- and/or N-induced alterations in mineral nutrition could also disrupt growing season energy relations (McLaughlin et al., 1993; Schaberg et al., 1997), which, in turn, could alter the adaptive significance of winter carbon capture. Furthermore, red spruce's limited genetic variability could greatly limit its ability to respond to changing pollution-induced stress (DeHayes and Hawley, 1992).

Possible Influence of Glacial Refugia

If indeed red spruce face a tradeoff between carbon capture and cold hardiness, then it is reasonable to question how selection may have influenced the initial development and continued presence of this response to thaw. One possibility is that this response is a remnant of selection during the last ice age. Compared with other conifers of the region, red spruce is thought to have had a restricted glacial refugia that extended from the Mid-Atlantic states eastward through an area that is now submerged continental shelf (White and Cogbill, 1992). Maritime influences likely moderated temperatures within the refugium, and may have reduced the adaptive advantage conveyed by the development of deep cold hardiness (White and Cogbill, 1992). Prolonged exposure to moderate coastal temperatures may also have promoted an extended period for carbon capture, including winter photosynthesis if the risk of freezing injury was low. Conifers from the maritime regions of the Pacific Northwest may provide a contemporary example of this; they are known to be photosynthetically active and have only limited cold tolerance during the mild winters typical of that region (e.g., Larcher and Bauer, 1981; Hawkins et al., 1995).

In addition to the potential impacts of refugia location, the small size of red spruce's glacial refugium may have reduced gene flows and increased levels of inbreeding, resulting in reduced genetic variability. Combined with the enhanced potential for environmental homogeneity (and thus more uniform selection pressure) within a restricted refugium, breeding limitations associated with small refugium size may help account for red spruce's low genetic variability relative to other conifers (White and Cogbill, 1992; Hawley and DeHayes, 1994). This reduction in variability may also be one reason why red spruce's physiological response to thaw has persisted over time: selective forces are now interacting with a more limited gene pool.

Acid Mist–Induced Modifications of Membrane-Associated Calcium

Considerable research has now established that exposure of red spruce foliage to both simulated (Fowler et al., 1989; Waite et al., 1994) and ambient acidic cloud water (DeHayes et al., 1991; Vann et al., 1992) during the growing season or autumn results in a significant reduction in late autumn and midwinter freezing tolerance of current-year needles (see Table 6.5). Such reductions are sufficient to explain the dramatic increase in freezing injury to red spruce observed in the northern montane forests over the past 40 years (Johnson et al., 1988, 1996; DeHayes, 1992) and growth and vigor losses typical of the decline (Wilkinson, 1990; Tobi et al., 1995). Although several studies have examined the role of specific pollutant ions in cold tolerance reductions (Cape et al., 1991; Sheppard, 1994), no empirically based physiological mechanism for this phenomenon has been established. Specific, highly repeatable findings that are critical to understanding the influence of acidic mist on red spruce freezing tolerance and that must be accounted for in any viable physiological mechanism are:

1. Freezing injury in red spruce is specific to current-year foliage, which is about 10°C less freezing tolerant in winter than year-old foliage.
2. Acid mist–induced reductions in cold tolerance persist throughout the winter even when the acid mist is removed.
3. Relatively short-term exposure to acidic mist in autumn only has an equal or greater influence on winter freezing tolerance than growing season exposure.
4. Simulated mists, which include a base solution with an ionic composition patterned after regional cloud chemistry analyses, reduce cold tolerance whether mist pH adjustments are made with HCl or H_2SO_4.

Potential Role of Calcium in Cold Tolerance

We have examined acid mist–induced alterations in Ca physiology as a potential mechanism for low temperature sensitivity and injury in red spruce foliage. Calcium is an abundant element in trees and a major cation in soil and surface waters and has been a focus of recent attention because of depletion resulting from acid deposition, diminished base cation deposition, foliar leaching, N saturation, and competitive interactions with Al (Joslin et al., 1988; Shortle and Smith, 1988; Lawrence et al., 1995; Hedin et al., 1994; Aber et al., 1995; Likens et al., 1996). Empirical information has demonstrated an inverse relationship between foliar Ca concentration, freezing tolerance, and freezing injury among native red spruce trees and experimental work has shown reduced foliar Ca in response to exposure to acidic mist (Joslin et al., 1988, DeHayes et al., 1991; McLaughlin et al., 1993). McLaughlin et al. (1991, 1993) have also implicated acid deposition-induced calcium deficiency as a potential causal factor in dark respiration increases and altered carbon metabolism associated with red spruce decline in the southern Appalachians, where ambient winter temperatures preclude freezing injury. McLaughlin et al. (1993) observed a partial reversal of acid mist–induced respiration increases through soil Ca fertilization, and suggest that foliar as well as soil-driven reactions may be important in altered carbon metabolism resulting from acid deposition.

Calcium is highly compartmentalized within cells and tissues and this partitioning is critical to its physiological function in plants. The major fraction of Ca in conifer needles is insoluble extracellular Ca oxalate and pectate crystals (Fink, 1991), while Ca ions associated with the plasma membrane region (including some free and displaced apoplastic Ca from the cell wall) are labile and of major physiological importance. This pool of mCa, although a relatively small fraction of total foliar Ca ion pools, strongly influences the response of cells to changing environmental conditions and stress (Hepler and Wayne, 1985; Dhindsa et al., 1993; McLaughlin et al., 1993). Membrane-associated Ca influences plasma membrane structure and function, stabilizing membranes and influencing permeability by bridging phosphate and carboxylate groups of membrane phospholipids and proteins (Palta and Li, 1978; Davies and Monk-Talbot, 1990; Steponkus, 1990). The plasma membrane plays a critical role in mechanisms of cold acclimation and low temperature injury in plants (Pomeroy and Andrews, 1985; Davies and Monk-Talbot, 1990; Steponkus, 1990). By influencing membrane architecture, mCa influences solution movement across membranes, the ability of cells to resist dehydration, extracellular ice damage, and perhaps intracellular freezing during cold acclimation (Guy, 1990). This labile mCa pool (Atkinson et al., 1990) may also play an important role as a second messenger ("messenger Ca") in the perception and transduction of stress signals, including low temperature signals (Dhindsa et al., 1993; Monroy et al., 1993; Crotty and Poole, 1995), across membranes by binding to proteins such as calmodulin (Hepler and Wayne, 1985). Calcium bridges can be broken in acidic environments resulting in foliar Ca leaching, which could potentially lead to destabilization of membranes, depletion of a pool of messenger Ca, and enhanced susceptibility to environmental stress such as low temperature.

Evidence of Acid-Induced Calcium and Cold Tolerance Perturbations

We have conducted a series of *in vivo* acidic mist experiments that consistently demonstrate significantly greater Ca leaching (Fig. 6.7a) from current-year needles of red spruce in response to pH 3.0 vs. 5.0 (base mist solution adjusted with HCl) mist throughout summer and autumn (DeHayes et al., 1999). Ca concentrations in foliar leachate range about 2 to 10 times greater in foliage exposed to the more acidic mist. Furthermore, an *in vitro* experiment designed to partition throughfall Ca has conclusively verified that most (~85%) of the acid-leached Ca is derived from needles rather than stems and nearly 50% more Ca is leached from current-year than year-old foliage (Fig. 6.7b). Analysis of throughfall

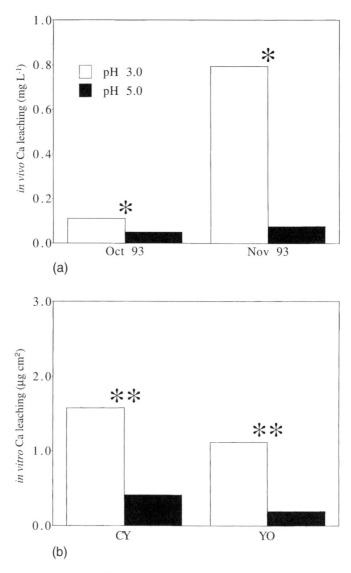

Figure 6.7. Impacts of acid mist treatment on (a) *in vivo* Ca leaching, (b) *in vitro* Ca leaching, (c) total foliar Ca, and (d) cold tolerance of foliage from red spruce. All data for current-year foliage unless indicated otherwise. CY, Current-year foliage; YO, year-old foliage. (∗ = $P \leq 0.05$; ∗∗ = $P \leq 0.01$).

also indicate nearly 60 times greater uptake of H^+ in response to pH 3.0 vs. 5.0 mist. Acid mist–induced Ca losses are accompanied by reductions in membrane stability (DeHayes et al., 1999) and a significant 4 to 10°C decrease in freezing tolerance during late autumn and winter subsequent to acid mist exposure (Fig. 6.7d).

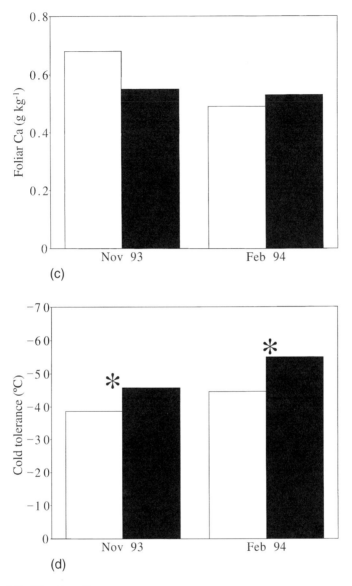

(c)

(d)

Figure 6.7. (*Continued*)

Despite up to 10-fold differences in foliar Ca leaching between pH treatments, a reduction in total foliar Ca pools of pH 3–treated foliage is typically not evident (Fig. 6.7c; see Table 6.4). It is likely that total foliar Ca estimates primarily reflect the dominant insoluble and immobile extracellular Ca pool, which may mask relatively subtle, but critical, shifts in the labile and environmentally sensitive mCa pool. If acid-induced foliar Ca leaching preferentially removes mCa, then membrane dysfunction and a concomitant increase in freezing injury susceptibility would be expected.

Using an analytical procedure allowing localization and quantification of subcellular Ca (Borer et al., 1997), we have examined the specific influence of acid deposition on the mCa pool. In a comprehensive experiment designed to examine freezing tolerance, mCa, total foliar Ca, and foliar Ca leaching responses to a factorial combination of Ca perturbation treatments, we demonstrated that acid rain displaces Ca ions specifically associated with the plasma membrane–cell wall compartment in red spruce mesophyll cells (Schaberg et al., 2000a). This membrane alteration is accompanied by a 4–10°C reduction in foliar freezing tolerance. Treatments were applied as soil Ca amendments, enhanced soil aluminum, and acid mist (base mist solutions adjusted with H_2SO_4) applications from June through September (pH 3.0 vs. 5.0). Although reduced soil Ca and enhanced soil Al treatments resulted in significant and predictable reductions in total foliar Ca, they had no consistent influence on foliar leaching, mCa, or freezing tolerance (see Table 6.4), which are critical red spruce decline precursors in northern montane forests. In contrast, pH 3.0 acid mist treatments applied in summer and early autumn resulted in significant foliar Ca leaching, and a consistent, significant, and parallel reduction in late fall and winter freezing tolerance and mCa in current-year needles of red spruce (see Table 6.4). Although relatively minor in late summer, acid mist-induced reductions in mesophyll cell mCa reached a maximum of about 35% between mist treatments by midwinter (see Table 6.4), even though the acid mist treatments had been removed months earlier.

The persistence of the acid mist influence on mCa and cold tolerance throughout winter and the responsiveness to short-term acid mist applications in autumn is a reflection of the lack of mobility of Ca in the phloem (Marschner, 1986) and the dependence upon xylem transport to replenish depleted mCa. It appears that the physiological implications of acid deposition are exacerbated in autumn when post–growing season losses of mCa cannot be replaced via the transpiration stream. These results conclusively demonstrate that acid deposition represents a unique environmental stress in that it preferentially removes mCa, which is not readily replaced in autumn, resulting in a mCa deficiency that may not be detectable by examination of total foliar Ca pools.

Other Evidence Implicating mCa in Red Spruce Freezing Injury

Considerable additional empirical data implicates mCa involvement or an association between mCa and freezing tolerance/injury in red spruce. For instance, the pronounced needle age class difference in both cold tolerance and freezing injury susceptibility is consistent with the pattern of developmental variation in mCa in current-year needles as well as age class differences in foliar Ca leaching. There is little or no detectable mCa in current-year needles of red spruce during early summer (Fig. 6.8; DeHayes et al., 1997) and mCa levels in current-year needles are inherently lower than in year-old needles throughout summer and early autumn when acid mist–induced Ca leaching is substantial. The substantial age class difference in foliar Ca leaching (see Fig. 6.7) further accentuates the mCa and cold tolerance differences between red spruce needle age classes and offers a viable explanation for the unique low temperature sensitivity of red spruce current-year foliage.

Furthermore, unlike total foliar Ca pools, mCa pools in current-year, but not year-old, needles are seasonally dynamic and responsive to temporal environmental changes that parallel seasonal changes in membrane structure related to cold acclimation. For example, late summer–early autumn increases in mCa reflect an increase in Ca ion exchange sites associated with short day–induced increases in membrane phospholipids, while apparent frost-initiated reductions in mCa are likely associated with changes in the fatty acid composition of membrane lipids (DeHayes et al., 1997) and perhaps a role for mCa in the perception and transduction of the low temperature cold acclimation signal as has been suggested for some agronomic crops (Dhindsa et al., 1993; Monroy et al., 1993; Crotty and Poole, 1995).

We have also documented abrupt, but temporary, mCa reductions in current-year needles of both red spruce seedlings and native trees in response to midwinter thaws (Table 6.6). Cold tolerance was examined during one of these thaws and verified that thaw-induced mCa reductions were accompanied by significant precocious dehardening over only 3 days (DeHayes et al., 1997; Strimbeck et al., 1995). After the thaw, both mCa and freezing tolerance returned to prethaw levels. Sequential changes in mCa and frost tolerance levels, coupled with the well documented changes in membrane structure in leaf tissue of north temperate conifers during cold acclimation (e.g., DeYoe and Brown, 1979; Senser and Beck, 1982, 1984), strongly support the contention that mCa is important to cold tolerance and that acid mist alterations of mCa cause membrane destabilization and a concomitant loss in freezing tolerance.

Alterations to mCa also seem pertinent to the documented physiological impairment and growth decline of red spruce in the southern Appalachians, which is attributed to acid mist–induced alterations in Ca nutrition and carbon relations (McLaughlin and Kohut, 1992; McLaughlin et al., 1991, 1993) rather than to freezing injury. Although they have not examined the mCa pool specifically, McLaughlin et al. (1993) suggest that tissue respiration increases in response to acid deposition-induced Ca losses may be associated with the critical regulatory role of Ca in membrane permeability. Acid deposition–induced alteration of mCa that we have documented and that leads to enhanced freezing injury

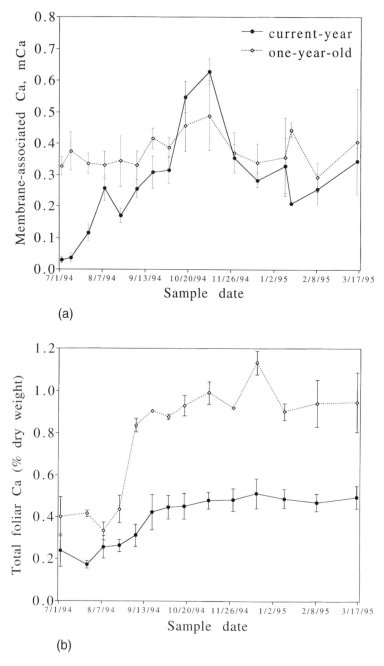

Figure 6.8. Seasonal patterns of (a) membrane-associated Ca (mCa) and (b) total foliar Ca of current-year and 1-year-old needles of red spruce seedlings. Error bars are ±1 SE from means. In some cases, error bars are obscured by symbols representing means.

Table 6.6. Mesophyll Cell Membrane-associated Ca (mCa) Levels in Current-year Foliage of Red Spruce Seedlings (1995) and Mature Trees (1997) before and after Winter Thaws

Year	Relative mCa Concentration		P
	Pre-thaw	Post-thaw	
1995	0.328	0.209	0.065
1997	0.232	0.116	0.001

susceptibility in the north would also be expected in response to acidic deposition in the southern Appalachians. As such, the acid deposition–mCa leaching–membrane alteration explanation for red spruce decline that we describe represents a comprehensive explanation for acid-induced red spruce decline throughout the montane forests of eastern North America.

Model for Enhanced Freezing Injury and Membrane-associated Calcium Alteration

Our results provide an empirically supported mechanistic explanation of acid mist– and winter thaw–induced reductions in freezing tolerance and overall health of red spruce forests (Fig. 6.9). Acid deposition–induced cation exchange results from H^+ replacement of Ca on exchange sites at the cell wall–membrane interface. This pool of mCa, unlike other insoluble extracellular Ca pools, is readily available for acid-induced cation exchange resulting in Ca leaching, membrane destabilization, depletion of a potential pool of messenger Ca, and significant reductions in the freezing tolerance of current-year needles of red spruce. Although SO_4^{2-} has been suggested as the critical pollutant ion in cold tolerance reductions (Shepherd, 1994; Cape et al., 1991), we have demonstrated similar cold tolerance reductions and significant mCa leaching in response to mists containing equalized SO_4^{2-} concentrations and with different anionic constituents (DeHayes et al., 1999; Schaberg et al., 2000a). This similarity in response across different anionic solutions coupled with evidence of reduced H^+, but not SO_4^{2-} (Joslin et al., 1988; McLaughlin et al., 1996), concentrations in throughfall suggest that acid mist-induced Ca leaching and cold tolerance reductions are likely the result of cation exchange driven by differential H^+ exposure. There is strong empirical support for most elements of the model and it effectively accounts for each of the aforementioned unique and specific highly repeatable factors that are integral components of the red spruce freezing injury and decline phenomenon.

Using this model, one would predict that sites exposed to high H^+ inputs and winter thaws followed by very low air temperatures would

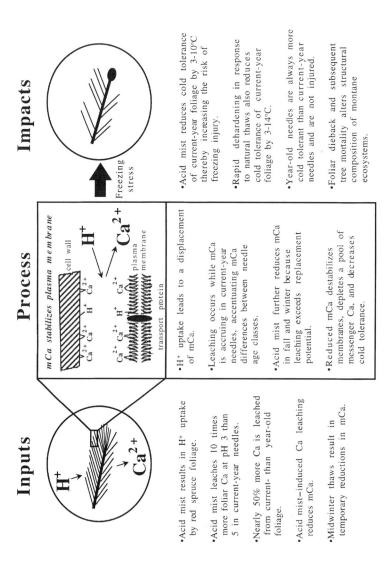

Inputs

- Acid mist results in H⁺ uptake by red spruce foliage.
- Acid mist leaches 10 times more foliar Ca at pH 3 than 5 in current-year needles.
- Nearly 50% more Ca is leached from current- than year-old foliage.
- Acid mist–induced Ca leaching reduces mCa.
- Midwinter thaws result in temporary reductions in mCa.

Process

mCa stabilizes plasma membrane

- H⁺ uptake leads to a displacement of mCa.
- Leaching occurs while mCa is accruing in current-year needles, accentuating mCa differences between needle age classes.
- Acid mist further reduces mCa in fall and winter because leaching exceeds replacement potential.
- Reduced mCa destabilizes membranes, depletes a pool of messenger Ca, and decreases cold tolerance.

Impacts

Freezing stress

- Acid mist reduces cold tolerance of current-year foliage by 3-10°C thereby increasing the risk of freezing injury.
- Rapid dehardening in response to natural thaws also reduces cold tolerance of current-year foliage by 3-14°C.
- Year-old needles are always more cold tolerant than current-year needles and are not injured.
- Foliar dieback and subsequent tree mortality alters structural composition of montane ecosystems.

Figure 6.9. Empirically based mechanistic model depicting acid mist–induced Ca perturbations, destabilization of plasma membranes, and reduction in red spruce foliar cold tolerance.

experience freezing injury with the greatest severity and frequency. Indeed, montane populations in the northeastern U.S. meet these criteria and commonly experience freezing injury (Friedland et al., 1984; Johnson et al., 1986; Peart et al., 1991). However, lower elevation forests are not without risk and currently experience some freezing injury (Morgenstern, 1969; Peart et al., 1991; DeHayes et al., 1990). In addition, continued H^+ inputs, predicted increases in the frequency and extent of winter thaws, and possibly even soil Ca depletion (because soils ultimately supply membranes with mCa) could be expected to increase the risk of freezing injury above current levels.

Although of some potential value in predicting trends in injury, the real importance of the model we present is its utility in synthesizing scientific information and thought regarding freezing injury of red spruce. We believe this model provides a physiological explanation for the dramatic increase in red spruce freezing injury over the past 40 years that coincided with increased pollution emissions and for the well documented decline of red spruce in northern montane forest ecosystems.

Conclusions

The unique freezing injury susceptibility of red spruce is, in part, a function of the modest depth of midwinter cold tolerance attained by the species, which appears to not be well adapted to northern montane temperature conditions. Historical analyses, however, indicate that freezing injury to red spruce occurred only sporadically during the late 19th century and first 50 years or so of this century and then increased dramatically over recent decades (Johnson et al., 1986). Because there is no evidence that winters in New England or New York have become colder (in fact, some evidence suggests a slight warming trend) during this recent period when freezing injury has increased (Hamburg and Cogbill, 1988), one must conclude that environmental changes that have impaired freezing tolerance during the approximate period 1950 to the present are primarily responsible for the enhanced freezing injury documented over the past several decades. A substantial body of evidence has demonstrated that both acid deposition and increasingly frequent winter thaws result in a consistent and significant reduction in midwinter cold tolerance, which dramatically increases the probability of freezing injury. Cold tolerance reductions averaging up to 10°C have been documented in response to either acidic deposition or thaw-induced precocious dehardening in red spruce trees under both ambient and experimental conditions. The extent and implications of interactions between these environmental conditions are not yet known, but are currently under study. Regardless of the relationship between acid- and temperature-impacts on cold tolerance, forests are exposed to such a heterogeneous mix of acid inputs and temperature perturbations (including thaws, rapid freezing, and cycles of thaw and freezing) that levels and patterns of injury predictably appear complex and temporally and/or spatially variable.

Equally as important to the understanding of red spruce freezing injury are environmental changes that have not resulted in cold tolerance impairments for this species. Based on the current state of knowledge, there appears to be no compelling evidence that exposure to high concentrations of tropospheric O_3 or Al in soil solution has any direct impact on the development or attainment of freezing tolerance in red spruce. Given the documented importance of freezing injury as an initiating and predisposing factor in red spruce decline in northern montane forests, it would appear that these potential stresses are not principally associated with the deterioration of red spruce forests. Although considerable empirical evidence has demonstrated either neutral or positive cold tolerance responses to N supplements, we are hesitant to discount a potential physiological impairment associated with atmospheric N deposition over the long-term, especially in high-elevation forests. Recent evidence has shown that long-term N supplements can negatively impact red spruce forest health and nutrient dynamics in ways that could conceivably alter plant cold tolerance development (Schaberg et al., 1997; Perkins et al., 2000).

Both acid mist-induced and thaw-induced reductions in red spruce cold tolerance appear to involve alterations to the structure and/or function of mCa in mesophyll cells. The physiologically critical and labile mCa pool, although a relatively small fraction of the total foliar Ca ion pool, strongly influences the response of cells to changing environmental conditions and stress. With respect to acidic mist, it appears that H^+ displaces Ca on exchange sites at the cell wall–membrane interface leading to membrane destabilization, depletion of a pool of messenger Ca, and enhanced susceptibility to low temperature stress. The extent to which acid mist alteration of mCa also influences the apparent hypersensitive red spruce dehardening response to winter thaws is not as yet known, but evidence indicates that mCa is involved in the dehardening response. It is important to emphasize that the commonly measured total foliar Ca ion pool is not correlated with or a meaningful indicator of the physiologically important mCa pool (DeHayes et al., 1997). In fact, it is likely that environmentally induced shifts in the critical mCa pool would not be detected in an analysis of total foliar Ca content. As such, it is critical that the mCa pool specifically be monitored in woody plant leaf tissue (Borer et al., 1997), especially in studies examining plant responses to pollution- or climate-induced environmental change.

Of greater concern relative to overall forest health is the strong likelihood that acid rain alteration of mCa and membrane integrity is not unique to red spruce, but simply exacerbated in this species because of its low temperature sensitivity. It is now well established that Ca plays a critical role in plant responses to numerous stresses, such as low temperature, salt, drought, and low light. (Hepler and Wayne, 1985; Dhindsa et al., 1993; Monroy et al., 1993; Sheen, 1996). It is also likely that the relatively small and environmentally sensitive mCa pool serves a critical and active role in such plant stress responses. Pollution-induced alteration to the mCa pool in forest trees would be expected to lead to physiological impairment of the plant stress response system and a predisposition to an array of environmental and biological stresses. The potential forest stress implications are further compounded by the lasting influence of acid deposition-induced leaching, depletion, and cycling disruption of Ca in forest soils. Estimates show that the pool of Ca in the soil complex in northeastern forests may have shrunk by as much as 50% during the past 45 years as a result of acidic deposition (Likens et al., 1996). As a result, the available pool of soil Ca for mCa replenishment and reconstruction of the plant stress response system may not be sufficient to stabilize healthy forests. Ironically, at a time of unprecedented abiotic and biotic stress loading from numerous sources, including atmospheric pollutants, changing climatic conditions, exotic pests, pathogens, and nutrient depletion, acidic inputs may be depleting a critical pool of Ca within plants that helps them respond to stress.

References

Aber JD, Magill A, McNulty SG, Boone RD, Nadelhoffer KJ, Downs M, Hallet R (1995) Forest biogeochemistry and primary production altered by nitrogen saturation. Water Air Soil Pollut 85:1665–1670.

Adams GT, Perkins TD (1993) Assessing cold tolerance in *Picea* using chlorophyll fluorescence. Environ Exp Bot 33:377–382.

Adams GT, Perkins TD, Klein RM (1991) Anatomical studies on first-year winter injured red spruce foliage. Am J Bot 78:1199–1206.

Atkinson MM, Keppler LD, Orlandi EW, Baker CJ, Mischke CF (1990) Involvement of plasma membrane calcium influx in bacterial induction of the K^+/H^+ and hypersensitive responses in tobacco. Plant Physiol 92: 215–221.

Borer CH, DeHayes DH, Schaberg PG, Cumming JR (1997) Relative quantification of membrane-associated calcium (mCa) in red spruce mesophyll cells. Trees 12:21–26.

Boyce RL (1995) Patterns of foliar injury to red spruce on Whiteface Mountain, New York, during a high-injury winter. Can J For Res 25:166–169.

Burns RM, Honkala BH (1990) *Silvics of North America. Vol. 1. Conifers.* Agric Handbook 654. United States Department of Agriculture (USDA) Forest Service, Washington, DC.

Campagna MA, Margolis HA (1989) Influence of short-term atmospheric CO_2 enrichment on growth, allocation patterns, and biochemistry of black spruce seedlings at different stages of development. Can J For Res 19:773–782.

Cape JN et al. (1991) Sulphate and ammonium in mist impair the frost hardiness of red spruce seedlings. New Phytol 125:119–126.

Crotty CM, Poole RJ (1995) Activation of an outward rectifying current by low temperature in alfalfa protoplasts. Plant Physiol 108:38.

Dalen LS, Johnsen O, Ogner G. (1977) Frost hardiness development in young *Picea abies* seedlings under simulated autumn conditions in a phytotron—effects of elevated CO_2, nitrogen and provenance. Plant Physiol 114:126.

Davies HW, Monk-Talbot LS (1990) Permeability characteristics and membrane lipid composition of potato tuber cultivars in relation to Ca^{2+} deficiency. Phytochem 29:2833–2835.

DeHayes DH (1992) Winter injury and developmental cold tolerance in red spruce. In: Eager C, Adams MB (eds) *The Ecology and Decline of Red Spruce in the Eastern United States*. Springer-Verlag, New York, pp 296–337.

DeHayes DH, Hawley GJ (1992) Genetic implications in the decline of red spruce. Water Air Soil Pollut 62:233–248.

DeHayes DH, Williams MW (1989) *Critical Temperature: A Quantitative Method of Assessing Cold Tolerance*. Gen Tech Rep NE-134. United States Department of Argiculture (USDA) Forest Service, Northeastern Forest Experiment Station, Broomall, PA.

DeHayes DH, Ingle MA, Waite CE (1989) Nitrogen fertilization enhances cold tolerance of red spruce seedlings. Can J For Res 19:1037–1043.

DeHayes DH, Waite CE, Ingle MA, Williams MW (1990) Winter injury susceptibility and cold tolerance of current and year-old needles of red spruce trees from several provenances. For Sci 36:982–994.

DeHayes DH, Thornton FC, Waite CE, Ingle MA (1991) Ambient cloud deposition reduces cold tolerance of red spruce seedlings. Can For Res 21:1292–1295.

DeHayes DH, Schaberg PG, Hawley GJ, Borer CH, Cumming JR, Strimbeck GR (1997) Physiological implications of seasonal variation in membrane-associated calcium in red spruce mesophyll cells. Tree Physiol 17:687–695.

DeHayes DH, Schaberg PG, Hawley GJ, Strimbeck GR (1999) Acid rain impacts calcium nutrition and forest health. BioScience 49:789–800.

DeYoe DR, Brown GN (1979) Glycerolipid and fatty acid changes in eastern white pine chloroplast lamellae during the onset of winter. Plant Physiol 64:924–929.

Dhindsa RS, Monroy A, Wolfraim L, Dong G (1993) Signal transduction and gene expression during cold acclimation in alfalfa. In: Li PH, Christersson L (eds) *Advances in Plant Cold Hardiness*. CRC Press, Boca Raton, FL, pp 57–72.

Federer CA, Hornbeck JW, Tritton LM, Martin CW, Pierce RS, Smith CT (1989) Long-term depletion of calcium and other nutrients in eastern US forests. Environ Manage 13:593–601.

Fink S (1991) The micromorphological distribution of bound calcium in needles of Norway spruce (*Picea abies* [L.] Karst.). New Phytol 119:33–40.

Fincher J, Cumming JR, Alscher RG, Rubin G, Weinstein L (1989) Long-term ozone exposure affects winter hardiness of red spruce (*Picea rubens* Sarg.) seedlings. New Phytol 113:85–96.

Fowler D, Cape JN, Deans JD, Leith ID, Murray MB, Smith RI, Sheppard LJ, Unsworth MH (1989) Effects of acid mist on the frost hardiness of red spruce seedlings. New Phytol 113:321–335.

Friedland AJ, Gregory RA, Karenlampi L, Johnson AH (1984) Winter damage to foliage as a factor in red spruce decline. Can J For Res 14:963–965.

Grusak MA, Minchin PEH (1989) Cold-inhibited phloem translocation in sugar beet. J Exp Bot 40:215-223.

Guy CL (1990) Cold acclimation and freezing stress tolerance: role of protein metabolism. Ann Rev Plant Physiol Plant Mol Biol 41:187–223.

Hadley JL, Amundson RG (1992) Effects of radiational heating at low air temperature on water balance, cold tolerance, and visible injury of red spruce foliage. Tree Physiol 11:1–17.

Hadley JL, Amundson RG, Laurence JA, Kohut RJ (1993) Physiological response to controlled freezing of attached red spruce branches. Environ Exper Bot 33:591–609.

Hadley JL, Friedland AJ, Herrick GT, Amundson RG (1991) Winter desiccation and solar radiation in relation to red spruce decline in the northern Appalachians. Can J For Res 21:269–272.

Hadley JL, Manter D, Herrick J (1996) The effects of post-freezing environment on freezing injury to red spruce: implications for cold tolerance testing in conifers. In: Bernier PY (ed) *Proceedings of the 14th North American Forest Biology Workshop. 16–20 June*, Laval University, *Quebec City, Canada*. p 106.

Hamburg SP, Cogbill CV (1988) Historical decline of red spruce populations and climatic warming. Nature 331:428–430.

Hanninen H (1991) Does climatic warming increase the risk of frost damage in northern trees? Plant Cell Environ 14:449–454.

Hawkins BJ, Davradou M, Pier D, Shortt R (1995) Frost hardiness and winter photosynthesis of *Thuja plicata* and *Pseudotsuga menziesii* seedlings grown at three rates of nitrogen and phosphorus supply. Can J For Res 25:18–28.

Hawley GJ, DeHayes DH (1994) Genetic diversity and population structure of red spruce (*Picea rubens*). Can J Bot 72:1778–1786.

Hedin LO, Granat L, Likens GE, Bulshand TA, Galloway JN, Butler TJ, Rodhe H (1994) Steep declines in atmospheric base cations in regions of Europe and North America. Nature 367:351–354.

Hepler PK, Wayne RO (1985) Calcium and plant development. Annu Rev Plant Physiol 36:397–439.

Jacobson JS, Heller LI, L'Hirondelle SJ, Lassoie JP (1992) Phenology and cold tolerance of red spruce (*Picea rubens* Sarg.) seedlings exposed to sulfuric and nitric acid mist. Scand J For Res 7:331–344.

Johnson AH, Cook ER, Siccama TG (1988) Climate and red spruce growth and decline in the northern Appalachians. Proc Natl Acad Sci USA 85:5369–5373.

Johnson AH, DeHayes DH, Siccama TG (1996) Role of acid deposition in the decline of red spruce (*Picea rubens* Sarg.) in the montane forests of Northeastern USA. In: Raychudhuri SP, Maramorosch K (eds) *Forest Trees and Palms: Disease and Control*. Oxford and IBH, New Delhi, India, pp 49–71.

Johnson AH, Friedland AJ, Dushoff JG (1986) Recent and historic red spruce mortality: evidence of climatic influence. Water Air Soil Pollut 30:319–330.

Johnson DW, Fernandez IJ (1992) Soil mediated effects of atmospheric deposition on eastern U.S. spruce-fir forests. In: Eager C, Adams MB (eds) *The Ecology and Decline of Red Spruce in the Eastern United States*. Springer-Verlag, New York, pp 235–270.

Joslin JD, Wolfe MH (1988) Response of red spruce seedlings to changes in soil aluminum in six amended forest soil horizons. Can J For Res 18:1614–1623.

Joslin JD, Wolfe MH (1992) Red spruce soil solution chemistry and root distribution across a cloud water deposition gradient. Can J For Res 22: 893–904.

Joslin JD, McDuffie C, Brewer PF (1988) Acidic cloud water and cation loss from red spruce foliage. Water Air Soil Pollut 39:355–363.

Kellomaki S, Hanninen H, Kolstrom M (1995) Computations on frost damage to Scots pine under climatic warming in boreal conditions. Ecol Applications 5:42–52.

Klein RM, Perkins TD, Myers HL (1989) Nutrient status and winter hardiness of red spruce foliage. Can J For Res 19:754–758.

Krauchi N (1993) Potential impacts of a climate change on forest ecosystems. Eur J For Path 23:28–50.

Larcher W, Bauer H (1981) Ecological significance of resistance to low temperature. In: Lange OL, Nobel PS, Osmond CB, Ziegler H (eds) *Encyclopedia of Plant Physiology N.S. Vol 12A. Physiological Plant Ecology I.* Springer, Berlin, Germany, pp 403–437.

Lawrence GB, David MB, Shortle WC (1995) A new mechanism for calcium loss in forest-floor soils. Nature 378:162–165.

L'Hirondelle SJ, Jacobson JS, Lassoie JP (1992) Acid mist and nitrogen fertilization effects on growth, nitrate reductase activity, gas exchange, and frost hardiness of red spruce seedlings. New Phytol 121:611–622.

Likens GE, Driscoll CT, Buso DC (1996) Long-term effects of acid rain: response and recovery of a forest ecosystem. Science 272:244–246.

Lindberg SE, Lovett GM (1992) Deposition and forest canopy interactions of airborne sulfur: Results from the integrated forest study. Atmos Environ 26a:1477–1492.

Lorius C, Jouzel J, Raynaud D, Hansen J, LeTrent H (1990) The ice-core record: climate sensitivity and future greenhouse warming. Nature 347:7–12.

Lund AE, Livingston WH (1998) Freezing cycles enhance winter injury in *Picea rubens*. Tree Physiol 19:65–69.

MacCracken M, Cubasch U, Gates WL, Harvey LD, Hunt B, Katz R, Lorenz E, Manabe S, McAvaney B, McFarlane N, Meehl G, Meleshko V, Robock A, Stenchikov G, Stouffer R, Wang W-C, Washington W, Watts R, Zebiak S (1991) A Critical Appraisal of Model Simulations. In: Schlesinger MF (ed) *Greenhouse-Gas-Induced Climatic Change: A Critical Appraisal of Simulations and Observations.* Developments in Atmospheric Science 19. Elsevier, Amsterdam, The Netherlands, pp 583–591.

Manter DK, Livingston WH (1996) Influence of thawing rate and fungal infection by *Rhizosphaera kalkhoffii* on freezing injury in red spruce (*Picea rubens*) needles. Can J For Res 26:918–927.

Margolis HA (1989) Influence of short-term atmospheric CO_2 enrichment on growth, allocation patterns, and biochemistry of black spruce seedlings at different stages of development. Can J For Res 19:733–782.

Margolis HA, Vezina L-P (1990) Atmospheric CO_2 enrichment and the development of frost hardiness in containerized black spruce seedlings. Can J For Res 20:1392–1398.

Marschner H (1986) *Mineral Nutrition of Higher Plants.* Academic Press, New York.

McLaughlin JW, Fernandez IJ, Richards KJ (1996) Atmospheric deposition to a low-elevation spruce–fir forest, Maine, USA. J Environ Qual 25: 248–259.

McLaughlin SB, Kohut RJ (1992) The effects of atmospheric deposition and ozone on carbon allocation and associated physiological processes in red spruce. In: Eager C, Adams MB (eds) *The Ecology and Decline of Red Spruce in the Eastern United States.* Springer-Verlag, New York, pp 338–382.

McLaughlin SB, Tjoelker MG, Roy WK (1993) Acid deposition alters red spruce physiology: laboratory studies support field observations. Can J For Res 23: 380–386.

McLaughlin SB, Anderson CP, Hanson PJ, Tjoelker MG, Roy WK (1991) Increased dark respiration and calcium deficiency of red spruce in relation to acidic deposition at high-elevation southern Appalachian Mountain sites. Can J For Res 21:1234–1244.

McNulty SG, Aber JD, Newman SD (1996) Nitrogen saturation in a high elevation New England spruce–fir stand. For Ecol Manage 84:109–121.

Mohnen VA (1992) Atmospheric deposition and pollutant exposure of eastern U.S. forests. In: Eager C, Adams MB (eds) *The Ecology and Decline of Red Spruce in the Eastern United States.* Springer-Verlag, New York, pp 64–124.

Monroy AF, Sarban F, Dhinza RS (1993) Cold-induced changes in freezing tolerance, protein phosphorylation and gene expression: evidence for a role of calcium. Plant Physiol 102:1227–1235.

Morgenstern EK (1969) Winter drying of red spruce provenances related to introgressive hybridization with black spruce. Bi-month Res Notes. Can For Serv 25:34–36.

Palta JP, Levitt J, Stadlemann EJ (1977) Freezing injury in onion bulb cells. Plant Physiol 60:393–397.

Palta JP, Li PH (1978) Cell membrane properties in relation to freezing injury. In: Li PH, Sakai A (eds) *Plant Cold Hardiness and Freezing Stress.* Academic Press, London, England, pp 93–115.

Peart DR, Jones MB, Palmiotto PA (1991) Winter injury to red spruce at Mt. Moosilauke, NH. Can J For Res 21:1380–1389.

Perkins TD, Adams GT (1995) Rapid freezing induces winter injury symptomatology in red spruce foliage. Tree Physiol 15:259–266.

Perkins TD, Adams GT, Klein RM (1991) Desiccation or freezing? Mechanisms of winter injury to red spruce foliage. Am J Bot 78:1207–1217.

Perkins TD, Adams GT, Lawson S, Hemmerlein MT (1993) Cold tolerance and water content of current-year red spruce foliage over two winter seasons. Tree Physiol 13:119–129.

Perkins TD, Adams GT, Lawson ST, Schaberg PG, McNulty SG (2000) Long-term nitrogen fertilization increases winter injury in montane red spruce (*Picea rubens*) foliage. J Sustain For 10:200–205.

Pomeroy MK, Andrews CJ (1985) Effects of low temperature and calcium on survival and membrane properties of isolated winter wheat cells. Plant Physiol 78:484–488.

Ramanathan V (1988) The greenhouse theory of climate change: a test by an inadvertent global experiment. Science 240:293–299.

Sakai A, Larcher W (1987) *Frost Survival of Plants. Responses and Adaptation to Freezing Stress.* Springer-Verlag, New York.

Schaberg PG, Shane JB, Hawley GJ, Strimbeck GR, DeHayes DH, Cali PF, Donnelly JR (1996) Physiological changes in red spruce seedlings during a simulated winter thaw. Tree Physiol 16:567–574.

Schaberg PG, DeHayes DH, Hawley GJ, Strimbeck GR, Murakami PF, Cumming JR, Borer CH (2000a) Acid mist, soil Ca and Al treatments alter the mineral nutrition and physiology of red spruce. Tree Physiol 20:101–106.

Schaberg PG, Strimbeck GR, Hawley GJ, DeHayes DH, Shane JB, Murakami PF, PerkinsTD, Donnelly JR, Wong BL (2000b) Cold tolerance and photosystem function in a montane red spruce population: physiological relationships with foliar carbohydrates. J Sustain For 10:225–230.

Schaberg PG, Perkins TD, McNulty SG (1997) Effects of chronic low-level N additions on foliar elemental concentrations, morphology, and gas exchange of mature montane red spruce. Can J For Res 27:1622–1629.

Schaberg PG, Shane JB, Cali PF, Donnelly JR, Strimbeck GR (1998) Photosynthetic capacity of red spruce during winter. Tree Physiol 18:271–276.

Schaberg PG, Wilkinson RC, Shane JB, Donnelly JR, Cali PF (1995) Winter photosynthesis of red spruce from three Vermont seed sources. Tree Physiol 15:345–350.

Senser M, Beck E (1982) Frost resistance in spruce (*Picea abies* [L.] Karst.). V. Influence of photoperiod and temperature on the membrane lipids of needles. Z Pflanzenphysiol B 108:71–85.

Senser M, Beck E (1984) Correlation of chloroplast ultrastructure and membrane lipid composition to the different degrees of frost resistance achieved in leaves of spinach, ivy, and spruce. J Plant Physiol 117:41–55.

Sheen J (1996) Ca^{2+}-dependent protein kinases and stress signal transduction in plants. Science 274:1900–1902.

Sheppard LJ (1994) Causal mechanisms by which sulphate, nitrate and acidity influence frost hardiness in red spruce: review and hypothesis. New Phytol 127:69–82.

Sheppard LJ, Cape JN, Leith ID (1993) Acid mist affects dehardening, budburst, and shoot growth in red spruce. For Sci 39:680–691.

Sheppard LJ, Smith RI, Cannell MGR (1989) Frost hardiness of *Picea rubens* growing in spruce decline regions of the Appalachians. Tree Physiol 5:25–37.

Shortle WC, Smith KT (1988) Aluminum-induced calcium deficiency syndrome in declining red spruce. Science 240:1017–1018.

Snyder MC (1990) Seasonal patterns of carbohydrate reserves within red spruce seedlings in the Green Mountains of Vermont. Masters thesis, University of Vermont, Burlington, VT.

Steponkus PL (1990) Cold acclimation and freezing injury from a perspective of the plasma membrane. In: Katterman F (ed) *Environmental Injury to Plants*. Academic Press, New York, pp 1–16.

Strimbeck GR (1997) Cold tolerance and winter injury of montane red spruce. Ph.D. Dissertation, University of Vermont, Burlington, VT.

Strimbeck GR, Johnson AH, Vann DR (1993) Midwinter needle temperature and winter injury of montane red spruce. Tree Physiol 13:131–144.

Strimbeck GR, Schaberg PG, DeHayes DH, Shane JB, Hawley GJ (1995) Midwinter dehardening of montane red spruce foliage during a natural thaw. Can J For Res 25:2040–2044.

Strimbeck GR, Vann DR, Johnson AH (1991) *In situ* experimental freezing produces symptoms of winter injury in red spruce foliage. Tree Physiol 9: 359–367.

Thornton FC, Pier PA, McDuffie C (1990) Response of growth, photosynthesis, and mineral nutrition of red spruce seedlings to O_3 and acidic cloud deposition. Environ Exp Bot 30:313–323.

Tobi DR, Wargo PM, Bergdahl DR (1995) Growth response of red spruce after known periods of winter injury. Can J For Res 25:669–681.

Vann DR, Strimbeck GR, Johnson AH (1992) Effects of ambient levels of airborne chemicals on freezing resistance of red spruce foliage. For Ecol Manage 51:69–79.

Waite CE, DeHayes DH, Rebbeck J, Schier GA, Johnson AH (1994) The influence of elevated ozone on freezing tolerance of red spruce seedlings. New Phytol 126:327–335.

Wareing PF, Phillips IDJ (1981) *Growth and Differentiation in Plants*. Pergamon, New York.

White GJ (1996) Effects of chronic ammonium sulfate treatments on forest trees at the Bear Brook Watershed in Maine. Ph.D. Dissertation, University of Maine, Orno, ME.

White PS, Cogbill CV (1992) Spruce–fir forests of Eastern North America. In: Eager C, Adams MB (eds) *The Ecology and Decline of Red Spruce in the Eastern United States.* Springer-Verlag, New York, pp 3–39.

Wilkinson RC (1990) Effects of winter injury on basal area and height growth of 30-year-old red spruce from 12 provenances growing in northern New Hampshire. Can J For Res 20:1616–1622.

7. Tree Health and Physiology in a Changing Environment

Walter C. Shortle, Kevin T. Smith, Rakesh Minocha,
Subhash Minocha, Philip M. Wargo,
and Kristiina A. Vogt

A tree is a large, long-lived, perennial, compartmented, woody, shedding, walling plant. This definition is based on new tree biology concepts (Shigo, 1986a,b, 1991) and explains much about how mature trees function through their unique structure. When the tree begins its life, it is mostly leaf in mass (Fig. 7.1a). As a tree grows in stature, it becomes mostly stem in mass and the foliage represents only a few percent of the total mass. Roots remain relatively constant at about one-fifth the total mass as a tree grows from a small sapling to a mature standard in the forest canopy. Branches represent only a small fraction of total mass, which decreases over time, as older branches are shed. Also shed are leaves, roots, and outer bark. However, aging wood cannot be shed, but dies internally as sapwood is transformed into a core of protection wood, often called "heartwood" (Fig. 7.1b,c; Table 7.1).

Annual rings are formed each growing season in the wood of trees in the temperate zone (Fig. 7.1b). The rings can be used in dendrochronology and dendrochemistry as a record of a changing external environment. These rings are further compartmented into cells. Large, dead, thick-walled cells function for transport and support. Small, live, thin-walled cells store food and defend against the spread of infection (Table 7.1). The

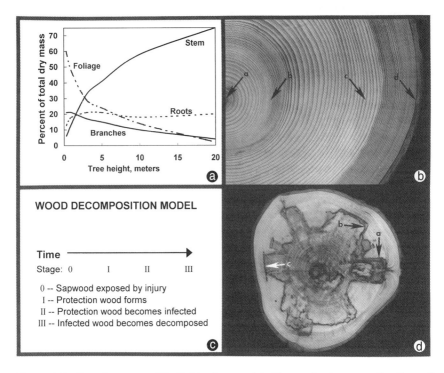

Figure 7.1. Development of individual trees. (a) Change in dry mass fraction of F = foliage, S = stem, R = roots, B = branches associated with increasing height growth in red spruce. (b) Record of diameter growth seen as annual growth rings in red spruce from pith (a) to cambium (d); see also outer band of living sapwood and inner core of protection wood. Note that segment a–b contains wood formation during the sapling stage of tree growth, b–c contains wood formed during the pole stage, and c–d wood formed as a mature standard. (c) Wood decomposition model (see Table 1). (d) Tree survival model illustrated in red maple, a = protective zone, b = protective zone incorporated into protective wood, and c = barrier zone (see Table 1).

small cells make up over 90% of the cell number, but only 10% of the mass. Most of the mass is composed of large, dead, woody cells that give stemwood its commercial value. When the live cell network dies due to age, injury, or infection, a protective tissue is formed to reduce the internal rate of infection and decomposition following exposure to the external environment by injury. The protective tissue may be of the protection wood type, which forms in the core of the stem, or it may be a protective zone, which forms in the peripheral sapwood to stop the spread of infection into the outer living wood, vascular cambial zone, and inner living bark (Fig. 7.1c,d; see Table 7.1). Infection in living trees will spread at highly variable rates through the dead core of wood exposed by

Table 7.1. Summary of Properties of Major Types of Wood Formed in the Stem of Living Trees

Wood type	Derivation	Composition	Characterization
Sapwood	Cell division by the vascular cambium	Mostly live cells	Stores food Defends against infection Conducts sap Supports crown
Protection wood	Forms from sapwood	Mostly dead cells	Preservatives formed from stored food Reduced rate of infection Dehydrated and deionized Partial support of crown
Infected wood	Forms in protection wood	Live cells of bacteria and fungi	Preservatives degraded Onset of decay process Ionized and rehydrated Partial loss of support
Rotted wood	Final stage of infection	Complex living community	Preservatives lost Cell wall structure lost Retains water and nutrients Humus-like
Protective zones	Forms from sapwood as a column boundary layer, or From the vascular cambium as a barrier zone and woundwood	Dead cells: hypersensitive reaction	Layer of highly preserved cells Blocks infection spread Low permeability High mineral content Results in tree survival, not disease resistance

shedding and injury. Trees may live a very long time with internal infections due to the formation of protective "walls" of altered tissue by a process called compartmentalization.

The use of dated stemwood tissue to study dendrochemistry as a record of external change requires knowledge about the internal development of sapwood, protection wood, infected wood, decomposing wood, and protective zones because these processes can alter the pattern of external chemical change recorded in functioning sapwood. Furthermore, differences in anatomical structure associated with the sapling, pole, and mature stages of trees (Fig. 7.1b) result in a greater density of binding sites in the core of the tree than in the periphery. The consequence of this structural change is a decreasing concentration of exchangeable divalent base cations from the core to the periphery under constant uptake conditions.

Taking a standard stocking chart for northern conifers as a model of how populations of trees develop (Fig. 7.2a), the number of trees per unit

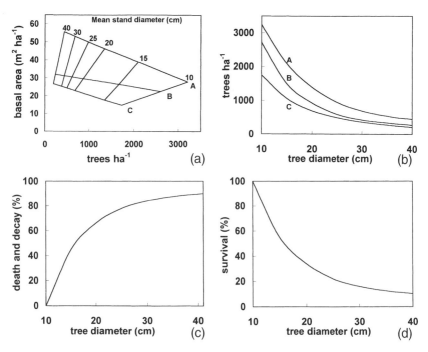

Figure 7.2. Development of tree populations as derived from a forest stocking table. (a) Standard stocking table for northern coniferous species. The A-line indicates overstocked stand, B-line indicates fully stocked stand, and C-line indicates understocked stand. (b) Decreased stand density associated with increased stem diameter (dbh). (c) Cumulative frequency of dead and decaying trees as trees progress from saplings (<10 cm dbh), poles (10 to 30 cm dbh), and mature standards (>30 cm dbh). (d) Survival frequency of individual trees through stand development. The time scale for stand development is highly variable from several decades to centuries.

area decreases at all stocking levels as trees increase in size (Fig. 7.2b). The greatest rate of loss of trees takes place during the early pole stage (trees 10 to 20 cm dbh). Losses continue through the late pole stage (20 to 30 cm dbh), but at a decreasing rate. Further decreases in mature standard trees (>30 cm dbh) will occur when various disturbances take place in the forest. Pests, pathogens, and declines caused by multiple stressors can cause tree loss to occur at rates observed in earlier stages of forest development.

By the time the average tree diameter reaches 40 cm dbh, the death and decay of 90% of the large sapling population (10 cm dbh) will have occurred (Fig. 7.2c). As in the individual tree, death and decay always accompany population growth. Forests are sustained through the survival to maturity of 10% of the initial sapling population (Fig. 7.2d). Survival occurs because of a highly developed defense system in these remarkable plants that sustain the forests of the terrestrial ecosystem. Trees cover only 10% of the land area, but produce 90% of the terrestrial biomass.

Human activities of forest clearing, tree harvesting, and air pollution have added a new dimension to the delicate balance of life and death that sustains forest life. Will trees adapt to modern environmental change to sustain forest health and productivity?

We have investigated the response of the northern spruce forests to environmental change during the 20th century, which has experienced marked increases in acidic deposition. We have looked for records of changing tree growth and changes in the base cation nutrition of the root zone in the stemwood of mature trees. We have related the current status of tree stress from below ground change to biochemical markers in the foliage. What have we learned?

Tree Growth and Dendrochronology

Assessments of tree growth are part of most investigations of tree health and environmental change. Increases in the number, girth, and extent of stems, branches, and roots occur as a tradeoff. Tree energy is partitioned between growth and defensive investments (Herms and Mattson, 1992; Loehle, 1988). A growth investment such as annual ring width is far easier to measure than any defensive investment. Constitutive defensive investments include the programmed cell death and chemical enrichment associated with the transformation from sapwood to heartwood. Induced defensive investments include the enrichment of wound-altered sapwood with phenolics and resins (Hillis, 1987; Smith, 1997). Enhanced wood production can be an induced defensive feature through the formation of abnormally thick woundwood or "callus" ribs near a wound (Mattheck et al., 1992).

Dendrochronology, the interpretation of patterns of precisely dated tree rings, can provide a reliable temporal record of tree growth as affected by environmental change. The decline of red spruce in the northeastern U.S. prompted a detailed application of dendrochronology to tree health and forest disturbance (Cook and Zedaker, 1992; Johnson et al., 1995). Primarily through research on climate and the dating of wooden cultural objects, basic principles of dendrochronology have been developed (Fritts, 1976). With some adaptation, these principles may be applied to investigations of tree health and physiology in a changing environment (Table 7.2).

The principle of integrative growth states that tree-ring characteristics, such as width and density, are a composite response to intrinsic and extrinsic factors. Intrinsic factors include the genetic capacity for growth of the individual tree, provenance, and species. Extrinsic factors that limit

Table 7.2. Revised Principles of Dendrochronology[a]

1.	Integrative growth
2.	Limiting factors
3.	Range sensitivity
4.	Site selection
5.	Crossdating
6.	Uniform linkage
7.	Signal and noise

[a] Modified from Fritts (1976).

growth include competition from other plants, biotic and abiotic stressors, edaphic conditions, and climate. This integration is rarely linear and additive, although it may be approximated with a linear model (Van Deusen, 1989).

To characterize growth in red spruce forest stands across the northeastern United States, we extracted two core samples from each of 36 trees at nine locations (Fig. 7.3). Although sampling was restricted to apparently healthy, canopy-dominant individuals, trees varied with respect to age and canopy crowding. The measured ring width series

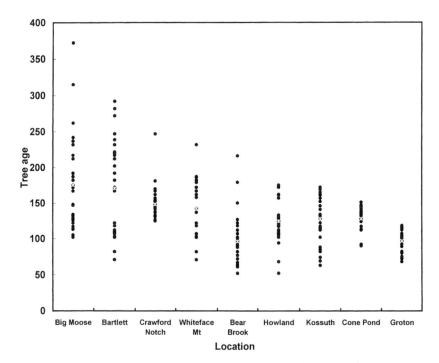

Figure 7.3. Tree age at breast height (solid circles) and location means (white crosses) for bored red spruce trees at the sampled locations across the northeastern U.S.

frequently appeared quite different from one another in general outline (Fig. 7.4a,b). Ring series were analyzed by two separate processes to examine (1) the high-frequency variation that may be attributable to annual climate variation and (2) the long-term growth trend.

In order to remove the medium- and long-term growth trend and to maintain as much as possible of the annual growth variation that may contain a year-to-year climate signal, the individual ring series were standardized. For standardization, each ring width series was fitted with a trend line calculated as a cubic spline (50% frequency cut-off of 30 years) using ARSTAN software (Cook, 1985; Holmes et al., 1986) (Fig. 7.4a,b). A standard index series was calculated by dividing the measured ring width by the trend line value for each year.

Patterns among index series (Fig. 7.4c) are more similar than they are among ring width series (Figs. 7.4a,b). The averaging of index series

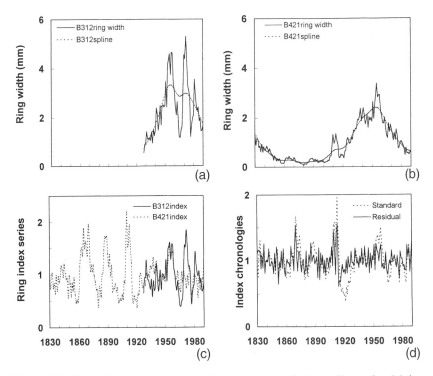

Figure 7.4. Tree ring data from red spruce sampled at Kossuth, Maine. (a,b) Splines representing individual growth trends were fitted to ring width series. (c) For dendroclimatic analysis, dimensionless index series were calculated by dividing the ring width by the trend for each year. (d) Index series were averaged to produce a standard chronology for each location. Each index series was also fitted with an autoregressive model and the residuals averaged to produce a residual chronology.

resulted in a standard index chronology for each location (Fig. 7.4d). One property of ring width series, index chronologies, and most naturally occurring time series is autocorrelation. Autocorrelation is the property that each value is dependent in part on the value for the preceding year or years. Given the physiological carryover of tree energy reserves, foliage production and retention (in the case of conifers), etc. autocorrelation may be expected. Autocorrelation violates the underlying assumption of independence for regression analysis. Additionally, as index chronologies may be more autocorrelated than climatic series, autocorrelation in tree rings may obscure sensitivity to climate. To reduce this effect, each index series was fitted with an autoregressive model. The mean of the residuals produced from autoregressive modeling was used to produce a residual chronology for each forest location. Although similar through most of the time period, the residual chronologies did diverge from the standard chronologies (Fig. 7.4d).

The principle of limiting factors states that although growth is integrative, growth is restricted by the essential factor or condition that is most limiting in supply. Ring width may be limited by damage to shoots caused by winter injury in the previous year or a shortened growing season due to adverse weather conditions. This principle is analogous to the "rate-limiting step" of chemical reactions, that the overall rate of a multi-part reaction is determined by the rate of the slowest component step. Limiting factors in common throughout a stand can result in the formation of "pointer years," an unusually narrow ring or series of rings that aids in the synchronization or crossdating of tree-ring series.

To determine the overall growth trend, tree-ring series were partitioned by location and age class. Bootstrap confidence intervals were calculated for mean ring widths to indicate variation about the mean (Cook, 1990). Trends in certain combinations of tree species and forest locations produce ring series with identifiable characteristics. For example, ring series of red spruce from Maine frequently contain growth suppression about 1920 (Fig. 7.5, plots for Howland, Kossuth, and Bear Brook). This suppression was likely due to defoliation by the spruce budworm followed by decreased availability of carbohydrate for growth. As this form of defoliation primarily takes place in the bud, and spruce typically retains functional needles for several years, the period of growth suppression can persist beyond the actual outbreak period. The general pattern for all age classes of red spruce sampled across the northeastern United States includes a peak in ring width about 1960 followed by a decline in radial growth (see Fig. 7.5).

The principle of range sensitivity states that trees sampled at or near the edge of their range are likely to contain a stronger common signal, especially of climate. Range in this case may be determined by latitude, elevation, edaphic, or climatic characteristics. This concept is related to

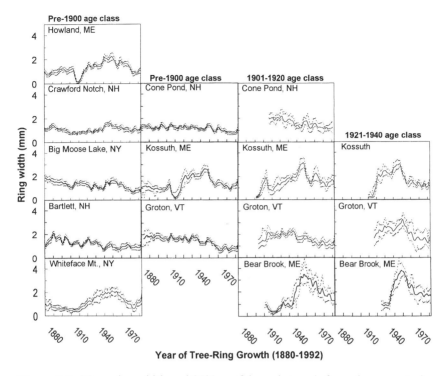

Figure 7.5. Mean ring width and 95% confidence intervals for red spruce at nine locations and various age classes. For each plot, the *x*-axis represents the years of growth from 1880–1993. The *y*-axis represents tree-ring width.

but not identical with ecological amplitude. Ecological amplitude is the capacity of a tree species to grow across a range of environmental conditions (Fritts, 1976). Ecological amplitude varies among species, from narrow, as in the case of coastal redwood, to broad, as in the case of red maple. Despite the breadth of the ecological amplitude, ring series from trees at the edge of the range are likely to contain a stronger common signal of environmental events.

The principle of site selection states that sampling sites can be identified as being likely to produce tree-ring series sensitive to the environmental variable being investigated. Investigations of tree health and forest decline may sample stands at various distances from the center of disturbance and decline (Bartholomay et al., 1997; McLaughlin et al., 1987).

The principle of crossdating is the process of matching patterns among several to many ring series. By including series with a known sampling date, the exact year of formation may be assigned to a particular ring. Crossdating is the rigorous process that moved tree-ring analysis from "ring counting" to dendrochronology. Crossdating is aided by the presence of a recurring environmental factor, such as drought or extreme cold, that periodically suppresses growth throughout the stand or region. In this manner, crossdating relies on common patterns of high frequency variation in ring width. Linkage of overlapping series contained in living trees, wooden structures, and subfossil remnants permits the construction of long tree-ring chronologies (Stokes and Smiley, 1968). Unfortunately, even long and perfectly constructed chronologies do not provide information on trends longer than the length of individual segments (the "segment length curse") (Cook et al., 1995).

Crossdating wooden timbers contained in old churches to existing chronologies allowed for the estimation of the timing of spruce budworm outbreaks for the 1700s and 1800s in Quebec (Krause, 1997). Krause (1997) constructed floating index chronologies (tree-ring chronologies not fixed to calendar dates) from structural members of buildings constructed of white spruce, a host species for the budworm, white cedar, a nonhost species, and other tree species. Crossdating with a local, recent spruce chronology allowed the assignment of calendar dates to both the white spruce and white cedar chronologies. Inspection of the crossdated spruce and cedar chronologies indicated periods of acute growth suppression in the spruce host chronologies that were not present in the cedar nonhost chronology. These periods of suppression were interpreted as indicators of episodic spruce budworm infestation (Krause, 1997).

Use of tree rings as a proxy record of climate or environmental disturbance depends on uniformity of linkage between external conditions and tree biology. This concept is an extension of the uniformitarian principle developed by 18th century naturalist James Hutton to describe geological processes. Briefly stated, "the present is the key to the past," those currently observable natural processes are the same processes that occurred in the past. Although this may seem to be self-evident, dendrochronology in a changing environment provides a cautionary note to the uncritical application of geological uniformitarianism to growth patterns of trees.

The decline of red spruce is still controversial with respect to causes and implications, although the dendrochronology of spruce decline has been well studied and reviewed (Cook and Zedaker, 1992; Johnson et al., 1995). The linkage at some locations between red spruce growth and climate markedly changed at about 1960 (Van Deusen, 1990). We tested this linkage at nine of our study locations. For all nine locations, significant regression models (nominal $P < 0.95$) of residual ring index chronologies were calibrated to climate variables for the 1896 to 1940. Although the true probability value was likely to have been less than the nominal value due to the multiple regression process, the probability values for the correlation coefficients determined for the verification periods could not have been similarly inflated. The portion of the index chronology for 1910 to 1929 was not included in the calibration of the Howland model due to previously discussed growth suppression. Each model was verified for two periods, 1941 to 1960 and 1971 to 1990. The endpoints for the verification period were chosen to test periods prior to and following the 1960s, the time of most rapid increases in environmental perturbation attributed to acidic atmospheric deposition. This procedure allowed model verification using independent data not used in the construction of the model. Significant models that related growth to temperature were used to predict ring indices for the verification periods. Total monthly precipitation had no significant effect on any of the constructed models. The predicted indices derived from the regression models were significantly correlated ($P \leq 0.05$) to actual indices for the 1941 to 1960 verification period at five locations (Fig. 7.6a–e). Actual and predicted indices were not significantly correlated for the 1941 to 1960 verification period at four locations (not shown). The lack of correlation for these four sites indicated that either there was no predictive relationship between index chronologies and mean monthly temperatures or that the nature of the predictive relationship differed between the calibration and the verification periods. For the 1971 to 1990 verification period, predicted indices were not correlated to the actual indices at any location. This suggests that at least five locations, the relationship of growth to temperature for the 1941 to 1960 verification period was similar to the 1896 to 1940 calibration period. However, growth in the 1971 to 1990 verification period was no longer responding to temperature as growth responded to temperature in the 1896 to 1940 calibration period.

In an earlier investigation of spruce decline, concentrating especially on montane trees, Johnson et al. (1988) found a similarly timed change in climate dependence. They hypothesized that this change may have been due to a decade of unusually cold winters that killed foliage and buds and shifted carbon allocation away from wood production and into producing new shoots and foliage (Johnson et al., 1988). In contrast with Johnson et al. (1988), monthly climate variables with a significant effect on growth were specific to each sampled location (see Fig. 7.3). This may reflect that trees in our investigation were further removed from the edge of their elevation range and may have been affected differently by the factors causing the change in climate dependence. Other changes in climate dependence dated to about 1960 have been reported for red pine and Norway spruce (Leblanc et al., 1987) and shortleaf pine (Grissino-Mayer and Butler, 1993).

As a matter of faith, uniformitarianism is quite appealing. When all else is equal, conditions that limit growth in the present are likely to have limited growth in the past. However, environmental change and associated changes in tree energy allocation could change the growth response of trees to climate, as has been shown in white pine exposed to atmospheric ozone (Bartholomay et al., 1997). This is a cautionary note for dendroclimatic reconstructions based on the premise that trees had

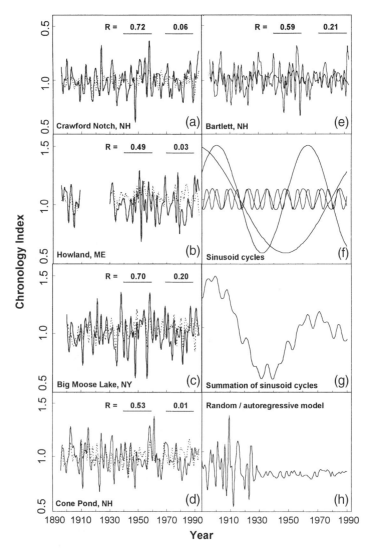

Figure 7.6. Red spruce and model growth series. (a–e) Residual index chronologies were compared with chronologies estimated by dendroclimatic models.
Correlation coefficients (R) for the observed and estimated chronologies for the
1941 to 1960 and 1971 to 1990 verification periods are listed. Correlations were
significant for all comparisons in the 1941 to 1960 period ($P \geq 0.95$). No
significant correlations were observed for comparisons in the 1971 to 1990 period.
Sinusoid waves (f) and their sum (g) approximate the appearance of tree ring data
(after Reams and Huso, 1993). (h) An autoregressive process acting on random
values assigned to 1890 to 1930 produces a smooth periodic series resembling
series derived from tree rings.

the same response to climate in the past as trees in the present (Smith et al., 1999). Fig. 7.6a–e argues against the acceptance of simple uniformitarianism.

The principle of signal and noise arises from the integrative growth of tree rings and the variety of research questions to which the tree-ring record is applied. The principle is sometimes stated as the replication principal, that multiple observations within and among trees will remove individual growth responses and better represent growth of the stand or region. However, even with adequate replication, the maximization of the desired information (signal) still needs to be separated from the information not believed to be useful (noise) in answering the research question.

An individual tree-ring series may contain trends, for example, of low frequency due to tree age and canopy closure, medium frequency due to local gap formation and recovery, and high frequency due to weather. Which of these trends is signal and which is noise is entirely dependent on the research question. Year-to-year variation can be maximized by detrending (Fig. 7.4a–d) or minimized through averaging by tree age class (see Fig. 7.5).

An individual tree-ring series typically contains cyclic as well as unique peaks and valleys. Model cyclic process can be represented by sinusoid waves of specific frequency and amplitude (Fig. 7.6f). The summation of these cyclic processes can yield a pattern suggestive of ring series (Reams et al., 1993) (Fig. 7.6g). Numerical techniques such as the Kalman filter (Van Deusen, 1990) may be able to increase our ability to precisely identify the timing of the change in climate dependence. However, purely mathematical approaches are not likely to identify the biological basis for growth cycles. Robust interpretation of ring patterns will require coming to grips with the mechanisms of the underlying cyclicity. Any suggested linking mechanism is likely to still leave doubt. Successive random events, entering autoperiodic processes such as ring formation, can themselves result in cycles or regular periodicity (Slutzky, 1937). In Fig. 7.6h, y-values for the 1890 to 1930 period were randomly drawn from a normal distribution. The y-values for 1931 to 1990 were calculated as a first-order autoregressive process seeded from the 1890 to 1930 period. The modeled values (1931 to 1990) quickly dampened the wide variation of the earlier, random portion of the series and appear to contain some pattern. Consequently, random series acted on by an autoregressive process can form the basis of an apparently periodic series.

Dendrochronology provides a record of tree growth and current and past environmental change. Interpreting that record requires a clearly specified hypothesis that may not be obvious from the record itself. The dendrochronology of red spruce indicates that across the northeastern U.S., growth has declined since the 1960s (Cook and Zedaker, 1992). The growth decline is apparent at all elevations and for all age classes. The change in climate dependence of red spruce and other species following 1960 may be of greater significance than the growth decline itself. These results are consistent with a widespread, fundamental change in the response of trees to the environment. The potential relationship of this change to perturbations in soil chemistry, land use patterns, and overall stress loading is unclear. Defining that relationship of biology and environment is made more difficult through the presence of natural cyclic processes in growth and climate. Identifying the cyclic processes is itself hindered through the natural integration of cycles of various frequencies and amplitudes and the generation of cyclic processes through the autoregressive process itself.

The body of principles for dendrochronology will continue to evolve. As more researchers apply the concepts to diverse problems, both the limitations and strengths of the concepts will become more apparent. This emphasis on the conceptual structure of the process should enhance the power of the tree-ring record to identify and explain environmental change.

Dendrochemistry: Change in Cation Mobility

Dendrochemistry is the interpretation of radial trends in the chemistry of dated wood. Dendrochemistry is based on the chemical relationship between the environment external to the tree and to the wood active in water conduction. For dendrochemistry, increment cores or stem sections are dated and samples of wood formed at different time periods are researched and analyzed. The application of dendrochemistry to environmental change requires distinguishing between the chemical characteristics of internal processes of maturation, defense, and infection from external environmental conditions (Smith and Shortle, 1996).

Process models predicted a mobilization of essential base cations in response to acidic deposition (Shortle and Smith, 1988; Lawrence et al., 1995). A signal of the mobilization of the essential bases calcium (Ca) and magnesium (Mg) appears in the dendrochemical record of red spruce (Shortle et al., 1997). Interpreting this signal of potential mobilization requires an understanding of transport processes of Ca and Mg. The binding of divalent Ca and Mg to ion exchange sites in the wood cell wall system follows the characteristics of a Donnan equilibrium (Momoshima and Bondietti, 1990). In brief, the amount of Ca and Mg in wood is a function of pH, the peak concentration of Ca and Mg in the surrounding sap, and the number of binding sites available for exchange. Under uniform soil and sap conditions, Ca and Mg concentrations tend to decrease with increasing radial distance from the pith, due to decreasing numbers of available binding sites (Momoshima and Bondietti, 1990).

In a dendrochemical analysis of red spruce across the northeastern U.S. (Shortle et al., 1997), wood formed in the 1960s was more frequently enriched in Ca (Fig. 7.7a) and Mg (Fig. 7.7b) than at any other time over the past eight decades ($P < 0.05$). The enrichment frequency of 28% for Ca and 52% for Mg in wood formed from 1961 to 1970 is remarkable because

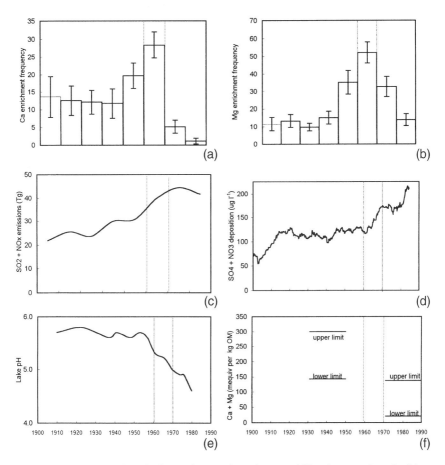

Figure 7.7. Dendrochemical markers of cation mobilization and coincident environmental changes. The vertical, dashed lines mark the years 1960 and 1970. (a) Percent frequency of Ca enrichment by decade. For example, the 1961 to 1970 bar indicates that the mean frequency of Ca enrichment relative to the preceeding decade was 28% across all sample locations (±SE). (b) Enrichment frequency for Mg was calculated as for Ca in (a). (c) Combined emissions of SO_2 and NO_x for the United States (NAPAP, 1993). (d) Combined deposition of nonmarine SO_4 and NO_3 in ice core 20D collected from south central Greenland (Mayewski et al., 1986). (e) Water pH at Big Moose Lake in the southwest Adirondacks of NY. Acidity was estimated through stratigraphy of diatoms in sediments (Charles, 1984). (f) Historical range of concentration of Ca and Mg in the forest floor of the northeastern U.S. according to available reports (Shortle and Bondietti, 1992).

the enrichment indicates a perturbation in the equilibrium between soil chemistry, sap chemistry, and cation binding sites in stemwood. The observed Ca and Mg enrichment is also remarkable because of its regional nature.

What internal or external factors could cause the enrichment in Ca and Mg for wood formed in the 1960s? Internal infection processes and the tree response to injury and infection can greatly alter Mg and Ca concentration in the wood of living trees. However, the careful screening of both candidate trees and cores argue against that explanation. In some tree species, especially those (unlike red spruce) that contain a highly colored heartwood, heartwood formation can affect wood chemistry. Simply because of tree age and the maturation process of red spruce, 31 of the 40 (78%) marked sapwood–heartwood boundaries occurred in wood formed in 1961 to 1970 (Shortle et al., 1997). However, the sapwood–heartwood boundary coincided with peak Ca enrichment in only 22 of the 40 cores (55%). We suggest that if Ca enrichment was a constitutive feature of heartwood formation in red spruce, essentially all, rather than the observed 55%, of the peak Ca enrichment would coincide with the sapwood–heartwood boundary.

The pattern of stemwood enrichment in Ca and Mg is consistent with the hypothesis that Ca and Mg were mobilized in the soil as acidic deposition increased, and transported in unusually high concentrations to stemwood via root uptake and sap transport (Shortle et al., 1997). This temporary increase in binding of Ca and Mg in wood formed in 1961 to 1970 coincided with (1) increased sulfur dioxide (SO_2) and oxides of nitrogen (NO_x) emissions (NAPAP, 1993; Fig. 7.7c), (2) increased deposition of sulfate (SO_4) and nitrate (NO_3) in the Greenland ice sheet (Mayewski et al., 1986; Fig. 7.7d), (3) decreased pH of lake water in the Adirondacks (Charles, 1984; Fig. 7.7e), and (4) reduced concentrations of Ca and Mg in the forest floor (Shortle and Bondietti, 1992; Fig. 7.7f).

If further reductions in the concentration of Ca and Mg from the forest floor were due to exchange of hydrogen (H) for Ca, either from plant growth (Johnson and Anderson, 1994) or from acidic deposition, then the concentration of exchangeable Ca and Mg ($[Ca + Mg]_{ex}$) would be negatively correlated with the concentration of exchangeable H ($[H]_{ex}$). Across our 12 study locations (Shortle et al., 1997), soil analysis indicated that $[Ca + Mg]_{ex}$ was not significantly related to $[H]_{ex}$ for untransformed or $\log_e$ transformed data from the forest floor ($P = 0.12$) or mineral soil ($P = 0.70$) (Fig. 7.8a). However, $\log_e [Ca + Mg]_{ex}$ was significantly and negatively correlated to $\log_e [Al]_{ex}$ in the forest floor, the principle rooting zone of red spruce (Fig. 7.8b, $P < 0.001$). Lawrence et al. (1995) have attributed this negative correlation between exchangeable Ca and exchangeable aluminum (Al) in the forest floor to increased translocation of Al from the mineral soil, where acidic deposition has increased Al dissolution. Acid-extractable Al in the forest floor increased over the past

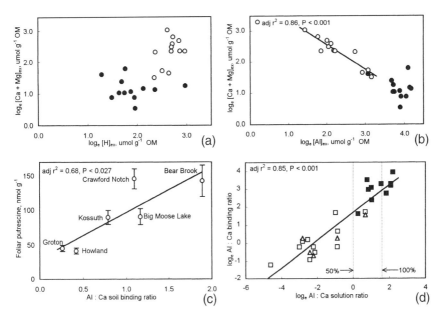

Figure 7.8. Relationships among major cations in forest soils and soil solutions (David and Lawrence, 1996; Lawrence et al., 1995) and the stress marker, putrescine. (a) Regression analysis indicates no significant relationship between $\log_e [Ca + Mg]_{ex}$ to $\log_e [H]_{ex}$ either in mineral soil (closed circles, $P < 0.70$) or forest floor (open circles, $P < 0.12$). (b) Regression analysis indicates a significant relationship between $\log_e [Ca + Mg]_{ex}$ and $\log_e [Al]_{ex}$ in the forest floor (open circles, $P < 0.001$) but no significant relationship in mineral soil (closed circles, $P < 0.36$). (c) Regression analysis indicates that concentrations of putrescine in red spruce foliage are significantly related to the Al/Ca binding ratio (molar charge ratio of exchangeable Al to exchangeable Ca) of the forest floor (open circles, $P < 0.027$). (d) Regression analysis indicates a significant relationship between $\log_e$ Al/Ca ratio in the forest floor soil solution to the $\log_e$ Al/Ca binding ratio in the forest floor (open squares) and fine root tips (triangles), and between $\log_e$ Al/Ca ratio in the mineral soil solution to the $\log_e$ Al/Ca binding ratio in mineral soil (closed squares, $P < 0.001$). Vertical, dotted lines indicate the suggested 50 and 100% risk level of forest damage due to adverse Al/Ca ratios.

two decades at the Hubbard Brook Experimental Forest, and ratios of Al to Ca in mineral soil solutions (but not forest floor solutions) were strongly correlated with exchangeable Al content in the forest floor. Mineral dissolution of Al in the forest floor was ruled out as a cause for this relation because mineral Al concentrations were unrelated to exchangeable Al concentrations (Lawrence et al., 1995). Leaching of Ca by SO_4^{2-} has directly contributed to depletion of forest floor Ca, but mobilization of Al in the mineral soil has also indirectly contributed to this process by (1) reducing uptake of Ca from the mineral soil (2) providing a supply of reactive Al that exchanges with Ca in the forest floor, enabling the leaching of Ca from the forest floor, and (3) increasing Al saturation so that the number of exchange sites available for adsorbing added Ca is reduced.

The foliar concentration of the stress marker putrescine in putatively healthy red spruce trees was significantly correlated ($P < 0.02$) to the Al/Ca binding ratio of the forest floor (Fig. 7.8c). This indicates that even healthy-appearing trees are stressed due to adverse Al/Ca ratios. Regression analysis indicates that $\log_e$ Al/Ca binding ratio was strongly related with the $\log_e$ Al/Ca ratio in the soil solution of both the forest floor and the mineral soil (Fig. 7.8d, $P < 0.001$). Nonwoody absorbing root tips in the forest floor appear to be binding Al in the same manner as the surrounding soil, at the expense of Ca and Mg (Fig. 7.8d; Shortle and Bondietti, 1992; Smith et al., 1995). The Al/Ca ratio in the soil solution (expressed as the reciprocal Ca/Al ratio) was described as an ecological indicator of approximate thresholds beyond which the risk of forest damage increased due to Al stress and nutrient imbalances (Cronan and Grigal, 1995). The risk of forest damage was assessed at 50% when the soil solution Al/Ca ratio was 2 and at 100% when the ratio was 5. At all locations in the regional red spruce study, Al/Ca ratios of mineral soil solutions exceeded the 50% risk level and half of the locations were above the 100% risk level (Shortle et al., 1997). Most forest floor solution ratios were below the 50% risk level and trees appeared generally healthy. The forest floor solution ratio was above the 50% risk level at only one location, Mt. Abraham, Vermont, where half of the canopy spruce were dead or dying. Because we sampled the most healthy portions of the red spruce forest at each location, the area we investigated at Whiteface Mountain, New York, did not have a particularly high Al/Ca binding ratio (0.8) in the forest floor. This particular area contained neither evidence of unusual mortality or current tree decline.

These results support and expand upon the model that Robert Hartig proposed in 1897 to relate sulfate deposition to damage ("smoke injury") of spruce–fir forests (Hartig, 1897). Hartig stated that essential base cations would be lost from the root zone of trees subjected to high inputs of sulfate. We suggest that mobilization of Al has accelerated the loss of essential bases. We make this suggestion because of the evidence of a general mobilization of base cations in the mid-20th century and the dynamics of Al behavior as described above. As Hartig described, the loss of essential bases would suppress root growth. Roots not able to grow into new areas of the forest soil would eventually starve and die, making the belowground parts of trees susceptible to attack by facultative pathogens. Aboveground parts would respond to the loss of roots by shedding needles. Most trees would survive but grow poorly, be more frequently infected by facultative pathogens and insects, and be less tolerant of abiotic stressors such as drought.

In conclusion, the patterns of the dendrochemical marker of Ca and Mg enrichment in red spruce stemwood is inconsistent with a stable chemical environment in forest soil for the 20th century in the northeastern United States. The enrichment in wood formed in the 1960s is consistent with mobilization of base cations in the mid-20th century. This hypothetical mobilization coincides with increases in the atmospheric emission and deposition of nitrates and sulfates. The biochemical marker, foliar putrescine concentration, indicated that even apparently healthy trees are under stress due to adverse Al/Ca ratios in the forest floor. The hypothetical mobilization of Al, which would result in greater Al/Ca ratios, would be enhanced by acidic deposition (Cronan and Goldstein, 1989; Driscoll et al., 1984; Lawrence and Fernandez, 1991). Spruce stands most vulnerable to multiple stressors, such as those at high elevations, have already been damaged. We anticipate that red spruce and other tree species growing under less harsh conditions will become stressed and more vulnerable with continued depletion of essential base cations and mobilization of Al.

Biochemical Markers: Indicators of Changing Stress

Trees, by their perennial nature, are constantly exposed to a variety of biological and environmental stresses. Plants, including forest trees, respond to these stresses with physiological and developmental changes that lead to either an avoidance or a tolerance of the stress factors. Depending upon the duration of exposure, plants employ either short-term responses that overcome stress or long-term adaptation to that stress by long-lived epigenetic changes. Some species have developed genetic mechanisms to adapt to harsh and variable environments. The complex response to stress begins with the perception of stress which, through a series of signal transduction pathways, leads to changes at the cellular, physiological, and developmental levels (Vernon et al., 1993). While some physiological responses may be specific to the stressor, other responses are generic. Also, whereas some of these responses may be localized to the tissue or organ that was stressed, others may be systemic responses of the whole tree. In all cases, stress-induced genes are activated, some of which lead to metabolic changes that can be used as physiological and biochemical markers of stress.

Whereas some of these metabolic changes are quantitative, for example, an increase or decrease in cell size or the cellular content of one or more metabolites, others are qualitative, for example, the presence or absence of certain metabolites or proteins. The first step in alleviating stress in plants is to detect and characterize the stress response at an early stage. Certain forms of stress (e.g., drought) occur intermittently and the stress periods are followed by recovery periods. Other forms of stress are more constant and require different adaptive mechanisms. For example, plants grown in high salinity develop mechanisms to exclude sodium (Na^+) or accumulate potassium (K^+) ions. On the other hand, plants exposed intermittently to salt may accumulate organic solutes, such as proline and polyamines, to stabilize cell membranes and achieve osmotic balance. In most cases these changes persist for some time even after the stress is temporarily removed. Thus, if these changes could be detected early, they could be useful indicators or markers of stress in visually nonsymptomatic trees.

The need to assess a number of early physiological and biochemical markers of stress in trees is obvious. With slow and long-term changes in the soil and environmental factors due to human activity, forest trees have been affected in many ways, leading to loss of productivity in the forest. Unfortunately, in most cases, symptoms of stress appear too late to manage or treat trees to reverse the effects of stress. Thus, the availability of markers that can assess the current status of stress in apparently healthy trees in a forest is crucial for planning a potential treatment or a management practice for either alleviating the deleterious effects of the stress or removing the cause of stress. To be useful, a physiological or biochemical marker should be a part of an adaptive mechanism and not a short-term response to applied stress. A useful marker is easily detected, quantified, and able to reliably estimate the health status of a stand.

The process to identify such markers involves (1) a comparison across a gradient of environmental change (e.g., acid precipitation, urban pollution, increased fertilization, etc.) and (2) experiments to study the effects of specific stressors on plant metabolism. While a few studies of the former type have been reported with forest trees, most of the experimental work has been done with either crop plants or young seedlings and tissue cultures of tree species.

Osmotic stress, salt stress, mineral deficiency, and increased solubilization of Al are some of the soil factors that have changed in recent years and have affected forest productivity. Among the important atmospheric factors are increased O_3, NO_x, sulfur oxides (SO_x), and particulate deposition. All of these factors can contribute to a decline in tree health and forest productivity. Metabolic changes associated with soil and environmental factors are changes in photosynthetic rates, foliage fluorescence characteristics, organic metabolites or inorganic ions, phytochelatins, activities of antioxidative enzymes, and the presence of certain stress-induced proteins. The following discussion examines these changes in plants, including trees, followed by a summary of our work on biochemical indicators of Al stress in woody plants.

Many stress factors, such as ozone, low temperature, and drought, induce the production of reactive oxygen species, such as oxygen radicals and hydrogen peroxide (H_2O_2). These oxidants damage cellular metabolism, particularly photosynthesis. To prevent oxidative damage, plant cells utilize a scavenging system consisting of low molecular weight antioxidants (e.g., ascorbic acid, glutathione, and vitamin E) and a set of inducible protective enzymes, including superoxide dismutase (SOD), glutathione reductase (GTR), dehydroascorbate reductase (DHAR), monodehydroascorbate reductase (MDHAR), and ascorbate peroxidase (APX). Thus, the constant presence of elevated levels of these oxygen-scavenging compounds can be indicative of one or more forms of stress, such as O_3 stress in birch and drought stress in oak and pine (Schwanz et al., 1996). Spruce showing symptoms of decline disease at high elevation have higher concentrations of antioxidants (Nageswara et al., 1991; Bermadinger-Stabentheiner, 1996). Molecular genetic techniques to affect stress tolerance through the modulation of cellular levels of these antioxidants and antioxygenic enzymes are being investigated in several species, including poplar (Foyer et al., 1995).

Polyamines (putrescine, spermidine, and spermine) are a group of open-chained polycations of low molecular weight that are found in all organisms and exist in both free and bound forms. They play an important role in the growth and development of all organisms. The biological function of the polyamines could be partly attributed to their cationic nature and their electrostatic interactions with polyanionic nucleic acids and negatively charged functional groups of membranes, enzymes, or structural proteins in the cell (Houman et al., 1991). Recently, considerable attention has been paid to the study of changes in polyamine metabolism in plants subjected to various kinds of environmental stress. External stress can result in an increase or a decrease in cellular polyamines, depending upon the type of stress, the plant species, and the time of stress application (Zhou et al., 1995). Abiotic stress conditions that affect polyamine concentration include low pH, SO_2, high salinity, osmotic shock, nutrient stress (e.g., K or Ca deficiency), low temperature, and ozone (Flores, 1991 and references therein), and Al (Minocha et al., 1992, 1996a,b, 1997). Polyamines generally show a reverse proportionality to ions such as Ca, Mg, manganese (Mn), and K in response to Al treatment (Minocha et al., 1992, 1997; Zhou et al., 1995). Houman et al. (1991) observed a 25-fold increase in putrescine in the leaves and an 80-fold increase in roots of *Populus maximowiczii* stem cuttings grown in low K. Spermidine and spermine content decreased under low K conditions. The increased levels of putrescine were sustained over several weeks, thus acting as useful markers of K deficiency.

In addition to polyamines that may stabilize cellular membranes, nucleic acids, and some proteins, another mechanism often used by plants to protect themselves under prolonged stress conditions is the accumulation of compatible solutes. These solutes are a group of compounds that are not highly charged, but are polar and highly soluble. They include proline, glycerol, glycine, betaine, and so forth. Besides osmoregulation, they perform the important function of osmoprotection by helping to preserve the native conformation of proteins (Rudolph et al., 1986;

Carpenter and Crow, 1989). While higher concentrations of inorganic ions may be inhibitory to enzyme activities, compatible solutes at higher concentrations are not inhibitory to most enzyme activities (Pavlicek and Yopp, 1982; Richter and Kirst, 1987). At present, the literature on the role of compatible solutes in osmoprotection in trees is scarce. An increase in foliar proline content has been shown in several species of oak in response to drought stress (Kim and Kim, 1994), and in spruce and *Eucalyptus* in response to elicitor-induced stress and insect herbivory, respectively (Lange et al., 1995; Marsh and Adams, 1995). Factors that affect fine root structure and function, which would in turn affect the water and nutrient status of the plant, could affect the levels of compatible solutes in the foliage.

Plants growing in soils rich in heavy metals are known to accumulate high concentrations of metal-binding peptides called phytochelatins (Grill et al., 1987). These peptides, which have the general structure (γ-Glu-Cys)$_n$-Gly ($n = 2$ to 10), are synthesized from glutathione by enzymatic peptide synthases and not by direct translation of messenger ribonucleic acid (mRNA). Phytochelatins have also been referred to occasionally and erroneously as metallothioneins. True metallothioneins are a group of small metal-binding proteins found mostly in bacteria and animals (Hamer, 1986). Metallothioneins are direct products of the genes synthesized through mRNA translation. More recently, metallothioneins have been located in plants. The correlation of metallothein to phytochelatin concentrations remain unknown. Phytochelatins bind to heavy metals, such as cadmium (Cd), lead (Pb), and chromium (Cr), and effectively reduce the cellular concentrations of free metals. The biosynthesis of phytochelatins is induced only by heavy metals and thus their presence in high quantities is a reliable and specific indicator of heavy metal stress (Grill et al., 1988). Gawel et al. (1996) have proposed that heavy metals may be the likely contributing factor in the decline of forests in the northeastern U.S. Their data showed that the phytochelatin concentrations were higher in red spruce, a species in decline in northeastern U.S., as compared with balsam fir, a species that is not. Phytochelatin concentrations increased with increasing altitude and had a positive correlation with decline. They also increased across the region in forest stands with increasing tree damage.

There are many classes of stress-inducible proteins, for example, heat-shock proteins (hsp), metallothionein-like proteins, late embryogenic abundant (lea) proteins, calmodulin, and several proline-rich cell wall proteins that have also been reported to be synthesized in stressed cells of several plant species. In most cases, specific functions of these proteins are not known. Likewise, increased cellular levels of salicylic acid, pathogen-related proteins (PRPs) and changes in foliage fluorescence have been suggested as potential markers of stress.

Biochemical Indicators of Aluminum Stress

The adverse effects of acidic deposition on soil productivity, due to the solubilization of Al and leaching of bases, are of major concern to forest land managers because such processes may impact growth over large areas. Monomeric Al is an important toxic species to plants, both in the rhizopshere and within the plant symplasm (Shortle and Smith, 1988; Cronan and Grigal, 1995; Kochian, 1995) Aluminum interferes with cation uptake and can cause damage to plant cells by interaction with sensitive macromolecules (Haug, 1984; Sucoff et al., 1990). Aluminum is known to affect the needle biomass, root growth, seedling height, and cellular inorganic ion content in plants (Schier et al., 1990). However, very little is known about the primary site(s) of action of Al and the chain of biochemical and molecular events associated with short- and long-term effects of Al. We have employed suspension tissue cultures of two woody plants, periwinkle (*Catharanthus roseus*) and red spruce as a model experimental system to study cellular responses to Al. We have also conducted field studies to correlate foliar polyamines and inorganic cations to soil Al/Ca ratios.

Tissue Culture

Cell cultures are a reliable tool for gaining insights into the effects of certain elements on plant cells under controlled conditions. This is a valuable approach as it is extremely difficult and expensive to create a controlled environment for mature trees. Model systems using cell cultures can identify biochemical and molecular processes that are early indicators of a particular stress, for example, Al exposure. Methods developed using the model system may then be applied to foliage or other tissues of mature trees collected across a natural or manipulated stress gradient to evaluate the current and future health status of forest stands.

Rapid methods for the extraction of free polyamines and exchangeable inorganic ions and for the quantitation of polyamines using high pressure liquid chromatography (HPLC) were developed in our laboratory using several herbaceous and woody plant tissues (Minocha et al., 1990, 1994; Minocha and Shortle, 1993). Preliminary work on Al stress was done using a woody plant model culture system of periwinkle until a suspension culture model system of red spruce became available in our laboratory.

Addition of 0.2, 0.5 or 1.0 mM aluminum chlorate (AlCl$_3$) to 3-day-old periwinkle cells showed a small but significant increase in cellular levels of putrescine at 4 hours followed by a sharp decline by 16 hours (Minocha et al., 1991). There was no further decline in putrescine during the next 32 hours. Spermidine levels did not change appreciably as compared with the control cultures. However, spermine levels increased by 2- to 3-fold at 24 and 48 hours. Whereas all concentrations of Al caused a slight decrease in total cell number, cell viability was affected only by 1.0 mM Al. There was a decrease in the cellular levels of Ca, Mg, Na, K, Mn, P, and iron (Fe) in the cells treated with Al at 4 hours but a significant increase by 16 and 24 hours. The results suggest that both the absolute amounts of Al and the length of exposure to it are important for cell toxicity (Minocha et al., 1992).

Similar to the results with periwinkle, treatment of 3-day-old red spruce cultures with 0.2, 0.5, and 1.0 mM AlCl$_3$ caused a dose-dependent inhibition of fresh weight production. These levels of Al also caused a significant increase in cellular putrescine concentrations. This increase was dose-dependent up to 2 days and could be observed as early as 4 hours (Fig. 7.9). Spermidine levels were either not affected or showed a slight increase, which was not always dose-dependent. Al generally caused a dose-dependent increase in spermine and a decrease in K, Ca, Mg, and Mn (Minocha et al., 1996a).

Aluminum tolerance varies considerably among species and even among genotypes within species (Delhaize and Ryan, 1995). The different responses observed for putrescine content during long-term incubation with Al for periwinkle (Minocha et al., 1992) and red spruce suspensions are probably due to their different sensitivities to Al. Periwinkle showed an increase in putrescine content in response to Al treatment during the first 4 hours, but an inhibition thereafter in response to Al treatment. In contrast, this stress response (increase in putrescine levels) lasted for several days in red spruce. Differences in the growth rates of cell cultures (i.e., periwinkle is faster growing than red spruce) may be partly responsible for this observed difference (Minocha et al., 1996a).

Under conditions of increased oxidative stress, such as low temperature and/or nutrient deficiencies, adjustments of enzymatic activities and antioxidant levels have been observed in herbaceous plants as well as spruce needles (Hendry and Broklebank, 1985; Price and Hendry, 1991; Baker, 1994). Aluminum has been reported to cause Ca deficiency in spruce. Thus Al may also induce cells to respond to nutrient deficiency by changing their antioxidant levels.

There is no available information in the literature on the effects of Al on antioxidants and their biosynthetic enzymes in plants. However, research done using rats, animal cell culture systems, and in vitro assays has shown that aluminum was responsible for altered GTR and SOD activities (Zaman et al., 1990; Shainkin-Kestenbaum et al., 1989). Our work with red spruce suspension cultures treated with Al showed a dose-dependent decrease in GTR, MDHAR, and APX. Thse changes in GTR were not consistent from experiment to experiment.

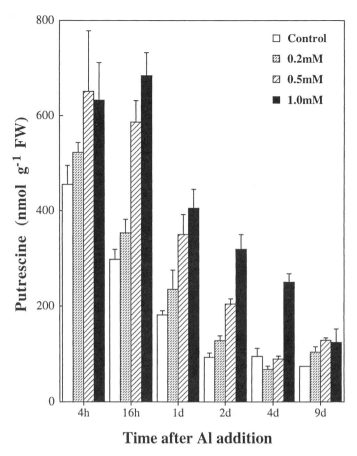

Figure 7.9. Effects of AlCl₃ on cellular putrescine. Al was added to 3-day-old red spruce cell suspension cultures of red spruce and subsequently analyzed. Values are mean ± SE of three replicates.

Recently, several researchers have postulated a protective role for organic acids (e.g., malate or citrate) in alleviating Al stress in herbaceous plants (Delhaize and Ryan, 1995 and references therein). The amount of malate released by an Al-tolerant genotype of wheat was 5- to 10-fold greater than the amount released from a near-isogeneic, but Al-sensitive genotype (Ryan et al., 1995). Following these suggestions we undertook a study (1) to determine which organic acids, if any, are accumulated or secreted by red spruce cultures in response to Al treatment and (2) to determine if exogenous addition of any of these organic acids along with Al would reverse the effects of Al on red spruce cultures. Our data show that addition of 0.5 and 1.0 mM AlCl₃ to 3-day-old suspension cultures caused a significant increase in the cellular content of succinate with a concomitant decrease in the concentration of oxalate at 48 hours.

However, both succinate and oxalate were present in significantly higher amounts in the spent media of Al-treated cell cultures. The cellular amounts of malate, ascorbate, and citrate were not affected by Al and they were not present in detectable quantities in the spent media. Exogenous addition of succinate alone or along with Al had no effect on cellular polyamine concentrations or the cell mass. Addition of either oxalate, ascorbate, or citrate alone to the cultures caused an increase in polyamine concentrations without any effects on cell mass. However, while the addition of malate by itself had no effect on cellular putrescine and cell mass, its addition along with Al caused a complete reversal of Al effects on putrescine metabolism and the growth rate of cells. The data support the postulated role for malate in alleviating Al stress.

Field Studies

The next step in the study was to validate observable changes in polyamine metabolism to Al exposure under field conditions. If valid, we could then use polyamine chemistry as an early biochemical marker of stress in apparently healthy trees (Minocha et al., 1996b, 1997). Six red spruce stands from the northeastern U.S. were selected for collection of soil and foliage samples. These stands all had soil solution pH values below 4.0 in the Oa horizon but varied in their soil Al concentrations. Some of these sites were apparently under some form of environmental stress as indicated by a large number of dead and dying red spruce trees. Samples of soil and needles (from apparently healthy red spruce trees) were collected from these sites four times during a 2-year period. The data showed a strong positive correlation between Ca and Mg content in the needles and that of the Oa horizon of the soil, the area of greatest occurrence of fine absorbing roots. Needles from trees growing on relatively Ca-rich soils with a low exchangeable Al concentration and a low Al/Ca soil solution ratio had significantly lower concentrations of putrescine and spermidine than those growing on Ca-poor soils with a high exchangeable Al concentration and a high Al/Ca soil solution in the Oa horizon. The magnitude of this change was several-fold higher for putrescine concentrations than for spermidine concentrations. Neither putrescine nor spermidine was correlated with soil solution chemistry (Ca, Mg, and Al concentrations) in the B horizon. The putrescine concentrations of the needles always correlated significantly with exchangeable Al ($r^2 = 0.73$, $P \leq 0.05$) and soil solution Al/Ca ratios (Fig. 7.10) ($r^2 = 0.91$, $P \leq 0.01$) of the Oa horizon. This suggests that in conjunction with soil chemistry, putrescine and/or spermidine may potentially be used as early indicators of Al stress before the appearance of visual symptoms in red spruce trees.

We conclude that similar inferences on the effects of Al may be drawn from cell cultures and the foliage of mature red spruce in the field (Minocha et al., 1996a). This indicates that the cell culture studies are

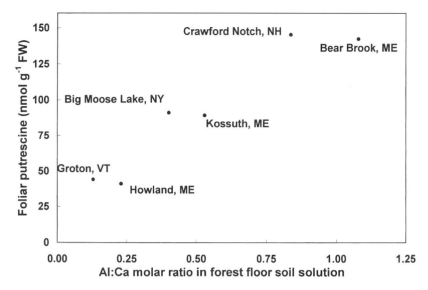

Figure 7.10. The relationship of foliar soluble putrescine to the Al/Ca molar ratio calculated for total monomeric Al and Ca in the soil solution of the forest floor. Foliar data are means of 40 replicate observations except for Kossuth, Maine, and Big Moose, New York, which are means of 20 replicate observations. Soil data are means of 12 replicate observations.

a valuable and reliable tool for investigating the effects of potential chemical stressors on trees. We have also demonstrated that an increase in foliar putrescine and/or spermidine concentrations in response to direct or indirect stress imposed on red spruce trees by Al exposure may possibly be used as an early warning tool for assessing and predicting tree health before the appearance of visual symptoms of damage.

Polyamines, in addition to a role in membrane stabilization, act as scavengers of free radicals, modulators of the metabolism of reduced N and modulators of free Ca in the cells (Slocum and Flores, 1991). Since their biosynthetic pathway competes with the biosynthesis of ethylene, a senescence hormone, they can also modulate ethylene metabolism. Thus, the elevated cellular polyamine content could help the plant in tolerating a variety of stresses that affect the above mentioned cellular functions.

Belowground Processes

Root Physiology and Pathology

Acidic deposition can deplete soil nutrients, promote soil acidification (Robarge and Johnson, 1992), and increase the concentration of anions.

Increased deposition of acidic anions increases the concentration of Al in the soil solution (Reuss, 1983). These changes adversely affect root viability and sustainability.

Root physiology and pathology are affected indirectly by acidic deposition primarily by two mechanisms. The first mechanism is a change in soil chemistry that modifies (1) root morphology, longevity, and turnover rates, (2) the mycorrhizal status of the root, and (3) root chemistry, metabolism, and ultimately the ability to take up nutrients. The second mechanism is a change in carbon allocation patterns to the roots caused by reductions in total tree carbon production due to reduced nutrient uptake, reduced photosynthetic area (foliage) related to dieback phenomena (e.g., winter injury in spruce), and increased root mortality. Change in carbon allocation reduces the available carbohydrate, required to maintain existing roots, replace dead and dying roots, and provide defense chemicals needed to protect roots against pathogenic fungi and other microorganisms (Bloomfield et al., 1996).

Ulrich (1983) proposed that the adverse effects of acidic deposition on forest ecosystems resulted from decreases in the base cations in soil solution, especially Ca levels, resulting in low Ca/Al ratios. Studies on fine root deterioration in conifer forests in Germany linked fine root death, decreases in growth, deteriorating crowns, and increased tree mortality to changes in the molar ratio of Al/Ca in the forest floor (Bauch and Schroeder, 1982; Bauch et al., 1985a,b; Stienen et al., 1984). These studies found that where decreased growth or tree mortality occurred, Al/Ca ratios in the cortex of fine roots were >1.

Higher levels of Al seem to decrease the uptake of Ca by tree roots eventually resulting in the loss of root integrity. Radioisotope studies showed that when equimolar concentrations of Al^{3+} and Ca^{2+} existed at pH 4, Ca uptake by the roots was reduced (Schroder et al., 1988). The Ca/Al molar ratio of the soil solution seems to be one of the prime mechanisms by which acidic deposition affects forest growth and it has been proposed as an important indicator of potential stress in forest ecosystems (Cronan and Grigal, 1995). This ratio is especially critical in the forest floor where the fine root systems of coniferous species are predominantly located and capture nutrients (Vogt et al., 1981).

Fine Root Distribution in the Forest Floor

The forest floor is an important source of nutrients for forest trees, especially for coniferous trees in cold temperate and boreal forests (Cole and Rapp, 1981; Vogt et al., 1986); the importance of this horizon as the dominant site for nutrient capture increases as the stand ages and develops. In the Pacific Northwest, Douglas fir (*Pseudotsuga menziesii* [Mirb.] Franco) forests about 70 years old obtained almost 100% of their nutrient requirements from the forest floor (Johnson et al., 1982b). At this stage of stand development, trees have fully occupied the site, and foliar and root biomasses have stabilized (Vogt et al., 1987). There is also some evidence that suggests that as the stand ages, fine roots concentrate in the forest floor. Fine roots in a 23-year-old stand of *Abies amabilis* Forbes in the Pacific Northwest was predominantly located in the E horizon while in an older stand (~180-year-old) root biomass was mostly located in the forest floor (Vogt et al., 1980).

The few studies conducted in forests of the northeastern U.S. show that fine roots of conifers occupy predominantly the forest floor, which in these forests can be quite deep (5 to 15 cm) (Fernandez, 1992; Joslin et al., 1992). In a survey of hardwood and spruce–fir forests in the Adirondack Mountains of NY, fine roots of red spruce (*Picea rubens* Sarg.) and balsam fir (*Abies balsamea* L.) occupied the forest floor predominantly, while fine roots of yellow birch (*Betula alleghaniensis* Britt.), sugar maple (*Acer saccaharum* Marsh.), and beech (*Fagus grandifolia* Ehrh.) were distributed throughout the forest floor and mineral horizons (Hopkins, 1939). For example, about 95% of balsam fir roots and 100% of red spruce roots were found in the forest floor horizons designated as the F or fermentation layer (Oe) and H or humus layer (Oa). In contrast, yellow birch had about half (40 to 50%) of its fine roots in the forest floor, while beech (37%) and sugar maple (12%) had a third or less of their roots in the forest floor. More recent studies in the same area, but over a greater range of sites than used in the Hopkins study (Hopkins, 1939), showed that 70 to 80% of the conifer fine root biomass (<2-mm diameter) occurred in the forest floor, while more than 50% of the fine roots of hardwood species were in the mineral horizons.

This preponderance of fine roots in the surface organic horizons has also been observed in spruce–fir forests in the southern Appalachians, where fine roots were almost exclusively limited to the Oa horizon and surface of the underlying mineral horizon (Johnson and Fernandez, 1992; Kelly and Mays, 1989). Both of these reports indicated that fine roots (<0.5 cm) were found in the lower depths of the mineral horizon but they represented only a very small fraction of the total fine roots.

In these northern forests, the restriction of conifer roots primarily to the forest floor is in strong contrast to the distribution of hardwood roots. In the Adirondacks, fine root abundance of yellow birch in the mineral horizon ranged from 32 to 100% depending on associated species composition and site type (Hopkins, 1939). In the same sites, total abundance of sugar maple and beech roots in the mineral horizon ranged from 88 to 100% and 63 to 100%, respectively. The same trend has also been recorded for mixed mesophytic forests in eastern Kentucky. Root abundance was consistently and significantly higher in the mineral soil and represented 79%, 83% and 88% of the total root abundance for oak, transitional, and beech forest types, respectively (Kalisz et al., 1987).

Soil Chemistry Effects on Fine Roots

The effects of acidic deposition on the humus–root relationships in sensitive forest sites and its role in subsequent localized damage was first proposed in the late 1800s in Germany for conifer sites apparently suffering from sulfur emissions (Hartig, 1897). Sulfur emissions were postulated to affect the soil, which led to suppressed root growth and nutrient uptake, and, eventually, dieback, decline, and mortality. However, it was not until the late 1970s and early 1980s that significant large-scale tree mortality was reported in Europe and stimulated research on its cause(s), which to some extent corroborated Hartig's hypothesis (Ulrich et al., 1979). This led to specific studies that showed significant disruption of root functions and damage to root tissues due to the mobilization of Al and its uptake by trees (Godbold et al., 1988).

In general, spruce on acidic "mor" soils carry larger bio- and necro-masses and have greater turnover rates than trees on more fertile soils (Kottke, 1987; Ulrich, 1987; Vogt et al., 1986). Thus, trees must allocate a larger percentage of their net primary production to root system growth to provide sufficient nutrients and water for growth (Rehfuess, 1989). Any significant reduction in root growth and turnover rates would have a significant adverse effect on above ground productivity.

The impact of acidic deposition on root growth and hence on overall productivity is mediated most likely through its effect in the rooting zone on the abundance of aluminum (Al), which interferes with and regulates processes at the soil–root interface (Cronan and Grigal, 1995). The effect may be manifested through two mechanisms. One mechanism is the competition by Al for nutrient uptake exchange sites, and subsequent physiological alterations to plant growth and resource acquisition. The other is the restriction of the rooting zone where roots can acquire resources. The first mechanism is related to the affinity of root exchange sites for Al at low pH and high Al/Ca ratios (Cronan, 1991; Dahlgren et al., 1991). This allows Al to significantly interfere with the uptake of other important cations (Ca and Mg), thus eventually causing nutrient imbalances within the whole plant, for example, the observed Mg deficiencies in Norway spruce in Germany (Schlegel and Hutterman, 1990). Also, increased uptake of Al by the fine roots directly and adversely affects their physiology, causing reduced water uptake and decreased fine-root biomass, branching, and terminal elongation (Cronan et al., 1989).

The second mechanism is related to the potential toxicity of Al in the rooting medium (Dahlgren et al., 1991). Organic horizons typically have high cation exchange capacities and low Al/Ca ratios; B horizons, in contrast, have lower exchange capacities, higher relative amounts of exchangeable acidity, and high Al/Ca ratios (Cronan, 1994; Dahlgren et al., 1991). Thus, as concentrations of anions from acidic deposition increase in the mineral soil, the concentration of Al (primarily Al^{3+}) in the soil solution increases making the mineral horizon an inhospitable rooting medium for roots of Al-sensitive species. Root concentrations of Al tracked closely with the Al^{3+} concentrations but not total aqueous Al concentrations showing that the form of Al is important to toxicity (Dahlgren et al., 1991). In addition, Al concentrations in the forest floor increased with increased depth. This relationship was due both to changes in the chemical affinity for Al and to the progressively greater incorporation of mineral soil and/or mineral soil aluminum with increasing depth into the organic horizon (Rustad and Cronan, 1995). This increase in Al concentrations in the B mineral horizon and lower depths of the organic horizon will decrease the availability of hospitable rooting zones and further decrease the ability of tree roots to capture necessary quantities of base cations other than Al.

The effects of these chemical changes on roots and their relationships with aboveground deterioration have been documented for several forest systems. Prolonged suppression of cambial growth and crown dieback and mortality were linked with high molar Al/Ca ratios in fine roots of red spruce in stands in New England (Shortle and Smith, 1988). Shortle and Bondietti (1992) showed that in stands receiving high acidic deposition, Ca/Al ratios in the root tips were low (0.3 to 1.1), root density was sparse, and tree conditions were poor, based on crown condition and abundance of mortality. The Ca/Al ratio was higher (2.5 to 10) in root tips of red spruce in stands receiving less acidic deposition and where tree condition was judged good. Similar relationships with acidic deposition were observed in high-elevation red spruce stands in the southern Appalachians (Joslin and Wolfe, 1992). Restricted root growth was attributed to an unfavorable soil chemical environment characterized by high Al concentrations in the soil solution and a low Ca/Al molar ratio. Fine roots of red spruce have a high affinity for Al at low soil pHs and especially at low Ca/Al ratios; Al is taken up in preference to "nontoxic" cations. In the root itself, Al is co-precipitated with phosphate (PO_4) and oxalate thereby decreasing the availability of P to the trees. This co-precipitation of Al with PO_4 decreases with depth and thereby decreases the ability of roots to avoid Al toxicity by co-precipitating Al (Dahlgren et al., 1991). Smith et al. (1995) showed that this preferential uptake can occur differently depending on the location of roots in the organic horizon. Root tips in the upper portion of the organic horizon (Oa, F) had higher Ca and lower Al concentrations than root tips in the lower portion of the organic horizon (Oa, H). Smith et al. (1995) also observed that the Al/Ca ratios in the root tips were consistently higher at higher-elevation sites compared with lower-elevation sites. This differential concentration of Ca and Al in the fine root tips corresponds to the observation that the Al concentrations are higher in the lower portion (Oa) of the organic horizon in these stands (Rustad and Cronan, 1995).

Mycorrhizal Relationships

Not a great amount of work has been done in the northern forest ecosystems on the relationship of mycorrhizae and acidic deposition. One study in West Virginia compared ectomycorrhizal-basidiomycete communities in red spruce with those in northern hardwood forests (Bills et al., 1986). Twenty-seven species of putative mycorrhizal partners were identified by sporocarp presence in red spruce stands, and thirty-five were found on northern hardwood stands; eight of the species were common to both areas. The spruce sites were on ridge crests between 1200 and 1350 m but there was no comparison with stands at higher elevations that are potentially more polluted. In a limited study of root pathology on red spruce on Mt. Abraham, Vermont, fewer mycorrhizae were found on fine roots of declining (crown deterioration—dieback, needle loss) trees than on healthy trees in a high elevation stand (Wargo et al., 1993). In addition to lower percentages of root tips that were mycorrhizal, declining trees had fewer mycorrhizae morphotypes (Glenn et al., 1991; Wargo et al., 1993).

In a more extensive study of root pathology on red spruce conducted on trees in plots above and below cloud base and on windward and leeward aspects on Whiteface Mountain, NY, no consistent relationships between mycorrhizal measurements and tree health were detected on random samples of fine roots from healthy and declining trees (Wargo et al., 1993). In general, percentages of mycorrhizae on the Whiteface Mountain site (avg. 27%) were significantly lower than those reported for other red spruce sites in mesic and wetland sites in New Jersey (avg. 89%) (Glen et al., 1991) and at high-elevation sites on Mt. Mitchell in North Carolina (avg. 35%) (Bruck, 1984).

Data on abundance of mycorrhizae on red spruce from the Whiteface Mountain study and studies at Mt. Mitchell (Bruck, 1984) suggest that trees in high-elevation sites, regardless of health, have fewer mycorrhizae. This could be a function of soil type, climate, or nutrient status. In the Whiteface Mountain study, the percentage of root tips that were mycorrhizal on randomly selected fine roots was significantly lower than on fine roots chosen for their healthy appearance. Percentages of mycorrhizae on these healthy-appearing roots were 2 to 3 times higher than on the randomly chosen fine-root systems and were similar to those found on healthy red spruce at lower-elevation sites on Mt. Mitchell, North Carolina (Bruck, 1984). This suggests that the turnover rate of mycorrhizal roots is higher in those high-elevation sites.

Studies in Bavaria related changes in mycorrhizae on declining Norway spruce (*Picea abies* L.) to soil chemistry (Meyer et al., 1988). In these Bavarian stands, mycorrhizae numbers were highly and positively correlated with the molar Ca/Al ratio in the mineral soil. The low incidence of mycorrhizae on red spruce at Whiteface Mountain sites also could be related to the Al/Ca soil chemistry. For example, total Ca and

Mg concentrations in the forest floor on Whiteface Mountain were higher at the low-elevation spruce site than at the higher-elevation site, and were lower at the single site on Mt. Abraham (Shortle and Bondetti, 1992) where root damage was greater than at both elevations on Whiteface Mountain (Wargo et al., 1993). Also, exchangeable Ca and Mg were considerably lower on Mt. Abraham (Shortle and Bondietti, 1992). In addition to the low Ca and Mg, both sites had relatively high Al/Ca ratios; it was 0.9 on Whiteface Mountain and 3.1 on Mt. Abraham. The value for Whiteface Mountain is just slightly above the Ca/Al ratio at which Cronan and Griegel (1995) estimate that there is a 50:50 risk of impact on tree growth or nutrition. The Ca/Al ratio at Mt. Abraham is well below that estimated to have a 75% chance of causing or resulting in damage.

Other studies have shown negative relationships between Al and degree of mycorrihzal abundance on fine-roottips. The degree of mycorrhizal infection on Norway spruce in Germany was positively related to the Ca/Al ratio in the mineral soil and increased as the ratio increased (Schneider et al., 1989). Mycorrhizal formation on balsam fir seedlings growing in environmental chambers was inhibited by Al at 50 and 100 mg Al g^{-1} substrate and at pH 3 and 4 (Entry et al., 1987). Reduction in mycorrhizal colonization on pitch pine seedlings also was caused by Al at 50 mg l^{-1} concentration (McQuattie and Schier, 1992). In addition, the mantle hyphae were devoid of cytoplasm, and cortical cells of the fine roots contained dense material and were often disrupted. Also, the zone of actively dividing cells in Al-treated seedlings was significantly less than controls, thus significantly reducing the amount of overall root growth.

Similar relationships of colonization rates by mycorrhizal fungi and Ca and Al properties of the soil have also been observed in northern hardwoods. A survey of 18 mature sugar maples showed that the frequency of endomycorrhizal formation was positively related to soil pH but negatively related to the level of H + Al held on the soil exchange sites (Ouimet et al., 1995). In this same study, the frequency of mycorrhizal formation also was positively correlated with the foliar and fine root Ca content. In contrast, evaluation of fine roots in sugar maple stands that had been treated with a base cation mixture showed that fertilization had no effect on the incidence of endomycorrhizae (Cooke et al., 1992; Moutoglis and Widden, 1996). However, this lack of response to fertilizer may reflect the short duration (1 year) between treatment and evaluation of mycorrhizal colonization of fine roots.

Fine Nonwoody Root Pathology

The basic processes of root turnover and senescence in forest trees have been recently reviewed (Bloomfield et al., 1996). However, the role of fine root pathogens in contributing to root turnover and senescence is not well understood, nor are the biotic and abiotic factors that influence the importance of fine-root pathogens in forests.

Primary fungal pathogens of fine nonwoody roots of trees, though common on tree seedlings in nurseries, have been rarely observed in natural northern forests. Most of the known aggressive fine root pathogens have been documented in the lower latitudes where soils are generally warm (Sinclair et al., 1987). Even fine root pathogens that develop on tree seedlings growing in nurseries fail to spread to other seedlings or trees when infected seedlings are transplanted to forest soils (Sinclair et al., 1987). It has been suggested that for those fine root fungal pathogens that do develop in forest settings, some form of stress may be required to predispose roots to infection by these fungi (Manion, 1981).

The relationship between stress and fine root pathogens has not been thoroughly studied. Woody roots in both deciduous and coniferous trees are commonly predisposed by stressors (e.g., drought, defoliators, temperature extremes, pollutants, etc.) to less aggressive root pathogens such as *Armillaria* spp. (Carey et al., 1984; Hudak and Singh, 1970; Wargo, 1977, 1981; Wargo and Houston, 1974). The same phenomenon probably occurs for fine roots; however, since these relationships have not been studied extensively, the pathogens have not been identified.

One potential stress-induced pathogen of fine roots of conifers is *Mycelium radicis atrovirens* Melin. This name was originally given to a sterile gray fungus that was isolated from fine roots of *Pinus sylvestris* L. and *Picea abies* Karst. (Melin, 1922). It was designated a pseudo-mycorrhiza that was potentially harmful to tree roots. Richard and Fortin (1973) observed the same fungus on fine roots of black spruce (*Picea mariana* [Mill] B.S.P.) and *M. r. atrovirens* has been isolated from fine roots of declining Norway spruce (*P. abies*) in Bavaria (Livingston and Blaschke, 1984). Subsequent work has shown that many isolates of dark-pigmented fungi that were designated as *M. r. atrovirens* actually represent many different fungi (Kowalski, 1973). Two such fungi have been identified as *Phialocephala demorphospora* Kendrick (Kendrick, 1961) and *Phialocephala fortinii* Wang and Wilcox (Wang and Wilcox, 1985). Both form pseudomycorrhizal associations with conifer roots (Wang and Wilcox, 1985). *Phialocephala fortinii* is considered to be a pathogen of fine roots and may be triggered to cause mortality of fine roots on trees weakened by stress that accelerates senescence and reduces resistance in root tissue. Recent studies have shown that *P. fortinii* was pathogenic on red pine seedlings in greenhouse experiments (Harney, 1994; Wang and Wilcox, 1985).

In studies on root vitality on declining red spruce, dematiaceous fungi were consistently isolated from putative dead, nonwoody roots from 10 sites in the states of New York, Vermont, New Hampshire, and Maine (Harney et al., 1995). Several of these isolates were identified by RFLP analysis of PCR amplified rDNA as being genetically related to *P. fortinii*

(Harney et al., 1995). These sites represented a putative west-to-east health and acid deposition gradient, and, consequently, a potential Al/Ca ratio gradient from a high deposition, poor health, central NY site, east to a low deposition, good health, ME site. There was a tendency for the sites with greater mortality or greater crown deterioration of red spruce to have greater levels of fine-root mortality but no correlation with these dematiaceous fungi have been verified as yet.

Tomato mosaic virus (ToMv) is another pathogen that has been found in fine roots of red spruce trees in areas potentially impacted by acid deposition. ToMv was initially detected in stream water exiting a deteriorating red spruce–balsam fir stand on Whiteface Mountain in New York (Jacobi and Castello, 1991). The virus was subsequently transmitted to herbaceous hosts. The virus also was detected in red spruce needles and subsequently transmitted to herbaceous hosts as well as red spruce, black spruce, and balsam fir seedlings (Jacobi et al., 1992). The virus was found in the roots of these infected seedlings, suggesting that the root system was the potential point of ToMv infection in infested soils in red spruce stands (Jacobi and Castello, 1992). Additional studies in windward and leeward red spruce stands, above and below cloud base (see Wargo et al., 1993), verified the presence of ToMv in fine roots of mature red spruce (Castello et al., 1995). The incidence and concentration of virus varied, but in general were greater in the healthier trees with little-to-moderate crown dieback than in declining trees with severe dieback. There was a higher percentage of virus-infected trees at the higher-elevation plots, that is, above cloud base on the windward and leeward aspects, but there was no aspect affect. Infected trees had smaller live crowns than uninfected trees, suggesting a long-term effect of virus infection on tree growth (Castello et al., 1995).

The role of this virus and other potential viruses in red spruce decline is unclear. The random pattern of virus occurrence in the fine nonwoody roots of red spruce (Castello et al., 1995) is coincident with a random pattern of fine root deterioration observed on these trees on Whiteface Mountain (Wargo et al., 1993). This coincidence suggests possible cause and effect. However, this random pattern of fine-root deterioration was in sharp contrast to apparent simultaneous and progressive deterioration that occurred on red spruce growing on Mt. Abraham in the Green Mountains of Vermont (Wargo et al., 1993). Root deterioration on trees on Mt. Abraham was related to low energy reserves (starch) in the woody root system and may indicate that these root systems were observed at a later stage of deterioration than those on trees on Whiteface Mountain.

The greater deterioration of fine-root systems on red spruce on Mt. Abraham also may be related to the chemical environment in which the roots were growing. The molar Al/Ca ratio of 3.1 in fine roots of red spruce in the humus layer on Mt. Abraham was significantly greater than the ratio of 0.9 measured in fine roots of red spruce on Whiteface Mountain (Shortle and Smith, 1988; Shortle and Bondietti, 1992). The soils on the two sites are also distinctly different, which helps explain the chemical differences in the fine roots. On Mt. Abraham plots, the forest floor rests directly on bedrock predominantly, while on Whiteface Mountain, plots had a well developed mineral soil (Wargo et al., 1993). Other studies on Whiteface Mountain on soil and foliar chemistry of red spruce have corroborated the lower Al to Ca ratio in soil, soil solution, and red spruce foliage on Whiteface Mountain (Johnson et al., 1994a,b).

Woody Root Pathology

No pathogens of the coarse woody root system have been consistently observed nor isolated from symptomatic trees and related to acidic deposition levels or putative gradients of acidic deposition in the northern forests. Armillaria root disease was reported on trees on Whiteface Mountain but the role of the fungus in the decline of red spruce was discounted because it was considered to be a secondary agent and not a direct cause of decline (Weidensaul et al., 1989). The secondary role of *Armillaria* seems to be exactly what its role in decline generally is, to attack trees after they have been weakened by stress agents (Wargo and Harrington, 1991). No information was given in the Weidensaul et al. report (1989) to determine if there was a relationship between *Armillaria* and decline severity or elevation. Lack of root pathogens as contributors to decline of red spruce and frazier fir also has been reported for the southern Appalachians (Bruck, 1989).

Where species of *Armillaria* have been observed and isolated from roots of declining red spruce, no consistent pattern or link between its presence and decline symptoms has been observed. In studies on red spruce on Mt. Abraham, Vermont and Whiteface Mountain, New York, *Armillaria* was found occasionally on some woody roots of trees classified to the severe decline category (Wargo et al., 1993). However, it was not on all severely declining trees and the fungus occurred only occasionally on trees with better crown conditions. In a survey to determine the incidence and severity of Armillaria root disease on red spruce at nine sites in the Northeast, the fungus was frequently found colonizing and killing trees at lower elevation sites (<800 m) (Carey et al., 1984). However, colonization by *Armillaria* was less severe and less frequent at elevations above 800 m. An aerial assessment of spruce mortality in New York, Vermont, New Hampshire, and the western mountains of Maine indicated that mortality was heaviest at elevations above 800 m (~2600 ft) (Miller-Weeks and Smoronk, 1993). The reasons for these differences by elevation were speculative. However, subsequent studies in seven sites (five in the Northeast and two in the southern Appalachians) on the distribution of rhizomorphs (the major infection and spreading structure of the fungus (Shaw and Kile, 1991) of *Armillaria* in soils in declining red spruce stands showed that both the frequency of occurrence and the abundance of rhizomorphs were significantly less in higher-elevation sites, thus decreasing the chances for root contact and infection (Wargo et al., 1987b). In this study on *Armillaria*, forest type or elevation, pH, and lead were significant variables that were related to rhizomorph density across all sites; all three variables were highly correlated among themselves. Lead concentration increases in the soil with elevation and has been related to atmospheric pollutant deposition (Johnson et al., 1982a). In addition, in vitro studies with *Armillaria* showed that concentrations of soluble lead as low as 10 ppm could significantly reduce thallus growth and inhibit rhizomorph formation and subsequent growth (Wargo et al., 1987a).

Species of *Armillaria* are not the only fungi that decrease in abundance with elevation in the northern forest ecosystems. Root and butt rot fungi in spruce–fir forests in New Hampshire decreased in importance as causes of gap-makers (trees that die and or snap-off and create forest gaps) as elevation increased (Worrall and Harrington, 1988). The relationship of acidic deposition or related factors with the reduction in fungi was not measured.

Conclusions

Few young trees survive to maturity. Tree survival depends on resisting the adverse effects of environmental stress. The stress to which an individual tree is subjected is the results of several to many stressors. Stressors include pests, pathogens, and adverse environmental conditions. Resistance to stress results from the application of metabolic energy within the genetic limits of the individual tree. Changes in landuse patterns, climate, soil nutrition, insect infestations, and the constant threat of disease all contribute to tree stress and all affect forest health and productivity.

Air pollution is an additional stressor to which trees have been constantly subjected throughout most of the 20th century. Air pollution comes in several forms, such as acids that alter the availability of essential elements in the soil, oxidants that disrupt leaf function, and heavy metals that block metabolism. Is there evidence that in the 20th century new stressors from human activity added to those already at work in the forest have had any impact on health and productivity?

We investigated the potential impact of acidic deposition on northern spruce forests. These forests naturally develop a thick raw or "mor" humus or organic layer in which the cation exchange capacity is highly sensitive to acid input. In this scenario, rapid increase in the atmospheric deposition of sulfate and nitrate in the middle of the 20th century mobilized essential Ca and Mg that were then available either for uptake by tree roots or for leaching from the rooting zone. This period of historical mobilization appears to be contained in the dendrochemical record of red spruce. Enhanced leaching may be responsible for the decreased storage of Ca and Mg in the soil as seen in the comparison with historical analyses. Acid deposition also mobilized Al from the mineral soil beneath the organic layer. Mobile Al ions displace Ca and Mg from the soil's organic layer and reduce the storage of Ca and Mg, which is essential for tree growth and development. In addition to displacing Ca and Mg from organic matter in the soil, Al binds to root cell walls, blocking the uptake of essential Ca and reducing root elongation. Roots unable to elongate and spread soon exhaust the local supply of essential elements and die. Dying roots are susceptible to attack by facultative root pathogens. Loss of roots leads to loss of crown and stem growth that results in reduced productivity and increased vulnerability to harmful abiotic and biotic stressors. Apparently healthy trees are stressed in relation to adverse Al:Ca ratios in the soil, as indicated by concentrations of putrescine, a biochemical marker in foliage.

These results taken together are consistent with widespread, fundamental change in the response of trees to the environment, which is due at least in part to acidic deposition. No one stressor, such as acid deposition, can be seperated from other perturbations of soil chemistry, landuse patterns, and overall stress loading, but the perturbation of cation cycling caused by acidic deposition can clearly be a significant contributing factor to the decline of productivity, and sometimes tree health, on base-poor forest sites.

In the 1980s Robert Hartig, the great German forest biologist, deduced that "smoke injury" (localized acidic deposition) reduced stem growth before the appearance of crown symptoms. Hartig hypothesized that loss of basic cations Ca and Mg from the root zone leads to root dysfunction, root death, and root infection. In the 1980s after modern technology had pushed acidic deposition to the regional level, Josef Bauch found that Al bound to the outer cortex of absorbing root tips reduced the uptake of Ca and Mg. The concept of aluminum-induced calcium deficiency syndrome was then developed to explain how acidic deposition can have an adverse effect on northern conifers growing on forest mor. Bauch further stated that the "bottom-up" disease described by Hartig coupled with "top-down" disease due to the direct action of air pollutants, such as ozone, on the crowns of trees could produce a "pincer effect." Either the bottom-up problem or the top-down problem alone caused growth suppression, but in combination caused mortality. Early detection, improved diagnosis, and management strategies designed to reduce levels of external stressors and enhance the effectiveness of trees to respond to stress will be needed to meet this challenge.

References

Baker, NR (1994) Chilling stress and photosynthesis. In: Foyer CH, Mullineaux PM (eds) *Causes of Photo-oxidative Stress and Amelioration of Defense Systems in Plants*. CRC Press, Boca Raton, FL, pp 127–154.

Bartholomay GA, Eckert R, Smith KT (1997) Reductions in tree-ring widths of white pine following ozone exposure at Acadia National Park, Maine, U.S.A. Can J For Res 27: 361–368.

Bauch J, Rademacher P, Berneike W, Kroth J, Michaelis W (1985a) Breite und Elementgehalt der Jahrringe in Fichten aus Waldschadensgebieten. In: *Waldschaden-Einflussfaktoren und ihre Bewertung*. VCI Berichte 560, Dusseldorf, Germany, pp 943–959.

Bauch J, Schroeder W (1982) Zellularer Nachweis von Elementen in den Feinwurzeln gesunder und erkrankter Tanne (*Abies alba* Mill.). Forstwiss Ctrblatt 101:285–294.

Bauch J, Stienen H, Ulrich B, Matzner E (1985b) Einfluss einer Kalkung bzw. Dungung auf den Elementgehalt in Feinwurzeln und das Dickenwachstum von Fichten aus Waldschadensgebieten. Allgem Forst Z 43:1148–1150.

Bermadinger-Stabentheiner E (1996) Influence of altitude, sampling year, and needle age class on stress-physiological reactions of spruce needles investigated on an alpine altitude profile. J Plant Physiol 148:339–344.

Bills GF, Holtzman GI, Miller OK Jr. (1986) Comparison of ectomycorrhizal-basidiomycete communities in red spruce versus northern hardwood forests of West Virginia. Can J Bot 64:760–768.

Bloomfield J, Vogt K, Wargo PM (1996) Tree root turnover and senescence. In: Waisel Y, Eshel A, Kafkafi U (eds) *Plant roots. The Hidden Half*. Marcel Dekker, New York, pp 363–381.

Bruck RI (1989) Survey of diseases and insects of Fraser fir and red spruce in the southern Appalachian Mountains. Eur J For Pathol 19:389–398.

Bruck RI (1984) Decline of montane boreal ecosystems in central Europe and the southern Appalachian Mountains. TAPPI Proc 159–163.

Carpenter JF, Crowe JH (1989) An infrared spectroscopic study of the interactions of carbohydrates with dried proteins. Biochemistry 28:3916–3922.

Carey AC, Miller EA, Geballe GT, Wargo PM, Smith WH, Siccama TG (1984) *Armillaria mellea* and decline of red spruce. Plant Dis 68(9):794–795.

Castello JD, Wargo PM, Jacobi V, Bachand GD, Tobi DR, Rogers MAM (1995) Tomato mosaic virus infection of red spruce on Whiteface Mountain, New York: prevalence and potential impact. Can J For Res 25:1340–1345.

Charles DF (1984) Recent pH history of Big Moose Lake (Adirondack Mountains, New York, USA) inferred from sediment diatom assemblages. Verh Int Verein Limnol 22:559–566.

Cole DW, Rapp M (1981) Elemental cycling in forest ecosystems. In: Reichle DE (ed) *Dynamic Properties of Forest Ecosystems*. International Biological Programme 23. Cambridge University Press, Cambridge, United Kingdom, pp 341–409.

Cook ER (1990) Bootstrap confidence intervals for red spruce ring-width chronologies and an assessment of age-related bias in recent growth trends. Can J For Res 20:1326–1331.

Cook ER (1985) *A Time Series Approach to Tree-ring Standardization*. PhD dissertation, University of Arizona, Tucson, AZ.

Cook ER, Briffa KR, Meko DM, Graybill DS, Funkhouser G (1995) The segment length curse in long chronology development for paleoclimatic studies. Holocene 5:229–237.

Cook ER, Zedaker SM (1992) The dendroecology of red spruce decline. In: Eagar C, Adams MB (eds) *Ecology and Decline of Red Spruce in the Eastern United States.* Springer-Verlag, New York, pp 92–231.

Cooke MA, Widden P, O'Halloran I (1992) Morphology, incidence and fertilization effects on the vesicular-arbuscular mycorrhizae of *Acer saccharum* in a Quebec hardwood forest. Mycologia 84(3):422–430.

Cronan CS (1994) Aluminum biogeochemistry in the Albios forest ecosystems: The role of acidic deposition in aluminum cycling. In: Godbold DL, Heutterman A (eds) Effects of acid rain on forest processes. Wiley-Liss, New York, pp 51–81.

Cronan CS (1991) Differential adsorption of Al, Ca, and Mg by roots of red spruce (*Picea rubens* Sarg.). Tree Physiol 8:227–237.

Cronan CS, April R, Bartlett R, Bloom P, Driscoll C, Gherini S, Henderson G, Joslin J, Kelly JM, Newton R, Parnell R, Patterson H, Raynal D, Schaedle M, Schofield C, Sucoff E, Tepper H, Thornton F (1989) Aluminum toxicity in forests exposed to acidic deposition: the ALBIOS results. Water Air Soil Pollut 48:181–192.

Cronan CS, Goldstein RA (1989) ALBIOS: a comparison of aluminum biogeochemistry in forested watersheds exposed to acidic deposition. In: Adrian DC, Hava M (eds) *Acidic Precipitation. Vol. 1.* Case Studies. Advances in Environmental Science. Springer-Verlag, New York, pp 113–135.

Cronan CS, Grigal DF (1995) Use of calcium/aluminum ratios as indicators of stress in forest ecosystems. J Environ Qual 24:209–226.

Dahlgren R, Vogt KA, Ugolini FC (1991) The influence of soil chemistry on fine root aluminum concentrations and root dynamics in a subalpine spodosol. Plant Soil 133:117–129.

Delhaize E, Ryan PR (1995) Aluminum toxicity and tolerance in plants. Plant Physiol 107:315–321.

Driscoll CT, van Breemen N, Mulder J (1984) Aluminum chemistry in a forested spodosol. Soil Sci Soc Am J 49:437–444.

Entry JA, Cromack K Jr., Stafford SG, Castellano MA (1987) The effect of pH and aluminum concentration on ectomycorrhizal formation in *Abies balsamea*. Can J For Res 17:865–871.

Fernandez IJ (1992) Characterization of eastern U.S. spruce–fir soils. In: Eager C, Adams MB (eds) *Ecology and Decline of Red Spruce in the Eastern United States.* Springer-Verlag, New York, pp 40–63.

Flores HE (1991) Changes in polyamine metabolism in response to abiotic stress. In: Slocum RD, Flores HE (eds) *Biochemistry and Physiology of Polyamines in Plants.* CRC Press, Boca Raton, FL, pp 213–228.

Foyer CH, Souriau N, Perret S, Lelandais M, Kunert K-J, Pruvost C, Jouanin L (1995) Overexpression of glutathione reductase but not glutathione synthetase leads to increases in antioxidant capacity and resistance to photoinhibition in poplar trees. Plant Physiol 109:1047–1057.

Fritts HC (1976) *Tree Rings and Climate.* Academic Press, New York.

Gawel JE, Ahner BA, Friedland A, Morel FMM (1996) Role for heavy metals in forest decline indicated by phytochelatin measurements. Nature 381:64–65.

Glenn MG, Wagner WS, Webb SL (1991) Mycorrhizal status of mature red spruce (*Picea rubens*) in mesic and wetland sites of northwestern New Jersey. Can J For Res 21:741–749.

Godbold DL, Fritz E, Huttermann A (1988) Aluminum toxicity and forest decline. Proc Natl Acad Sci USA 85:3888–3892.

Grill E, Winnacker E-L, Zenk MH (1987) Phytochelatins, a class of heavy metal–binding peptides from plants, are functionally analogous to metallothioneins. Proc Natl Acad Sci USA 84:439–443.

Grill E, Winnacker E-L, Zenk MH (1988) Occurrence of heavy metal binding phytochelatins in plants growing in a mining refuse area. Experientia 44:539–540.

Grissino-Mayer HD, Butler DR (1993) Effects of climate on growth of shortleaf pine in northern Georgia: a dendroclimatic study. South Geogr 33:65–81.

Hamer DH (1986) Metallothionein. Annu Rev Biochem 55:913–951.

Harney SK (1994) *Dematiaceous Endophytes from Plant Roots: Molecular Characterization and Interactions*. PhD thesis, State University of New York, Syracuse, NY.

Harney SK, Wentworth TS, Wargo PM (1995) *Phialocephala fortinii*, a potential fine root pathogen, isolated from red spruce. Phytopathology 85:1141.

Hartig R (1897) Ueber den Einfluss des Hütten- und Steinkohlenrauches auf den Zuwach der Nadelholzbäume. Forst Naturwiss Z 6:49–60.

Haug A (1984) Molecular aspects of aluminum toxicity. Crit Rev Plant Sci 1:345–373.

Hendry GAF, Broklebank KJ (1985) Iron-induced oxygen radical metabolism in waterlogged plants. New Phytol 101:199–206.

Herms DA, Mattson WJ (1992) The dilemma of plants: to grow or defend. Quart Rev Biol 67:283.

Hillis WE (1987) *Heartwood and tree exudates*. Springer-Verlag, Berlin, Germany.

Holmes RL, Adams RK, Fritts HC (1986) *Tree-ring Chronologies of Western North America: California, Eastern Oregon, and Northern Great Basin, with Procedures Used in the Chronology Development Work Including User's Manuals for Computer Programs COFECHA and ARSTAN*. Chronology Series VI. Laboratory of Tree-Ring Research, University of Arizona, Tucson, AZ.

Hopkins HT Jr. (1939) *The Root Distribution of Forest Trees in the Central Adirondack Region*. PhD thesis, Cornell University, Ithaca, NY.

Houman F, Godbold DL, Majcherczyk A, Shasheng W, Huttermann A (1991) Polyamines in leaves and roots of *Populus maximowiczii* grown in different levels of potassium and phosphorus. Can J For Res 21:1748–1751.

Hudak J, Singh P (1970) Incidence of Armillaria root rot in balsam fir infested by balsam woolly aphid. Can Plant Dis Surv 50:99–101.

Jacobi V, Castello J (1992) Infection of red spruce, black spruce, and balsam fir seedlings with tomato mosaic virus. Can J For Res 22:919–924.

Jacobi V, Castello JD (1991) Isolation of tomato mosaic virus from waters draining forest stands in New York State. Phytopathology 81:1112–1117.

Jacobi V, Castello JD, Flachmann M (1992) Isolation of tomato mosaic virus from red spruce. Plant Dis 76:518–522.

Johnson AH, Anderson SB (1994) Acid rain and soils of the Adirondacks. I. Changes in pH and available calcium, 1930–1984. Can J For Res 24:39–45.

Johnson AH, Cook ER, Siccama TG (1988) Climate and red spruce growth and decline in the northern Appalachians. Proc Natl Acad Sci USA 85:5369–5373.

Johnson AH, Cook ER, Siccama TG, Battles JJ, McLaughlin SB, LeBlanc DC, Wargo PM (1995) Comment: Synchronic large-scale disturbances and red spruce growth decline. Can J For Res 25:851–858.

Johnson AH, Schwartzman TN, Battles JJ, Miller R, Miller EK, Friedland AJ, Vann DR (1994b) Acid rain and soils of the Adirondacks. II. Evaluation of calcium and aluminum as causes of red spruce decline at Whiteface Mountain, New York. Can J For Res 24:654–662.

Johnson AH, Friedland AJ, Miller EK, Siccama TG (1994a) Acid rain and soils of the Adirondacks. III. Rates of soil acidification in a montane spruce–fir forest at Whiteface Mountain, New York. Can J For Res 24:663–669.

Johnson AH, Siccama TG, Friedland AJ (1982a) Spatial and temporal patterns of lead accumulation in the forest floor in the northeastern United States. J Environ Qual 11:577–580.

Johnson DW, Cole DW, Bledsoe CS, Cromack K, Edmonds RL, Gessel SP, Grier CC, Richards BN, Vogt KA (1982b) In: Edmonds RL (ed) *Nutrient Cycling in Forests of the Pacific Northwest. Analysis of Coniferous Forest Ecosystems in the Western United States.* Hutchinson Ross, Stroudsburg, PA.

Johnson DW, Fernandez IJ (1992) Soil-mediated effects of atmospheric deposition on eastern spruce–fir forests. In: Eager C, Adams MB (eds) *Ecology and Decline of Red Spruce in the Eastern United States.* Springer-Verlag, New York, pp 235–270.

Joslin JD, Kelly JM, Van Miegroet H (1992) Soil chemistry and nutrition of North American spruce–fir stands: evidence of recent change. J Environ Qual 21: 12–30.

Joslin JD, Wolfe MH (1992) Red spruce soil solution chemistry and root distribution across a cloud water deposition gradient. Can J For Res 22:893–904.

Kalisz PJ, Zimmerman RW, Muller RN (1987) Root density, abundance, and distribution in the mixed mesophytic forest of eastern Kentucky. Soil Sci Soc Am J 51:220–225.

Kelly JM, Mays PA (1989) Root zone physical and chemical characteristics in southeastern spruce–fir stands. Soil Sci Soc Am J 53:1248–1255.

Kendrick WB (1961) The *Leptographium* complex *Phialocephala* gen. nov. Can J Bot 39:1079–1085.

Kim JW, Kim JH (1994) Comparison of adjustments to drought stress among seedlings of several oak species. J Plant Biol 37:343–347.

Kochian LV (1995) Cellular mechanisms of aluminum toxicity and resistance in plants. Annu Rev Plant Physiol Plant Mol Biol 46:237–260.

Kottke J (1987) Zusammenfassung und Wertung zum Themenbereich Mykorrhiza/rhizosphare. Tagungsbericht Statusseminar Kfa Julich 30.3.-3.4: 342–345.

Kowalski S (1973) Mycorrhiza forming properties of various strains of the fungus *Mycelium radicis atrovirens* Melin. Bull Acad Polon Sci, Ser Sci Biol 21: 767–770.

Krause C (1997) The use of dendrochronological material from buildings to get information about past spruce budworm outbreaks. Can J For Res 27:69–75.

Lange BM, Lapierre C, Sandermann H Jr. (1995) Elicitor-induced spruce stress lignin: Structural similarities to early developmental lignins. Plant Physiol 108:1277–1287.

Lawrence GB, David MB, Shortle WC (1995) A new mechanism for calcium loss in forest-floor soils. Nature 378:162–165.

Lawrence GB, Fernandez IJ (1991) Biogeochemical interactions between acidic deposition and a low-elevation spruce–fir stand in Howland, Maine. Can J For Res 21:867–875.

Leblanc DC, Raynal DJ, White EH (1987) Acidic deposition and tree growth. II. Assessing the role of climate in recent growth declines. J Environ Qual 16:334–340.

Livingston WH, Blaschke H (1984) Deterioration of mycorrhizal short roots and occurrence of *Mycelium radicis atrovirens* on declining Norway spruce in Bavaria. Eur J For Pathol 14:340–348.

Loehle C (1988) Tree life history strategies: the role of defenses. Can J For Res 18:209–222.

Manion PD (1981) Fungi as agents of tree diseases: root rot. In: Manion PD (ed) *Tree Disease Concepts.* Prentice-Hall, Englewood Cliffs, NJ, pp 295–309.

Marsh NR, Adams MA (1995) Decline of *Euclyptus teriticornis* near Bairsdale, Victoria: insect herbivory and nitrogen fractions in sap and foliage. Aust J Bot 43:39–49.

Matthek C, Gerhardt H, Breloer H (1992) VTA: visual tree defect assessment based on computer simulation of adaptive growth. In: Little C (ed) *Experimental Mechanics*. Elsevier, London, United Kingdom, pp 109–120.

Mayewski PC (1986) Sulfate and nitrate concentrations from a south Greenland ice core. Science 232:975–977.

McLaughlin SB, Dowing DJ, Blasing TC, Cook ER, Adams HS (1987) An analysis of climate and competition as contributors to decline of red spruce in high elevation Appalachian forests of the eastern United States. Oecologia 72:487–501.

McQuattie CJ, Shier GA (1992) Effect of ozone and aluminum on pitch pine (*Pinus rigida*) seedlings: anatomy of mycorrhizae. Can J For Res 22:1901–1916.

Melin E (1922) On the mycorrhizas of *Pinus silvestris* L. and *Picea abies* Karst. A preliminary note. J Ecol 9:254–257.

Meyer J, Schneider BU, Werk K, Oren R, Schulze ED (1988) Performance of two *Picea abies* (L.) Karst. stands at different stages of decline. V. Root tip and ectomycorrhiza development and their relations to above ground and soil nutrients. Oecologia 77:7–13.

Miller-Weeks M, Smoronk D (1993) *Aerial Assessment of Red Spruce and Balsam Fir Condition*. NA-TP-16-93. United States Department of Agriculture (USDA) Forest Service, Northeastern Area, Radnor, PA.

Minocha R, Minocha SC, Komamine A, Shortle WC (1991) Regulation of DNA synthesis and cell division by polyamines in *Catharanthus roseus* suspension cultures. Plant Cell Rep 10:126–130.

Minocha R, Minocha SC, Long SL, Shortle WC (1992) Effects of aluminum on DNA synthesis, cellular polyamines, polyamine biosynthetic enzymes and inorganic ions in cell suspension cultures of a woody plant, *Catharanthus roseus*. Physiol Plant 85:417–424.

Minocha R, Shortle WC (1993) Fast, safe, and reliable methods for extraction of major inorganic cations from small quantities of woody plant tissues. Can J For Res 23:1645–1654.

Minocha R, Shortle WC, Coughlin DJ, Minocha SC (1996a) Effects of Al on growth, polyamine metabolism, and inorganic ions in suspension cultures of red spruce (*Picea rubens*). Can J For Res 26:550–559.

Minocha R, Shortle WC, Lawrence GB, David MB, Minocha SC (1996b) Putrescine: a marker of stress in red spruce trees. In: Hom J, Birdsey R, O'Brian K (eds) *Proceedings, 1995 Meeting of the Northern Global Change Program, 14–16 March 1995, Pittsburgh, PA*. Gen Tech Rep NE-214. United States Department of Agriculture (USDA) Forest Service, Northeastern Forest Experiment Station, Radnor, PA, pp 119–130.

Minocha R, Shortle WC, Lawrence GB, David MB, Minocha SC (1997) Relationships among foliar chemistry, foliar polyamines, and soil chemistry in red spruce trees growing across the northeastern United States. Plant Soil 191:109–122.

Minocha SC, Minocha R, Robie CA (1990) High-performance liquid chromatographic method for the determination of dansyl-polyamines. J Chromatogr 511:177–183.

Momoshima N, Bondietti EA (1990) Cation binding in wood: applications to understanding historical changes in divalent cation availability to red spruce. Can J For Res 20:1840–1849.

Moutoglis P, Widden P (1996) Vesicular-arbuscular mycorrhizal spore populations in sugar maple (*Acer saccharum* Marsh.) forests. Mycorrhiza 6:91–97.

National Acid Precipitation Assessment Program [NAPAP] (1993) *1992 Report to Congress*. NAPAP, Washington, DC.

Nageswara R, Madamanchi R, Hausladen A, Alscher R, Amundson RG, Fellows S (1991) Seasonal changes in antioxidants in red spruce (*Picea rubens* Sarg.) from three field sites in the Northeastern United States. New Phytol 118:331–338.

Ouimet R, Camire C, Furlan V (1995) Endomycorrhizal status of sugar maple in relation to tree decline and foliar, fine-roots, and soil chemistry in the Beauce region, Quebec. Can J Bot 73:1168–1175.

Pavlicek KA, Yopp JH (1982) Betaine as a compatible solute in the complete relief of salt inhibition of glucose-6-phosphate dehydrogenase from a halophilic blue-green alga. Plant Physiol 69:58.

Price AH, Hendry GAF (1991) Iron catalyzed oxygen radical formation and its possible contribution to drought damage in nine native grasses and three cereals. Plant Cell Environ 14:477–484.

Reams GA, Nicholas NS, Zedaker SM (1993) Two hundred year variation of southern red spruce radial growth as estimated by spectral analysis. Can J For Res 23:291–301.

Rehfuess KE (1989) Acidic deposition—extent and impact on forest soils, nutrition, growth and disease phenomena in central Europe: A review. Water Air Soil Pollut 48:1–20.

Reuss JO (1983) Implications of the calcium–aluminum exchange system for the effect of acid precipitation on soils. J Environ Qual 12:591–595.

Richard C, Fortin JA (1973) The identification of *Mycelium radicis atrovirens* (*Phialocephala dimorphospora*). Can J Bot 51:2247–2248.

Richter DFE, Kirst GO (1987) D-Mannitol dehydrogenase and mannitol-1-phosphate dehydrogenase in *Platymonas subcordiformis*: some characteristics and their role in osmotic adaptation. Planta 170:528–534.

Robarge WP, Johnson DW (1992) The effects of acidic deposition on forested soils. Advan Agron 47:1–83.

Rudolph AS, Crowe JH, Crowe LM (1986) Effects of three stabilizing agents: proline, betaine, and trehalose on membrane phospholipids. Arch Biochem Biophys 245:134–143.

Rustad LE, Cronan CS (1995) Biogeochemical controls on aluminum chemistry in the O horizon of a red spruce (*Picea rubens* Sarg.) stand in central Maine, USA. Biogeochemistry 29:107–129.

Ryan PR, Delhaize E, Randall PJ (1995). Characterization of Al stimulated efflux of malate from the apices of Al-tolerant wheat roots. Planta 196:103–110.

Schier GA, McQuattie CJ, Jensen KF (1990) Effect of ozone and aluminum on pitch pine (*Pinus rigida*) seedlings: growth and nutrition relations. Can J For Res 20:1714–1719.

Schlegel H, Huttermann A (1990) Identification of ion stress in roots of forest trees. In: Persson H (ed) *Above- and Below-ground Interactions in Forest Trees in Acidified Soils*. Air Pollut Res Rep 32. Environmental Research Programme of the Commission of the European Communities, Brussels, Belgium. pp 110–118.

Schneider BU, Meyer J, Schulze E-D, Zech W (1989) Root and mycorrhizal development in healthy and declining Norway spruce stands. In: Schulze E-D, Lange OL, Oren R (eds) *Forest Decline and Air Pollution: A Study of Spruce (Picea abies) on Acid Soils*. Ecological Studies, Vol. 77, Springer-Verlag, Berlin, Germany, pp 370–391.

Schroeder WH, Bauch J, Endeward R (1988) Microbeam analysis of Ca exchange and uptake in the fine roots of spruce: influence of pH and aluminum. Trees 2:96–103.

Schwanz P, Picon C, Vivin P, Dreyer E, Guehl J-M, Polle A (1996) Responses of antioxidative systems to drought stress in pedunculate oak and maritime pine as modulated by elevated CO_2. Plant Physiol 110:393–402.

Shaw CG III, Kile GA (eds) *Armillaria Root Disease*. Agric Hndbk 691. United States Department of Agriculture (USDA) Forest Service, Washington, DC.

Shainkin-Kestenbaum R, Adler AJ, Berlyne GM, Caruso C (1989) Effect of aluminum on superoxide dismutase. Clin Sci 77:463–466.

Shigo AL (1986a) *A New Tree Biology*. Shigo and Trees, Durham, NH.

Shigo AL (1986b) *A New Tree Biology Dictionary*. Shigo and Trees, Durham, NH.

Shigo AL (1991) *Modern Arboriculture*. Shigo and Trees, Durham, NH.

Shortle WC, Bondietti EA (1992) Timing, magnitude, and impact of acidic deposition on sensitive forest sites. Water Air Soil Pollut 61:253–267.

Shortle WC, Smith KT (1988) Aluminum induced calcium deficiency syndrome in declining red spruce. Science 240:1017–1018.

Shortle WC, Smith KT, Minocha R, Lawrence GB, David MB (1997) Acidic deposition, cation mobilization, and biochemical indicators of stress in healthy red spruce. J Environ Qual 26:871–876.

Sinclair WA, Lyon HH, Johnson WT (1987) Diseases caused by *Phytophthora* species. In: *Diseases of Trees and Shrubs*. Cornell University Press, Ithaca, NY, pp 284–291.

Slocum RD, Flores HE (1991) (eds) Biochemistry and physiology of polyamines. CRC Press, Boca Raton, FL, p 228.

Slutzky E (1937) The summation of random causes as the source of cyclic processes. Econometrica 5:105–146.

Smith KT (1997) Phenolics and compartmentalization in the sapwood of broad-leaved trees. In: Dashek WV (ed) *Methods in Plant Biochemistry and Molecular Biology*. CRC Press, Boca Raton, FL, pp 189–198.

Smith KT, Cufar K, Levanic T (1999) Temporal stability and dendroclimatology in silver fir and red spruce. Phyton 39(3):43–54.

Smith KT, Shortle WC (1996) Tree biology and dendrochemistry. In: Dean JS, Meko DM, Swetnam TW (eds) *Tree Rings, Environment and Humanity*. Radiocarbon, Tucson, Arizona, pp 629–635.

Smith KT, Shortle WC, Ostrofsky WD (1995) Aluminum and calcium in fine root tips of red spruce collected from the forest floor. Can J For Res 25:1237–1242.

Stienen H, Barckhausen R, Schaub H, Bauch J (1984) Mikroskopische und rontgenenergiedispersive Untersuchungen an Feinwurzeln gesunder und erkrankter Fichten (*Picea abies* [L.] Karst.) verschiedener standarte. Forstwiss Ctrblatt 103:262–274.

Stokes MA, Smiley TL (1996) *An Introduction to Tree-Ring Dating*. University of Arizona Press, Tucson AZ. [Reprint of 1968 edition]

Sucoff E, Thornton F, Joslin JD (1990) Sensitivity of tree seedlings to aluminum. I. Honeylocust. J Environ Qual 19:163–171.

Ulrich B (1987) Raten der Deposition, Akkumulation und des Austrags Toxischer Luftverunreinigungen als Mass der Belastung und Belastbarkeit von Waldo-kosystemen. Tagungsbericht Statusseminar KfA Julich 30.3.-3.4:277–278.

Ulrich B (1983) Soil acidity and its relations to acid deposition. In: Ulrich B, Pankrath J (eds) *Effects of Accumulation of Air Pollutants in Forest Ecosystems*. D Reidel, Boston, MA, pp 127–146.

Ulrich B, Mayer R, Khanna PK (1979) Deposition von Luftverunreinigungen und ihre Auswirkungen in Waldoekosystemen im Solling in Schriften. Forstl Fak Univ Goettingen 58:1–291.

Van Deusen PC (1990) Evaluating time-dependent tree ring and climate relationships. J Environ Qual 19:481–488.

Van Deusen PC (1989) A model-based approach to tree ring analysis. Biometrics 45:763–779.

Vernon DM, Osterm JA, Bohnert HJ (1993) Stress perception and response in a facultative halophyte: the regulation of salinity-induced genes in *Mesembryanthemum crystallinum*. Plant Cell Environ 16:437–444.

Vogt KA, Edmonds RI, Grier CC (1981) Seasonal changes in biomass and vertical distribution of mycorrhizal and fibrous-textured conifer fine roots in 23- and 180-year-old subalpine *Abies amabilis* stands. Can J For Res 11:223–229.

Vogt KA, Edmonds RL, Grier CC, Piper SR (1980) Seasonal changes in mycorrhizal and fibrous root growth in 23- and 180-year-old Pacific silver fir stands in western Washington. Can J For Res 10:523–529.

Vogt KA, Grier CC, Vogt DJ (1986) Production, turnover and nutrient dynamics of above- and belowground detritus of world forests. Adv Ecol Res 15:303–377.

Vogt KA, Vogt DJ, Moore EE, Fatuga BA, Redlin MR, Edmonds RL (1987) Douglas-fir overstory and understory live fine root biomass in relation to stand age and productivity. J Ecol 75:857–870.

Wang CJK, Wilcox HE (1985) New species of ectendomycorrhizal and pseudo-mycorrhizal fungi: *Phialophora finlandia*, *Chloridium paucisporum*, and *Phialocephala fortinii*. Mycologia 77(6):951–958.

Wargo PM (1981) Defolation, dieback and mortality. In: Doane CC, McManus ML (eds) *The Gypsy Moth: Research toward Integrated Pest Management*. Tech Bull 1584. United States Department of Agriculture (USDA) Animal and Plant Health Inspection Service, Washington, DC, pp 240–248.

Wargo PM (1977) *Armillariella mellea* and *Agrilus bilineatus* and mortality of defoliated oak trees. For Sci 23:485–492.

Wargo PM, Bergdahl DR, Tobi DR, Olson CW (1993) *Root Vitality and Decline of Red Spruce*. Biologia Arborum. Ecomed, Munich, Germany.

Wargo PM, Carey AC, Geballe GT, Smith WH (1987a) Effects of lead and trace metals on growth of three root pathogens of spruce and fir. Phytopathology 77:123.

Wargo PM, Carey AC, Geballe GT, Smith WH (1987b) Occurrence of rhizomorphs of *Armillaria* in soils from declining red spruce stands in three forest types. Plant Dis 71:163–167.

Wargo PM, Harrington TC (1991) Host stress and susceptibility. In: Shaw CG III, Kile GA (eds) *Armillaria root disease*. Agric Hndbk 691. United States Department of Agriculture USDA Forest Service, Washington, DC, pp 88–101.

Wargo PM, Houston DR (1974) Infection of defoliated sugar maple trees by *Armillaria mellea*. Phytopathology 64:817–822.

Weidensaul TC, Fleck AM, Hartzler DM, Capek CL (1989) *Quantifying Spruce Decline and Related Forest Characteristics at Whiteface Mountain, New York*. Summary Report. Ohio Agricultural Research and Development Center, Ohio State University, Wooster, OH.

Worrall JJ, Harrington TC (1988) Etiology of canopy gaps in spruce–fir forests at Crawford Notch, New Hampshire. Can J For Res 18:1463–1469.

Zaman K, Miszta H, Dabrowski Z (1990) The effect of aluminum on the activity of selected bone marrow enzymes in rats. Folia Haematol 117:447–452.

Zhou X, Minocha R, Minocha SC (1995) Physiological responses of suspension cultures of *Catharanthus roseus* to aluminum: changes in polyamines and inorganic ions. J Plant Physiol 145:277–284.

8. Atmospheric Deposition Effects on Surface Waters, Soils, and Forest Productivity

Gregory B. Lawrence, Kristiina A. Vogt, Daniel J. Vogt,
Joel P. Tilley, Philip M. Wargo,
and Margaret Tyrrell

When acid rain was discovered to be a regional problem in North America in the 1970s, initial concerns focused on surface-water acidification. Some of the earliest acidic deposition research found that fish populations in some lakes and streams in the Adirondack Mountains of NY had been eliminated by acidification (Schofield, 1976). In the 1980s, research in the U.S. expanded greatly through the National Acid Precipitation Assessment Program (NAPAP) to include soils and forests, as well as aquatic ecosystems. Because little environmental monitoring had been done before the start of NAPAP, however, information on changes that led to the conditions observed in the 1970s and 1980s was limited. As a result, NAPAP research focused on assessments of current conditions, short-term experimental manipulations, reconstructions from paleolimnological evidence, and mathematical modeling, to investigate past and possible future changes caused by acidic deposition. This program yielded conclusive evidence that acidic deposition had acidified poorly buffered surface waters, resulting in the loss of fish populations and other aquatic organisms, although uncertainties remained about the extent of these effects (NAPAP, 1991). The NAPAP Integrated Assessment Report (NAPAP, 1991) also concluded that acidic deposition may have affected soil chemistry but effects on forest health were not apparent, except for high-elevation spruce–fir forests where stand dieback was attributed to acidic deposition.

Although considerable uncertainty about the effects of acidic deposition remained at the close of the intensive research phase of the NAPAP program in 1990, a wealth of information was obtained through NAPAP-sponsored research. This research greatly increased our understanding of natural processes within forest and aquatic ecosystems, and provided much information on the effects of acidification on various components of these ecosystems. The NAPAP program also produced baseline information on the magnitude, composition, and location of atmospheric deposition within the U.S. boundaries, which indicated that the northeastern U.S. and southeastern Canada were regions receiving the highest inorganic chemical deposition in North America (Sisterson, 1991). "Median annual precipitation-weighted concentrations were shown to be higher in North America than in remote regions of the world by factors of 9 for SO_4^{2-}, 14 for NO_3^-, 7 for NH_4^+, and 5–6 for H^+" (Sisterson, 1991). Baseline deposition data, in conjunction with initiation of monitoring programs, and data collected through assessments in the 1970s and 1980s, have now provided the opportunity to study ecosystem responses to changes in atmospheric deposition rates that have occurred over the past three decades. These changes include declines in the atmospheric deposition of sulfate (SO_4^{2-}, 37% between 1964 and 1992), hydrogen ions (H^+, 40% between 1964 and 1992), and calcium (Ca^{2+}, 75% between 1963 and 1992), measured at the Hubbard Brook Experimental Forest in New Hampshire (Likens and Bormann, 1995). Perhaps as significant, atmospheric deposition of nitrogen (N) has not shown a trend between 1964 and 1992 at this site.

The purpose of this chapter is to synthesize the current knowledge (as of 1998) of how atmospheric deposition has affected forested watersheds in the northeastern U.S. Although the main topic of this book is forests, information on surface-water chemistry is included in this chapter to provide additional insight into how acidic deposition affects soils and trees. The following discussion first reviews the most significant findings of research conducted before and during the NAPAP era, which is considered in this chapter to have ended in 1990, the year in which NAPAP ceased funding large-scale research (although some results were published after 1990). This discussion is followed by descriptions of the specific advances in the 1990s that were funded through other programs. The current state of knowledge and remaining uncertainties are then synthesized from an ecosystem perspective.

What Was Learned from the NAPAP Era

To review the research conducted before and during the NAPAP era, this section has been divided into the areas of surface-water chemistry, soil chemistry, and forest productivity. The order in which these areas are presented is based on the chronology of acid rain research. Acidic deposition (commonly referred to as acid rain) was discovered to be a regional phenomenon in the early 1970s (Cogbill and Likens, 1974). This information, plus evidence of reduced fish populations in the Adirondack Mountain region of New York, triggered studies to determine the effects of acid rain on surface-water chemistry and fish mortality. Studies of soils and forests followed, but progress was slow because the many interacting processes that affect soils and forest ecosystems occur over long time scales.

Surface Waters

Effects of acidic deposition on aquatic ecosystems was one of the primary components of the NAPAP program. Although this topic is too broad to be covered in this section, information on the specific areas of surface-water chemistry and hydrological processes relevant to soil processes has been included.

Water Chemistry

Initial concerns about the effects of acidic deposition on surface waters focused on pH decreases resulting from sulfuric acid inputs. In a survey of 214 lakes in the Adirondack Mountains, Schofield (1976) found that 52% of these lakes above 610 m elevation had a pH less than 5.0. Burns et al. (1981) also documented that pH was significantly lower in 38 low-order New Hampshire streams in 1979 than in the period between 1936 through 1939. Quantifying the changes in surface-water chemistry that occurred since the onset of acidic deposition was not possible in the Northeast, however, because continuous monitoring of stream chemistry before the 1960s was extremely rare. Further analyses of surface-water chemistry revealed that lakes with pH < 5.0 often had aluminum (Al) concentrations that were toxic to fish, and that inorganic forms of Al were the most toxic (Driscoll et al., 1980). Elevated concentrations of inorganic Al in surface waters were attributed to the mobilization of Al in the soil by acidic deposition (Cronan and Schofield, 1979).

Through the 1970s and 1980s, acidity in precipitation was found to be comprised of approximately two-thirds sulfuric acid (H_2SO_4) and one-third nitric acid (HNO_3) (Cogbill and Likens, 1974). In surface waters, SO_4^{2-} concentrations (as eq l^{-1}) during this period were at least an order of magnitude greater than nitrate (NO_3^-) concentrations, but during spring snowmelt and other high-flow periods, NO_3^- concentrations increased by a factor of 2 or more (Driscoll and Schafran, 1984). In an analysis of data from the United States Environmental Protection Agency (USEPA) Eastern Lake Survey, Driscoll et al. (1989a) found that organic acids also contributed to low acid-neutralizing capacity (ANC) of surface waters.

However, organic anions were shown to be less than 25% of the total anion concentrations, whereas SO_4^{2-} comprised 40 to 80% of total anion concentrations in Adirondack lakes with pH < 6.5.

One of the most significant findings of surface-water investigations in the NAPAP era was the identification of episodic acidification, a process characterized by short-term decreases in ANC associated with increases in flow (Wiggington et al., 1990). These decreases in ANC were usually accompanied by increases in Al concentrations and decreases in pH. The transient nature of episodic acidification, however, made measurements of this process difficult to quantify. For this reason, most of the surface-water investigations during the NAPAP era focused on chronic acidification. One exception was the USEPA-sponsored Episodic Response Program, which was started in 1988. Data analysis for this program, however, was only partially completed at the time of publication of the NAPAP Integrated Assessment Report. As a result, episodic acidification was listed as one of the major remaining uncertainties in this report.

To evaluate trends in surface water chemistry, the USEPA initiated the Long-Term Monitoring Project (LTM) in the early stages of NAPAP. During the 1980s, 7 of the 17 Adirondack lakes monitored in this program showed a decline in ANC; 1 lake showed an increase (Driscoll et al., 1991). Concentrations of SO_4^{2-} declined in 11 lakes (no increases were observed), whereas 4 of the lakes showed an increase in NO_3^- concentrations (no decreases were observed). Five of the lakes showed an increase in base cation concentrations (Ca^{2+}, magnesium [Mg^{2+}], sodium [Na^+], and potassium [K^+]), and none of the lakes showed a decrease in base cations. Catskill Mountain streams monitored in this program (Murdoch and Stoddard, 1993) showed general decreases in SO_4^{2-} concentrations, but increases in NO_3^- concentrations between 1983 and 1989. The significance of these trends in relation to decreases in acidic deposition was uncertain because of the short length of these records. Long-term monitoring at the Hubbard Brook Experimental Forest, however, showed similar trends with a 25 year record; with the notable exception of a decreasing trend in base cation concentrations in stream water (Driscoll et al., 1989a). This decrease was attributed to a declining trend in atmospheric deposition of base cations and reduced leaching of base cations from soil caused by the decline in SO_4^{2-} deposition.

Hydrological Effects on Surface-Water Chemistry

Strong relationships between stream flows and chemical concentrations were identified before the discovery of acidic deposition in the Northeast. By modeling flow-related changes in stream chemistry as a mixture of rain with soil water, Johnson et al. (1969) concluded that chemical concentrations of soil water from the upper profile were likely to be different from those of the lower profile. In a hydrological assessment of two Adirondack

Mountain lakes, Chen et al. (1984) found that during high flows, stream water was partially comprised of soil solution originating from shallow, acidic soil horizons, whereas, during base flows, stream water was primarily derived from deeper soil or till that tended to be less acidic. Through a survey of Adirondack lakes, Driscoll and Newton (1985) determined that the amount of till underlying the soil was a key factor in determining the sensitivity of watersheds to acidic deposition. Watersheds with little till tended to produce acidic surface waters.

Further hydrological analyses (Lawrence et al., 1988) showed that elevational variability within a watershed plays an important role in controlling temporal variations in stream chemistry. Because neutralization of acidic deposition decreases as elevation increases (Johnson et al., 1981), contributions to stream flow from upslope areas tend to be more acidic than contributions from lower elevations in the watershed. During high-flow conditions, flowing water tends to extend upward in the watershed, thereby increasing the amount of water from upper elevations that becomes surface flow. Because these upslope contributions to flow tend to more acidic than contributions from lower in the watershed, stream chemistry at the base of the watershed becomes more acidic as flow increases. Models developed during the NAPAP era to predict the effect of variations in acidic deposition on surface-water chemistry, however, focused on time scales of seasons or longer and did not incorporate the effects of areal variability on storm-related changes in stream chemistry (Gherini et al., 1985; Cosby et al., 1985).

Soils

Evaluation of acidic deposition effects on soils was limited at the start of the NAPAP era by a lack of monitoring data and an incomplete understanding of the natural processes integral to soil development in forests. These factors complicated efforts to identify and quantify changes in soil properties resulting from acidic deposition throughout the 1980s.

Natural Acidification Processes

In the earliest stages of acid rain research, it was already known that forest soils can be acidified through natural processes (Schnitzer and Kahn, 1972). Early work also showed that quantitatively differentiating natural sources of acidity from acidity due to acidic deposition inputs is complicated by the many competing chemical reactions that take place in forest soils (van Breemen et al., 1983). The most common natural processes that generate H^+ in the soil can be generalized into the following processes: (1) carbonic acid production, (2) organic acid production, (3) nitrification, and (4) accretion of living plant biomass and their associated uptake of cations. Each of these processes is described in the following paragraphs.

Carbon dioxide (CO_2) is released into the soil atmosphere through metabolic processes during the growth and maintenance of plant root tissues and their symbiotic associations, and decomposition of organic detritus (Schlesinger, 1977). The buildup of CO_2 in the soil leads to the formation of carbonic acid, which can dissociate into H^+ and bicarbonate, a process that lowers the pH of soil solution and facilitates the leaching of base cations (van Breemen et al., 1983). Because carbonic acid does not dissociate to a large degree below pH 5.0 ($pK_a = 6.3$), typical CO_2 partial pressures in soils are generally not high enough to lower soil solution pH below 5.0 (Reuss and Johnson, 1985).

Production of organic acids, as end-products of incomplete decomposition of detrital materials, had long been recognized as a key factor in the pedogenic process of podzolization (Schnitzer and Delong, 1955), and as a result received considerable attention during the 1980s as a possible cause of surface-water acidification. Quantifying organic acidity, however, was difficult because of the molecular complexity and diversity of naturally derived organic acids. This, and the limited information on how organic and inorganic acids interact, made the importance of soil-derived organic acids in surface-water acidification the most contentiously debated issue during the NAPAP era (Krug and Frink, 1983; Driscoll et al., 1989a). The development of combined analytical and modeling approaches reduced the uncertainty in quantifying organic acidity, which led to the general acknowledgment that organic acidity plays an important role in the acidification of soils (Johnson et al., 1991) and some surface waters (Baker et al., 1990b).

Nitrification (microbial conversion of ammonium [NH_4^+] to NO_3^-) represents another important natural source of H^+. During nitrification, NH_4^+ is converted to NO_3^-, which results in a net production H^+ ion. Nitrification is common in forests dominated by N-fixing plants, such as red alder (*Alnus rubra*), where N builds up in the soil. In a stand dominated by red alder for 50 years, pH in the surface soil horizon was 0.4 unit less than in the same soil horizon in an adjacent conifer stand (Van Miegroet and Cole, 1985).

Net forest growth results in the acidification of soil through generation of H^+ if cations are taken up by vegetation at a greater rate than anions (Nye, 1981). The predominant N form used by most evergreen plants growing in a forest of low site quality is NH_4^+, which results in soil acidification from root uptake of N. The generation of H^+ occurs because plants have to maintain electrical neutrality at their membrane surfaces and the uptake of excess positively charged ions necessitates the release of equivalent positive charge to maintain electroneutrality (Nye, 1981).

These natural process of soil acidification have been shown to be important factors in determining the exchangeable base status of soil (Alban, 1982), therefore, efforts were made to compare inputs of H^+ from acidic deposition to internal ecosystem inputs of H^+ (Driscoll and Likens,

1982; van Breemen et al., 1984). That research elucidated the key components of H^+ budgets, but the difficulty of accurately measuring fluxes and pool sizes (Binkley and Richter, 1987) precluded a definitive comparison between the sources of acidity. At the close of the NAPAP era, the relative importance natural biological processes and acidic deposition to soil acidification was not fully resolved.

Anion Mobility

During the NAPAP era, acidification of soils and surface waters was considered in large part to be a function of the mobility of acid anions. When an anion derived from acid dissociation moves from the soil to surface waters, it provides countercharge, which facilitates leaching of H^+ or other cations that have been mobilized by H^+, such as Ca^{2+}. Surface-water acidification results if H^+ is leached with this anion, whereas soil acidification occurs if Ca^{2+} is leached with this anion.

Because SO_4^{2-} is the predominant anion in precipitation and most acidified surface waters, considerable research was done during the NAPAP era on the mobility of this anion within soils. Laboratory experiments demonstrated that retention of SO_4^{2-} within the soil can occur through adsorption, which increases as pH decreases (Nodvin et al., 1986), and biological assimilation that converts SO_4^{2-} to organic sulfur (S) (David et al., 1987). An assessment of watershed S budgets for the eastern U.S., however, showed that atmospheric inputs of S were approximately balanced by leaching of S into surface waters in the Northeast (Rochelle et al., 1987). The SO_4^{2-} adsorption capacity of northeastern soils was found to be low (Fuller et al., 1985), in part a result of the most recent glaciation, which ended approximately 12,000 years ago (Rochelle et al., 1987). Biological assimilation of S was also shown to be ineffective as a process for retaining SO_4^{2-} within the soil, because of high net S mineralization rates, which were found to exceed inputs of S in wet deposition (David et al., 1987).

The discovery of an association between surface-water acidification and elevated NO_3^- concentrations during spring snowmelt led to investigations to determine the mechanism through which NO_3^- was transported to surface waters. Because the snowpack was found to contain high concentrations of NO_3^- (Mollitor and Berg, 1980), and meltwater was shown to be enriched in NO_3^- relative to the snowpack during the initial phases of snowmelt (Cadle et al., 1985), surface-water acidification during the early spring was directly attributed to release of NO_3^- from the snowpack (Galloway et al., 1980). This hydrochemical mechanism provided an explanation for how acidic deposition could acidify surface waters during snowmelt without acidifying soils. A paper by Rascher et al. (1987), however, questioned whether the soil could also be a source of NO_3^- in the early spring. In this paper, concentrations of NO_3^- in solutions collected below the forest floor were 5-fold higher than concentrations collected directly below the snowpack. Rascher et al. (1987), however, could not rule out the possible effect of disturbance caused by the installation of the forest floor lysimeters on nitrification rates in their study.

Driscoll et al. (1989b) also presented evidence that N deposited in snow did not pass through the soil without being altered. Measurements in the western Adirondack Mountains showed that as meltwater from the snowpack passed through the Oa horizon, the total flux of N did not change, but the proportion of NH_4^+ decreased and the proportion of NO_3^- increased. The total flux of N in stream water was also found to be less than half the total N flux leaving the Oa horizon. This type of mass balance approach, however, did not identify the source of NO_3^- entering streams. Thus, the role of snowmelt in delivering NO_3^- to surface waters was not resolved.

The mobility of organic anions in soil received considerable attention before and during the NAPAP era through investigations of podzolization. Results showed that given the opportunity to pass through mineral horizons, organic anions tended to readily adsorb to particle surfaces (McDowell and Wood, 1984). Increases in organic matter content in the mineral soil, however, can decrease retention of organic anions, thereby increasing transport into surface waters (Lawrence et al., 1986). During storms, organic anions can also be transported from soils to surface waters through shallow flow paths that tend to have higher organic carbon (C) concentrations than deeper soils (Lawrence et al., 1988). This information indicated the need to consider the mobility of organic anions in assessments of soil and surface-water acidification.

Base Cation Leaching

The potential for acidic deposition to cause or contribute to the depletion of exchangeable base cations in soil through leaching by SO_4^{2-} and NO_3^- was recognized early in the NAPAP era (Cowling and Dochinger, 1980). Conclusive evidence of base cation depletion in the Northeast resulting from acidic deposition or any other cause, however, was not found in the 1980s. Acid addition experiments such as those described in David et al. (1990) documented increased base cation leaching associated with increased acid inputs. However, treatment levels were considerably higher than rates of ambient deposition, and soil disturbance effects caused by the experiments could not be completely ruled out.

The lack of data on the potential causes of base cation depletion resulted in opposing interpretations from the limited information that was available. For example, Johnson et al. (1982a) concluded that the release of exchangeable bases was generally sufficient to neutralize acid inputs, and that the buffering capacity provided by exchange reactions and weathering was exceeded by acid inputs only in rare circumstances. In contrast, van Breemen et al. (1984) concluded that in the Northeast, proton loading from atmospheric deposition exceeded internal proton inputs, and that neutralization of this additional acidity required dissolution of Al. Contrasting viewpoints continued to be published through the end of the NAPAP era. In a literature review, Johnson and Fernandez (1992) wrote that soils in eastern spruce–fir forests were naturally acidic and therefore unlikely to be further acidified by acidic deposition. In a second literature review, Joslin et al. (1992) wrote that current rates of Ca and Mg leaching in soils of red spruce forests were substantial in relation to exchangeable pools of these elements in both organic and mineral horizons, and that acidic deposition probably caused more than half of the leaching losses.

The role of base cation leaching was evaluated by the NAPAP-sponsored Direct–Delayed Response Project (DDRP) to assist in predicting future trends of surface-water chemistry that would result from various acidic deposition scenarios. This effort relied on cation-exchange models to determine the effect of leaching on base saturation and leachate chemistry. The approach was considered by the report authors to be a worst-case scenario because insufficient information was available to include resupply of base cations from weathering in the models (Church et al., 1989). Results indicated that acidic deposition would cause significant depletion of exchangeable base cations in the absence of resupply by weathering.

Interpretations of the effects of acidic deposition on exchangeable base cation pools varied because natural processes, such as mineral weathering, plant uptake, and organic acid leaching, were known to affect leaching rates of exchangeable base cations. However, the effects of these processes were not accurately measured during the NAPAP era. To simplify these measurement complications, Federer et al. (1989) focused on total rather than available soil Ca pools, and estimated that leaching, including that caused by acidic deposition, would decrease total Ca pools in New England from 11 to 23% over 120 years.

Aluminum Mobilization

Research on forest soils demonstrated that organic acids mobilized Al in the upper soil, which was subsequently precipitated lower in the profile through the natural process of podzolization (DeConink, 1980). Transport of Al into surface waters, however, suggested that podzolization was altered by inputs of H_2SO_4 and HNO_3 to the soil. Cronan and Scofield (1979) documented that Al concentrations in soil solutions increased with depth in the soil profile at Mt. Moosilauke, New Hampshire. Early analysis of surface waters indicated that Al concentrations were close to equilibrium with Al-trihydroxide solubility (Johnson et al., 1981). In a detailed characterization of soil solution chemistry, however, David and

Driscoll (1984) showed that organically bound Al was the predominant form of dissolved Al in all horizons, and that dissolved Al concentrations were undersaturated with respect to mineral forms of Al. Results of this study also suggested that vegetation played a role in transporting Al into the forest floor.

Cronan et al. (1986) developed an empirical model to quantify interactions between C and Al. This model successfully describes Al solubility as a function of the bound Al ratio (equivalents of adsorbed Al per mole of carboxyl groups) and pH. This, along with other field measurements (Lawrence et al., 1988), provided conclusive evidence that organic interactions, not kinetic constraints on Al dissolution, were most commonly responsible for undersaturation with respect to mineral solubility. The effect of acidic deposition on mobilization of organic Al, however, was not established during the NAPAP era. As a result, Al-trihydroxide solubility equations continued to be used in hydrochemical models of acidic deposition effects on surface-water chemistry through the end of the NAPAP era.

Interactions between Al and base cation concentrations were studied to a limited extent during the NAPAP era. Reuss (1983) showed that if soil base saturation was above 20%, soil solution Al concentrations were not strongly related to base saturation. Below a base-saturation value of 20%, however, soil solution Al concentrations were strongly related to base saturation as a result of high concentrations of exchangeable Al adsorbed to soil surfaces. An inverse correlation between soil base saturation and Al concentrations in soil water was later shown by an extensive field study that included sites in both Europe and North America (Cronan and Goldstein, 1989). Reuss et al. (1990) also developed the first model to link Al solubility to Ca–Al exchange. The model was ineffective, however, if Al solubility did not fit an Al-trihydroxide formulation, which was generally the case in Spodosols in the Northeast (David and Driscoll, 1984; Driscoll et al., 1985; Walker et al., 1990). As in the case with organic Al, the effect of acidic deposition on exchangeable Al concentrations was not resolved during the NAPAP era.

Increased attention was focused on Al–Ca interactions in the late stages of the NAPAP era as a result of a paper by Shortle and Smith (1988). This paper indicated that high Al concentrations in soil solutions might inhibit uptake of Ca by trees and lead to spruce dieback, a phenomenon commonly observed at high elevations in the late 1980s.

The Role of Soil in Surface-Water Acidification

Soils research on acidic-deposition effects began in the early 1980s from the perspective that soils were generally well buffered, both in terms of pH and base saturation (Johnson et al., 1982a; Krug and Frink, 1983). The question soon arose, however, as to how surface waters could become acidified without acidifying soils. The conservative transport of HNO_3 from the snowpack into surface waters provided a partial explanation but didn't address chronic acidification associated with SO_4^{2-}, nor episodic acidification at other times of the year. Reuss and Johnson (1985) published the only study during the NAPAP era that directly addressed this question. Through equilibrium modeling, they showed that the addition of acidic deposition could increase ionic strength sufficiently to shift soil solution ANC from positive to negative values. In this circumstance, acidic deposition could lower soil solution pH by 0.3 units, whereas surface-water pH would be lowered by a full unit. Degassing of soil solutions as they discharge at the surface increases pH, but only if ANC is positive. This mechanism was only operable, however, if soils derived most of their ANC from carbonic acid, and had a low but positive ANC before being impacted by acidic deposition. David and Vance (1989), the only workers to test the Reuss and Johnson model, found that carbonic acid did affect ANC, but to achieve the results predicted by the model, CO_2 concentrations had to be raised to unrealistically high levels (David and Vance, 1989). This mechanism, therefore, did not provide an explanation for how surface waters could become acidified without accompanying soil acidification.

During the NAPAP era, considerable attention was given to an alternative explanation for how changes in surface-water chemistry could be attributed to acidic deposition (Krug and Frink, 1983). These authors contended that extensive logging followed by forest fires in the early 1900s resulted in significant increases in soil and surface-water pH, which gradually reversed as forests regrew. The natural increase in soil acidity (in the form of organic acids) caused surface-water acidification that became apparent in the 1970s. Soil acidification caused by these large-scale disturbances coincided with increased industrialization and accompanying acidic deposition after World War II. Krug and Frink (1983) attributed elevated surface-water concentrations of SO_4^{2-} and NO_3^- to the immobilization of organic anions caused by inputs of H^+ from acidic deposition; the end result being replacement of organic acidity by inorganic acidity without a significant change in pH or leaching of cations. Experiments such as those done by James and Riha (1986) and Vance and David (1989) demonstrated that organic anions did have the capacity to buffer acid inputs to the soil. Results from an analysis of lake chemistry by Driscoll et al. (1989a) were also qualitatively consistent with a shift from organic to inorganic acidity. For no net change in the transport of acidity from soils to surface waters, however, the immobilization of organic anions would have to be assumed to be equal to the input of SO_4^{2-} and NO_3^-, on an equivalent basis. This assumption was proven to be invalid, however, when Vance and David (1991) showed that the addition of 200 µeq l^{-1} of H_2SO_4 reduced leaching of organic anions by a maximum of 75 µeq l^{-1} (assuming a charge density of 5 µeq l^{-1}, Cronan and Aiken, 1985).

Krug and Frink (1983) explained elevated concentrations of Al in surface waters by arguing that Al immobilization associated with podzolization was often ineffective, thereby allowing naturally derived, organically complexed Al to enter surface waters. These authors also contended that organic-Al complexation decreased substantially below pH 4.5, thereby allowing inorganic Al to naturally enter the most acidic surface waters. These explanations, however, were later shown to be inconsistent with field measurements. Surface waters with pH 4.5 to 5.0 were found to have inorganic Al concentrations that were over 50% of total Al concentrations in streams at the Hubbard Brook Experimental Forest (Lawrence et al., 1986), and organically complexed Al was shown to be the predominant Al species in soil waters with pH < 4.5 in both the Adirondack Mountains (David and Driscoll, 1984) and White Mountains (Driscoll et al., 1985).

Forest Productivity

In the 1980s, studies of pollution effects on forests focused on single or multiple causal mechanisms to explain individual tree mortality and decreased growth of trees. Research was not designed to use forest productivity measurements as an indication of how the system was responding to acidic deposition. In fact, few complete C budgets of forest ecosystems existed during this time for natural or human-manipulated sites worldwide. In 1991, relatively complete biomass budgets (above- and belowground) were found for only 10 forested ecosystems around the world, whereas 21 had complete productivity data (Vogt, 1991). The use of forest productivity data has been further complicated by the input of N as part of acidic deposition because this nutrient has generally been considered to be growth limiting (Vogt et al., 1990).

However, prior to this time, most of the research focused on understanding what controlled the growth of aboveground parts of trees and how fertilization could be used to increase aboveground tree growth under non–acidic deposition conditions (Ballard, 1979; Vogt et al., 1986; Vogt, 1991). Trying to understand the management impacts of N additions as fertilizers on tree growth was a focus of research that predated interest in pollution inputs of nutrients (Albrektson et al., 1977). This research increased knowledge of aboveground responses to excess nutrients but generally involved one-time applications of a limiting nutrient in varying amounts to identify optimum levels of fertilization. The focus of these fertilizer studies was the tree rather than the ecosystem because the primary goal of management was to increase tree growth rates. At the end of the NAPAP era, it was recognized that imbalanced input of nutrients, as occurs with acid deposition and fertilization, can have very negative impacts on tree growth even if that nutrient limits tree growth (Ingestad and Ågren, 1992).

During the early 1980s, several studies were also published that drew attention to the importance of belowground processes, and the fact that the belowground could not be understood from measurements of aboveground variables (Grier et al., 1981; Vogt et al., 1982; Santantonio and Hermann, 1985). The importance of studying the belowground, however, was not well documented until the late 1980s and early 1990s (Vogt et al., 1993, 1996). Therefore, the type of data necessary to study the effect of pollution on forest ecosystem productivity was not collected during the 1980s.

Although studies in the 1980s were not specifically designed to investigate the effects of atmospheric deposition on forest productivity, information related to productivity was gained from studies of pollution effects on individual tree species, and basic research on nutrient cycling, C allocation, and organic matter accumulation.

Tree Species Responses

During the NAPAP era, growth reductions and branch mortality of red spruce (*Picea rubens* Sarg.) (Eager and Adams, 1992) and sugar maple (*Acer saccharum* Marsh) (Allen et al., 1992) were hypothesized to be caused by atmospheric deposition. Atypical prevalence of pest and pathogen outbreaks were also documented (Wargo et al., 1993), but because pathogens are secondary agents that affect trees already under stress, the role of acid deposition was not clear. Further discussion of pathogens is given in Chapter 4.

Research also suggested that the atmospheric deposition of N was much higher at high elevations (Lovett et al., 1982) where dieback of red spruce stands was extensive (Siccama et al., 1982; Scott et al., 1985). Areas that were not polluted by atmospheric deposition typically had N addition rates of $<5\,\mathrm{kg\,ha^{-1}\,yr^{-1}}$ whereas areas with pollution inputs could vary from 20–25 to $>50\,\mathrm{kg\,ha^{-1}\,yr^{-1}}$ (Driscoll et al., 1989b). The existence of N as a component of acid deposition stimulated research on the impacts of increased inputs of N into forests. Waring (1987) hypothesized that coniferous species growing in boreal and subalpine zones would be unable to utilize additional inputs of N, especially as NO_3^-. In general, coniferous species were believed to be adapted to acquiring NH_4^+ and organic N, which were considered the dominant forms available in these ecosystems (Waring, 1987). Friedland et al. (1985) also suggested that red spruce foliage could be damaged by low winter temperatures if excess N levels prevented needles from hardening off.

Red spruce growing at high elevations of New York, Vermont, and New Hampshire showed significant declines in basal area and density that started about 1960 and continued through the 1980s (Johnson and Siccama, 1983) . The NAPAP Integrated Assessment concluded that the high-elevation red spruce forest was the only forest ecosystem that could be "confidently identified" as showing growth declines caused by acid deposition (NAPAP, 1991). These growth declines were attributed to inputs of S and N that "reduced midwinter cold tolerance by 4 to 10°C, compared with trees growing at the same location but protected from the acidic cloud water and rain" (NAPAP, 1991).

Red spruce and black spruce (*P. mariana* [Mill.] B.S.P.) inhabit seemingly similar environments at upper elevations in the Northeast, but red spruce has a lower tolerance for the cold and moisture stress (Hart, 1959). In spruce–fir stands in ME, spruce populations shifted from black spruce to black/red spruce hybrids, to red spruce, as soil fertility and drainage improved. Speculation had been published that drought contributed to the growth declines experienced by red spruce (Johnson and Siccama, 1983), but the relative significance of factors that could cause growth declines were not determined. Other studies of high-elevation red spruce showed the strongest evidence linking acid deposition to changing soil solution chemistry and nutrient availabilities to plants. For example, Shortle and Smith (1988) directly linked the increased Al cycling due to acid deposition to Ca deficiencies in red spruce, which indirectly decreased the resilience of red spruce to general stresses.

Sugar maple was the only other species identified during the NAPAP era to have unusually high rates of branch mortality and decreased growth (Allen et al., 1992; Millers et al., 1989). The observed decline of sugar maple resulted in the formation of the North American Sugar Maple Decline Project (NAMP) in 1987. Because air pollution was considered the causal agent for this decline, this research project was incorporated as part of NAPAP. Field studies were not sufficiently complete to determine the role of acid deposition in the decline of sugar maple by the close of NAPAP.

Nutrient Cycles

Prior to the NAPAP era, atmospheric N deposition was strictly viewed as a fertilization process that was beneficial to forest productivity. Research had established that most trees and decomposer microbes in forests had their growth limited by a low availability of N (Tamm, 1976). Losses of N from the ecosystem were much less than N inputs from atmospheric deposition, so ecosystems were assumed to benefit from sequestering this additional N in the soil and tree biomass (Tamm, 1976). The amount of N being added annually from the atmosphere was typically about one-tenth that added in single fertilizer applications as part of timber management in Washington and Oregon (Van Miegroet and Cole, 1985).

During the 1980s, however, this view gradually expanded to include the possibility that atmospheric N deposition may cause eutrophication of some forest ecosystems that could lead to decreased forest growth and productivity (Friedland et al., 1985; Aber et al., 1989). Experimental investigations to test this hypothesis had not been completed by the end of the NAPAP era, however.

Although N saturation of forest ecosystems was not documented during the NAPAP era, Heil et al. (1988) reported that increased N inputs in the Netherlands caused declines in some grassland species. This study suggested that plant species that are less effective at utilizing increased amounts of available N may be out-competed by other species more capable of using these inputs to increase their growth rates. Other effects of atmospheric N inputs were also hypothesized. Because atmospherically deposited N was known to be present in a form that foliage could directly assimilate (Waring, 1987), availability of N for plant growth could potentially become less dependent on decay rates of surface organic layers (Vogt et al., 1986). The relative significance of foliar and root uptake of N in forests, and the effect of atmospheric deposition on this relation, however, was not determined during the NAPAP era.

The effects of seasonal variation of N deposition on forest ecosystems was also identified as a potentially important area of investigation. Mineralization of N in the soil was shown to be greatest during the summer (Melillo, 1977; Thorne, 1985) or early autumn (Vogt, 1987), whereas atmospheric inputs of N to the soil were shown to be greatest during spring snowmelt (Driscoll et al., 1989b). The relation of seasonal variations of N inputs to seasonal variations in the capacity of the ecosystems to retain N, however, was an additional area not investigated during the NAPAP era.

Knowledge of excess nutrient effects on productivity was advanced by fertilization experiments on individual species, which showed that excess input of one nutrient could significantly stress a plant by causing nutrient imbalances that resulted in decreased plant growth rates, increased susceptibility to pests and pathogens, and changes in carbon allocation patterns (Ingestad and Ågren, 1992). In a study of loblolly pine (*Pinus taeda* L.), N additions to P-limited trees caused reduced leaf areas, decreased volume growth, and extremely high foliar N levels and N/P levels (Vose, 1987). Experiments also showed that the amount of stemwood production per unit of leaf area could be manipulated by low level additions of N fertilizer but was adversely affected at higher application levels (Vose, 1987). Belowground biomass and production were not estimated in the loblolly pine studies, however, so effects on C allocation could not be determined. Nevertheless, results did support the findings of seedling experiments which also showed that nutrient imbalances could result from application of high levels of a single nutrient (Ingestad and Ågren, 1992), as well as the suggestion that nutrient deficiencies could be symptomatic of acidic deposition (Bondietti et al., 1990).

During the 1970s and 1980s, studies conducted in forest ecosystems not impacted by acid deposition helped to define some of the causal mechanisms through which acidic deposition could decrease the availability of nutrients in soils (Johnson et al., 1977; Johnson and Cole, 1980; van Miegroet and Cole, 1984, 1985; Dahlgren et al., 1991; Vogt et al., 1987a,b). This research demonstrated that decreased nutrient availability could occur naturally on a site due to the ability of some plant species to increase the cycling of Al (e.g., Al-accumulating species, Vogt et al., 1987a,b) and NO_3^- (e.g., N-fixing plants, van Miegroet and Cole, 1984, 1985)—both elements whose availability could be increased by acid deposition, as well. For example, mountain hemlock (*Tsuga mertensiana* [Bong.] Carr.) increases the cycling of Al and its potential toxicity to plant communities by accumulating Al at significantly higher levels in its foliage and fine roots than other tree species that grow in similar environments (Vogt et al., 1987a,b; Dahlgren et al., 1991). Despite the fact that mountain hemlock comprised only 20% of the basal area at one site, it contributed more than half of the Al being added annually to detritus as part of litter inputs to this site (Vogt et al., 1987b).

Trees have some degree of tolerance for Al by either preventing root uptake through symbiotic associations or by taking up Al and then sequestering it in nontoxic complexes within plant tissues (Vogt et al., 1987b; Dahlgren et al., 1991). A key factor in Al effects is the threshold at which plant tolerance to Al is reached and whether acid deposition has increased Al availability beyond this threshold, as was suggested by Shortle and Smith (1988). Other research in the 1980s also suggested that those species, such as red spruce, which showed growth declines or mortality as a result of acidic deposition were not well adapted to preventing Al uptake or the accumulation of Al within roots (Persson, 1990). If so, these species would have a low threshold of tolerance to Al and probably be inefficient at assimilating Ca, which is needed to maintain the membrane integrity within roots (Persson, 1990).

The effects of increased nitrification on the availability of exchangeable bases were also evaluated in the absence of acidic deposition. This was accomplished in a study that compared differences in soil chemistry between a 40-year-old red alder stand that was adjacent to a 40-year-old Douglas fir (*Pseudotsuga menziesii* [Mirb.] Franco.) stand in Washington (van Miegroet and Cole, 1984, 1985). Because red alder fixes N and increases N availability, nitrification rates and subsequent NO_3 leaching in this stand were much higher than in the Douglas fir stand. Production of HNO_3 through nitrification increased leaching of bases from soil exchange sites (Johnson and Cole, 1980), resulting in base saturation values that were 50% lower in soils of the red alder site than the Douglas fir site. Up to 14% of the exchangeable Ca, Mg and K were exported annually through NO_3-mediated leaching (Van Miegroet and Cole, 1984). This study showed how increasing the amount of NO_3 in the soil (an effect that can also result from atmospheric deposition of N) increases the leaching of bases from soil exchange sites.

Studies in the 1980s in forest ecosystems unimpacted by acidic deposition demonstrated the need to (1) understand how tree species influence nutrient availability through natural acidification processes, and (2) determine how acidic deposition affects growth and productivity of the forest.

Carbon Allocation

Prior to the NAPAP era, the importance of C allocation as a functional adaptive response of plants to their environment had been identified (Mooney, 1972). In herbaceous plants, substantial shifts in carbon allocation were found to be associated with changes in biotic and abiotic site factors, such as nutrient concentrations, soil moisture, light, and temperature (Penning de Vries, 1975; Wareing and Patrick, 1975; Brix, 1971; Pate and Layzell, 1981; Keyes and Grier, 1981; Waring et al., 1985; Chapin et al., 1987). Field studies of trees presented evidence, although inconclusive, which similarly suggested that biotic and abiotic variables affected allocation of carbon to woody tissues, roots, foliage, and defensive secondary chemicals (Grier et al., 1981; Coley et al., 1985; Vogt et al., 1986). Laboratory studies also suggested that photosynthate was preferentially translocated to those tissues actively acquiring the resource that is most limiting to growth at that time. Therefore, when N is limiting growth, plants will increase C allocation to roots to increase N uptake, whereas a plant limited by light may increase its leaf area.

Research showed that N fertilizer applications generally increased growth rates, if the availability of water was not limiting (Grier and Running, 1977; Gholz, 1982). Numerous field studies also showed a strong positive relationship between aboveground production and nutrient availability (Mitchell and Chandler, 1939; Lea et al., 1980; Chapin et al., 1987). This relationship was supported by physiological investigations, which showed that 75% of the leaf N may be located in chloroplasts, which are involved in C fixation (Chapin et al., 1987). Field research by Gholz (1982) and Vose (1987) suggested a strong coupling of moisture, nutrients, and aboveground growth response for conifers.

Organic Matter Accumulation

Surface organic matter accumulation and carbon accumulation within the soil was not a major area of research in the Northeast during the NAPAP era, because soil organic matter was assumed to turn over in 100-year cycles and therefore would not respond to several decades of pollution in a measurable way. This perception began to change, however, with the finding that soil organic matter existed in several different forms that vary in lability (Strickland and Sollins, 1987). The response of microbial activity to increased nutrient inputs was also shown to be an important factor because fungal tissues play an important role in the cycling of available nutrients. Bååth and Söderström (1979) calculated that up to 19.6% of the N and up to 18.2% of the P in the soil was immobilized in fungal hyphae in the soil of coniferous forests in Sweden. Further research in Swedish forests showed that a one-time application of N fertilizer caused major changes in the populations of decomposer organisms responsible for breakdown of leaf materials (Bååth et al., 1981). In that study, application of 150, 300, and $600\,kg\,ha^{-1}$ of ammonium nitrate (NH_4NO_3) resulted in decreased fungal and bacterial biomass, as well as decreased respiration rates in soil and surface organic horizons five years after application (Bååth et al., 1981). The cause of these reductions in microbial activity could not be deduced from the study, but results did suggest that increased N additions from acid deposition would have an effect on the decomposition rate of litter and, therefore, the availability of other nutrients needed by plants.

The role of surface organic accumulations as sites of nutrient availability was extensively studied in many forest types around the world before and during the NAPAP era (Vogt et al., 1986). Under unpolluted natural conditions, the forest floor was shown to be more important for nutrient conservation in high-elevation and high-latitude forests than at lower elevations or in hardwood forests (Vogt et al., 1986). In high-elevation balsam fir forests in New Hampshire forest floor masses were reported as high as $117\,Mg\,ha^{-1}$ (Vitousek et al., 1982). A direct correlation between the mass of the forest floor and the content of N and P in this horizon was also shown (Vogt et al., 1986), indicating that high-elevation forests have a greater proportion of available nutrients in the forest floor than do low-elevation forests. It was believed during this period that, under natural conditions, little NO_3^- is generated in the forest floor, and nutrient cycling is tightly controlled by the decomposition rates of surface organic horizons and the subsequent mineralization of nutrients from this horizon. The hypothesis of N saturation, developed in the later stages of NAPAP, also raised the possibility that increased leaching of nutrient cations could occur if atmospheric deposition increased the availability of N and stimulated nitrification rates (Aber et al., 1989).

In forests with thick organic matter accumulations, plants are highly dependent on mycorrhizal associations, which function as nutrient-acquiring organs for trees (Harley and Smith, 1983). Acid deposition could alter the functioning of decomposers and mycorrhizal symbionts, thereby decoupling plants from their soil environment by impeding the uptake of nutrients. The effect of acid deposition on mycorrhizas began to receive attention near the end of the NAPAP era, following the suggestion of Arnolds (1991) who stated that dramatic declines in mycorrhizal populations were occurring in Europe due to acid deposition.

Summary

Results of NAPAP research indicated that acidic deposition had chronically acidified surface waters in geologically sensitive areas of the

Northeast (Sullivan et al., 1990), causing loss of fish populations and disruption of aquatic ecosystems (Baker et al., 1990a). Naturally acidified surface waters were also identified, however, and considerable uncertainty remained regarding the regional extent of chronic and episodic acidification of surface waters resulting from acidic deposition. Research results on soils indicated that acidic deposition enhanced leaching of Al and base cations, but the effect of increased leaching on the soil pool sizes and availability of these cations was not quantified. Evidence of surface-water acidification was recognized as a possible, but unproven, indication of soil base depletion and decreased soil pH, and the relative significance of acidic deposition and natural processes as causes of soil acidification remained unresolved at the close of NAPAP.

From the NAPAP forest effects research it was concluded that multiple stressors (which included ozone and acidic deposition) were still threatening "the long-term structure, function, and productivity of many sensitive ecosystems by changing their chemical composition and nutrient cycling" (NAPAP, 1993). NAPAP research, however, did not document any forest health effects that could be directly attributed to acidic deposition, other than red spruce dieback at high elevations (NAPAP, 1993). Research did show that acidic deposition could potentially decrease nutrient retention in forest ecosystems and cause imbalances in the availability of nutrients, which may have contributed to the growth declines of red spruce observed in the 1980s. Research conducted during the NAPAP era also indicated that further research was needed in the following areas: (1) the effect of multiple stressors on tree growth, (2) the thresholds at which additional inputs of nutrients will cause the ecosystem to degrade, (3) interactions between nutrients and Al, and the nutritional imbalances that may occur, (4) understanding the ecosystem-level impacts of acid deposition and the feedbacks that exist at different spatial and temporal scales, (5) identification of ecological indicators that can be used to determine if a system is degrading before visual symptoms are expressed, and (6) the influence of natural disturbance cycles (e.g., droughts) on the expression of acid deposition effects.

Research Advances in the 1990s

The end of the intensive research phase of the NAPAP program brought to a close much of the acidic deposition research throughout the country. Research at a greatly reduced scale continued in the Northeast, however, through the continuation of monitoring programs, further interpretation of data collected during the NAPAP era, and the initiation of other projects, such as those funded through the USDA Forest Service Global Change Research Program. The extensive databases and information collected through NAPAP provided the basis from which interdisciplinary studies could be designed with modest project resources. Acidic deposition researchers in the 1990s were also able to address the issue of recovery following 30 years of declining acidic deposition rates. As a result, questions could be addressed from an ecosystem perspective in a manner not possible during the NAPAP era. Significant advances were therefore achieved despite the substantial reduction in resources that were committed to the problem. The following section summarizes the most important research on surface water, soils, and forest productivity published from late 1992 through 1998.

Surface Waters

Investigations of acidic deposition effects on surface-water chemistry continued through the the 1990s, primarily through extensions of monitoring programs established before or during the NAPAP era, as well as some new hydrological investigations. Trends identified through these programs provided important information for assessing effects of declining trends of acidic deposition.

Surface-Water Chemistry

Before the discovery of acidic deposition, little monitoring of surface-water chemistry was being done in regions where surface waters were poorly buffered, with two important exceptions. Records of stream-water chemistry in the Catskill Mountains of New York showed that NO_3^- concentrations in stream water of protected forest watersheds in the New York City water supply system had shown overall increases during this century (Murdoch and Stoddard, 1992). This paper drew considerable attention, because it was the first to show trends in stream chemistry that were consistent with the theory of N saturation as introduced by Aber et al. (1989). The New York City program also provided data that showed stable concentrations of Ca^{2+} plus Mg^{2+} from 1952 through 1972, but declining Ca^{2+} concentrations from 1970 to 1996 (although the Ca^{2+} record included a period from 1975 to 1983 when data were unavailable, Lawrence et al., 1999).

The other important record of stream chemistry that predated the discovery of acidic deposition in the Northeast was developed at the Hubbard Brook Experimental Forest, beginning in 1963. Up through 1993, concentrations of SO_4^{2-} and basic cations (including Ca^{2+}) steadily decreased in stream water (Likens et al., 1996). Concentrations of NO_3^- also showed an overall decrease, but the trend was less consistent than for SO_4^{2-} or basic cations; pH increased slightly during this period (Likens et al., 1996).

By the mid-1990s, NAPAP-initiated monitoring had also accumulated records of 12 years or longer in a number of locations. Driscoll et al.

(1995) found that from 1982 to 1994, concentrations of SO_4^{2-} and base cations in 16 Adirondack lakes had decreased markedly, whereas pH, ANC, and concentrations of NO_3^- did not exhibit any systematic trends. For the same period, Stoddard et al. (1998) found a small decrease in ANC (-0.6 µeq l^{-1}, $P < 0.05$) in 13 Adirondack lakes and a small increase in ANC for 23 lakes in Vermont and Maine ($+0.8$ µeq l^{-1}, $P < 0.001$). Concentrations of SO_4^{2-} declined steadily in lakes in both the Adirondack and New England data sets. In Catskill streams, Lawrence et al. (1999) found consistent decreases in Ca^{2+} concentrations and ANC from 1984 to 1997, and Murdoch et al. (1998) found that NO_3^- concentrations showed no clear trend from 1984 to 1995.

Trends in surface-water chemistry over the past 10 to 30 years can be summarized for acid-sensitive regions of the Northeast as follows: (1) consistent decreases in concentrations of SO_4^{2-} and base cations, (2) minimal changes in ANC and pH, and (3) an increase in NO_3^- concentrations in the 1980s followed by a decrease in the 1990s that resulted in little or no overall trend. Despite the steady decrease in concentrations through this period, SO_4^{2-} remained the predominant anion in most surface waters in the mid-1990s (e.g., Stoddard, 1994; Driscoll et al., 1995; Lawrence et al., 1999). In an analysis of data (predominantly baseflow samples) from 159 streams draining small watershed in Massachusetts, Vermont, New Hampshire, and Maine, Hornbeck et al. (1997) showed that 50% of the streams had SO_4^{2-} concentrations greater than 55 µeq l^{-1}, whereas 50% of the streams had NO_3^- concentrations greater than only 5 µeq l^{-1}. In this analysis, 50% of the streams were also found to have pH < 6.0 and Ca concentrations <100 µeq l^{-1}.

The declining trend in SO_4^{2-} concentrations through the 1980s and 1990s, without a corresponding NO_3^- decline, did increase the relative importance of NO_3^- somewhat as a contributor to chronic surface-water acidification. Sullivan et al. (1997) contended that NO_3^- concentrations in Adirondack lakes were sufficient to account for acidic-deposition–induced pH declines determined from paleolimnological analyses. The most significant aspect of NO_3^-, however, was proven to be its role in episodic acidification, which was documented in detail by Wiggington et al. (1996). During high flows, NO_3^- concentrations were shown to increase by an order of magnitude, whereas SO_4^{2-} concentrations remained fairly stable. An exception to this relationship was shown by data from Pennsylvania, where NO_3^- concentrations were continuously low and SO_4^{2-} concentrations increased with increases in flow (Wiggington et al., 1996). A significant, but secondary role of organic acids in both chronic (Munson and Gherini, 1993; Driscoll et al., 1994) and episodic (Wiggington et al., 1996) surface-water acidification was also confirmed by further work in the 1990s. Little new information on Al concentrations in northeastern surface waters was published in the 1990s.

Hydrological Effects on Surface-Water Chemistry

Results of the NAPAP-initiated Episodic Response Program, completed in the early 1990s, provided detailed documentation of the frequency and severity of stream acidification associated with high flows at sites in the Adirondack Mountains and Catskill Mountains, New York, and western Pennsylvania (Wiggington et al., 1996). Further hydrological investigations in the 1990s helped to clarify mechanisms through which episodic acidification occurred. These studies focused on the effects of watershed scale and areal variability; a significant shift from NAPAP-era research, which was conducted almost entirely in small watersheds, with an emphasis on vertical variability of subsurface flowpaths. Lawrence et al. (1996) identified a positive correlation between the extent of channelized overland flow within a watershed and the total flow at the base of the watershed ($P < 0.01$; $R^2 = 0.88$). Upslope extension of channelized flow resulted primarily from return flow water that had previously infiltrated the soil surface before discharging to channels. This study quantified the relationship between variable source-area contributions to streamflow and episodic acidification, which was inferred in the study of Lawrence et al. (1988). Further research in the Catskill Mountains showed that neutralization of acidic stream water during base flow increased longitudinally from watersheds that ranged in drainage area from 0.2 to 166 km^2, in a manner consistent with an increase in the value of a topography-based index of subsurface contact time (Wolock et al., 1997).

These studies demonstrated that (1) precipitation has considerable interaction with soil even in headwater drainages, (2) the chemical characteristics of soil in variable source areas influence episodic variations in stream chemistry, and (3) the effectiveness of acid neutralization processes within the soil and subsoil is strongly related to surface topography. These findings indicate the necessity of incorporating areal variations of watershed characteristics to extend watershed hydrochemistry models to large watersheds. Evaluating the recovery potential of acidified stream reaches will also require spatial characterization of the biogeochemical processes that control acid neutralization.

Soils

Progress in understanding the effects of acidic deposition on soils accelerated after the close of NAPAP through the application of new investigative approaches, which showed that soil properties were more sensitive to change than was generally thought during the NAPAP era.

Nitrogen Retention and Release

In the early stages of NAPAP research, SO_4^{2-} was identified as the primary anionic component of acidic deposition and acidified surface waters. As a result, the focus throughout the program was on SO_4^{2-} as the primary acidifying agent. The NAPAP Integrated Assessment Report acknowledged, however, that the role of NO_3^- in episodic acidification warranted further study (Sullivan et al., 1990). This conclusion, and the paper by Aber et al. (1989), which described how atmospheric deposition could potentially overfertilize forest ecosystems, contributed in large part to a redirection of research toward N cycling investigations in the 1990s. The large number of post-NAPAP N cycling studies will not be summarized in this section, but are discussed in the later section of this chapter on Forest Productivity, and in Chapter 10. The following section summarizes results of studies that have addressed N mobility from a watershed perspective.

Stoddard (1994) published an exhaustive review of available information on watershed retention of N in which he concluded that high rates of atmospheric N deposition have contributed to the degradation of water quality in the Northeast, primarily through episodic acidification. In this analysis, he developed a system to classify the degree of watershed N saturation from seasonal patterns of NO_3^- concentrations in surface waters. From this analysis he also concluded that N saturation could not be viewed solely as a function of atmospheric deposition rates, but that others factors also affected N loss from watersheds in ways that were not well understood.

Burns et al. (1998) identified a hydrologic mechanism in Catskill watersheds that complicated the use of stream chemistry to assess N saturation. Elevated concentrations of NO_3^- observed in streams during the growing season were initially assumed to be an indication that N was not growth-limiting in these watersheds. Analysis of subsurface flow paths, however, showed that stream concentrations during the growing season were not always a good indication of biotic N demands during this time of year. Streams that were fed by a groundwater flow system maintained moderate levels of NO_3^- concentrations through the summer, but streams that relied on shallow sources of subsurface water exhibited NO_3^- concentrations that approached undetectable values. Isotope tracer measurements showed that the groundwater system was recharged primarily during the non–growing season, and that discharged water was 4 to 8 months of age. Growing-season concentrations of NO_3^- in streams therefore reflected soil water concentrations during the non–growing season.

A watershed experiment at Bear Brook, Maine, demonstrated that addition of ammonium sulfate ($[NH_4]_2SO_4$) at a rate approximately twice ambient N deposition substantially increased NO_3^- concentrations in the stream (Norton et al., 1994). Analysis of ^{15}N distribution, which was added with the $(NH_4)_2SO_4$, indicated, however, that most of the NO_3^- in stream water was not derived from nitrification of $^{15}NH_4^+$. The treatment apparently stimulated nitrification of pre-existing N. This effect, however, was inconsistent with results from studies in the Catskill Mountains.

Murdoch et al. (1998) found that NO_3^- concentrations in stream water
were not related to atmospheric N deposition rates on either a seasonal or
an annual basis, but were strongly related to average annual air
temperature. A positive correlation between soil temperature and net
nitrification suggested that microbial dynamics were controlling release of
NO_3^- to streams. Lawrence et al. (1999) also found that the elevational
trend of NO_3^- concentrations during snowmelt in a Catskill stream was the
reverse of the elevational trend of atmospheric deposition. This relation
was attributed to an increase in retention of N in the soil as elevation
increased, temperature decreased, and soil organic matter increased.
Despite high levels of deposition, annual loss of N in stream water in this
watershed was 15% of atmospheric inputs and 0.2% of the total N pool in
the Oa horizon (Fig. 8.1).

   The effect of climate on watershed release of N was also documented by
Mitchell et al. (1996). High concentrations of NO_3^- in stream water
following unusually low winter temperatures were measured at Bear
Brook watershed Maine, the Hubbard Brook Experimental Forest,
Arbutus watershed in the Adirondack Mountains, and Biscuit Brook
watershed in the Catskill Mountains. This response was attributed to soil
freezing and thawing, which released N previously immobilized by
microbes.

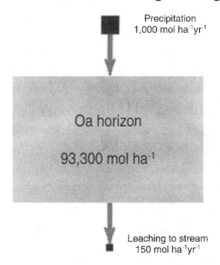

Figure 8.1. Nitrogen fluxes (black boxes) and pool size of the Oa soil horizon
(gray box) in Winnisook watershed, Catskill Mountains, NY. Box sizes are
proportional by their mol ha^{-1} value.

Calcium Cycling

Research during the NAPAP era indicated that acidic deposition probably enhanced leaching of basic cations, but the effect of increased leaching on pool sizes of available base cations was unknown, primarily because weathering rate estimates were highly uncertain. The first evidence of a decrease in the size of basic cation pools in the Northeast was shown by Shortle and Bondietti (1992) in a comprehensive data review. Although variation in methods and stand characteristics precluded direct comparisons, concentrations of exchangeable Ca^{2+} plus Mg^{2+} measured before 1950 were generally twice the concentrations measured after 1970.

Johnson et al. (1994a) were the first to document long-term changes in concentrations of exchangeable Ca, the predominant exchangeable base cation in most forest soils of the Northeast. In this study, 48 soil profiles in the Adirondack Mountains, first sampled in 1930 to 1932, were resampled in 1984 by the same methods of collection and analysis. Results showed that extractable Ca concentrations were significantly less in 1984 than in 1930 to 32. Estimates of net forest growth developed by Johnson et al. (1994a) for 16 sites suggested that Ca uptake by trees during the approximate 50-year interval was almost equal to the loss of available Ca from the soil. Detailed Ca budgets developed at Whiteface Mountain, New York, by Johnson et al. (1994b), however, indicated that the loss of forest floor Ca from 1986 to 1990 significantly exceeded the loss rate of the previous 50 years. In this study, the authors concluded that the increased loss rate of Ca from the forest floor was most likely the result of a declining trend in atmospheric deposition of Ca (Hedin et al., 1994). In the mineral soil, weathering inputs of Ca (estimated to be 4% annually of the exchangeable Ca pool and a fraction of a percent of the total Ca pool) were assumed to counter leaching losses.

The use of strontium (Sr) isotopes as an analog for Ca provided a new method in the 1990s to estimate Ca weathering rates in forest ecosystems. Development of this method represented a significant advance over the commonly used watershed mass balance approach, which required the assumption that the pool size of exchangeable basic cations was static. In the first application of this approach, Miller et al. (1993) found that 50 to 60% of the Sr (and presumably Ca) in vegetation and forest floor exchangeable pools at Whiteface Mountain originated from the atmosphere. The discovery that trees were utilizing Ca deposited by the atmosphere, as well as Ca released from weathering, was significant in light of the regionwide declining trend of atmospheric Ca deposition discovered by Hedin et al. (1994). Bailey et al. (1996) also used Sr isotopes at Cone Pond watershed, New Hampshire, to show that 32% of the Sr in vegetation and forest floor exchangeable pools originated from the atmosphere and that the ecosystem was experiencing a net depletion of Ca.

Results from the Whiteface Mountain and Cone Pond studies were extremely useful in providing detailed information on Ca availability and ecosystem cycling; however, the regional significance of this information was unknown because the limited number of studies conducted at other sites had used a variety of methods that were not directly comparable. To address this problem, Lawrence et al. (1997) conducted a soil survey in 12 red spruce stands that were selected to encompass the range of environmental conditions experienced by this species in the Northeast. The data collected in this survey were related to other studies by conducting comparisons of chemical analysis methods. Results showed that exchangeable Ca concentrations in the Oa horizon varied by an order of magnitude among the 12 sites, which were located from the western Adirondack Mountain region to eastern Maine. These authors concluded that the primary factor causing the variability in the Oa horizon was mineralogy of parent material, but that exchangeable Ca concentrations in the mineral soil were probably reduced by acidic deposition at all sites. A strong relationship was observed between relative Ca weathering rates (estimated from bedrock and surficial mineralogy) and Ca saturation (exchangeable Ca concentration expressed as a percentage of cation-exchange capacity) in the Oa horizon, whereas in the mineral soil, this relation was weakly expressed (Fig. 8.2). The difference in the relation between these two horizons was by explained by the depletion of exchangeable Ca in the mineral soil below the threshold at which soil solution becomes strongly buffered by exchangeable Al (Reuss, 1983). This type of buffering by exchangeable Al minimizes further leaching of base cations, which caused the differences in parent material mineralogy to be masked. Depletion of Ca was less advanced in the Oa horizons, where high root activity helped reduce Ca loss through vegetative recycling. This study provided additional evidence of the increasing importance of the forest floor in supplying Ca for root uptake. Results of the study of Lawrence et al. (1997) supported the findings of Shortle and Bondietti (1992), Johnson et al. (1994a,b) and Bailey et al. (1996), which in sum, indicated that a regional decline in available Ca concentrations had occurred in the Northeast in the second half of the 20th century.

Likens et al. (1996) published the first paper to link long-term trends in stream chemistry to soil base depletion in the Northeast. Although steadily decreasing acidic deposition rates have been measured at the Hubbard Brook Experimental Forest over the past 30 years, ANC of stream water at this site has shown only slight increases. These authors contended that depletion of available soil bases was preventing a strong recovery of surface-water chemistry on the basis of a decrease in the ratio of base cations to acid anions in stream water. This study, however, was unable to provide soil data that could be directly linked to acidic deposition or stream chemistry.

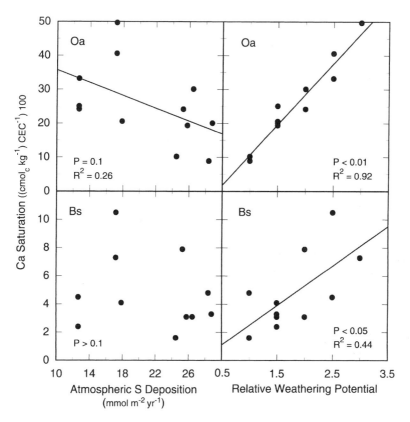

Figure 8.2. Ca saturation as a function of atmospheric deposition and relative weathering potential in Oa and Bs horizons in northeastern red spruce forests. (Modified from Lawrence et al., 1997.)

A subsequent study by Lawrence et al. (1999) in the Neversink River Basin, in the Catskill Mountains of New York, utilized an elevational gradient of atmospheric deposition to investigate soil base leaching. Exchangeable base cation concentrations, in both native soil and homogeneous soil (the upper 10 cm of the B horizon collected from a 3-m^2 area then thoroughly mixed) that was incubated in mesh bags in the field, decreased with increasing elevation in accordance with an elevational increase in atmospheric deposition (Fig. 8.3). The concentration ratio of base cations to acid anions in stream water also decreased with increasing elevation in this watershed. To further investigate the relation between soil base depletion and stream chemistry, these workers conducted laboratory leaching experiments to quantify the effect of base saturation and strong acid inputs on Ca leaching (Fig. 8.4). Results showed that decreasing either acid inputs or base saturation lowered Ca concentrations in leachates, and that the slope of the leachate Ca–acid anion relation (see

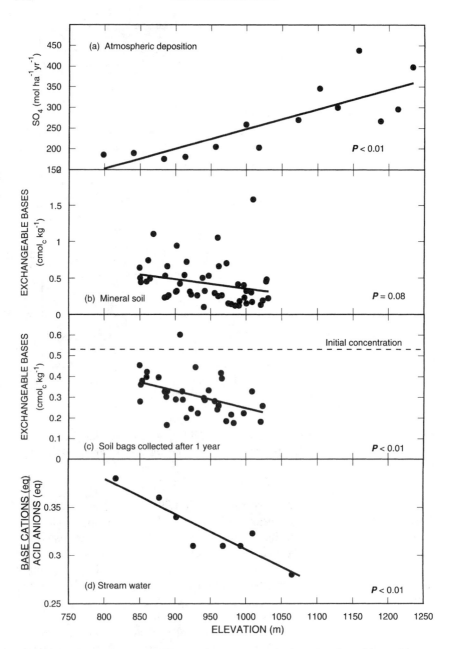

Figure 8.3. Elevational gradients of (a) atmospheric SO_4 deposition, (b) exchangeable base concentrations (Ca, Mg, Na, K) in the mineral soil, (c) exchangeable base concentrations in mineral soil bags after 1 year, and (d) base cation to acid anion ration in stream water in Winnisook watershed, Catskill Mountains, NY. (Modified from Burns et al., 1998.)

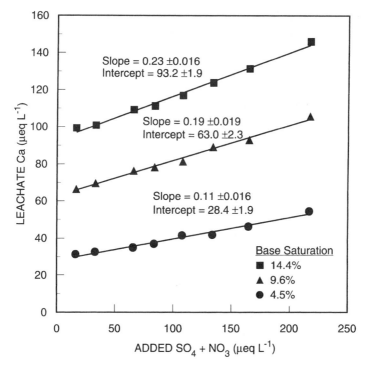

Figure 8.4. Concentration of Ca in solutions leached through mineral soil with a vaccum extractor, as a function of SO₄ plus NO₃ concentrations in the added solution. Results of duplicate leachings are shown. 95% confidence intervals of regression statistics are given. (Modified from Lawrence et al., 1999.)

Fig. 8.4) decreased as base saturation decreased. This experiment supported the interpretation of Likens et al. (1996) that a decreasing ratio of base cation to acid anion concentrations in drainage waters resulted from a decrease in available soil bases. Trends of stream water ANC indicated that recovery of soil base saturation from declining acidic deposition was not occurring in the Neversink River Basin.

Aluminum Mobilization

Because little was known about the chemistry of Al in natural systems before the NAPAP era, much of the research conducted during this period focused on characterizing Al speciation in soils and waters. By building on the findings of the NAPAP era research, some progress was made in the 1990s despite substantially reduced research on Al. Rustad and Cronan (1995), the first to quantify Al cycling in the forest floor, showed that Al leached out of the forest floor could account for about 80% of the total mass of Al entering surface waters. This study also showed that biocycling of Al through litterfall, and possibly root translocation, were significant pathways for inputs of Al to the forest floor, as was admixing with mineral soil horizons. The largest fraction of Al in the forest floor was identified as primary and secondary minerals, but control of solution Al concentrations was attributed to complexation with organic matter.

In a study of forest-floor Al by Lawrence et al. (1995), exchangeable Ca concentrations were found to be unrelated to exchangeable H concentrations but inversely related to exchangeable Al concentrations. Analysis of archived soil samples from the Hubbard Brook Experimental Forest also showed that concentrations of exchangeable and extractable Al increased in the forest floor from about 1970 to 1990, whereas concentrations of exchangeable and extractable Ca decreased. These results were attributed to transport of Al from the mineral soil to the forest floor through biocycling and water movement (a rising water table or capillary movement), which displaced Ca due to the high affinity of Al for organic binding sites. Ratios of inorganic Al to Ca in soil solution of the mineral soil were found to be positively related to the exchangeable Al content of the forest floor (Fig. 8.5). Although acidic deposition has not lowered the pH of the forest floor, it probably has lowered the pH of the mineral soil, which has led to the mobilization of Al within this section of the profile. Subsequent transport of this Al from the mineral soil into the forest floor results in reduced storage of available Ca.

The effects of pH and organically complexed Al on dissolved Al concentrations in soils were established during the NAPAP era through empirical modeling (Cronan et al., 1986), but the chemical reactions that controlled these relations remained undefined. In an effort to theoretically describe the effect of pH and organic complexation on Al solubility, Tipping et al. (1995) developed an equilibrium model of humic ion binding. Dissolved Al concentrations predicted by this model were inconsistent with concentrations predicted by the empirical model of Walker et al. (1990) under conditions common in northeastern forest soils. The one similarity in these approaches was that concentrations of organically bound Al were assumed to remain constant. Results of the manipulation experiment conducted at Bear Brook, Maine (Norton et al., 1994), however, indicated that organically bound Al concentrations are changed by inputs of strong acids (Lawrence and David, 1997). In this experiment dry $(NH_4)_2SO_4$ was added at a rate of $1800\,eq\,ha^{-1}\,yr^{-1}$ from 1989 to 1995 to a red spruce stand in eastern Maine. The addition of $(NH_4)_2SO_4$ elevated nitrification, which produced HNO_3 and lowered solution pH of the Oa horizon. Concentrations of copper chloride $(CuCl_2)$-extractable Al (which represents organically complexed Al) were significantly higher in the Oa horizon of the treated stand than in this horizon of the reference stand, but Al saturation with respect to mineral solubility was approximately the same in both stands. These results were explained by mobilization of Al through accelerated weathering of mineral

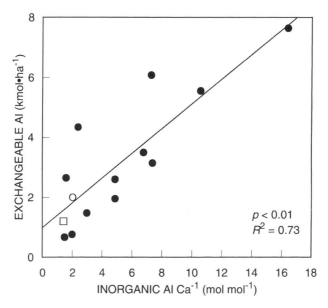

Figure 8.5. Exchangeable-Al content of Oa horizons of red spruce stands in New York, Vermont, New Hampshire, and Marine as a function of the molar concentration ratio of inorganic Al and Ca in B-horizon soil solution. Each filled circle represents the mean of 18–36 soil–solution samples (combined into 6–12 samples before analysis) collected at each of 12 sites. The open circle represents the mean of 68 soil samples and 31 seep-water samples, collected in a mixed hardwood forest in Winnisook watershed, Catskill Mountains, NY. The open square represents soil and seep-water concentrations measured at Tunk Mountain, ME by Rustad (1988). (Modified from Lawrence et al., 1995.)

particles caused by lowered pH in the treated stand. Much of this mobilized Al bonded with organic matter, causing higher $CuCl_2$-extractable Al concentrations in the treated stand than in the reference stand. Lawrence and David (1997) concluded that dissolved Al concentrations in these soils were the result of complex mechanisms through which mineral matter, organic matter, and pH interact to control Al solubility; mechanisms that are only partially understood, and are not accurately represented in current Al solubility models.

Forest Productivity

In the 1990s, measurements of forest productivity and biomass increased substantially as a result of a growing interest in the effects of atmospheric CO_2 increases on forest ecosystems (and vice versa). By 1997, at least 200 complete data sets of organic matter biomass and productivity (above- and belowground vegetative, forest floor, and soil) were available from different forests around the world (Vogt et al., 1996). Research conducted in the 1990s also contributed to understanding the effects of pollutants on ecosystems through holistic studies that included the biology of the belowground. Pre-NAPAP studies had typically focused on the effects of pollutants on one or two characteristics of individual species, under tightly controlled experimental conditions. The shift in focus to field investigations of ecosystems provided more realistic process information, although lack of experimental control increased the challenge of isolating the effects of interacting factors (Vogt et al., 1997).

Post-NAPAP era studies of atmospheric deposition effects on forest health and productivity in the Northeast gained direction from earlier studies of sugar maple and red spruce. These investigations also benefited from new information on soil chemical processes obtained in the 1990s, which improved the current understanding of atmospheric deposition effects on nutrient balances and availability. The following two sections summarize results from fertilization studies designed to evaluate the effects of soil base depletion on sugar maple growth and the effects of N saturation and base depletion on spruce–fir forests. The third section is a case study that presents previously unpublished results of a fertilization experiment designed to study interacting effects of N and Ca availability on red spruce–dominated ecosystems.

Sugar Maple, Climate, and Nutrients

Sugar maple decline in the mid- to late 1980s was often associated with acid deposition (Millers et al., 1989). In 1987, the North American Sugar Maple Decline Project was initiated to identify the cause of of sugar maple decline in the U.S. and Canada. Most of the research conducted on sugar maple focused on possible links between Ca deficiency and crown dieback (Ellsworth and Liu, 1994; Wilmot et al., 1995; Zahka et al., 1995). Forest management may also have affected sugar maple growth because many of stands that originally were comprised of ~4 to 52% sugar maple were converted to >90% sugar maple to increase sap production. Conversion of these stands may have occurred on sites that did not favor sugar maple because soils were drought prone.

Wilmont et al. (1995) measured annual growth rates of apparently healthy sugar maple stands and those that exhibited significant crown dieback in northern Vermont during 1989 to 92. They found that in the stands with crown dieback the average annual growth rates from 1953 to 1992 were approximately half the rates measured in the stands with healthy crowns during 1989 to 1992. Those stands exhibiting crown dieback also had lower concentrations of exchangeable soil bases, lower soil pH, higher exchangeable soil Al concentrations, and lower foliar Ca concentrations than stands with healthy crowns. However, after three years of lime additions to the stands with crown dieback, they observed increases in growth rates, soil pH, and concentrations of exchangeable Ca, and decreases in exchangeable Al concentrations; responses that suggested crown decline was caused by Ca deficiency in these stands (Wilmot et al., 1996). Although this study could not directly link changes in soil chemistry to crown dieback at this site, results nevertheless identified a strong relationship between soil Ca availability and long-term growth rates, which suggested that declines of available soil Ca, such as those identified in the Adirondack Mountains (Johnson et al., 1994a,b), Catskill Mountains (Lawrence et al., 1999) and White Mountains (Likens et al., 1996; Bailey et al., 1996) may have reduced sugar maple growth rates within the region where moderately fertile sites have degraded to marginal sites.

Sugar maple decline since the mid-1980s has also been documented in northwestern Pennsylvania, where crown dieback has led to extensive mortality. In a fertilization experiment begun in 1985, Long et al. (1997) found that liming significantly increased diameter growth, improved crown vigor and increased flower and seed crops of overstory sugar maple in these declining stands. Liming also increased exchangeable base–cation concentrations in the soil and decreased concentrations of exchangeable Al. This study, however, did not show that deficiency of nutrient bases was the direct cause of decline because other factors such as drought and insect defoliation were also evident (Kolb and McCormick, 1993). The study did, however, provide further evidence of the importance of Ca and Mg nutrition in maintaining the health of sugar maple.

Although fertilization studies showed strong relationships between sugar maple health and nutrient base availability, these studies did not isolate pollution effects from other interacting factors. Therefore, the causes of sugar maple decline in the Northeast continued to be debated past the mid-1990s. Evidence that past declines were generally followed by periods of recovery (Robitaille et al., 1995) and that past management strategies resulted in establishment of sugar maple stands on marginal sites may have also contributed to sugar maple declines in the Northeast (USDAFS, 1993). Evidence that both natural and pollution-related processes have been involved in past sugar maple declines suggests that future research should focus on the multiple interactive factors that control the long-term productivity of sugar maple, rather than single mechanisms that trigger periodic declines.

Fertilization Effects on Red Spruce

Experiments to evaluate the effects of increased N and Ca availability on red spruce growth were conducted during the 1990s in both the Northeast and Southeast. From 1988 to 1994, N additions were made to a spruce–fir stand in southeastern Vermont to evaluate the effects of varying N availability on this ecosystem (McNulty et al., 1996). From 1988 to 1990 plots that received either 15.7 or $19.5\,\mathrm{kg\,N\,ha^{-1}\,yr^{-1}}$ had significantly higher growth rates for red spruce, maple (*Acer* spp.), birch (*Betula* spp.) and balsam fir (*Abies balsamea* [L.] Miller) than control plots, whereas plots that received either 25.5 or 31.4 kg N ha^{-1} yr^{-1} showed little or no growth increase. From 1991 to 1994, all species, at all treatment levels, showed significantly lower growth rates and higher mortality than in control plots. Authors of this study concluded that N fertilization had resulted in ecosystem saturation that was causing a possible shift in species composition. Although growth of all measured species had decreased, and mortality increased as a result of N additions, spruce and fir were not regenerating, but deciduous stump sprouts and seedlings were observed. These responses led the authors to propose that additional inputs of N would result in the replacement of spruce and fir by birch and maple as the dominant species in this stand. The more effective utilization of N by the deciduous species than the spruce and fir could give them a competitive advantage under conditions of high N availability. These results provided experimental support of the theory developed during the NAPAP era, which stated that forest ecosystem stress and lower productivity could result from eutrophication (Ingestad and Ågren, 1992).

Experiments to evaluate the effects of Ca fertilization on red spruce were conducted in the early 1990s in Great Smokey National Park to determine if low concentrations of available Ca and Mg in soil limited growth rates. Results of an experiment by van Miegroet et al. (1993) showed that needle weight of spruce saplings with initial needle concentrations of 1700 μg Ca g^{-1} increased significantly after two applications (over two years) of either CaCl$_2$ or CaCl$_2$ plus MgCl$_2$. Saplings with initial needle concentrations of 1940 μg Ca g^{-1} did not show a response to fertilization, suggesting that Ca deficiency may be indicated by a needle concentration of 1700 μg Ca g^{-1} or less.

A similar Ca fertilization experiment conducted by Joslin and Wolfe (1994) found that the same amounts of Ca and Mg additions also increased needle weight of mature, co-dominant red spruce trees on Whitetop Mountain, VA, and supported the suggestion of van Miegroet et al. (1993) that a needle concentration of less than 1700 μg Ca g^{-1} indicates Ca deficiency. This study also observed no foliar growth response to additions of N, indicating that N availability was not limiting growth rates at this site.

Laboratory experiments with seedlings were conducted by McLaughlin et al. (1993) to investigate acidic deposition effects on carbon metabolism. This study showed that the ratio of net photosynthesis to dark respiration was positively correlated with foliar Ca concentrations. The effect of acid mist and rain (pH 3.0 and 3.8, respectively) on this ratio was also decreased by addition of Ca to the soil.

Results of van Miegoet et al. (1993), Joslin and Wolfe (1994) and McLaughlin et al. (1993) provided valuable information by identifying a connection between Ca availability and growth of red spruce. However, because these experiments tested the response of a single species to short-term manipulations, without evaluating ecosystem interactions, the application of the results to the issue of long-term productivity of red spruce stands is uncertain. By maintaining manipulations over six years and collecting a variety of ecosystem measurements, the McNulty et al. (1996) study provided a more holistic perspective of the changes that may be taking place as a result of N saturation. The Ca fertilization studies described above indicate that the type of ecosystem investigation conducted by McNulty et al. (1996) is the next step in furthering our understanding of how Ca availability and Ca–N interactions affect forest productivity. Some progress in this area has been made, as described in the following case study, which presents preliminary results of the first experiment specifically designed to address these questions in northeastern forests.

Ecosystem Dynamics in Mixed Spruce–Hardwood Stands as Affected by Nitrogen and Calcium Additions: A Case Study in Stand and Species Response

The effects of increasing Ca, N and Ca + N levels have been studied in spruce stands located in the Hubbard Brook Experimental Forest, New Hampshire, Groton State Forest, Vermont, and a stand near Big Moose Lake in the Adirondack Mountains, New York, from 1992 to 1997. At all sites, red spruce was the most common tree species. A series of twelve $30 \, m \times 30 \, m$ plots were established at Hubbard Brook and Big Moose Lake, whereas six plots were established at Groton. Three replicate plots per treatment plus three controls were established at each site. Each year the Hubbard Brook and Big Moose Lake plots received three applications of N fertilizer (NH_4NO_3) totaling to $100 \, kg \, N \, ha^{-1} \, yr^{-1}$. All three sites received Ca fertilizer each year as a 1:1 ratio of calcium sulfate ($CaSO_4$) and calcium chloride ($CaCl_2$) at a total rate of $160 \, kg \, Ca \, ha^{-1} \, yr^{-1}$. The Ca + N plots at Hubbard Brook and Big Moose Lake received $100 \, kg \, ha^{-1} \, yr^{-1}$ of N fertilizer and $160 \, kg \, Ca \, ha^{-1} \, yr^{-1}$. As part of this research, litterfall mass, above- and belowground tree productivity and biomass, soil nutrient concentration, litterfall nutrient concentration, nutrient concentrations in tree components, soil organic matter and its fractionation into labile and recalcitrant forms, susceptibility to root pathogens, mycorrhizal colonization, and plant secondary chemical production have been measured. In this section, radial growth increment data are presented.

Results and Discussion

Results of this manipulation have been evaluated in terms of stand and species growth responses, carbon allocation, and soil organic matter accumulation. Although the three sites were similar in terms of stand composition, radial growth variations in the response to fertilizer additions indicated distinct differences in how these spruce stands responded to increased inputs of N, Ca, or N + Ca. These results suggest that the effects of nutrient addition will vary across the landscape and that a uniform response to higher Ca or N should not be expected at the species level. How spruce and associated species responded to the higher inputs of N, Ca, or both nutrients varied depending on where the site existed along a gradient of health (determined from an index of how many pathogens were present on the site), patterns of detrital decomposition (whether decomposition was complete to CO_2 and water or to CO_2, water, and organic acids) and how much a particular nutrient limited growth at each site.

Growth Responses

Application of N fertilizer caused varying diameter growth responses depending on the health of the spruce trees (Table 8.1). For example, the spruce forests at Hubbard Brook, which are healthy on the basis of a minor presence of fungal pathogens (see Chapter 4) showed significant growth increment increases after four years of N applications. These results indicated that N addition from atmospheric deposition was not impairing the growth of red spruce at this site. The significant radial growth increases due to N applications in the spruce forests in New Hampshire were also observed at the Big Moose Lake site for some hardwood species, such as yellow birch (Table 8.2). At this site, however, spruce mortality was an ongoing process, fungal pathogens were prevalent

Table 8.1. Mean Radial Increment of All Tree Species from 1991 to 1996, Located at Hubbard Brook, NH, Groton, VT and Big Moose, NY, and Grouped by Nutrient Addition Treatment Which Was Applied Annually Starting in 1992. Nutrient Addition Rates Were Calcium = $160\,kg\,Ca\,ha^{-1}\,yr^{-1}$; Nitrogen = $100\,kg\,N\,ha^{-1}\,yr^{-1}$; Ca/N at $160\,kg\,Ca\,ha^{-1}\,yr^{-1}$ + $100\,kg\,N\,ha^{-1}\,yr^{-1}$; n = Number of Trees Measured

Site	Control Group	Treatment Groups		
		Ca additions	N additions	Ca + N additions
	Mean Radial Increment, mm (n)			
Hubbard Brook, NH	0.85 a[a]	0.76 a	1.04 b	1.12 b
	(284)	(294)	(284)	(318)
Groton, VT	0.72 a	0.78 b	—	—
	(204)	(210)		
Big Moose, NY	0.69 a	0.94 b	0.84 b	0.88 b
	(228)	(244)	(244)	(234)

[a] Within the same site location (row) radial increment means followed by a similar lower case letter are not significantly different (Tukeys standardized range, $P < 0.10$).

Table 8.2. Mean Radial Increment for Individual Tree Species on Spruce-Dominated Sites Receiving Ca Additions ($160\,kg\,ha^{-1}\,yr^{-1}$), N Additions ($100\,kg\,ha^{-1}\,yr^{-1}$), Ca + N Additions ($160\,kg\,Ca\,ha^{-1}\,yr^{-1}$ + $100\,kg\,N\,ha^{-1}\,yr^{-1}$) for Five Consecutive Years and No Inputs (control) for Plots at Hubbard Brook, NH, Groton, VT, and Big Moose, NY (n = number of trees measured)

Site and Species	Treatment Groups			
	Control Group	Ca additions	N additions	Ca + N additions
	Mean Radial Increment mm (n)			
Hubbard Brook, NH				
Spruce	0.95 abA[a]	0.87 aA	1.00 abA	1.19 bB
	(84)	(102)	(79)	(96)
Balsam fir	0.76 abA	0.68 aA	1.01 bcA	1.31 cB
	(68)	(66)	(61)	(72)
Yellow Birch	0.76 abA	0.64 aA	1.02 bA	0.83 abA
	(60)	(60)	(60)	(60)
Mt. Ash	1.04 aA	0.58 bA	0.93 abA	0.86 abA
	(36)	(30)	(42)	(42)
Maple	0.76 aA	0.98 abA	1.28 bA	1.25 bB
	(36)	(36)	(42)	(48)
Groton, VT				
Spruce	0.68 aAB	0.77 aB	—	—
	(90)	(72)		
Balsam fir	0.43 aA	0.43 aA	—	—
	(12)	(24)		
Yellow Birch	0.81 aB	0.93 aB	—	—
	(54)	(54)		
Mt. Ash	0.76 aAB	0.81 aB	—	—
	(48)	(60)		
Big Moose, NY				
Spruce	0.47 aA	0.54 aB	0.57 aA	0.66 aB
	(60)	(78)	(90)	(84)
Balsam fir	0.99 aB	1.23 aA	1.29 aB	1.30 aA
	(78)	(72)	(54)	(54)
Yellow Birch	0.36 aA	0.84 bA	0.84 bB	0.62 abB
	(42)	(46)	(60)	(48)
Mt. Ash	0.78 aB	1.26 aA	0.85 aB	1.08 aA
	(48)	(48)	(54)	(48)

[a] Within the same species (row) radial increment means followed by a similar lower case letter are not significantly different. Within each site location/treatment combination (column), means followed by the same uppercase letter are not significantly different (Tukeys standardized range, $P < 0.10$).

(see Chapters 4 and 7) and red spruce trees did not respond to N addition. The patterns recorded during this study closely followed those proposed by Tamm (1976), in which initial increases of N act as a fertilizer, but as N availability continues to increase, growth responses to further N inputs cease. Results of the N fertilization suggest that N saturation had not occurred at the ecosystem level at the New York site, but that certain tree species had reached a threshold of response, whereas others had not; results similar to those observed by McNulty et al. (1996).

Identifying these thresholds at the tree species or ecosystem level is difficult in field experiments because different species in a given environment may accumulate nutrients at different ratios, a process typically controlled by genetics (Westman, 1985). Most of these studies, therefore, have been conducted under controlled conditions using smaller sized plants (Vogt et al., 1997). Results of these studies do show a general pattern of slower growth at low nutrient levels, exponential growth when nutrient availabilities are optimal, luxury uptake as nutrient levels increase beyond levels required for growth, and eventual toxicity as the levels increase further. However, utilization of this type of information to understand the effects of acidic deposition has been complicated by genetic differences in abiotic resource utilization at the species level (Johnson et al., 1996) and the difficulty in determining the level of exposure of a particular pollutant (Vogt et al., 1997).

The stands at the Big Moose site showed a stronger response to Ca additions than N additions, although the effects of the different fertilizer treatments were not significantly different (see Table 8.1). This contrasted with the Hubbard Brook site, where no response to Ca inputs was observed, which suggests that Ca was not limiting growth at this site. Radial growth of the stands at the Groton site was increased by Ca fertilization, although the increase was not as large as the responses recorded for N additions at the other sites. Thus, Ca availability appeared to be limiting the growth of trees at both Groton and Big Moose but not at Hubbard Brook. These results demonstrate that responses are dependant on site characteristics that could include factors such as the initial nutrient status and the developmental stage of the stand, which are directly influenced by geology, atmospheric deposition, and land use history.

The growth responses by individual tree species to the treatments at the sites in New York, Vermont and New Hampshire were generally nonsignificant, but consistent with patterns observed for the stands (see Tables 8.1 and 8.2). We had hypothesized that Ca deficiency was responsible for the patterns of reduced growth of red spruce in the Northeast, which may have resulted from the tendency of this species to concentrate root activity (and therefore their zone of nutrient uptake) in the forest floor. Increases in radial growth increment of red spruce that were observed in response to Ca additions at Groton and Big Moose Lake, however, were not significant (see Table 8.2), and red spruce at Hubbard Brook showed a nonsignificant decrease in response to Ca additions. Only the negative response of mountain ash at Hubbard Brook and the positive response of yellow birch at Big Moose Lake were statistically significant (see Table 8.2). Perhaps the presence of hardwood species enhances litter decay, and hence Ca cycling rates (Gosz et al., 1973), to a level that satisfies

Table 8.3. Mean Radial Increment Grouped by All Species and All Treatments by Year at Hubbard Brook, NH, Groton, VT, and Big Moose, NY

Year	Hubbard Brook		Groton		Big Moose	
	Radial Increment (mm)	n	Radial Increment (mm)	n	Radial Increment (mm)	n
1991	1.00 a[a]	196	0.79 ab	69	0.87 a	159
1992	0.99 ab	196	0.86 a	69	0.89 a	159
1993	0.98 ab	196	0.86 a	69	0.85 a	159
1994	0.92 ab	196	0.75 abc	69	0.80 a	159
1995	0.91 ab	197	0.67 bc	69	0.88 a	159
1996	0.88 b	199	0.58 c	69	0.76 a	159

[a] Lower case letters used to show significant radial increment differences within columns (Tukeys standardized range, $P < 0.10$).

the Ca requirements of spruce. The tendency for rooting in the forest floor by red spruce (see Chapter 6) may also enable this species to acquire sufficient Ca from decomposition of hardwood leaf litter that occurs within the forest floor. There is also a possibility that spruce growth rates are being controlled by Al levels in the rooting zone for which Ca fertilization is unable to compensate. Aluminum-induced Ca deficiency may also be impairing uptake of other nutrients if Ca levels are insufficient to maintain root integrity (Persson, 1990).

There are year-to-year variations in the response of forest stands to pollution that have to be considered when assessing the role of pollutants in affecting the long-term growth of forests. Studies, usually conducted during a 3- to 4-year period (often less), can be complicated by inter-annual variations. The variation of dominant factors controlling the growth of stands will also vary by location, as there are no true replicates in nature. Forest stands at Big Moose Lake, for example, showed no significant change in annual radial increment between 1991 to 1996 (Table 8.3). However, forest stands at the Hubbard Brook site had significantly higher growth in 1991 compared with 1996, and the Groton site had significantly higher radial growth rates in 1991, 1992, and 1993 than in 1996 (see Table 8.3). This type of year-to-year variability highlights the necessity for long-term studies of ecosystem processes and the importance of understanding that generalizations of patterns should not be inferred from data collected at one site.

Carbon Allocation

There are many indirect effects of changing the N status of a site that can stress plants and result in reduced growth, such as shifts in growth and metabolic activity of fine roots, branches, and foliage. Part of the increased aboveground production that occurs with N fertilization is due to increased foliage biomass and photosynthetic efficiency (Brix, 1971; Albrektson et al., 1977). However, little information is available on the structural, functional, and stress resistance properties of the increased foliage area. Increased shoot growth has been shown to be associated with decreased allocation of C to roots, which results in increased C availability for other tree functions (Gower et al., 1992). In a fertilized young Scots pine (*Pinus sylvestris* L.) stand, increased production occurred in foliage, foliage-bearing branches, and coarse roots, while allocation to fine roots decreased from 41 to 17.5% of total annual production (Axelsson, 1981).

At the Big Moose Lake and Groton sites, addition of N caused similar decreases in allocation to fine root biomass. These decreases in fine root biomass have implications for the uptake of other nutrients because the surface area for nutrient uptake is reduced. This reduced fine root surface area also has strong implications for the ability of plant roots to avoid Al toxicity (Dahlgren et al., 1991). For silver fir (*Abies amabilis* [Dougl.] Forb.) in the WA, the large allocation of C to maintaining fine root biomass was hypothesized to be a mechanism of this species that diminishs the negative impacts that Al can have on root growth and nutrient uptake (Vogt et al., 1987b). Both factors may contribute to decreased Ca uptake, which could feedback to decreased fine root membrane integrity, eventually resulting in decreased tree growth on these sites. Studies with red spruce have shown that there is a strong relationship between the vitality of fine roots and the condition of the crown (Wargo et al., 1988), and that Al can induce Ca deficiencies in spruce that appear to reduce cambial growth and ultimately cause trees to shed part of the crown (Shortle and Smith, 1988).

Nitrogen fertilization can also influence C allocation by (1) causing decreased storage of carbohydrates in foliage and boles (Ericsson, 1979; Vogt et al., 1985), (2) increasing protein plus free amino acid-N and chlorophyll content of foliage (Barnes and Bengtson, 1968), and (3) decreasing production of secondary plant chemicals in foliage (Horner, 1987). Increased N availability may initially result in higher levels of carbohydrate reserves in foliage than in nonfertilized trees (Ericsson, 1979), but these reserves are depleted rapidly during the growing season (Vogt et al., 1985). Changes in the amount of stored carbohydrates are important because these metabolites are utilized in secondary chemical production (e.g., lignin, phenolics, terpenoids) and may be related to root production and longevity (Wargo, 1972; Vogt et al., 1985; Marshall and Waring, 1985). These changes in C allocation may be reflected in a decreased ability of trees to protect themselves against pathogens, but at present no direct link between pathogens and acid deposition has been shown (see Chapter 4). Pathogens have been considered as secondary agents responding to stressed plants, but it is important to determine whether pathogen activity exceeds levels which would be expected from normal development of these forests.

Litterfall

As part of the experimental manipulations of N and Ca levels in New York, Vermont, and New Hampshire, litterfall inputs were dramatically changed, but responses varied by site and foliage type (conifer or deciduous). If changes occur in the magnitude of litterfall inputs, ecosystem level cycling and availability of nutrients will change. As expected, litterfall of senesced conifer needles in treatment plots was not significantly higher than in the control plots at the Big Moose Lake and Hubbard Brook sites, reflecting the small change in conifer radial growth increments recorded at these sites. The amount of green needle litterfall, however, did increase 30-fold in response to Ca additions at Big Moose, which suggests that increased Ca availability was increasing the amount of winter loss of foliage. This higher input of green litterfall results in increased nutrient inputs to the forest floor at this site because labile nutrients have not been translocated out of the foliage prior to being dropped. At Groton, Ca additions also increased the amount of conifer litter, but only by a factor of 2. This increase in coniferous litterfall, however, was not accompanied by an increase in radial increment of the coniferous species, although the stand did show a significant radial growth increase in response to Ca fertilization (see Table 8.1).

Although an increase in coniferous green litterfall was measured, the proportion of total litterfall in deciduous tissues increased in response to the addition of Ca and also Ca + N. This increased cycling of decomposable litterfall with a high Ca content could help to ameliorate the negative impacts of Al toxicity if base saturation of the soil is increased (Vogt et al., 1986).

Preliminary data from the fertilization experiments also suggest that additional input of N and Ca may increase the role of the soil organic matter (SOM) in retaining nutrients. It appears that there is a seasonal change in SOM that may be related to the role of the microbial carbon pool in carbon sequestration. This effect was significant at the Hubbard Brook site, but less so at the other two sites, which suggests that C sequestration is dependant on the type and degree of litter decomposition. Increased N additions could possibly increase C sequestration. These results support the earlier finding that there is a strong positive relationship between the amount of organic matter that accumulates in the soil and the N content of this organic matter (Vogt et al., 1995). If higher levels of nutrients increase net productivity, the system may be able to retain more C in the soil and thereby improve the water retention capacity of the soil and also the availability of soil nutrients for plants.

Conclusions

Despite decreased levels of funding, significant progress in understanding the environmental effects of atmospheric deposition continued through the 1990s. Much of this progress was the result of an increased emphasis on interdisciplinary investigations that addressed the linkages among forests, soils, and surface waters. These studies were able to build upon the extensive data collection programs and basic process studies of the NAPAP era to address fundamental questions about soil acidification. Through an expanded knowledge of soil processes, progress in the evaluation of forest responses was accelerated.

A Conceptual Model of Acidic Deposition Effects on Soils

One of the most important advancements in the 1990s was an increased understanding of how trends in surface-water chemistry reflected interactions between acidic deposition and soil chemistry. The study of Johnson et al. (1994a) was able to compare soil chemistry in 1984 to soil chemistry 50 years earlier, but there were no long-term records of soil chemistry that could be directly related to trends in acidic deposition. Long-term records of stream chemistry were available, however, from which inferences could be drawn regarding trends in soil chemistry; as shown by the papers of Likens et al. (1996) and Lawrence et al. (1999). By combining this information with other important advances in the 1980s and 1990s, a conceptual model of soil processes can be developed that characterizes the distinct differences in how organic and mineral soil horizons respond to acidic deposition.

Results of NAPAP-era research showed that soil chemistry is controlled by the interactions of complex processes; therefore, the effects of acidic deposition on soils could not be viewed as merely a lowering of pH. In northeastern forest soils, H is generated naturally within the forest floor through decomposition, resulting in solution pH values in the forest floor as low as 3.3 (David and Lawrence, 1995). Natural acidity in these soils is primarily associated with organic anions, which, after leaching from the forest floor, tend to be readily sorbed to surfaces in the mineral soil as described by podzolization theory (DeConink, 1980). As a result, mineral soils have solution pH values that are typically 0.5 to 1.0 pH unit higher than forest floors (David and Lawrence, 1996). Removal of organic acids from solution depends on the extent to which subsurface flow paths intersect mineral soil or till before discharging to surface waters. Watersheds, with steep slopes that have little or no mineral soil and/or wetlands with thick accumulations of organic matter, allow significant amounts of organic acidity to reach surface waters, whereas watersheds with well developed mineral soils (0.5 m or more thick) are generally effective at removing most of the organic acidity generated in the forest floor (Lawrence et al., 1986; David et al., 1992).

Acidic deposition introduces H in association with SO_4^{2+} and NO_3^-, which are less effectively retained in the mineral soil than organic anions (Cronan et al., 1978; Fernandez et al., 1995). Exchange of H for base cations, or consumption of H through weathering, releases base cations to the soil solution. This process results in leaching of base cations out of the solum, facilitated by SO_4^{2+} and NO_3^- that provide countercharge to maintain solution electroneutrality. As base saturation of the mineral soil is lowered, leaching of base cations decreases and neutralization of H through mobilization of Al increases as discussed by van Breemen et al. (1984), and observed in field experiments by Lawrence et al. (1999). The end result is acidification of the mineral soil, reflected by a decrease in base saturation and an increase in Al and H saturation. This process results from the introduction of acidity associated with mobile anions, not merely an addition of H to the already abundant sources of natural acidity.

Direct effects of acidic deposition on the forest floor are limited because the pH of soil solution in this horizon is typically less than acidic deposition as a result of organic acids (Johnson and Fernandez, 1992; Lawrence et al., 1995). Mobilization of Al in the mineral soil that results from a lowering of pH by mobile acids (H_2SO_4, HNO_3), however, increases the ratio of Al to Ca in soil solution within the mineral profile. When this solution is transported into the forest floor, Al is more effectively retained than Ca owing to the high affinity of Al for organic binding sites; a process that increases Al saturation of the forest floor. An increase in the ratio of Al to Ca in mineral soil solutions also enhances uptake of Al relative to Ca by roots in the mineral soil horizons (Cronan and Grigal, 1995), which causes increased biocycling of Al into the forest floor. Because this process involves replacement of exchangeable Ca by Al, soil and soil solution pH in the forest floor is not decreased by acidic deposition but can potentially be increased if Al replaces a significant amount of exchangeable H as well as Ca (Ross et al., 1996; Skyllberg, 1994).

Reduction of Ca availability in the forest floor by acidification of the mineral soil is compounded by the decreasing trend of atmospheric deposition of Ca that has extended into the 1990s (Hedin et al., 1994). The utilization of atmospherically deposited Ca by trees (Miller et al., 1993; Bailey et al., 1996) may reflect increased atmospheric deposition of Ca in past decades or the effects of mineral soil acidification on Ca uptake from the mineral soil. In either case, the current high Al to Ca ratios in the mineral soil suggest that trees will have difficulty compensating for lower atmospheric inputs by increasing uptake from the mineral soil (Cronan and Grigal, 1995). These factors suggest that increased Ca demand by trees will reduce Ca leaching, thereby limiting neutralization of drainage waters.

Future Research Needs

Despite steady declines in atmospheric deposition since the mid-1960s, ANC of surface waters in the Northeast has shown minimal improvement. Recent progress in understanding acidic deposition effects on soils suggests that recovery of surface waters is being limited by current levels of soil base saturation. Sustained recovery of surface waters will likely require recovery of ecosystem base status. To achieve a net ecosystem increase in base cation content, inputs of bases through weathering and atmospheric deposition must exceed leaching losses, and net gains must be stored in a root-available form. Large uncertainties regarding weathering rates, controls of leaching losses and storage of available nutrient cations, however, make current assessments of soil recovery potential uncertain.

Weathering fluxes in the Northeast have been estimated without the assumption of constant exchangeable base pools at only Whiteface Mountain, New York (Miller et al., 1994), Cone Pond Watershed, New Hampshire (Bailey et al., 1996), and the Hubbard Brook Experimental Forest (Likens et al., 1996). An indication of past changes in soil base saturation is also provided by sites with long-term records of surface-water chemistry, but current base saturation trends and ecosystem base status for the overall region remain largely unknown. Development of non-steady state methods for estimating weathering fluxes is a highly significant advancement, but further work is needed to reduce the uncertainty of these methods. Furthermore, development of an approach to efficiently apply these methods to a large number of sites will be essential for conducting a regional assessment of soil recovery potential.

Because recovery of soil base saturation is dependent on leaching losses as well as weathering inputs, trends in forest nitrogen cycling affect soil base saturation through the release of NO_3 to soil solution. Manipulation studies suggest that increased inputs of N cause increased leaching of NO_3 (Norton et al., 1994). Temporal and spatial studies of watershed NO_3^- release, however, indicate that, on a year-to-year basis, leaching of NO_3^- is more closely tied to climate than atmospheric deposition (Murdoch et al., 1998; Lawrence et al., 1999). Uncertainty of the role acidic deposition plays in NO_3^- leaching is unclear, because annual rates of atmospheric N inputs are at least two orders of magnitude smaller than ecosystem N pools. Effects of acidic deposition and climate variation on release of NO_3^- from large pools of soil N need to be better understood to improve assessments of base cation status.

In addition to increasing leaching losses of base cations, acidic deposition has also reduced the capacity of the soil to store base cations by increasing Al saturation in both the mineral soil and forest floor. The mechanism through which Al is mobilized remains only partially understood, however, so a quantitative understanding of processes that would decrease Al saturation has not been established. Research to date does indicate that reversing accumulation of reactive Al will be hindered by high exchangeable Al/Ca ratios in the mineral soil, the tendency for Al to strongly adsorb to organic matter, and continued decreases in atmospheric deposition of Ca. Further research is needed to define and quantify Ca–Al–organic matter interactions in both the mineral soil and forest floor.

Research on forest effects in the 1990s emphasized the need to understand the role of multiple factors in the declines recorded during the 1970s, 1980s and 1990s. Reoccurring droughts that occurred over these decades reflect the fact that these declines are probably phenomena related to long-term cycles of drought and associated insect outbreaks in the regions (Allen et al., 1992) that are exacerbated by air pollution (Long and Horsley, 1997). Drought appears to aggravate the response of certain tree species to their environment and appears to push them toward a threshold where higher cycles of mortality and growth decreases will be measured. Depletion of soil Ca and increased N availability are likely to have lowered this threshold in some areas of the Northeast. Air pollution effects on forest productivity therefore must be studied within the context of long-term climatic patterns, as in the study of McLaughlin and Downing (1995).

Research also needs to tease apart how different species respond to acid deposition in terms of cycling limiting resources because ecosystem-level responses to pollution may be controlled by species-level responses. Preliminary results suggest that the ability of spruce to grow on sites where Ca availability has decreased due to lower atmospheric inputs and changing soil chemistries may depend on the presence of hardwood species capable of maintaining higher Ca levels in the system.

Future research also needs to characterize acidic deposition effects at a landscape level, recognizing that the response of particular ecosystem may not transfer to other similar ecosystems (Vogt et al., 1997). The legacy of past land use, ecosystem succession, and disturbance history will influence nutrient availability and the sensitivity of the site to disturbances, which determine the overall health of that system. To address these factors, more sites will need to be studied to identify the variables that control how species and ecosystems responds to acidic deposition. It will be important to determine which variables are transferable to other sites and which factors modify the expression of some of the site specific attributes. The lack of change in radial growth of spruce in response to Ca fertilization at the Hubbard Brook site coupled with significant increases in radial increment at Groton and Big Moose sites demonstrates that Ca is not a universally limiting factor controlling the growth of spruce.

Ecosystem sensitivity to acidic deposition also needs to be determined to assess whether forest health is changing. For example, changes in soil chemistry, such as increased prevalence of reactive Al, may not impact forest health if the plant species are adapted to these changes. It will be important to link the common variables that have been used to study acidic deposition effects to determine the thresholds of ecosystem and/or species sustainability.

References

Aber JD, Nadelhoffer KJ, Steufler P, Melillo JM (1989) Nitrogen saturation in northern forest ecosystems. BioScience 39:378–386.

Alban DH (1982) Effects of nutrient accumulation by aspen, spruce, and pine on soil properties. Soil Sci Soc Am J 46:853–861.

Albrektson A, Aronsson A, Tamm CO (1977) The effect of forest fertilization on primary production and nutrient cycling in the forest ecosystems. Silva Fenn 11:233–239.

Allen DC, Bauce E, Barnett JC (1992) Sugar maple declines—causes, effects, and recommendations. In: Manion PD, Lachance C (eds) Forest Decline Concepts. APS Press, New York, pp 123–125.

Arnolds E (1991) Decline of ectomycorrhizal fungi in Europe. Agric Ecosyst Environ 35:209–244.

Axelsson B (1985) Increasing forest productivity and value by manipulating nutrient availability. In: Ballard R, Farnum P, Ritchie GA, Winjum JK (eds) Weyerhaeuser Science Symposium 4. Forest Potentials. Productivity and Value. Weyerhaeuser Co, Washington, DC, pp 5–37.

Axelsson B (1981) Site Differences in Yield-differences in Biological Production or in Redistribution of Carbon within Trees. Ecol Environ Rep 9. University of Sweden, Agricultural Science Department, Uppsala, Sweden, p 11.

Bååth E, Lundgren B, Söderström B (1981) Effects of nitrogen fertilization on the activity and biomass of fungi and bacteria in a podzolic soil. Zbl Bakt Hyg, I Abt Orig C 2:90–98.

Bååth E, Söderström B (1979) Fungal biomass and fungal immobilization of plants nutrients in Swedish coniferous forest soils. Rev Ecol Biol Sol 16: 477–489.

Bailey SW, Hornbeck JW, Driscoll CT, Gaudette HE (1996) Calcium inputs and transport in a base-poor forest ecosystem as interpreted by Sr isotopes. Water Resour Res 32:707–719.

Baker JP, Bernard DP, Christensen SW, Sale MJ, Freda J, Heltcher KJ, Marmorek DR, Rowe L, Scanlon PF, Suter GW II, Watten-Hicks WJ, Welbourn PM (1990a) Biological Effects of Changes in Surface water Acid–Base Chemistry. NAPAP (National Acid Precipitation Assessment Program) Report 13. Acidic Deposition: State of Science and Technology. NAPAP, Washington, DC.

Baker LA, Kaufmann PR, Herlihy AT, Eilers JM (1990b) Current Status of Surface Water Acid–Base Chemistry. NAPA (National Acid Precipitation Assessment Program) Report 9. Acidic Deposition: State of Science and Technology. NAPAP, Washington, DC.

Ballard R (1979) Use of fertilizers to maintain productivity of intensively managed forest plantations. In: Proceedings, Impact of Intensive Harvesting on Forest Nutrient Cycling. State University of New York, Syracuse, New York. 13–16 August 1979. United States Department of Agriculture (USDA) Forest Service, Northeast Forest Experiment Station, Broomall, PA, pp 321–342.

Barnes RL, Bengtson GW (1968) Effects of fertilization, irrigation and cover cropping on flowering and on nitrogen and soluble sugar composition of Slash pine. For Sci 14:172–180.

Binkley D, Richter D (1987) Nutrient cycles and H^+ budgets of forest ecosystems. Adv Ecol Res 16:1–51.

Bondietti EA, Momoshima N, Shortle WC, Smith KT (1990) A historical perspective on divalent cation trends in red spruce stemwood and the hypothetical relationship to acidic deposition. Can J For Res 20:1850–1858.

Brix H (1983) Effects of thinning and nitrogen fertilization on growth of Douglas fir: relative contribution of foliage quantity and efficiency. Can J For Res 13:167–175.

Brix H (1971) Effects of nitrogen fertilization on photosynthesis and respiration in Douglas fir. For Sci 17:407–414.

Burns DA, Galloway JN, Hendrey GR (1981) Acidification of surface waters in two areas of the eastern United States. Water Air Soil Pollut 16:277–285.

Burns DA, Lawrence GB, Murdoch PS (1998) Streams in Catskill Mountains still susceptable to acid rain. EOS 79:197, 200–201.

Burns DA, Murdoch PS, Lawrence GB (1998) Effect of groundwater springs on nitrate concentrations during summer in Catskill Mountains streams. Water Resour Res 34:1987–1996.

Cadle SH, Dasch JM, Grossnickle NE (1985) Retention and release of chemical species by a northern Michigan snowpack. Water Air Soil Pollut 22:303–319.

Chapin FS III, Bloom AJ, Field CB, Waring RH (1987) Plant responses to multiple environmental factors. BioScience 37:49–57.

Chen CW, Gherini SA, Peters NE, Murdoch PS, Newton RM, Goldstein RA (1984) Hydrologic analyses of acidic and alkaline lakes. Water Resour Res 20:1875–1882.

Church MR, Thornton KW, Shaffer PW, Stevens DL, Rochelle BP, Holdren GR, Johnson MG, Lee JJ, Turner RS, Cassell DL, Lammers DA, Campbell WG, Liff CI, Brandt CC, Liegel LH, Bishop GD, Mortenson DC, Pierson SM, Shmoyer DD (1989) *Direct/Delayed Response Project: Future Effects of Long-Term Sulfur Deposition on Surface Water Chemistry in the Northeast* and *Southern Blue Ridge Province, Level I and Level II Analysis*. EPA/600/3-89/061b. National Acid Precipitation Program (NAPAP), Washington, DC.

Cogbill CV, Likens GE (1974) Acid precipitation in the northeastern United States. Water Resour Res 10:1133–1137.

Coley PD, Bryant JP, Chapin FS III (1985) Resource availability and plant antiherbivore defense. Science 230:895–899.

Cosby BJ, Hornberger GM, Galloway JN (1985) Modeling the effects of acid deposition: assessment of a lumped parameter model of soil water and streamwater chemistry. Water Resour Res 21:51–63.

Cowling EB, Dochinger LS (1980) *Effects of Acidic Presipitation on Health and Productivity of Forests*. Tech Rep PSW-43, United States Department of Agriculture (USDA) Forest Service, pp 165–173.

Cronan CS, Aiken GR (1985) Chemistry and transport of soluble humic substances in forested watersheds of the Adirondack Park, New York. Geochim Cosmochim Acta 49:1697–1705.

Cronan CS, Goldstein RA (1989) ALBIOS: a comparison of aluminum biogeochemistry in forested watersheds exposed to acidic deposition. In: Adriano DC, Havas M (eds) *Acidic Precipitation. Vol. 1. Case Studies*. Advances in Environmental Science, Springer-Verlag, New York, pp 113–135.

Cronan CS, Grigal DF (1995) Use of calcium/aluminum ratios as indicators of stress in forest ecosystems. J Environ Qual 24:209–226.

Cronan CS, Reiners WA, Reynolds RC, Lang GE (1978) Forest floor leaching: contributions from mineral, organic, and carbonic acids in New Hampshire subalpine forests. Science 200:309–311.

Cronan CS, Schofield CL (1979) Aluminum leaching response to acid precipitation: effect on high-elevation watersheds in the Northeast. Science 204:305–306.

Cronan CS, Walker WJ, Bloom PR (1986) Predicting aqueous aluminum concentrations in natural waters. Nature 324:140–143.

Dahlgren RA, Vogt KA, Ugolini FC (1991) The influence of soil chemistry on fine root aluminum concentrations and root dynamics in a subalpine Spodosol, Washington State, USA. Plant Soil 133:117–129.

David MB, Driscoll CT (1984) Aluminum speciation and equilibria in soil solutions of a Haplrthod in the Adirondack Mountains (New York, USA). Geoderma 33:297–318.

David MB, Fuller RD, Fernandez IH, Mitchell MH, Rustad LE, Vance GF, Stam AC, Nodvin SC (1990) Spodosol variability and assessment of response to acidic deposition. Soil Sci Soc Am J 54:541–548.

David MB, Lawrence GB (1996) Soil and soil solution chemistry under red spruce stands across the northeastern U.S. Soil Sci 161:314–328.

David MB, Mitchell MJ, Scott TJ (1987) Importance of biological processes in the sulfur budget of a northern hardwood ecosystem. Biol Fertil Soils 5: 258–264.

David MB, Vance GF (1989) Generation of soil solution acid-neutralizing capacity by addition of dissolved organic carbon. Environ Sci Tech 23: 1021–1024.

David MB, Vance GF, Kahl JS (1992) Chemistry of dissolved organic carbon and organic acids in two streams draining forested watersheds. Water Resour Res 28:389–396.

DeConinck F (1980) Major mechanisms in formation of spodic horizons. Geoderma 24:101–363.

Driscoll CT, Baker JP, Bisogni JJ, Schofield CL (1980) Effect of aluminum speciation on fish in dilute acidified waters. Nature 284:161–164.

Driscoll CT, Lehtinen MD, Sullivan TJ (1994) Modeling the acid–base chemistry of organic solutes in Adirondack, New York, lakes. Water Resour Res 30: 297–306.

Driscoll CT, Likens GE (1982) Hydrogen ion budget of an aggrading forested ecosystem. Tellus 34:283–292.

Driscoll CT, Likens GE, Hedin LO, Eaton JS, Bormann FH (1989a) Changes in the chemistry of surface waters. Environ Sci Tech 23:137–143.

Driscoll CT, Newton RM (1985) Chemical characteristics of Adirondack lakes. Environ Sci Tech 19:1018–1024.

Driscoll CT, Newton RM, Gubala CP, Baker JP, Christensen S (1991) Northeast overview. In: Charles DF (ed) *Acidic Deposition and Aquatic Ecosystems*: *Regional Case Studies*. Springer-Verlag, New York, pp 129–132.

Driscoll CT, Postek KM, Kretser W, Raynal DJ (1995) Long-term trends in the chemistry of precipitation and lake water in the Adirondack Region of New York, USA. Water Air Soil Pollut 85:583–588.

Driscoll CT, Schaefer DA, Molot LA, Dillon PJ (1989b) Summary of North American data. In: Malanchuk JL, Nilsson J (eds) *The Role of Nitrogen in the Acidification of Soils and Surface Waters*. Nordic Council of Ministers, Copenhagen, Denmark, 6-1 to 6-44.

Driscoll CT, Schafran GC (1984) Short-term changes in the base neutralizing capacity of an acidic Adirondack, New York, lake. Nature 310:308–310.

Driscoll CT, van Breeman N, Mulder J (1985) Aluminum chemistry in a forested Spodosol. Soil Sci Soc Am J 49:437–444.

Eager C, Adams MB (eds) (1992) *Ecology and Decline of Red Spruce in the Eastern United States*. Springer-Verlag, New York.

Ellsworth DS, Liu X (1994) Photosynthesis and canopy nutrition of four sugar maple forests on acid soils in northern Vermont. Can J For Res 24:2118–2127.

Federer CA, Hornbeck JW, Tritton LM, Pierce RS, Smith CT (1989) Long-term depletion of calcium and other nutrients in eastern U.S. forests. Environ Manage 13:593–601.

Fernandez IJ, Lawrence GB, Son Y (1995) Comparisons of soil-solution chemistry by horizons and over time under a low-elevation spruce–fir forest. Water Air Soil Poll 84:129–145.

Friedland AJ, Hawley GJ, Gregory RA (1985) Investigations of nitrogen as a possible contributor to red spruce (*Picea rubens* Sarg.) decline. In: *Symposium: Effects of Air Pollutants on Forest Ecosystems, 8–9 May 1985*. University of Minnesota Press, Minneapolis, MN, pp 95–106.

Fuller RD, David MB, Driscoll CT (1985) Sulfate adsorption relationships in forested spodosols of the northeastern U.S. Soil Sci Soc Am J 49:1034–1040.

Galloway JN, Schofield CL, Hendrey GR, Peters NE, Johannes AJ (1980) Sources of acidity in three lakes acidified during snowmelt. In: Drablos D, Tollan A (eds) *Ecological Impact of Acid Precipitation*. SNSF Project, Oslo, Norway, pp 264–265.

Gherini SA, Mok L, Hudson RJM, Davis GF, Chen C, Goldstein RA (1985) The ILWAS model: formulation and application. Water Air Soil Pollut 26: 95–113.

Gholz HL (1982) Environmental limits on above ground net primary production, leaf-area, and biomass in vegetation zones of the Pacific Northwest. Ecology 63:469–481.

Gosz JR, Likens GE, Bormann FH (1973) Nutrient release from decomposing leaf and branch litter in the Hubbard Brook Forest, New Hampshire. Ecol Monogr 43:173–191.

Gower ST, Vogt KA, Grier CC (1992) Above- and belowground carbon dynamics of Rocky Mountain Douglas fir: influence of water and nutrient availability. Ecol Monogr 62:43–65.

Grier CC, Running SW (1977) Leaf area of mature northwestern coniferous forests: relation to site water balance. Ecology 58:893–899.

Grier CC, Vogt KA, Keyes MR, Edmonds RL (1981) Biomass distribution and above- and belowground production in young and mature *Abies amabilis* zone ecosystems of the Washington Cascades. Can J For Res 11:155–167.

Harley JL, Smith SE (1983) *Mycorrhizal Symbiosis*. Academic Press. London.

Hart AC (1959) *Silvical Characteristics of Red Spruce* (Picea rubens). NE Paper 124. United States Department of Agriculture (USDA) Forest Service, Northeastern Forest Experiment Station, Radnor, PA.

Hedin LO et al. (1994) Steep declines in atmospheric base cations in regions of Europe and North America. Nature 367:351–354.

Heil GW, Werger MJA, de Mol W, van Dam D, Heijne B (1988) Capture of atmospheric ammonium by grassland canopies. Science 239:764–765.

Hornbeck JW, Bailey SW, Buso DC, Shanley JB (1997) Streamwater chemistry and nutrient budgets for forested watershed in New England: variability and management implications. For Ecol Manage 93:73–89.

Horner JD (1987) A preliminary investigation of the role of phenolic compounds in ecosystem processes. Ph.D. dissertation, University of New Mexico, Albuquerque, NM.

Ingestad T, Ågren GI (1992) Theories and methods on plant nutrition and growth. Physiol Plant 84:177–184.

James BR, Riha SJ (1986) pH buffering in forest soil organic horizons: relevance to acid precipitation. J Env Qual 15:229–234.

Johnson AH, Anderson SB, Siccama TG (1994a) Acid rain and soils of the Adirondacks. I. Changes in pH and available calcium. Can J For Res 24: 39–45.

Johnson AH, Friedland AJ, Miller EK, Siccama TG (1994b) Acid rain and soils of the Adirondacks. III. Rates of soil acidifiation in a montane spruce–fir forest at Whiteface Mountain, New York. Can J For Res 24:663–669.

Johnson AH, Siccama TG (1983) Acid deposition and forest decline. Environ Sci Tech 17:294A–305A.

Johnson CE, Litaor MI, Billet MF, Bricker OP (1994) Chemical weathering in small catchments: climatic and anthropogenic influences. In: Moldan B, Cerny J (eds) *Small Catchments: A tool for Environmental Studies*. John Wiley, New York, pp 229–254.

Johnson DW, Cole DW (1980) Anion mobility in soils: relevance to nutrient transport from forest ecosystems. Environ Internat 3:79–90.

Johnson DW, Cole DW, Bledsoe CS, Cromack K, Edmonds RL, Gessel SP, Grier CC, Richards BN, Vogt KA (1982b) Nutrient cycling in forests of the Pacific Northwest. In: Edmonds RL (ed) *Analysis of Coniferous Forest Ecosystems in the Western United States*. US/IBP Synthesis Series 14. Hutchinson Ross, Stroudsburg, PA, pp 186–232.

Johnson DW, Cresser MS, Nilsson SI, Turner J, Ulrich B, Binkley D, Cole DW (1991) Soil changes in forest ecosystems: evidence for and probable causes. In: Last FT, Watling R (eds) *Acidic Deposition Its Nature and Impacts*, Royal Society of Edinburgh, Edinburgh, United Kingdom, pp 81–116.

Johnson DW, Fernandez IJ (1992) Soil-mediated effects of atmospheric deposition on eastern U.S. spruce–fir forests. In: Eager C, Adams MB (eds) *Ecology and Decline of Red Spruce in the Eastern United States*. Springer-Verlag, New York, pp 235–270.

Johnson DW, Todd DE (1989) Nutrient cycling in forests of Walker Branch Watershed: Roles of uptake and leachng in causing soil change. J Environ Qual 19:97–104.

Johnson DW, Turner J, Kelly JM (1982a) The effects of acid rain on forest nutrient status. Water Resour Res 18:449–461.

Johnson NM, Driscoll CT, Eaton JS, Likens GE, McDowell WH (1981) 'Acid rain,' dissolved aluminum and chemical weathering at the Hubbard Brook Experimental Forest, New Hampshire. Geochim Cosmochim Acta 45: 1421–1437.

Johnson NM, Likens GE, Bormann FH, Fisher DW, Pierce RS (1969) A working model for the variation in stream water chemistry at the Hubbard Brook Experimental Forest, New Hampshire. Water Resour Res 5:1353–1363.

Johnson KH, Vogt KA, Clark HJ, Schmitz OJ, Vogt DJ (1996) Biodiversity and the productivity and stability of ecosystems. Tree 11:372–377.

Joslin JD, Kelly JM, Van Miegroet H (1992) Soil chemistry and nutrition of North American spruce–fir stands: evidence for recent change. J Environ Qual 21:12–30.

Joslin JD, Wolfe MH (1994) Foliar deficiencies of mature southern Appalachian red spruce determined from fertilizer trials. Soil Sci Soc Am J 58: 1572–1579.

Keyes MR, Grier CC (1981) Below- and above-ground biomass and net production in two contrasting Douglas fir stands. Can J For Res 11:599–605.

Kolb TE, McCormick LH (1993) Etiology of sugar maple decline in four Pennsylvania stands. Can J For Res 23:2395–2402.

Krug EC, Frink CR (1983) Acid rain on acid soil: a new perspective. Science 221:520–525.

Lawrence GB, David MB (1997) Response of Al solubility to elevated nitrogen saturation in soil of a red spruce stand in eastern Maine. Environ Sci Tech 31:825–930.

Lawrence GB, David MB, Bailey SW, Shortle WC (1997) Assessment of calcium status in soils of red spruce forests in the northeastern United States. Biogeochemistry 38:19–39.

Lawrence GB, David MB, Lovett GM, Murdoch PS, Burns DA, Baldigo BP, Thompson AW, Porter JH, Stoddard JL (1999) Calcium status and the response of stream chemistry to changing acidic deposition rates in the Catskill Mountains of New York. Ecol Applic 9:1059–1072.

Lawrence GB, David MB, Shortle WC (1995) A new mechanism for calcium loss in forest-floor soils. Nature 378:162–164.

Lawrence GB, Driscoll CT, Fuller RD (1988) Hydrologic control of aluminum chemistry in an acidic headwater stream. Water Resour Res 24:659–669.

Lawrence GB, Fuller RF, Driscoll CT (1986) Spatial relationships of Al chemistry in streams of the Hubbard Brook Experimental Forest, New Hampshire. Biogeochemistry 2:115–135.

Lawrence GB, Lovett GM, Baevsky YH (1999) Atmospheric deposition and watershed nitrogen export along an elevational gradient in the Catskill Mountains, New York. Biogeochemistry 51:119–228.

Lawrence GB, Wolock DM, Bailey SW, Hornbeck JW (1996) Variable-source-areas as controls of flow-related changes in stream chemistry. EOS, Transactions, AGU 77:226.

Lea R, Tierson WC, Bickelhaupt DH, Leaf AL (1980) Differential foliar response of northern hardwoods to fertilization. Plant Soil 54:419–439.

Likens GE, Bormann FH (1995) *Biogeochemistry of a Forested Ecosystem*. 2nd ed. Springer-Verlag, New York.

Likens GE, Driscoll CT, Buso DC (1996) Long-term effects of acid rain: response and recovery of a forest ecosystem. Science 272:244–246.

Long RP, Horsley SB, Lilja PR (1997) Impact of forest liming on growth and crown vigor of sugar maple and associated hardwoods. Can J For Res 27:1560–1573.

Lovett GM, Reiners WA, Olson RK (1982) Cloud droplet deposition in a subalpine balsam fir forest: hydrological and chemical inputs. Science 218:1303–1304.

Marshall JD, Waring RH (1985) Predicting fine root production and turnover by monitoring root starch and soil temperature. Can J For Res 15:791.

McDowell WH, Wood T (1984) Podzolization: soil processes control dissolved organic carbon concentrations in stream water. Soil Sci 137:23–32.

McLaughlin SB, Downing (1995) Interactive effects of ambient ozone and climate measured on growth of mature forest trees. Nature 374:252–254.

McLaughlin SB, Tjoelker MG, Roy WK (1993) Acid deposition alters red spruce physiology: laboratory studies support field observations. Can J For Res 23:380–386.

McNulty SG, Aber JD, Newman SD (1996) Nitrogen saturation in a high elevation New England spruce–fir stand. For Ecol Manage 84:109–121.

Melillo JM (1977) *Mineralization of Nitrogen in Northern Forest Ecosystems*. Ph.D. dissertation, Yale University. New Haven, CT.

Miller EK, Blum JD, Friedland AJ (1993) Determination of soil exchangeable-cation loss and weathering rates using Sr isotopes. Nature 362:438–441.

Millers I, Shriner DS, Rizzo D (1989) *History of Hardwood Decline in the Eastern United States*. Gen Tech Rep NE-126. United States Department of Agriculture (USDA) Forest Service, Northeastern Forest Experiment Station, Radnor, PA.

Mitchell HL, Chandler RF (1939) *The Nitrogen Nutrition and Growth of Certain Deciduous Trees of the Northeastern United States*. Black Rock For Bull 11.

Mitchell MJ, Driscoll CT, Kahl JS, Likens GE, Murdoch PS, Pardo LH (1996) Climatic control of nitrate loss from forested watersheds in the northeast United States. Environ Sci Tech 30:2609–2612.

Mollitor AV, Berg KR (1980) Effects of acid precipitation on forest soils. In: Raynal DJ, Leaf AL, Manion PD, Wang CJ (eds) *Actual and Potential Effects of Acid Precipitation on a Forest Ecosystem in the Adirondack Mountains*. New York State Energy Research and Development Authority (NYSERDA) 80-28. pp 3.1–3.88l.

Mooney HA (1972) The carbon balance of plants. Ann Rev Ecol Syst 3:315–346.

Munson RK, Gherini SA (1993) Influence of organic acids on the pH and acid-neutralizing capacity of Adirondack Lakes. Water Resour Res 29:891–899.

Murdoch PS, Burns DA, Lawrence GB (in review) Relation of climate change to the acidification of surface waters by nitrogen deposition. Environ Sci Tech 32:1642–1647.

Murdoch PS, Stoddard JL (1993) Chemical characteristics and temporal trends in eight streams of the Catskill Mountains, New York. Water Air Soil Pollut 67:367–395.

Murdoch PS, Stoddard JL (1992) The role of nitrate in the acidification of streams in the Catskill Mountains of New York. Water Resour Res 28:2707–2720.

[NAPAP] National Acid Precipitation Assessment Program (1993) *NAPAP 1992 Report to Congress*. NAPAP, Washington, DC, pp 37–49.

[NAPAP] National Acid Precipitation Assessment Program (1991) *NAPAP 1990 Integrated Assessment Report*. S/N040-000-00560-9. NAPAP, Washington, DC.

Nodvin SC, Driscoll CT, Likens GE (1986) The effect of pH on sulfate adsorption by a forest soil. Soil Sci 142:69–75.

Norton SA, Kahl JS, Fernandez IJ, Rustad LE, Schofield JP, Haines TA (1994) Response of the West Bear Brook Watershed, Maine, USA, to the addition of $(NH_4)_2SO_4$: 3-year results. For Ecol Manage 68:61–73.

Nye PH (1981) Changes of pH across the rhizosphere induced by roots. Plant Soil 61:7–26.

Pate JS, Layzell DB (1981) Carbon and nitrogen partitioning in the whole plant—a thesis based on empirical modeling. In: Bewley JD (ed) *Nitrogen and Carbon Metabolism*. Martinus Ninjhoff/Dr. W. Junk, The Hague, Netherlands.

Penning de Vries FWT (1975) The cost of maintenance processes in plant cells. Ann Bot 39:77–92.

Persson H (ed) (1990) *Above- and Belowground Interactions in Forest Trees in Acidified Soils. Air Pollution Research Report 32*. Commission of the European Communities. Directorate-General for Sciences, Research and Development, Environment Research Programme, Brussels, Belgium.

Rascher CM, Driscoll CT, Peters NE (1987) Concentration and flux of solutes from snow and forest floor during snowmelt in the West-Central Adirondack region of New York. Biogeochemistry 3:209–224.

Reuss JO (1983) Implications of the Ca–Al exchange system for the effect of acid precipitation on soils. J Environ Qual 12:591–595.

Reuss JO, Johnson DW (1985) Effect of soil processes on the acidification of water by acid deposition. J Environ Qual 14:26–31.

Reuss JO, Walthall PM, Roswall EC, Hopper RWE (1990) Aluminum solubility, calcium–aluminum exchange, and pH in acid forest soils. Soil Sci Soc Am J 54:374–380l.

Robitaille B, Boutin R, Lachance D (1995) Effects of soil freezing stress on sap flow and sugar content of mature sugar maples (*Acer saccharum*) Can J For Res 25:577–587.

Rochelle BP, Church MR (1987) Regional patterns of sulfur retention in watersheds of the eastern U.S. Water Air Soil Pollut 36:61–73.

Ross DS, David MB, Lawrence GB, Bartlett RJ (1991) Exchangeable hydrogen explains the pH of Spodosol Oa horizons. Soil Sci Soc Am J 60:1926–1932.

Rustad L, Cronan C (1995) Biogeochemical controls on aluminum chemistry in the O horizon of a red spruce (*Picea rubens* Sarg.) stand in central Maine, USA. Biogeochemistry 29:107–129.

Santantonio D, Hermann RK (1985) Standing crop, production, and turnover of fine roots on dry, moderate, and wet sites of mature Douglas fir in western Oregon. Ann Sci For 42:113–142.

Schlesinger WH (1977) Carbon balance in terrestrial detritus. Annu Rev Ecol Syst 8:51–81.

Schnitzer M, Kahn SU (1972) *Soil Organic Matter*. Elsevier, Amsterdam, Netherlands.

Schnitzer M, Delong WA (1955) Investigation on the mobilization and transport of iron in forested soils II. The nature of the reaction of leaf extracts and leachates with iron. Soil Sci Soc Am Proc 19:363–368.

Schofield CL (1976) Acid precipitation: effects on fish. Ambio 5:228–230.

Scott JT, Siccama TG, Johnson AH, Breisch AR (1985) Decline of red spruce in the Adirondacks, New York. Bull Torrey Bot Club 111:438–444.

Shortle WC, Bondietti EA (1992) Timing, magnitude, and impact of acidic deposition on sensitive forest sites. Water Air Soil Pollut 61:253–267.

Shortle WC, Smith KT (1988) Aluminum-induced, calcium deficiency syndrome in declining red spruce. Science 240:1017–1018.

Shortle WC, Smith KT, Minocha R, Lawrence GB, David MB (1997) Acidic deposition, cation mobilization, and stress in healthy red spruce trees. J Environ Qual 26:871–876.

Siccama TG, Bliss M, Vogelmann HW (1982) Decline of red spruce in the Green Mountains of Vermont. Bull Torrey Bot Club 109:162–168.

Sisterson DL (1991) Report 6. Deposition monitoring: methods and results. In: Irving PM (ed) *Acidic Deposition: State of Science and Technology. Summary Report of the U.S. National Acid Precipitation Assessment Program.* NAPAP Compendium of Summaries. U.S. Government Printing Office. Washington, DC, pp 65–74.

Skyllberg U (1994) Aluminum associated with a pH-increase in the humus layer of a boreal podzol. Interciencia 19(6):356–365.

Stoddard JL (1994) Long-term changes in watershed retention of nitrogen. In: Baker LA (ed) *Environmental Chemistry of Lakes and Reservoirs*. Am Chem Soc Adv Chem Ser 237:223–284.

Stoddard JL, Driscoll CT, Kahl S, Kellog J (1998) Linking rates and effects of acidic deposition through intensive monitoring and regional interpretation. Environ Monit Assess 51:399–413.

Strickland TC, Sollins P (1987) Improved method for separating light- and heavy-fraction organic material from soil. Soil Sci Soc Am J 51:1390–1393.

Sullivan TJ (1990) *Historical Changes in Surface Water Acid–Base Chemistry in Reponse to Acidic Deposition. National Acid Precipitation Assessment Program Report 11. Acidic Deposition: State of Science and Technology.* National Acid Precipitation Assessment Program (NAPAP), Washington, DC.

Sullivan TJ, Eilers JM, Cosby BJ, Vache KB (1997) Increasing role of nitrogen in the acidification of surface waters in the Adirondack Mountains, New York. Water Air Soil Pollut 95:313–336.

Tamm CO (1976) Acid precipitation: biological effects in soil and on forest vegetation. Ambio 5:235–238.

Thorne JF (1985) *Nitrogen Cycling in a Base-poor and a Relatively Base-rich Northern Hardwood Forest Ecosystem.* Ph.D. dissertation, Yale University, New Haven, CT.

Tipping E, Berggren D, Mulder J, Woof C (1995) Modelling the solid-solution distributions of protons, aluminum, base cations and humic substances in acid soils. Europ J Soil Sci 46:77–94.

[USDAFS] United States Department of Agriculture, Forest Service (1993) *Northeastern Area Forest Health Report.* Radnor, PA, NA-TP-03-93.

van Breemen N, Driscoll CT, Mulder J (1984) Acidic deposition and internal proton sources in acidification of soils and waters. Nature 307:599–604.

van Breemen N, Mulder J, Driscoll CT (1983) Acidification and alkalinization of soils. Plant Soil 75:283–308.

van Miegroet H, Cole DW (1985) Acidification sources in red alder and Douglas fir soils—importance of nitrification. Soil Sci Soc Am J 49:1274–1279.

van Miegroet H, Cole DW (1984) The impact of nitrification on soil acidification and cation leaching in a red alder ecosystem. J Environ Qual 13:586–590.

van Miegroet H, Johnson DW, Dodd DE (1993) Foliar response of red spruce saplings to fertilization with Ca and Mg in the Great Smoky Mountains National Park. Can J For Res 23:89–95.

Van Praag HJ, Sougnez-Remy S, Weissen F, Carlette G (1988) Root turnover in a beech and a spruce stand of the Belgian Ardennes. Plant Soil 105:87–103.

Vance GF, David MB (1989) Effect of acid treatment on dissolved organic carbon retention by a spodic horizon. Soil Soc Am J 53:1242–1247.

Vance GF, David MB (1991) Forest soil response to acid and salt additions of sulfate: III. Solubilization and composition of dissolved organic carbon. Soil Sci. 151:297–305.

Vitousek PM, Gosz JR, Grier CC, Melillo JM, Reiners WA (1982) A comparative analysis of potential nitrification and nitrate mobility in forest ecosystems. Ecol Monogr 52:155–177.

Vogt DJ (1987) *Douglas Fir Ecosystems in Western Washington: Biomass and Production as Related to Site Quality and Stand Age.* Ph.D. dissertation, University of Washington, Seattle, WA.

Vogt KA (1991) Carbon cycling in forest ecosystems. Tree Physiol 9:69–86.

Vogt KA, Gordon JC, Wargo JP, Vogt DJ, Asbjornsen H, Palmiotto PA, Clark HJ, O'Hara JL, Keeton WS, Patel-Weynand T, Witten E (1997) *Ecosystems: Balancing Science with Management.* Springer-Verlag, New York.

Vogt KA, Dahlgren R, Ugolini F, Zabowski D, Moore EE, Zasoski R (1987a) Aluminum, Fe, Ca, Mg, K, Mn, Cu, Zn and P in above- and belowground biomass. I. *Abies amabilis* and *Tsuga mertensiana.* Biogeochemistry 4:277–294.

Vogt KA, Dahlgren R, Ugolini F, Zabowski D, Moore EE, Zasoski R (1987b) Aluminum, Fe, Ca, Mg, K, Mn, Cu, Zn and P in above- and belowground biomass. II. Pools and circulation in a subalpine *Abies amabilis* stand. Biogeochemistry 4:295–311.

Vogt KA, Vogt D, Brown S, Tilley J, Edmonds R, Silver W, Siccama T (1995) Dynamics of forest floor and soil organic matter accumulation in boreal, temperate, and tropical forests. In: Lal R, Kimble J, Levine E, Steward B (eds) *Soil Management and Greenhouse Effect.* CRC Press, Boca Raton, FL, pp 159–178.

Vogt KA, Grier CC, Meier CE, Edmonds RL (1982) Mycorrhizal role in net primary production and nutrient cycling in *Abies amabilis* ecosystems in western Washington. Ecology 63:370–380.

Vogt KA, Grier CC, Vogt DJ (1986) Production, turnover, and nutrient dynamics of above- and belowground detritus of world forests. Adv Ecol Res 15:303–377.

Vogt KA, Publicover DA, Bloomfield J, Perez JM, Vogt DJ, Silver WL (1993) Belowground responses as indicators of environmental change. Environ Exp Bot 33:189–205.

Vogt KA, Publicover DA, Vogt DJ (1991) Integration of ectomycorrhizae to ecosystem ecology. Agric Ecosys Environ 35:171–190.

Vogt KA, Vogt DJ, Gower ST, Grier CC (1990) Carbon and nitrogen interactions for forest ecosystems. In: Persson H (ed) *Above- and Below-ground Interactions in Forest Trees in Acidified Soils. Air Pollution Report 32.* Commission of the European Communities. Directorate-General for Sciences, Research and Development, Environment Research Programme, Brussels, Belgium, pp 203–235.

Vogt KA, Vogt DJ, Moore EE, Littke W, Grier CC, Leney L (1985) Estimating Douglas fir fine root biomass and production from living bark and starch. Can J For Res 15:177–179.

Vogt KA, Vogt DJ, Palmiotto PA, Boon P, O'Hara J, Asbjornsen H (1996) Review of root dynamics in forest ecosystems grouped by climate, climatic forest type and species. Plant Soil 187:159–219.

Vose JM (1987) *Effects of Increased Nutrient Supply on Loblolly Pine Stand Leaf Area, Stemwood Growth, and Crown Architecture.* Ph.D. dissertation, North Carolina State University, Raleigh, NC.

Walker WJ, Cronan CS, Bloom PR, (1990) Aluminum solubility in organic soil horizons from northern and southern forested watersheds. Soil Sci Soc Am J 54:369–374.

Wareing PF, Patrick J (1975) Source-sink relations and the partition of assimilates in the plant. In: Cooper JP (ed) *Photosynthesis and Productivity in Different Environments.* International Biological Programme 3. Cambridge University Press, Cambridge, United Kingdom, 481–499.

Wargo PM (1972) Defoliation-induced chemical changes in sugar maple roots stimulated growth of *Armillaria mellea*. Phytopathology 62:1278–1283.

Wargo PM, Bergdahl DR, Olson CW, Tobi DR (1988) Carbohydrate and nitrogen content of roots of declining red spruce trees. (Abstr.) Phytopathology 78:1533.

Wargo PM, Bergdahl DR, Tobi DR, Olson CW (1993) *Root Vitality and Decline of Red Spruce. Contributiones Biologiae Arborum, Vol. 4.* (Fuhrer E, Schutt P, eds) Ecomed, Landsberg/Lech, Germany.

Waring RH (1987) Nitrate pollution: a particular danger to boreal and subalpine coniferous forests. In: Fujimore T, Kimura M (eds) *Human Impacts and Management of Mountain forests.* Forestry and Forest Products Research Institute, Ibaraki, Japan, pp 93–105.

Waring RH et al. (1985) Differences in chemical composition of plants grown at constant growth rates with stable mineral nutrition. Oecologia (Berl.) 66: 157–160.

Westman WE (1985) *Ecology, Impact Assessment, and Environmental Planning.* John Wiley, New York.

Wiggington PJ, Baker JP, DeWalle DR, Kretser WA, Murdoch PS, Simonin HA, Van Sickle J, McDowell MK, Peck DV, Barchet WR (1996) Episodic acidification of small streams in the northeastern United States: Episodic Response Project. Ecol Applic 6:374–388.

Wiggington PJ, Davies TD, Tranter M, Eshleman KN (1990) *Episodic Acidification of Surface Waters Due to Acidic Deposition. National Acid Precipitation Assessment Program Report 12. Acidic Deposition: State of Science and Technology.* National Acid Precipitation Assessment Program (NAPAP), Washington, DC.

Wilmot TR, Ellsworth DS, Tyree MT (1996) Base cation fertilization and liming effects on nutrition and growth of Vermont sugar maple stands. For Ecol Manage 84:123–134.

Wilmot TR, Ellsworth DS, Tyree MT (1995) Relationships among crown condition, growth, and stand nutrition in seven northern Vermont sugarbushes. Can J For Res 25:386–397.

Wolock DM, Fan J, Lawrence GB (1997) Effects of basin size on low-flow stream chemistry and subsurface contact time in the Neversink River Watershed, New York. Hydrol Process 11:1273–1286.

Zahka GA, Baggett KL, Wong BL (1995) Inoculum potential and other VAM fungi parameters in four sugar maple forests with different levels of stand dieback. For Ecol Manage 75:123–134.

3. Ecosystem-Scale Interactions with Global Change

9. Nitrogen Saturation in Experimental Forested Watersheds

Ivan J. Fernandez and Mary Beth Adams

Daniel Rutherford is credited with the discovery of nitrogen (N) in 1772. A.L. Lavoisier, the 18th century French chemist, called this element "azote" which translates to "without life" because free N is incapable of supporting life: a fitting name for this element given the paradoxes it presents to society and its role in global ecology. Despite being the most abundant element in the atmosphere, usually quoted as 78% by volume, primarily as N_2, life on the planet does not rely on either the uptake or release of N for the two most important complimentary biological processes, namely respiration and photosynthesis. Nonetheless, N has long been heralded as the most commonly limiting mineral nutrient for plant growth around the globe, and indeed is essential for the formation of amino acids, and thus proteins, for all life forms. No other element has been studied more extensively in the context of agricultural and forest ecosystems than N, yet at no time in the past has there been greater interest in research addressing information needs regarding N.

To add to the irony, N is the most commonly used fertilizer in crop and forest management, with increasing amounts used to maintain production goals. Yet in the past several decades, we have realized that N has its dark side in the environment. Fossil fuel combustion typically releases N gases (nitrogen dioxide [NO_2], nitrogen oxide [NO]) into the atmosphere that can be transported long distances and, along with sulfur dioxide (SO_2), create strong mineral acids in the atmosphere that return to terrestrial ecosystems in wet and dry forms as acid deposition. Atmospheric N also contributes to the formation of tropospheric ozone (O_3) or "smog." Ironically, N fertilizers have a high energy requirement for their fabrication. This means that high fossil fuel consumption is usually necessary to produce the electricity needed to manufacture a fertilizer form of N to be added to the landscape, while at the same time also contributing to the long-range transport of N in the atmosphere. Vitousek et al. (1997) recently discussed the indisputable human influence on global N cycles, emphasizing the serious and long-term environmental consequences we can expect as a result of our perturbation of this system on a global scale. Galloway (1998) estimated that, in 1997, human activities contributed 150 Tg N to the global terrestrial environment, of which 80% was due to food production (with twice as much attributable to fertilizer production as biological N fixation through cultivation) and 20% due to energy production.

During the past two decades there has been an emerging concern for the consequences of increased atmospheric deposition of N to forested ecosystems, despite the continued interest in certain regions in forest fertilization with N to promote growth. Early concerns were focused on the possibility of N deposition having direct effects on tree foliage and physiology, which later shifted to a focus on soil acidification and related indirect effects on forest health. More recently, it has become evident that chronic N deposition to forests may result in complex responses at both the organismal and ecosystem levels. In the mid-1980s two important experiments were initiated in the eastern U.S. using whole-watershed chemical manipulations to determine the effects of chronic N and sulfur (S) deposition on forested ecosystems. The focus of these studies has been to investigate the biogeochemical response of forested stream catchments to experimental treatments, and more recently research has included other aspects of ecosystem response. These studies have focused on the interaction between forested watersheds and the chemical climate, and the long-term nature of these studies is increasingly providing insight on interactions with physical climatic factors as well. This chapter discusses some of the highlights of these whole-watershed studies during the first six to seven years of experimental manipulation.

Nitrogen Deposition and Nitrogen Saturation

The concept of nitrogen saturation was developed in the early 1980s in Europe by scientists concerned with atmospheric deposition of N. In 1981, Swedish scientist T. Ingestad first presented a model that defined the initial concept (Aber, 1992). Many other European scientists have since worked on the N saturation concept. One noteworthy example is the "ammonium hypothesis" offered by Nihlgård (1985), based on observations of very high deposition of ammonium (NH_4) in the Netherlands. Ågren and Bosatta (1988) discussed the role of plant vs. soil processes in the evolution of N saturation, concluding that the soil subsystem must saturate first as the condition of N saturation develops.

Europeans have been known for their work in developing the concept of "critical loads" for N and S (Bull, 1991). This concept was defined in a Swedish workshop as "a quantitative estimate of an exposure to one or more pollutants below which significant harmful effects on specified sensitive elements of the environment do not occur according to present knowledge" (Nilsson and Grennfelt, 1988). This definition is more conceptual than the earlier definitions of N saturation, but specific policy goals have resulted from the critical loads approach in Europe. Nilsson (1986) summarized the outcome of a Nordic working group on critical loads who concluded approximately 10 to $20 \, kg \, N \, ha^{-1} \, yr^{-1}$ was the critical load for forests under conventional management. Ågren and Bosatta (1988) concurred with this estimate based on their own evaluation. Løkke et al. (1996) recently discussed some of the uncertainties that exist with the widespread utilization of this approach in regulating transboundary air pollution in Europe.

In the U.S., the concept of "nitrogen saturation" has received the greatest amount of attention as a condition of "excess" N in the ecosystem. Aber (1992) points out that several definitions of N saturation exist depending on whether this condition is defined as (1) the absence of a growth response in vegetation to N additions, (2) the initiation of NO_3 leaching, or (3) the lack of a net N accumulation in ecosystems as evidenced by an equivalence between inputs and losses. He included in his article a useful table shown here as Table 9.1 that suggests key mechanisms involved with ecosystem response to N, and the nature of those changes.

Several authors have discussed the evolution of N saturation in forested ecosystems and provide a useful framework not only for understanding the processes controlling these phenomena, but also for designing hypothesis-driven research to define mechanisms essential for accurate predictions of future response. Aber et al. (1989) defined N saturation as the availability of NH_4 and nitrate (NO_3) in excess of total combined plant and microbial nutrient demands (and excluding their use as a substrate for denitrification). They described the stages of the development of N saturation we can summarize as Stage 0, N limitations to biological processes ranging from tree growth to litter decomposition; Stage 1, initial increases in foliar N concentrations followed by increased biomass; Stage 2, N saturation where N no longer limits biological function; and finally, Stage 3, forest decline due to a range of potential mechanisms impacting forest health. Aber et al. (1998) recently refined their conceptual

Table 9.1. Characteristics of N-limited and N-saturated Forest Ecosystems (Aber, 1992)

Characteristic	N-limited	N-saturated
Form of N cycled (net, as plant uptake)	100% NH_4	25–50% NO_3, 50–75% NH_4
Soil DOC concentration	High	Low
Ratio of gross NO_3 immobilization to gross nitrification	Near 100%	Near 0%
Ratio of gross NH_4 immobilization to gross mineralization	High (90–95%)	Low (50%?)
Fraction of soil fungi that are mycorrhizal	High	Low
Nitrate loss during snow melt	Low	High
Nitrate loss at base flow	Zero	High
Foliar lignin concentration	High	Low
Foliar N concentration	Low	High
Foliar free amino acid (e.g., arginine) concentration	Zero	High
Soil C/N ratio	High	Low
N_2O production	Zero	High
CH_4 production	High	Low (zero?)

model of N saturation based on recent research. They included in their concept of the evolution of forest ecosystems toward N saturation a decline in both calcium (Ca)/aluminum (Al) and magnesium (Mg)/N ratios in foliage, and a decline in N mineralization in Stage 2, the latter not yet well understood. In this article they point out the paradox that forest studies using N manipulations have shown, and that is the strong retention of added N to forests despite the limited carbon (C) pool in soils to facilitate free-living microbial fixation. They discuss a number of possible mechanisms to explain this phenomenon. Their conclusion is that mycorrhizal conversion of N to organic forms may offer the most plausible mechanism since mycorrhizae have direct access to photosynthate without requiring its conversion to plant biomass and subsequent microbial degradation. This promising hypothesis remains to be tested.

Stoddard (1994) has been widely credited with defining the evolution of N saturation based on watershed retention of N and surface-water chemical response. Accordingly, he defined Stage 0, net N retention due to microbial and forest uptake with the exception of spring snowmelt; Stage 1, as with Stage 0 but with amplified export of NO_3 during snowmelt or spring episodes; Stage 2, decreased control of biological uptake of N and an increase in base flow NO_3 concentrations; and Stage 3, no biological N deficiency is evident and the watershed becomes a net source of N. Both approaches have provided conceptual templates for understanding the process of N saturation and the important terrestrial–aquatic linkages involved. Jeffries and Maron (1997) question the adequacy of either the N saturation or critical loads approach to defining ecosystem response to N enrichment alone, and argue for the addition of a third approach, that of careful monitoring of ground-layer species and mosses sensitive to soil N status to fulfill a critical gap in our understanding of the mechanisms of N transfer within and between ecosystems. Their recommendation seems sound, and would provide another tool to evaluate ecosystem condition that uses factors that integrate many processes, much as we used stream chemistry in watershed studies.

The greatest concern for N saturation in the U.S. has been in the northeastern states, and particularly at high elevations where deposition rates are greatest. Driscoll and Van Dreason (1993) reported increasing NO_3 concentrations in nine lakes of the Adirondack mountains of New York, despite the lack of any significant temporal change in N deposition throughout this region. They also noted 13 of the 17 lakes they studied showed declines in SO_4 concentrations commensurate with the documented declines in SO_4 deposition throughout the northeastern U.S. Sullivan et al. (1997) evaluated several extensive data sets on surface-water chemistry for this region, and concluded that the focus on S in the past may have underestimated the contribution of N to historical acidification, and that ignoring the role of N in surface-water acidification in the region may result in serious overestimations of recovery. McNulty et al. (1990) examined properties of forest soils in high-elevation spruce–fir sites throughout the Northeast from New York to Maine. They reported a spatial coincidence between the gradient of N deposition throughout the Northeast corridor, and N mineralization and nitrification potentials. They concluded that indicators of high rates of N cycling and available N in the systems receiving the higher atmospheric inputs of N might be evidence for the migration of these ecosystems to a condition of N saturation. McNulty et al. (1993, 1996) used N amendments to spruce–fir plots at a high-elevation site in southeastern Vermont to evaluate the mechanisms of response to treatments. They showed evidence that N enrichment resulted in changes in foliar N concentrations, basal area growth, nutrient balances, N cycling, and species composition based on initial differences in regeneration success on the treated plots. These results support and refine the model presented above for the terrestrial response to N saturation as described by Aber et al. (1989). Other N-saturated systems in the U.S. include the pine forests of southern California, unusual because of the very high levels of dry N deposition (Fenn et al., 1996), and high-elevation catchments in the Rocky Mountains, saturated because of very low uptake by biota (Williams et al., 1996). Land use history (e.g., fire, harvesting, blowdown) also appear to have a lasting effect on the susceptibility of forests to N saturation and influences ecosystem C storage in soils (Aber and Driscoll, 1997). It should be noted that the net effects of atmospheric N deposition on C sequestration in northern forests remains uncertain. Nadelhoffer et al. (1999) recently suggested that

N deposition is unlikely to explain the reported C sink in midlatitude forests attributing <20% of current C uptake by forests to N.

In order to study the effects of chronic elevated N deposition on forested ecosystems in situ, we can either (1) watch the evolution of existing ecosystems, over long periods of time, that are subjected to elevated or changing atmospheric deposition of N, (2) study differences among forest sites along carefully delineated spatial gradients of N deposition on the landscape, or (3) experimentally manipulate the deposition of N to the ecosystem. Sullivan (1997) points out the dangers of utilizing numerical simulation models based on an imperfect understanding of processes in ecosystems, and underscores the value of ecosystem manipulation experimentation as a means of testing these models. Plot-scale studies, such as the ones discussed above by McNulty and Aber (1993), are excellent and cost-effective ways to address specific hypothesis regarding mechanisms of response. However, this scale of study is not adequate for defining whole ecosystem responses, particularly with respect to biogeochemical processes that are strongly related to ecosystem hydrological properties. Nor do plot-scale studies allow us to study how different ecosystem components are integrated across landscapes. It is for this reason that watershed studies have been so valuable in understanding the integration of processes at the whole-ecosystem scale. Readers are encouraged to visit special issues of the journal Forest Ecology and Management, Vol. 71, 1/95 and Vol. 101, 2/98 (Wright and van Breeman, 1995; Wright and Rasmussen, 1998) for a collection of papers on the NITREX project, where whole catchments or large forest stands at 12 sites across a modern gradient of N deposition in Europe were used to study N deposition effects on forested ecosystems. Throughout these sites they employed a range of techniques to both add N as well as remove ambient levels of N to study both the risk of N saturation and the characteristics of recovery.

Two whole-watershed experimental N manipulation programs have also been conducted in forested ecosystems in the eastern U.S. since the 1980s, in Maine and West Virginia, specifically designed to study the effects of elevated N and S deposition. We know of no other comparable studies on the continent, and the following sections describe some of the findings to date.

Experimental Watersheds

Watershed Descriptions

The Bear Brook Watershed in Maine (BBWM) is located in eastern Maine (lat 44°52'N, long 68°6'W) approximately 50 km from the Atlantic Ocean. The site lies on the southeastern slope of Lead Mountain, with total relief of 210 m and a maximum elevation of 475 m. Two nearly perennial, low dissolved organic carbon (DOC), low acid neutralizing capacity (ANC)

Table 9.2. General Characteristics of the Study Sites

	BBWM	FEF
Watershed Pair	East Bear (Reference)	WS4 (Reference)
	West Bear (Treated)	WS3 (Treated)
Watershed Area	10.2/10.9 ha	34/39 ha
Relief	265–475 m	735–860 m
Soil Types	Typic Haplorthods	Typic Dystrochrepts
Soil Depths	<1 m	<2 m
Soil Texture	Fine sandy loam	silt loam
Bedrock	Siltstone/granite	sandstone/shale
Stream Yield	65%	45%
Annual Temp	5°C	10°C
Annual Precip	130 cm	145 cm
Forest Types	Northern hardwood/ spruce	Central Appalachian hardwoods

streams (East Bear Brook and West Bear Brook) drain 10.2 and 10.9 ha contiguous watersheds (Table 9.2). Vegetation at the site is dominated by northern hardwoods (*Fagus grandifolia* Ehrb., *Acer rubrum* L., *Acer saccharum* Marsh., *Betula alleghaniensis* Britt., *Betula papyrifera* Marsh., and *Acer pennsylvanicum* Marsh.) with stands of softwoods (*Picea rubens* Sarg., *Abies balsamea* Mill., and *Tsuga canadensis* [L.] Carr.) at higher elevations. Soils are coarse, loamy, mixed, frigid Typic Haplorthods developed on till. Bedrock consists of quartzites and meta-pelites intruded locally by granite.

Chemical additions of N and S, as ammonium sulfate ([NH$_4$]$_2$SO$_4$), have been delivered bimonthly via helicopter to the West Bear watershed beginning in November 1989 and continuing through the period of data reported and remain ongoing. The adjacent untreated East Bear watershed serves as a reference. Total experimental loadings were 1800 eq ha^{-1} yr^{-1} (~25.2 kg N ha^{-1} yr^{-1} and ~28.8 kg S ha^{-1} yr^{-1}), which is ~3× and ~2× the ambient-wet–plus–estimated-dry deposition of N and S, respectively (Rustad et al. 1994). Total N deposition was approximately 1.5× that of the highest estimated wet-plus-dry deposition in the U.S. (NADP, 1990), but was less than 70% of the total N deposition in areas of central Europe (Dise and Wright, 1992). Additional details on the BBWM site can be found in Kahl et al. (1993), Norton et al. (1994) and Norton and Fernandez (1999).

The Fernow Experimental Forest, located near Parsons, West Virginia (lat 39°3′15″N, long 79°42′15″W) is situated on the unglaciated Allegheny Plateau in the Allegheny Mountain Section, within the Central Appalachian Broadleaf Forest Province (Bailey et al., 1994). The two watersheds (Watershed 3 (WS3) = treatment watershed; Watershed 4 (WS4) = reference watershed) are drained by second-order streams. Precipitation is distributed evenly between dormant and growing seasons, and averages 145 cm per year. Average annual precipitation pH is 4.2 but lower values are common in summer (Adams et al., 1994). The predominant soils on both watersheds are loamy-skeletal, mixed mesic Typic Dystrochrepts. Elevation on the watershed ranges from 735 to 870 m.

Watershed 3, the treatment watershed (34 ha, see Table 9.2), contains a young stand that originated in 1969 from natural regeneration following clearcutting. Dominant tree species include *Prunus serotina* Ehrh. *Acer rubrum* L., *Betula lenta* L., and *Fagus grandifolia* Ehrh. Watershed 4, the control watershed (39 ha), contains a relatively undisturbed second growth stand of central Appalachian hardwoods. The current stand is approximately 90 years old, but some residual trees may be up to 200 years old (J.N. Kochenderfer, unpublished data). Dominant tree species include *Acer saccharum* Marsh., *Acer rubrum* L., *Fagus grandifolia* Ehrh., and *Quercus rubra* L.

Chemical additions of $(NH_4)_2SO_4$ fertilizer have been applied to WS3 since January 1989. Fertilizer was applied at a rate double the ambient N and S throughfall inputs, which was considered to approximate the combined N and S inputs of wet and dry deposition. Because ambient deposition varied seasonally, three applications were made each year, in March, July, and November. Yearly rates were $\sim40\,kg\,S\,ha^{-1}$ and $\sim35\,kg\,N\,ha^{-1}$. Additional details can be found in Adams et al. (1997).

Streams as Integrators of Response

Our discussion focuses on the behavior of these forested watersheds in response to experimentally increased N deposition. By definition, a watershed represents the land area contributing to a specific point of hydrological export, and integrates the inputs and losses of water from that system. Likewise, watersheds integrate biogeochemical processes reflecting the input and transformation of materials within that same hydrologically defined system.

Fig. 9.1 shows the long-term time series of stream water nitrate (NO_3) concentrations at both BBWM and Fernow. Measurements began in 1987 at BBWM, and, although a longer record of measurements is available for Fernow (Adams et al., 1994), our analyses begin with 1984 to facilitate a comparison between the two experimental watershed study programs. Treatments of $(NH_4)_2SO_4$ were initiated in both watersheds in 1989, starting in November for BBWM and in January at Fernow. Thus, data from 1988 and before shows how comparable the two streams were to each other at each site relative to stream water NO_3 concentrations prior to the onset of experimental manipulations. Although variability exists in these data, evidence presented suggests that the reference watersheds chosen behave nearly identical to the treatment watersheds prior to the commencement of treatments. A noteworthy difference between BBWM and Fernow is that a clear seasonal pattern in NO_3 concentrations exists at BBWM even before the treatments, and that the concentrations of NO_3

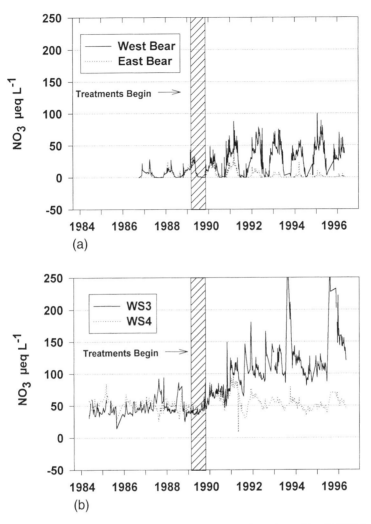

Figure 9.1. Stream NO_3 concentration time series for (a) BBWM and (b) Fernow.

are low and often below detection. This is the behavior we would expect in Stage 0 as described by Stoddard (1994) where N is biologically limiting and only during spring snowmelt, prior to vegetative uptake, are significant losses of inorganic N detected. The situation at Fernow is markedly different. Even before treatment began, there is only a vague representation of a seasonal pattern in stream water NO_3 concentrations, with a relatively constant stream water concentration of $\sim$50 µeq l^{-1} NO_3. This loss of seasonality in NO_3 concentration is indicative of the lack of biological controls on an increasingly N-enriched ecosystem referred to by both the Stoddard (1994) or Aber et al. (1989) models as Stage 2 in the progression to N saturation. Peterjohn et al. (1996) examined the 23-year record of stream water NO_3 concentrations in WS4 at Fernow. They were able to define three distinct periods of increasing NO_3 export and decreasing seasonality: an initial period of low variability, a period of high peak concentrations and high variability, and, most recently, a period of high concentrations and low variability. While these patterns of response to elevated N deposition seem well demonstrated in forested watersheds, Peterjohn et al. (1996) pointed out the need to better define the mechanisms controlling them. Indeed, as seen later in this chapter, even the treated watersheds exhibiting classic N saturation characteristics are still retaining most of the annual inputs of N. Ecosystem mechanisms controlling the retention of N and the evolution of N saturation characteristics remain poorly understood, with some of the most recent speculation being discussed earlier such as that of Aber et al. (1998).

Both the treated watersheds at BBWM (West Bear) and at Fernow (WS3) responded to the NH_4 amendments with an increase in stream water NO_3 concentrations. This indicates that the added NH_4 was either rapidly nitrified or treatments increased the rate of nitrification of native soil N pools. There is reason to believe that both processes are responsible, but there is evidence to suggest that initial increases in NO_3 export were largely attributable to native soil N (Nadelhoffer et al., 1999). Increased stream-water NO_3 in West Bear retained a strong seasonality, with increasing magnitude of peak concentrations as compared with the reference watershed best representing the characteristics of the response. The time series shows some evidence at BBWM of an increasing length of the period of elevated NO_3, but during the first three years the duration of low NO_3 concentrations during the annual cycle remained constant. In Fig. 9.1 it should be noted that while West Bear appears to show little increase in maximum NO_3 concentrations during the treatment period, the period between 1990 and 1993 actually had an elevated NO_3 export that was partially masked by declining ambient NO_3 export evident in the reference East Bear stream NO_3 concentrations. This pattern of elevated NO_3 export during the early 1990s was evident in forested watersheds throughout the northeastern U.S. Regional NO_3 stream concentrations then appeared to decline and returned to essentially zero in the reference watershed at BBWM. Mitchell et al. (1996) attributed this period of regionwide elevated surface-water NO_3 concentrations to an anomalous cold period in December of 1989, triggering several years of elevated but declining surface-water NO_3 concentrations. At BBWM, overall declining NO_3 concentrations in the reference East Bear watershed between 1990 and 1995 mean that net relative increases continued to occur in the treated watershed during this period when compared with the reference East Bear watershed. This pattern of surface-water NO_3 potentially attributable to relatively short-term climatic trends also points out the value and need for long-term, whole-ecosystem studies such as these to ascertain the role of climate on these ecosystems.

Fernow watersheds did not exhibit the period of elevated NO_3 export seen further north in the Northeast during the early 1990s, and NO_3 concentrations appear to show a continuous increase over time in response to treatments. Since biological control on stream-water NO_3 patterns are minimal due to excess N deposition at this site, it is logical that increased exports in WS3 beyond the reference WS4 would also not reveal seasonal patterns. There appears to be occasional instances of unusually high NO_3 concentrations but the data and analysis presented here do not attempt to determine their relationship to climatic patterns or antecedent conditions. It is interesting to note that Fig. 9.1 shows a 10-year record of NO_3 concentrations in WS4, with little evidence for increasing or decreasing trends. This may indicate that $\sim 50\,\mu eq\,l^{-1}$ NO_3 represents steady state with current levels of N deposition and stand development in this watershed. The time series suggests that stream NO_3 concentrations in WS3 are becoming more variable, but with the overall trend of steadily increasing concentrations that are twice or more the concentrations in WS4 after six years of treatment.

A great deal of literature has described the importance of atmospheric deposition on soil cation processes, with typical concerns being accelerated base cation leaching and aluminum (Al) mobilization (Likens et al., 1996; Lawrence et al., 1995, 1997; Shortle and Smith, 1988; Shortle et al., 1995). As with ambient deposition, the treatments at both BBWM and Fernow include a source of both N and S, and the effects of each element individually can not be definitively separated from the other in these studies. However, concern for net base cation losses due to chronic atmospheric deposition, and increasing concern for N deposition, has resulted in greater attention to the relationship between N saturation phenomena and base cation depletion. Both BBWM and Fernow are initiating research to specifically address some of these concerns. There is no question that $(NH_4)_2SO_4$ amendments to acid forest soils can result in accelerated depletion of Ca and the mobilization of Al depending on the rate of application. Carnol et al. (1997) recently demonstrated this phenomena in a Norway spruce (*Picea abies*) stand in England where $(NH_4)_2SO_4$ treatments caused soil water concentrations to "switch" from Ca to Al leaching, with relatively low Ca/Al ratios. They concluded that the most sensitive soils to acidification were those with low base saturation, typical of many northeastern forest soils in the U.S.

Fig. 9.2 shows the time series for Ca concentrations in the streams. The graph for BBWM shows the divergence between West Bear and East Bear as streams respond to the treatments. An apparent decline in Ca concentrations in East Bear over the period may reflect the declining contributions of NO_3 in stream water due to the climate-induced regional trends discussed earlier, and thus an overall decrease in the anion export.

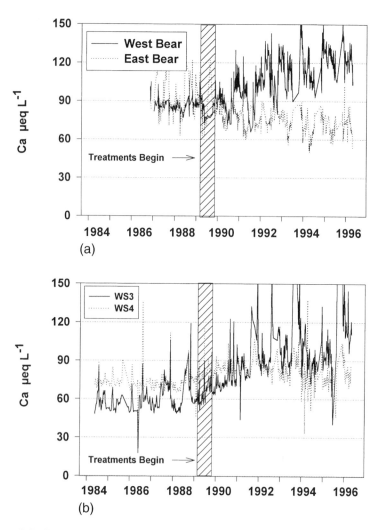

Figure 9.2. Stream Ca concentration time series for (a) BBWM and (b) Fernow.

West Bear clearly is showing increased export of Ca with treatment. This accelerated base cation leaching is partly attributable to NO_3 as evidenced by the parallels between Ca and NO_3 concentrations' seasonal patterns, and due to simple ion balance calculations discussed elsewhere (Norton et al., 1994). At Fernow, WS3 appears to have had a lower Ca concentration prior to treatments compared with WS4, but Ca concentrations have increased with the onset of treatment, largely attributable to N export in this highly N-enriched forested ecosystem.

As expected, both NO_3 and SO_4 are highly and significantly correlated with Ca leaching in these watersheds based on correlation analyses of the stream data presented here. At BBWM, simple correlation coefficients suggest that both of these strong mineral acid anions are governing Ca export. The correlation with Ca is stronger for SO_4 ($r = 0.83$) than for NO_3 ($r = 0.65$) in West Bear, but NO_3 is dramatically less important to Ca export in East Bear ($r = 0.59$ and $r = 0.17$ for Ca with SO_4 and NO_3, respectively). At Fernow, strong soil adsorption of SO_4 minimizes the effects of SO_4 on cation export but at this site even the reference watershed has high background NO_3 concentrations. Therefore, stream-water concentration data show that the correlation with Ca at Fernow in the treated WS3 is weaker for SO_4 ($r = 0.67$) than for NO_3 ($r = 0.91$). Both anions are much more weakly correlated with Ca in the reference WS4 watershed, which were $r = 0.20$ and $r = 0.38$ for SO_4 and NO_3, respectively.

Fig. 9.3 is a scatter plot of the relationships between stream concentrations of Ca and NO_3. Fig. 9.3a shows that a relationship appears to exist between stream concentrations of this cation–anion pair, although only in WS3 at Fernow is NO_3 apparently a dominant factor governing Ca concentrations in streams as inferred by correlations. In both reference watersheds, East Bear and WS4, there is little evidence of a strong relationship between Ca and NO_3, largely because of the lack of NO_3 in East Bear and the limited range of NO_3 concentrations in WS3 despite the elevated baseline NO_3 concentrations. It should also be noted that soil fertility, as indicated by soil exchangeable Ca concentrations, can influence nitrification and subsequent leaching. Willard et al. (1997) studied N mineralization and nitrification in nine mid-Appalachian forested watersheds including WS4 at Fernow. They concluded high soil fertility promoted nitrification due to significant positive correlations between soil exchangeable Ca and NO_3 production.

Table 9.3 and Fig. 9.4 present N mass balances for these watersheds. Table 9.3 shows that untreated East Bear has nearly total retention of ambient and treatment N in the final years of data presented here (exports of $<0.5 \, kg \, N \, ha^{-1} \, yr^{-1}$), while treatments have driven West Bear toward a condition similar to the reference watershed WS4 at Fernow with respect to N mass balance. Fernow clearly receives much greater inputs of atmospheric N compared with BBWM (approximately 2.3× using annual precipitation N data in this chapter), and has a greater export of N as a result (Fernow is approximately 4.8× BBWM using annual stream-water export N data in this chapter). Fig. 9.4 shows net N retention (%) for the study period at the treated watersheds. The most striking feature of these data is the relatively high retention of N in all watersheds despite the increasingly N-saturated character of WS3 and, perhaps most recently, West Bear. The mechanisms responsible for this continued net retention of N and the consequences for both forest management and recovery from N saturation are critically important areas for research. Even at WS3, where long-term N saturation processes are well advanced compared with

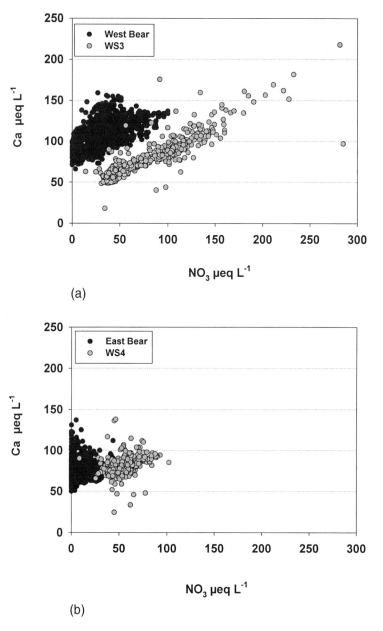

Figure 9.3. Scatter diagram for the relationship between stream NO_3 and Ca concentrations in the (a) treated and (b) reference watersheds at BBWM and Fernow.

BBWM, most N is retained by the ecosystem. As aboveground biota attain and surpass conditions of biological N limitation, it is logical that soil sequestration is largely responsible for the persistent N retention. Yet

Table 9.3. Inorganic Nitrogen Import–Export Budget for BBWM and Fernow from 1988 to 1995 (kg ha^{-1} yr^{-1})

	Wet Deposition (+ Treatment)[a]		Stream Export	
	East Bear	West Bear	East Bear	West Bear
1988	2.8	2.8	1.5	1.7
1989	4.5	8.7	2.7	3.1
1990	4.6	29.8	1.5	3.1
1991	3.7	28.9	2.2	7.0
1992	3.6	28.8	0.7	5.5
1993	3.1	28.3	0.4	6.0
1994	3.9	29.1	0.2	5.7
1995	3.8	29.0	0.2	5.8
	WS4	WS3	WS4	WS3
1988	8.0	8.2	4.2	3.5
1989	10.0	45.1	6.3	6.1
1990	9.3	44.4	8.6	10.3
1991	8.3	43.4	3.8	6.6
1992	8.1	43.2	4.3	8.8
1993	8.9	44.0	6.3	13.7
1994	10.3	45.4	3.7	7.8
1995	11.2	46.5	5.4	14.6

[a] Wet deposition is the sum of precipitation inputs of NH_4-N plus NO_3-N for the year in the reference watersheds (East Bear, WS4), and the sum of precipitation inputs plus N amendments in the treated watersheds (West Bear, WS3).

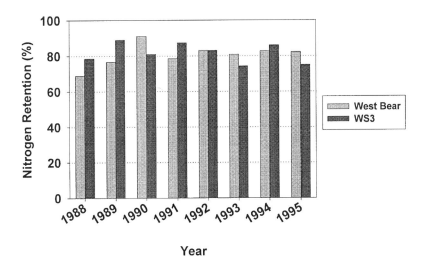

Figure 9.4. Nitrogen retention between 1988 and 1995 for BBWM and Fernow shown as a percentage of total calculated N input retained in the treated watersheds at BBWM and Fernow.

the mechanisms of this phenomena and our ability to define and predict both response and recovery remain incomplete at best.

An important question that has been raised is whether we really understand N mass balances given the difficulties in measuring all phases of N deposition, and the paucity of data on organic N export in streams. Certainly there are weaknesses in the current data from the literature. During the field seasons of 1996 and 1997, a pilot study was conducted that sampled streams from both BBWM and Fernow six times to gather preliminary data on the importance of other forms of N export besides inorganic NH_4 and NO_3. Fig. 9.5 shows the mean concentrations for those six collections, that included dissolved NH_4 and NO_3, NH_4 adsorbed on particulates, and total organic nitrogen on the unfiltered stream samples (includes dissolved and particulate forms of organic N). While these are only a few samples from a pilot study, they suggest that the majority of the stream export response in the treated watersheds at both BBWM and Fernow is in the form of NO_3. Even though the treatment is in the form of NH_4, no evidence of increased NH_4 export was found in the data. This reflects the ability of soil colloids to adsorb the NH_4 cation, and the subsequent nitrification potential of these soils. At Fernow, in both the treated and reference streams, NO_3 accounted for >96% of the N in streams. At BBWM the data were more variable, with organic N accounting for an average of 64% of the total stream N export in East Bear. It is important to note that this high percentage is of a very low concentration since in East Bear almost no N is leaving the ecosystem. In West Bear, these data averaged only 27% organic N with the majority of the treatment response expressed as increased NO_3 concentrations. These preliminary data suggest that organic forms of N

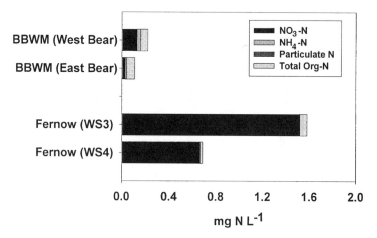

Figure 9.5. Means for the concentrations of NO_3, NH_4, particulate NH_4, and organic N from 1996 and 1997 at BBWM and Fernow ($n = 6$).

can be relatively important, particularly where N export is low. However, most of the response to N amendments in these experimentally manipulated watersheds during the initial years of treatment was clearly expressed as NO_3 increases in streams.

The Role of Soils and Vegetation

The focus of research at both BBWM and Fernow has been on maintaining a schedule of treatments to the watersheds, and tracking the integrated response of these treatments through stream-water monitoring. A number of studies are now in progress designed to evaluate the response of these watersheds to treatments that will provide insight on ecosystem response mechanisms. Typically, the first step in research at the ecosystem-level dealing with biogeochemical information needs is to treat the ecosystem as a "black box" and define inputs and outputs. Once the basic input–output budgets are defined, we then investigate what is taking place "inside the box."

One question about the fate of N in these ecosystems focuses on the role of forest vegetation and whether increased growth or N concentrations can explain any of the net N retention we see in these watersheds. We have initial insights from several studies that suggest that vegetative uptake likely plays a role in the N retention mechanism at both sites. At BBWM, White et al. (1999) showed that the treated West Bear watershed had significantly greater foliar concentrations of N in four of the major tree species, but there was no evidence of significant radial growth increases after four years of treatment in these species. They also reported declines in foliar Ca and increases in foliar Al concentrations on the treated watersheds. Similarly, Gillam et al. (1995) and Adams et al. (1993) reported increased concentrations of foliar N and declines in foliar Ca and Mg for major tree species and understory foliage concentrations at Fernow, but no change in diameter growth. Gilliam et al. (1995) also reported no significant effects on foliar phosphorus (P) concentrations that could be attributed to treatments. Dewalle et al. (1999) studied tree ring cation concentrations at both the West Virginia and Maine experimental watershed study sites. They included in their study a second watershed also treated with $(NH_4)_2SO_4$ in West Virginia (Clover Run), along with the Maine site. They reported significant decreases in Ca and/or Mg for all species in both treated West Virginia watersheds when compared with the reference watersheds, along with significant increases in manganese (Mn). A similar trend was reported for the Maine site, but the base cation declines were not statistically significant at the confidence level they chose. They suggested that sapwood Ca/Mn or Mg/Mn molar ratios were better indices of soil acidification than Ca/Al.

Weber and Wiersma (1997) studied the chemical composition of mosses at BBWM and reported significantly higher concentrations of N in the treated watershed in *Bazzania trilobata* and *Dicranum fulvum*, with significantly lower concentrations of Ca, Mg, potassium (K) and some trace elements, and found no treatment effects on tissue P concentrations. Magill et al. (1996) reported increases in foliar N in hardwood stands in a plot experiment adjacent to the BBWM watershed that was subjected to wet N treatments. They also reported no net increase in increment growth, but continuously increasing foliar N concentrations with time due to treatments. The limited data to date on these patterns suggest that some sequestration of increased N inputs is possibly attributable to above-ground vegetative uptake, but the studies have not been carried out long enough to see how these differences may result in growth changes by stand or species. There has been no quantification to date at either site of total aboveground biomass differences between treated and reference water-sheds that would define whether changes were taking place in minor stand components or understory species.

The other ecosystem compartment that plays a critical role in N accumulation and transformation is the soil. Limited intensive soils research has been conducted to date in these watersheds, although this is the focus of current studies. As yet, there appears to be no evidence for major shifts in forest floor total C and N pools or C/N ratios attributable to the treatments. Both the Maine and West Virginia sites show a significant increase in N mineralization and nitrification in the forest floor in response to elevated N (Wang and Fernandez, 1998; Gilliam et al., 1995). High rates of N mineralization were reported at Fernow in both WS3 and WS4 (6.7 and $7.7 \, \mathrm{g\,m^{-2}\,yr^{-1}}$, respectively) with between 90 to over 100% of this N release accounted for in net nitrification measure-ments. No comparable studies have been completed at BBWM, but data reported by Wang and Fernandez (1998) suggests much less nitrification in the BBWM ecosystems. Where increases in N mineralization or nitrifi-cation potentials in the forest floor due to treatments occurred at BBWM, they were largely confined to hardwood stands in the watershed, with little response in the softwoods. The influence of vegetation type, particularly the contrast between hardwood and softwood stands due to the influence on litter quality and, therefore, soil properties, should be further investigated and is likely critical in considering the relevance of these studies to landscapes.

Soil solutions at both BBWM and Fernow show differences due to treatments consistent with the changes in stream-water chemistry over time. Adams et al. (1997) described the treatment response in soil solutions at Fernow, with elevated NO_3 concentrations in WS3 soil solutions in the A horizon, diminishing with depth in the soil. They reported that both SO_4 and NO_3 contributed to increased Ca leaching losses, with little influence on Mg. Fernandez et al. (1999) also reported similar patterns in soil solutions of West Bear, but these changes were largely confined to the hardwood stands within the West Bear watershed.

They, too, reported increased leaching of Ca along with increased Al in soil solutions as a result of the treatments.

The Reversibility Question

The phenomenon referred to as N saturation, or the chronic effects of elevated atmospheric N deposition on forested ecosystems, has the benefit of drawing on a vast scientific literature of N in forested ecosystems either from the forest fertilization or nutrient cycling literature, or more broadly from studies in the ecological and agricultural sciences. What we do not have is a robust literature on the consequences of long-term, chronic N amendments to humid, temperate climate forested ecosystems. More importantly, we have little information on how these ecosystems respond to decreased N inputs following long-term elevated N exposures. Indeed, there is much we do not know, nor can adequately predict, about how forest ecosystems "recover" from N saturation. There is ample evidence in the literature to suggest that soil export of NO_3 in soil solutions rapidly declines following short-term experimental applications of N, such as reported by Rustad et al. (1993, 1996). However, data reported here on BBWM and Fernow show that the majority of N deposition even in N "saturated" ecosystems is retained. Although a significant amount of the retained N could be sequestered in an aggrading forest, a major proportion must be retained in the soil. Therefore, while the symptoms of N saturation may be quickly ameliorated by reduced inputs as judged by the geochemical signal in surface waters, we expect that the forest ecosystem will take much longer to recover, on the order of decades to centuries. Indeed, even the concept of "recovery" from N accumulations could be debated as to meaning and demonstration. Stand growth and composition, litter quality, and subsequent rates of decomposition and nutrient cycling, and short- and long-term soil pools of N are likely to respond slowly to major shifts in N inputs, with many of the controlling mechanisms in this context poorly understood.

Conclusions

The BBWM and Fernow whole-watershed experimental N enrichment studies provide a unique opportunity to study, in situ, the mechanisms governing forest ecosystem response to elevated N deposition and N saturation phenomena. Both sites are paired watershed studies offering a powerful tool to examine long-term changes in ecosystem structure and function, particularly with respect to the biogeochemistry of N. Our understanding of forest ecosystem response to N allows us to predict some changes in these watersheds, but our knowledge is lacking on mechanisms governing these responses and on the temporal development of the changes expected. Both sites together offer an interesting spectrum of forested stream watersheds ranging from little or no inorganic N export and modest deposition, to chronically elevated N export and rates of N inputs that rival some of the high N deposition European landscapes. Yet in all cases a majority of the added N is retained by the watersheds.

On the other hand, even in the reference watershed in Maine where no N saturation exists and tree growth remains N-limited, small increases in N inputs had almost immediate consequences expressed as increased stream export of NO_3. We suggest that these sites become more valuable each year because of their record of treatment and response, but neither of the treated watersheds has yet attained a new equilibrium with the elevated N inputs from treatments. Some have suggested they never will, others believe a new equilibrium can be attained in a practical time frame. Research on the biological and geochemical mechanisms controlling these response characteristics, and how they will respond in the future to potential changes in N deposition, should be a high priority for the national research agenda and contribute to our basic understanding of ecosystem function regardless of current policy issues. This becomes even more important in a changing climate that couples warming-induced changes in the rates of biogeochemical cycling with chronic exposure to elevated N deposition.

References

Aber JD, Driscoll CT (1997) Effects of land use, climate variation, and N deposition on N cycling and C storage in northern hardwood forests. Global Biogeochem Cycl 1:639–648.

Aber J, McDowell W, Nadelhoffer K, Magill A, Berntson G, Kamakea M, McNulty S, Currie W, Rustad L, Fernandez I (1998) Nitrogen saturation in temperate forest ecosystems—hypothesis revisited. BioScience 48:921–943.

Aber J (1992) Nitrogen cycling and nitrogen saturation in temperate forest ecosystems. Tree 7:220–223.

Aber JD, Nadelhoffer KJ, Steudler P, Melillo JM (1989) Nitrogen saturation in northern forest ecosystems. BioScience 39:378–386.

Adams MB, Angradi TR (1996) Decomposition and nutrient dynamics of hardwood leaf litter in the Fernow Whole-Watershed Acidification Study. For Ecol Manage 83:61–69.

Adams MB, Angradi TR, Kochenderfer JN (1997) Stream water and soil solution responses to 5 years of nitrogen and sulfur additions at the Fernow Experimental Forest, West Virginia. For Ecol Manage 95:79–91.

Adams MB, Edwards PJ, Wood F, Kochenderfer JN (1993) Artificial watershed acidification on the Fernow Experimental Forest, USA. J Hydrol 150:505–519.

Adams MB, Kochenderfer JN, Wood F, Angradi TR, Edwards PJ (1994) *Forty Years of Hydrometeorological Data from the Fernow Experimental Forest, West Virginia.* Gen Tech Rep NE-184. United States Department of Agriculture (USDA) Forest Service, Northeastern Forest Experiment Station, Radnor, PA.

Ågren GI, Bosatta E (1988) Nitrogen saturation of terrestrial ecosystems. Environ Pollut 54:185–197.

Bailey RG, Avers PE, King T, McNabb WH (1994) *Ecoregions and Subregions of the United States*. United States Department of Agriculture (USDA) Forest Service, Washington, DC.

Bull KR (1991) The critical loads/levels approach to gaseous pollutant emission control. Environ Pollut 69:105–123.

Carnol M, Ineson P, Dickinson AL (1997) Soil solution nitrogen and cations influenced by $(NH_4)_2SO_4$ deposition in a coniferous forest. Environ Pollut 97:1–10.

DeWalle DR, Tepp JS, Swistock BR, Sharpe WE, Edwards PJ (1999) Tree-ring cation response to experimental watershed acidification in West Virginia and Maine. J Environ Qual 28:299–309.

Dise NB, Wright RF (1992) *The NITREX Project* (*Nitrogen Saturation Experiments*). *Ecosystem Research Report 2*. Commission of European Communities, Brussels, Belgium.

Dise NB, Wright RF (1995) Nitrogen leaching from European forests in relation to nitrogen deposition. For Ecol Manage 71:153–161.

Driscoll CT, Van Dreason R (1993) Seasonal and long-term temporal patterns in the chemistry of Adirondack lakes. Water Air Soil Pollut 67:319–344.

Fenn ME, Poth MA, Johnson DW (1996) Evidence for nitrogen saturation in the San Bernardino Mountains in southern California. For Ecol Manage 82: 211–230.

Fernandez IJ, Rustad LE, David MB, Nadelhoffer KJ, Mitchell MJ (1999) Mineral soil and solution responses to experimental N and S enrichment at the Bear Brook Watershed in Maine (BBWM). Environ Monit Assess 55: 165–185.

Galloway JN (1998) The global nitrogen cycle: changes and consequences. Environ Pollut 102:15–24.

Gilliam FS, Adams MB, Yurish BM (1995) Ecosystem nutrient responses to chronic nitrogen inputs at Fernow Experimental Forest, West Virginia. Can J For Res 26:196–205.

Jeffries RL, Maron JL (1997) The embarrassment of riches: atmospheric deposition of nitrogen and community and ecosystem processes. Tree 12:74–77.

Johnson D (1992) Nitrogen retention in forest soils. J Environ Qual 21:1–12.

Kahl SJ, Norton SA, Fernandez IJ, Nadelhoffer KJ, Driscoll CT, Aber JD (1993) Experimental inducement of nitrogen saturation at the watershed scale. Environ Sci Tech 27:565–568.

Lawrence GB, David MB, Shortle WC (1995) A new mechanism for calcium loss in forest-floor soils. Nature 378:162–165.

Lawrence GB, David MB, Vailey SW, Shortle WC (1997) Assessment of calcium status in soils of red spruce forests in the northeastern United States. Biogeochemistry 38:19–39.

Likens GE, Driscoll CT, Busco DC (1996) Long-term effects of acid rain: response and recovery of a forest ecosystem. Science 272:244–246.

Løkke H, Bak J, Falkengren-Grerup U, Finlay RD, Ilvesniemi H, Nygaard PH, Starr M (1996) Critical loads of acidic deposition for forest soils: Is the current approach adequate? Ambio 25:510–516.

Magill AH, Downs MR, Nadelhoffer KJ, Hallet RA, Aber JD (1996) Forest ecosystem reponse to four years of chronic nitrate and sulfate additions at Bear Brooks Watershed, Maine, USA. For Ecol Manage 84:29–37.

McNulty S, Aber JD (1993) Effects of chronic nitrogen additions on nitrogen cycling in a high-elevation spruce–fir stand. Can J For Res 23:1252–1263.

McNulty S, Aber JD, McLellan TM, Katt SM (1990) Nitrogen cycling in high elevation forests of the northeastern US in relation to nitrogen deposition. Ambio 19:30–40.

McNulty S, Aber JD, Newman SD (1996) Nitrogen saturation in a high elevation New England spruce–fir stand. For Ecol Manage 84:109–121.

Mitchell MJ, Driscoll CT, Kahl JS, Likens GE, Murdoch PS, Pardo H (1996) Climatic control on nitrate loss from forested watersheds in the northeast United States. Environ Sci Technol 30:2609–2612.

Nadelhoffer K, Downs M, Fry B, Aber J (1999) Controls on N retention and exports in a forested watershed. Environ Monit Assess 55:187–210.

Nadelhoffer KJ et al. (1999) Nitrogen deposition makes a minor contribution to carbon sequestration in temperate forests. Nature 398:145–148.

NADP (1990) National Atmospheric Deposition Program (IR-7)/National Trends Network: 1990. Annual Report: NADP/NTN Coordination Office. Natural Resource Ecology Laboratory, Colorado State University, Fort Collins, CO.

Nihlgård B (1985) The ammonium hypothesis—an additional explanation to the forest dieback in Europe. Ambio 14:2–8.

Nilsson J (1986) *Critical Loads for Nitrogen and Sulphur*. Nordsk ministerrånd. Miljørapport 11. Copenhagen, Denmark.

Nilsson J, Grennfelt P (1988) *Critical Loads for Sulphur and Nitrogen. Report from a Workshop Held at Skokloster, Sweden, 19–24 March 1988*. Miljørapport 1988:15, Nordic Council of Ministers, Copenhagen, Denmark.

Norton SA, Fernandez IJ (eds) (1999) *The Bear Brook Watershed in Maine: A Paired Watershed Experiment—The First Decade (1987–1997)*. Kluwer Academic, Boston, MA.

Norton SA, Kahl JS, Fernandez IJ, Rustad LE, Scofield JP, Haines TA (1994) Response of the West Bear Brook Watershed, Maine, USA, to the addition of $(NH_4)_2SO_4$: 3-year results. For Ecol Manage 68:61–73.

Peterjohn WT, Adams MB, Gilliam FS (1996) Symptoms of nitrogen saturation in two central Appalachian hardwood forest ecosystems. Biogeochemistry 35:507–522.

Rustad LE, Fernandez IJ, David MB, Mitchell MJ, Nadelhoffer KJ, Fuller RB (1996) Experimental soil acidification and recovery at the Bear Brook Watershed in Maine. Soil Sci Soc Am J 60:1933–1943.

Rustad LE, Fernandez IJ, Fuller RB, David MB, Nodvin SC, Halteman WA (1993) Soil solution response to acidic deposition in a northern hardwood forest. Agric Ecosyst Environ 47:117–134.

Rustad LE, Kahl JS, Norton SA, Fernandez IJ (1994) Underestimation of dry deposition by throughfall in mixed northern hardwood forests. J Hydrol 162:319–336.

Shortle WC, Smith WC (1988) Aluminum-induced calcium deficiency syndrome in declining red spruce. Science 240:1017–1018.

Shortle WC, Smith WC, Minocha R, Alexeyev VA (1995) Similar patterns of change in stemwood calcium concentrations in red spruce and Siberian fir. J Biogeogr 22:467–473.

Stoddard JL (1994) Long-term changes in watershed retention of nitrogen: its causes and aquatic consequences. In: Baker LA (ed) *Environmental Chemistry of Lakes and Reservoirs*. American Chemical Society, Washington, DC, pp 223–284.

Sullivan TJ, Eilers JM, Cosby BJ, Vaché KB (1997) Increasing role of nitrogen in the acidification of surface waters in the Adirondack Mountains, New York. Water Air Soil Pollut 95:313–336.

Sullivan TJ (1997) Ecosystem manipulation experimentation as a means of testing a biogeochemical model. Environ Manage 21:15–21.

Vitousek PM (1997) Human alteration of the global nitrogen cycle: sources and consequences. Ecol Applic 7:737–750.

Wang Z, Fernandez IJ (1998) Soil type and forest vegetation influences on forest floor nitrogen dynamics in the Bear Brook Watershed in Maine (BBWM). Environ Monit Assess 55:221–234.

Weber KL, Wiersma GB (1997) Trace element concentrations in mosses collected from a treated experimental forest watershed. Toxic Environ Chem 65:17–29.

White GJ, Fernandez IJ, Wiersma GB (1999) Impacts of ammonium sulfate treatment on foliar chemistry of forest trees at the Bear Brooks Watershed in Maine. Environ Monit Assess 55:235–250.

Willard KW, DeWalle DR, Edwards PJ, Schnabel RR (1997) Indicators of nitrate export from forested watersheds of the mid-Appalachians, United States of America. Global Biogeochem Cycl 11:649–656.

Williams MW, Baron JS, Caine N, Sommerfeld R, Sanfrod R Jr. (1996) Nitrogen saturation in the Rocky Mountains. Environ Sci Technol 30(20):640–646.

Wright RF, van Breeman R (1995) The NITREX Project: an introduction. For Ecol Manage 71:1–5.

Wright RF, Rasmussen L (eds) (1998) The whole ecosystem experiments of the NITREX and EXMAN projects. For Ecol Manage Vol. 101.

10. Effects of Soil Warming on Carbon and Nitrogen Cycling

Lindsey E. Rustad, Jerry M. Melillo, Myron J. Mitchell,
Ivan J. Fernandez, Paul A. Steudler,
and Patrick J. McHale

Atmospheric concentrations of carbon dioxide (CO_2), methane (CH_4), and nitrous oxide (N_2O) have increased dramatically since the beginning of the industrial revolution, largely due to human activities such as fossil fuel combustion and land use change. These gases are currently accumulating in the atmosphere at the rate of approximately 0.40% yr^{-1} for CO_2, 0.60% yr^{-1} for CH_4, and 0.25% yr^{-1} for N_2O (Houghton et al., 1996). Because these gases have the capacity to retain heat in the atmosphere by trapping infrared radiation, considerable concern has been raised that increased concentrations of these gases will result in higher mean global temperatures, or the "greenhouse effect." Currently, the radiative forcing of climate by greenhouse gases is predicted to increase global mean annual temperature by 1 to 3.5°C in the next 100 years, with a greater warming occurring in the higher latitudes than at the equator (Houghton et al., 1996).

The consequence of this potential increase in temperature for northern U.S. forest soils needs to be evaluated. These soils serve as large reservoirs for nutrients necessary for forest growth and productivity as well as major sinks for organic carbon (C) (Turner et al., 1995). It has been hypothesized that the predicted increases in mean annual temperature will accelerate rates of soil organic matter decomposition, resulting in a significant increase in the release of C (as CO_2) from the soil to the atmosphere. This would, in turn, exacerbate increases in atmospheric CO_2, and provide a positive feedback to global warming. Elevated temperature could either enhance soil efflux or uptake of CH_4, resulting in either an augmentation or amelioration of the greenhouse effect, depending on the direction of flux. Rates of nitrogen (N) mineralization may also increase in response to elevated temperature, leading to greater availability of inorganic N (ammonium [NH_4^+] plus nitrate [NO_3^-]). For those forested sites where N is a limiting nutrient, increased N availability may have a positive effect on forest productivity (Melillo et al., 1995). However, at sites already enriched in N, additional, NH_4^+ or NO_3^- generated by soil warming could contribute to "N saturation," resulting in forest nutrient imbalances (Shortle and Smith, 1988), decreased growth (McNulty et al., 1996), and accelerated leaching of N (particularly NO_3^-) to surface water and groundwater (Aber et al., 1989; McNulty et al., 1996; Fenn et al., 1998).

In this chapter, we evaluate the potential consequences of climatic warming for belowground C and N cycling in northern U.S. forest soils. We focus here on soil C and N because (1) changes in C and N storage and cycling rates could have direct effects on the direction and magnitude of climate change, (2) soil C and N represent the limiting energy and nutrient resources in these soils and the cycling of these elements is directly regulated by microbial communities that are typically sensitive to temperature, and (3) forest soil organic matter quality is often defined by C and N composition , which then affects the cycling rates of other soil nutrients. Although the discussion will draw from the extensive literature on the temperature dependence of soil processes governing C and N cycling, we will highlight results from three in situ temperature manipulation experiments and two geographical gradient studies that have been conducted in the northeastern and north central U.S. during the last decade. Although these studies do not directly simulate the long-term and complex physical (e.g., temperature, moisture) and chemical (e.g., increased CO_2, changes in N availability) shifts predicted for global change, such experimental manipulations and gradient studies, particularly when considered together, provide powerful tools to test predictive relationships between soil temperature and soil processes, and allow us to scale-up results from the laboratory or microcosm to larger, more relevant ecological landscape units.

Soil Warming Experiments and Gradient Studies

Three soil warming experiments were initiated in the northeastern U.S. during the last decade at the Harvard Forest in Petersham, Massachasetts, the Howland Integrated Forest Study site in Howland, Marine, and the Huntington Forest near Newcomb, New York (Fig. 10.1). The Harvard

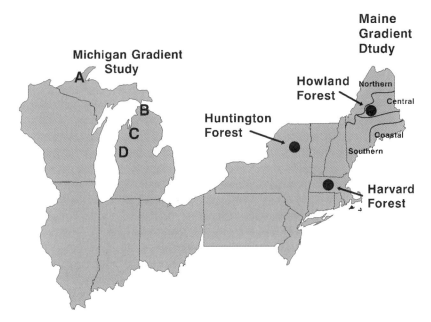

Figure 10.1. Location of soil warming and gradient studies in the north-eastern U.S.

Forest and the Huntington Forest are both northern hardwood forest ecosystems and the Howland Forest is a low elevation spruce–fir forest ecosystem. Elevated soil temperatures were maintained at all three sites using buried resistance heating cables. At the Harvard Forest, eighteen 6×6-m experimental plots were established in 1991, and one of three treatments was randomly assigned to one plot in each of six blocks. Treatments consisted of (1) heated plots in which soil temperature was raised 5°C above ambient using resistance heating cables buried at a depth of 10 cm in the Ap horizon, with 20 cm horizontal spacing, (2) disturbance control plots that were identical to the heated plots except that they received no power, and (3) undisturbed control plots. The plots were heated year round. Although the experiment is ongoing, results reviewed here are from the first six months of the study (Peterjohn et al., 1994) and the first two years of the study (Melillo et al., 1995). At the Howland Forest, six 15×15-m plots were established in 1992, and one of three treatments was randomly assigned to two plots each. The treatments were the same as for the Harvard Forest, except that the cables were buried ~ 2 cm below the surface of the O horizon. The plots were heated from May through November in 1993, 1994, and 1995. At the Huntington Forest, four 10×10-m plots were established in 1992. Three of the four plots were heated at 2.5, 5.0, and 7.5°C above ambient, respectively, using resistance heating cables buried at a depth of 5 cm in the O-horizon. The fourth plot served as the reference plot. The plots were heated from May through September in 1993 and 1994. For purposes of comparison, we only review data from the 5.0°C temperature treatment at the Huntington Forest. Field sampling at these three sites included measurements of nutrient cycling parameters identified as important to C and N dynamics in forested ecosystems. Further details on these heating experiments can be found in Peterjohn et al. (1994) for the Harvard Forest, in Rustad and Fernandez (1998a,b) for the Howland Forest, and in McHale et al. (1998) for the Huntington Forest.

The two gradient studies that we will review are the Michigan Gradient Study and the Maine Gradient Study (see Fig. 10.1). The Michigan Gradient Study was initiated in 1987 and includes four intensively monitored northern hardwood study sites (with three 30 × 30-m plots at each site) located along a 600-km climatic and pollutant gradient extending from northwestern to southern Michigan. From north to south along the gradient, mean annual air temperature increased from 4.2 to 7.6°C, mean field season soil temperature (mid-March to mid-November) increased from 10.0 to 13.0°C, and total annual wet-plus-dry deposition of NO_3-N decreased from 8 to 4 kg ha^{-1} yr^{-1}. The Maine Gradient Study was initiated in 1993 and included a network of 16 northern hardwood sites across four climatic regions in Maine identified by Briggs and Lemmin (1992). From north to south, mean annual air temperature increased from 2.0 to 6.2°C, mean annual soil temperature increased from 5.2 to 7.3°C, and mean annual precipitation increased from 90 to 140 cm. In 1995, the Maine Gradient Study was expanded to include nine additional softwood-dominated sites, which were also distributed across the climatic gradient in Maine. Further details on these gradient studies can be found in Burton et al. (1998) for Michigan and in Simmons et al. (1996) for Maine.

Carbon Cycling

The main sources of C inputs to forest soils are above- and belowground detritus. Carbon entering the soil is partially decomposed releasing CO_2 with the residual C converted to soil organic matter of which humus represents the most recalcitrant form. Root respiration is another major source of C loss from soil with leaching of carbonate ions (HCO_3^-) and dissolved organic carbon (DOC) generally of secondary importance.

Soil Respiration

Soil respiration represents the combined respiration of roots, mycorrhizae, and the soil microflora and fauna. Worldwide, soil respiration releases an estimated 68 to 100 Pg C yr^{-1} from the soil to the atmosphere (Raich and Schlesinger, 1992; Musselman and Fox, 1991), making it one of the largest fluxes in the global C cycle. Even a small increase in soil respiration could thus equal or exceed the estimated 7Pg C released annually from the combination of tropical deforestation and fossil fuel combustion (Lal et al., 1995), and could significantly exacerbate atmospheric increases in CO_2. For example, Jenkinson et al. (1991) predicted that a $0.03°\text{C yr}^{-1}$ increase in mean global temperature could result in a net release of 60Pg C from the soil to the atmosphere during the next 60 years due to accelerated soil organic matter decomposition.

Studies have long demonstrated that soil temperature is one of the main drivers of soil respiration. From a global perspective, Raich and Schlesinger (1992) and Raich and Potter (1995) demonstrated significant correlations between soil respiration and mean annual air temperature and mean annual precipitation, with air temperature being the single best climatic predictor of respiration. Raich and Schlesinger (1992) reported a median Q_{10} value of 2.4 for temperature influences on soil respiration based on in situ studies. Additional in situ studies demonstrating significant positive relationships between temperature and respiration include those reported by Witkamp (1969), Edwards (1975), Singh and Gupta (1977), Schleser (1982), Schlenter and Van Cleve (1985), Peterjohn et al. (1993) and Pinol et al. (1995). All investigators reported positive slopes for the temperature response function, but relationships between soil respiration and temperature have shown considerable variability due to interacting influences of moisture (including drought, flood, wetting and drying cycles), litter quantity, litter quality, freeze–thaw cycles, and the relative contribution of roots vs. microbes to total respiratory flux. In an attempt to eliminate some of this variability, Kirschbaum (1995) reviewed the temperature dependence of soil respiration in controlled laboratory studies where moisture was not limiting, and found that soil respiration consistently increased with soil temperature and the magnitude the response was greater at lower temperatures. This is consistent with observations by both Schleser (1982) and de Jong et al. (1974), and has led to the suggestion that global warming could result in an unprecedented release of C from soils, especially in high latitude ecosystems.

Soil respiration was measured at all three experimental soil warming sites and across all sites in the two gradient studies using the static chamber technique. For the soil warming studies, mean rates of CO_2 efflux in the reference plots were $33 \text{ mg CO}_2\text{-C m}^{-2} \text{ hr}^{-1}$ at the Howland Forest (Rustad and Fernandez, 1998a), $71 \text{ mg CO}_2\text{-C m}^{-2} \text{ hr}^{-1}$ at the Huntington Forest (McHale et al., 1998), and $80 \text{ mg CO}_2\text{-C m}^{-2} \text{ hr}^{-1}$ at the Harvard Forest (Melillo et al., 1995). These values are within the range of 5 to $410 \text{ mg CO}_2\text{-C m}^{-2} \text{ hr}^{-1}$ reported for North American forest soils (Edwards, 1975; Crill, 1991; Fernandez et al., 1993; Toland and Zak, 1994). Soil respiration was significantly correlated with soil temperature at all three sites, with temperature explaining 35, 53, and 92% of the observed variation of soil respiration at the Howland Forest (Rustad and Fernandez, 1998a), Huntington Forest (McHale et al., 1998), and

Harvard Forest (Peterjohn et al., 1994), respectively. Differences in both the absolute rates of respiration at these three sites as well as the "goodness of fit" of the relationship between soil respiration and soil temperature may be attributed to differences in soil organic C chemistry and N availability (Randlett et al., 1996; Zogg et al., 1996), with soil respiration being more responsive to temperature at the two hardwood sites, which had more labile C, higher soil N concentrations, and lower C/N ratios.

A 5°C increase in soil temperature resulted in a significant increase in CO_2 efflux at all three soil warming sites, with mean increases of 25, 26, and 33% at the Howland Forest (Rustad and Fernandez, 1998a), Harvard Forest (Melillo et al., 1995), and Huntington Forest (McHale et al., 1998), respectively. Temporal patterns of response varied among the sites. At the two hardwood sites (Harvard and Huntington Forests) considerably more CO_2 was evolved in response to warming during the first year compared with the second year, that is, the increase in CO_2 efflux in response to the first year of warming was 40 and 48% at the Harvard and Huntington Forests, respectively, whereas the increase in CO_2 efflux in response to the second year of warming was only 12 and 19% for the Harvard and Huntington Forests, respectively (Melillo et al., 1995; McHale et al., 1998). In contrast, CO_2 efflux showed no significant response to warming during the first year of manipulation at the Howland Forest, whereas CO_2 efflux increased significantly by 26% in both the second and third years, respectively (Rustad and Fernandez, 1998a). These differences are likely due to differences in organic matter content and quality, with the hardwood sites likely having larger initial pools of labile C than the conifer site. Once this initial pool has been exhausted, decomposition of more recalcitrant C pools would proceed more slowly.

For the Maine Gradient Study, mean soil respiration in hardwood sites ranged from 46 mg CO_2-C m^2 hr^{-1} in the northern region to 55 mg CO_2-C m^{-2} hr^{-1} in the southern region, or a 17% increase along the climatic gradient (Simmons et al., 1996). Normalized to an increase in respiration per °C increase in soil temperature, this represents an increase of 4.5 mg CO_2-C m^{-2} hr^{-1} per 1°C increase in temperature, which compares to increases of 1.6 to 6.6 mg CO_2-C m^{-2} hr^{-1} for the three soil warming studies. For the Maine Gradient Study, temperature explained 33% of the observed variation in the natural logarithm of soil respiration.

For the Michigan Gradient Study, Zogg et al. (1996) demonstrated significant positive relationships between fine root respiration and soil temperature, with the northern-most site having predictably the lowest fine root respiration rate (3.4 μmol O_2 kg^{-1} s^{-1}) compared with the other three sites (mean = 4.2 μmol O_2 kg^{-1} s^{-1}). Root tissue N concentration was also shown to be an important determinant of root respiration rate, with the combination of root tissue N concentration and soil temperature explaining 65% of the variability in fine root respiration rates. Burton et al. (1998) further demonstrated the importance of soil moisture in tempering these interactions, with root respiration rates being lower than predicted during periods of drought. MacDonald et al. (1995) also reported significant increases in microbial respiration with increasing temperature, based on laboratory incubations of soil from the four sites along the gradient. Soil organic C and incubation temperature together explained 90% of the variation in microbial respiration in this laboratory study.

These studies provide strong evidence that an increase in temperature will increase the efflux of CO_2 from the soil through soil respiration. The extent of this increase will likely be influenced by the availability of labile C, soil moisture, and soil N and/or other limiting nutrients, and will likely be at least initially greater in northern hardwood stands than in softwood stands due to the latter producing litter more recalcitrant to decomposition.

Litter Decomposition

The decomposition of litter release nutrients that are taken up by forest vegetation and the micro- and macrobiota, and the availability of these nutrients affects ecosystem production and litter inputs. The net balance between the rates of litter inputs and decomposition has a major influence on the accumulation of soil organic matter. Temperature has long been recognized to play a critical role in regulating rates of litter decay, with decay rates generally increasing with increasing temperature in the range 5 to 30°C (Daubenmire and Prusson, 1963; Witkamp, 1966; Alexander, 1977; Meentemeyer, 1978; Swift, 1979; Jansson and Berg, 1985; Moore, 1986; Van Cleve et al., 1990; Ruark, 1993; Hobbie, 1996). Thus, it is reasonable to hypothesize that an increase in mean global temperature will result in increased rates of litter decomposition. Critical questions in global change research are whether increased rates of litter decomposition due to elevated temperature will equal or exceed increased rates of NPP due to elevated CO_2, and whether future net ecosystem C exchange will be positive or negative. Wofsy et al. (1993), using measurements of whole ecosystem C exchange, have reported that the Harvard Forest was a net sink for atmospheric C ($\sim -2\,Mt\,C\,ha^{-1}\,yr^{-1}$), whereas Goulden et al. (1998) using similar techniques reported that the Boreas sites in eastern Canada are near neutral or a slight source for atmospheric C.

The effect of soil warming on litter decomposition was evaluated at the Howland and Huntington Forest sites using the litter bag technique (Rustad and Fernandez, 1998b; McHale et al., 1998). The response to the 5°C increase in ambient temperature varied both by site and species. At the Howland Forest, red maple (*Acer rubrum* L.) litter lost 27% more mass and 33% more C during the first 6 months of decay in the heated plots compared with the controls. After 30 months of decay, significant treatment effects were no longer evident for red maple litter. Red spruce (*Picea rubens* Sarg.) litter at the Howland Forest showed few treatment effects during the initial 18 months of decay. However, after 30 months of decay, red spruce litter in the heated plots had lost 19% more mass and C than red spruce litter in the control plots. At the Huntington Forest, American Beech (*Fagus grandifolia* Ehrh.) litter in the heated plots had lost significantly more mass (19%) and C (16%) in the heated plots compared with the controls after one year of decay, and significantly more C (19%) after two years of decay. Red maple litter, however, decayed at comparable rates in both the heated and control plots during the two years of the study at the Huntington site. Verburg et al. (1999) also reported no effect of a 3 to 5°C increase in temperature on decomposition of birch (*Betula* spp.) litter at the CLIMEX soil warming study in southern Norway, and Robinson et al. (1995) showed lower rates of litter decomposition in response to a 1°C increase in temperature in a high arctic polar semi-desert and a sub-arctic dwarf shrub heath in Sweden. Both studies attribute the lack of a positive response to temperature to reductions in soil moisture, with consequent desiccation of the litter. Similar reductions in soil moisture in soil warming experiments have been reported by Rustad and Fernandez (1998b) for the Howland Forest, Peterjohn et al. (1994) for the Harvard Forest, Harte et al. (1995) for a montane meadow in Colorado, Hantschel et al. (1995) for an agroeco-system in Germany, and Hartely et al. (1998) for a sub-arctic dwarf shrub tundra site in Sweden. The reduction of soil moisture may have reduced or confounded the effects of elevated temperature on decomposition as well as on other biologically mediated processes.

For the Maine Gradient Study, Delaney et al. (1996) found no significant differences in initial (6-month) decay rates between climatic regions for either red maple or white pine (*Pinus strobus* L.) foliar litter, despite higher temperature and greater precipitation at the southern sites compared with the northern sites. They attribute this lack of response to (1) the short duration of the study, (2) high variability, and (3) the relatively small regional temperature differences along the gradient (i.e., mean May to October O-horizon temperatures ranged from 10.3 at the northern sites to 13.4°C at the southern sites).

We conclude from these studies that a 3 to 5°C increase in mean annual soil temperature can result in subtle but significant increases in litter decomposition, which are consistent with the observation that soil C accumulation is generally inversely related to mean annual temperature at landscape and continental scales. However, litter quality and soil moisture also play important roles in determining the characteristics of the temperature effect on litter decomposition. Further, our current ability to detect changes in decomposition for more modest temperature increases (i.e., $< +3.0$°C) is limited and warrants further study. Although an increase in decomposition has the potential to release large amounts of C from the soil to the atmosphere, enhanced decomposition will also result in increased availability of nutrients, which may stimulate plant growth, particularly in a higher CO_2 environment. The net effect could be to shift C from soil organic matter (with typically low C/nutrient ratios) to plant biomass (with typically higher C/nutrient ratios), thereby increasing total C sequestration in forest ecosystems (Melillo et al., 1995).

Methane Flux

Estimates of global flux of CH_4 to the atmosphere range from 410 to 660 Tg CH_4 yr^{-1} (Houghton et al., 1996). Although this is considerably less than the flux of CO_2, CH_4 is an important greenhouse gas because CH_4 has the ability to trap 25 times more radiant energy than CO_2 on a molar basis (Rodhe, 1990). Soils serve as both a source and a sink for atmospheric CH_4, with the direction of flux depending largely on the oxygen status of the soils. Under anaerobic conditions, typical of swamps, bogs, fens, paddies, and other wetland soils, methanogenic bacteria use CO_2 as an electron acceptor, reducing CO_2 to CH_4. Under aerobic conditions that are more typical of forest and grassland soils, methano-trophic bacteria use CH_4 as a substrate for growth, ultimately converting CH_4 to CO_2 (Alexander, 1977). On a global scale, anaerobic soils produce an estimated 82 to 132 Tg CH_4 yr^{-1} (Cicerone and Ormeland, 1988) whereas aerobic forest soils consume 15 to 45 Tg CH_4 yr^{-1} (Houghton et al., 1996).

Not surprisingly, numerous studies have shown that CH_4 flux from soils is largely controlled by soil moisture (Steudler et al., 1989; Castro et al., 1994; Castro et al., 1995; Torn and Harte, 1996). Soil moisture affects both soil O_2 status as well as the diffusion of atmospheric CH_4 into the soil, as studies have shown that methanotrophs are typically located well below the soil surface (Crill, 1991; King, 1997). The importance of temperature in mediating CH_4 production and oxidation has also been demonstrated (Keller et al., 1983; Crill et al., 1988, 1991; Born et al., 1990; Adamsen and King, 1993; Koschorreck and Conrad, 1993; Peterjohn et al., 1993; Yavitt et al., 1995; Whalen and Reeburgh, 1996). For CH_4 oxidation, results from Crill (1991), King and Adamsen (1992), and Castro et al. (1995) suggest that CH_4 oxidation increases with increasing temperature between −5 and 10°C, with no further response at temperatures above 10°C. Site fertility, as defined by N availability, also affects CH_4 consumption, with greater consumption reported for more fertile sites, likely due to more active microbial populations (Castro et al., 1995). This does not, however, hold true for N-fertilized sites, where high N input rates may suppress CH_4 consumption (Crill, 1994; Castro et al., 1995).

Methane flux was measured at all three of the soil warming experiments using static chambers. Mean CH_4 flux in reference plots were −3, −56 and −100 µg CH_4-C m^{-2} hr^{-1} for the Howland Forest (Rustad and Fernandez, 1998a), Huntington Forest (McHale et al., 1998), and Harvard Forest (Peterjohn et al., 1994), respectively. These results show that oxidation was the primary process controlling CH_4 flux in these aerobic forest soils. These values are within the range of approximately -10 to $-145\,\mu g$ CH_4-C $m^{-2}\,hr^{-1}$ reported for temperate forest soils (Castro et al., 1995; King, 1997). Brief periods of CH_4 production were also reported at all three sites, typically in the spring when soils were saturated, indicating the presence and activity of methanogenic bacteria in these soils as well. Linear regression showed a significant positive correlation ($r^2 = 0.46$) between CH_4 uptake and soil temperature at the Harvard Forest (Peterjohn et al., 1994), and a weaker but still significant correlation ($r^2 = 0.12$) between CH_4 flux and soil temperature in heated plots only at the Huntington Forest (McHale et al., 1998). No relationship between CH_4 flux and soil temperature was observed at the Howland Forest (Rustad and Fernandez, 1998a). Relationships between CH_4 uptake and moisture were stronger than relationships between CH_4 uptake and temperature at all three sites (Peterjohn et al., 1994; McHale et al., 1998; Rustad and Fernandez, 1998a), suggesting that moisture plays a more important role in CH_4 dynamics than temperature for these sites.

The $5°C$ increase in soil temperature resulted in an $\sim14\%$ greater uptake of CH_4 in the heated plots compared with the controls at the Harvard Forest (Peterjohn et al., 1994). No significant differences in CH_4 uptake due to warming were observed at either the Huntington or Howland Forest experiments. Torn and Harte (1996) also reported no significant effect of soil warming ($+1.0°C$) on CH_4 flux for a montane meadow experiment and Christensen et al. (1997) reported no significant effect of soil warming ($+1.3°C$) on CH_4 flux from a sub-arctic heath site in northern Sweden.

These results suggest that an increase in mean annual temperature may increase the consumption of atmospheric CH_4 by some freely drained, aerobic forest soils in the northern U.S. Enhancement of CH_4 uptake will likely be greater at sites with higher N availability at higher latitudes (where mean summertime soil temperatures are typically below $10°C$), and the effect will be due, in part, to longer growing seasons during which methanotrophic bacteria are active, and will certainly be amplified if warmer temperatures are accompanied by decreases in summer precipitation. Temperature effects on CH_4 consumption at other sites, perhaps more typical of northern U.S. forests, will likely be minimal.

Unfortunately, there are relatively little data available on the effect of increased temperature on CH_4 production from wetlands. Wetlands account for an estimated one-fifth of global CH_4 emissions (Cicerone and Ormeland, 1988), and any change in CH_4 emissions from wetlands, either an increase due to elevated temperature and/or increased precipitation or a decrease due to increased frequency of summer drought or increased evapotranspiration, could significantly affect atmospheric CH_4 concentrations and supersede any potential changes in CH_4 oxidation by forest soils.

Soil Solution Dissolved Organic Carbon

The export of DOC in leachate is typically a small component of a forest's C cycle, particularly when compared to respiratory C losses. However, because DOC largely consists of organic acids, changes in soil solution DOC concentrations and flux can affect soil base cation leaching, metal dissolution, mineral weathering, and the acid–base chemistry of associated surface waters (Cronan and Aikens, 1985; McColl and Pohlman, 1986; Pohlman and McColl, 1988; Kahl et al., 1989; Schiff et al., 1990). If decomposition increases under a warmer climate, as has been suggested, it is possible that the concentration and flux of DOC in soil solutions would also change, either by increasing due to a greater production of DOC in upper soil horizons or decreasing due to greater biotic and abiotic uptake within the mineral soil. Although significant positive relationships between DOC leaching and soil temperature have been demonstrated in a few studies (e.g., Cronan and Aiken, 1985; Dalva and Moore, 1991), the response of DOC to elevated temperature is largely unknown.

To address this information need, the effect of a 5°C increase in soil temperature on soil solution export of DOC and associated cations was evaluated at the Howland Forest. Mean DOC concentrations at 5-cm depth in the B horizon were approximately 68% higher ($P < 0.07$) in the heated plots (mean $\pm$ SD $= 1866 \pm 265\,\mu$mol Cl^{-1}) than in the controls plots ($1106 \pm 370\,\mu$mol Cl^{-1}). This may be attributed, in part, to a concentration effect due to greater evaporation in the heated plots compared with the controls, but may also reflect a greater rate of DOC production in the O horizon and uppermost mineral soil. As solutions passed through the next 20 cm of the B horizon, a numerically greater amount of DOC was either oxidized, adsorbed, or precipitated in the heated plots compared with the controls, such that mean DOC concentrations at 25 cm depth in the B horizon were 56% lower in the heated plots ($177 \pm 52\,\mu$mol Cl^{-1}) compared with the controls ($402 \pm 217\,\mu$mol Cl^{-1}). Trends in total aluminum (Al) concentrations were similar to those observed for DOC. These results are consistent with an acceleration of the natural soil forming process of podzolization, where C is mobilized from upper eluvial soil horizons, typically in association with Al and iron (Fe) and then illuviated in lower B horizons.

Liechty et al. (1995) evaluated forest floor DOC dynamics at the northern- and southern-most sites of the Michigan Gradient Study. They demonstrated that forest floor DOC flux and concentration were both significantly correlated with mean soil temperature ($r^2 > 0.48$; $P < 0.035$). Consistent with the Howland soil warming study, they also showed that forest floor DOC flux and concentration were numerically greater at the southern site than the northern site. No data are presented for DOC concentrations and flux deeper in the soil profile.

These two studies, in concert with the limited data available in the literature, suggest that DOC concentrations in upper eluvial soil horizons will increase with increasing temperature. Research is needed to (1) verify this hypothesis, (2) determine whether DOC eluviated from upper horizons is redistributed deeper within the soil profile or exported entirely from the ecosystem, (3) determine the duration of accelerated DOC production in response to increased temperature, and (4) evaluate the consequences of DOC transport on cation mobilization, mineral weathering, and the acid–base chemistry of associated surface waters. Such results will be important in evaluating the linkages between upland forests and recently shown changes of DOC in surface waters (Schindler and Curtis, 1997).

Nitrogen Cycling

Nitrogen is generally considered to be the key limiting nutrient in many north temperate forested ecosystems (Vitousek et al., 1982; Waring and Schlesinger, 1985; Tamm, 1991). An important source of N to these forests is wet and dry atmospheric deposition, originating largely from fossil fuel combustion and agriculture. Atmospheric N deposition ranges from $<5.0\,\mathrm{kg}\ \mathrm{N}\ \mathrm{ha}^{-1}\ \mathrm{yr}^{-1}$ in relatively pristine areas upward to $50\,\mathrm{kg}\ \mathrm{ha}^{-1}\ \mathrm{yr}^{-1}$ or more in the more heavily polluted areas of the northeastern U.S. and Europe (Gunderson and Bashkin, 1994; Vitousek et al., 1997). Although high levels of atmospheric N deposition can serve to enhance net primary productivity (NPP) and ecosystem C storage, N deposition in excess of an ecosystem's ability to take up the added N can lead to symptoms of N saturation, including plant tissue nutrient imbalances, forest decline, and elevated concentrations of inorganic N in drainage waters (Aber et al., 1989; Vitousek et al., 1997; Fenn et al., 1998). Concerns have been raised that warmer temperatures could stimulate internal N cycling, particularly N mineralization, resulting in a greater production of inorganic N, which could exacerbate conditions of N saturation at sites already experiencing high N inputs from anthropogenic sources.

Inputs of N to the soil include above- and belowground detritus, precipitation, throughfall and stemflow, and in some ecosystems, N fixation. Within the soil, N is mineralized to NH_4^+ and nitrified to NO_3^-, both of which can then be taken up by plants or microbes, or exported in leachate. Nitrate can be reduced to gaseous NO, N_2O, or N_2. Large amounts of N are also stored in the forest floor and mineral soil in highly recalcitrant humic ring structures (Swift et al., 1979).

Nitrogen Mineralization

Nitrogen mineralization is the process by which organic N is oxidized first to NH_4^+ (ammonification) and then to nitrite (NO_2^-) and NO_3^- (nitrification) by soil microorganisms. Although studies have shown that soil properties such as total N, labile C, and soil texture, as well as overstory vegetation can influence annual rates of N mineralization (Pastor et al., 1984; Reich et al., 1997), temperature is also widely recognized as an important driver of N mineralization (Van Cleve et al., 1983; Emmer and Tietema, 1990; Bonan and Van Cleve, 1991; Gonncalves and Caryle, 1994; MacDonald et al., 1995; Hobbie, 1996; Reich et al., 1997). Thus, climatic warming may increase rates of N mineralization and nitrification in northern U.S. forest soils.

Net N mineralization (defined here as net ammonification plus net nitrification) was evaluated at the Harvard and Howland Forest soil warming experiments using the in situ buried soil bag technique (Eno, 1960). At the Harvard Forest, average reference plot net N mineralization rates in the O-horizon and upper 10 cm of the Ap-horizon were 1.02 and 0.09 mg N kg soil^{-1} day^{-1}, respectively (Melillo et al., 1995). These compare favorably with average net N mineralization rates in the reference plot O-horizon and upper 10 cm of the B-horizon at the Howland Forest, which were 1.19 and 0.02 mg N kg soil^{-1} day^{-1}, respectively. Net mineralization increased linearly with soil temperature in the reference plots at both sites (Peterjohn et al., 1994). Net nitrification was not observed at either site.

Warming the soils by 5°C significantly increased net N mineralization rates at both study sites. However, the magnitude and temporal pattern of response differed between them. At the Harvard Forest, N mineralization responded almost immediately to the manipulation with an approximate doubling of N mineralization rates in both the O- and Ap-horizons during the first and second years of the study (Melillo et al., 1995). At the Howland Forest, no treatment effects were observed on N mineralization rates in the B-horizon. In the O-horizon, a delayed response was observed, that is, no treatment effect was observed for O-horizon N mineralization rates during the first year of manipulation, whereas O-horizon mineralization rates increased by an average of 24 and 18% ($P < 0.06$) in the heated plots compared with the controls during the second and third years of the study, respectively. Similar delays in the response of N mineralization to experimental soil warming have been reported by Verburg et al. (1999) for the CLIMEX soil warming study in southern Norway and by Hartely et al. (1999) for a soil warming study in a subarctic dwarf shrub tundra in Sweden. These lags may reflect an initial immobilization of excess available N at these N-limited sites, as the in situ buried bag technique provides a measure of net N mineralization (defined here as N mineralized minus N immobilized). Following the initial lag in response, N mineralization increased significantly by 18% during the second year of the CLIMEX experiment (Verburg et al., 1999) and by 114% in the second and third years of the Swedish experiment (Hartley et al., 1998).

In the Maine Gradient Study, forest floor samples were collected from all 16 hardwood stands and 9 softwood stands in 1995, and analyzed for potential N mineralization and nitrification using a 28-day laboratory incubation following the methods of McNulty et al. (1990). No significant differences in net N mineralization were observed between stand types and no regional differences in net N mineralization were observed for the softwood sites. However, contrary to expectations, hardwood sites in the northern region had 2 to 4 times the potential net N mineralization compared with central, southern, and coastal regions (mean 57.9 mg N $kg^{-1} da^{-1}$ compared to means 30.7, 20.5, and 15.1 mg $N kg^{-1} da^{-1}$, respectively). These results are consistent with the significantly higher rates of in situ net N mineralization observed for hardwood sites in the northern region (mean 7.0 mg $N kg^{-1} da^{-1}$) compared with other regions (mean 2.9 to 3.8 mg $N kg^{-1} da^{-1}$) during the 2-year period 1993 to 1994. The higher rates of both potential and in situ net N mineralization at the northern sites corresponded to higher soil N concentrations and lower C/N ratios, and indicate that site quality, as defined here by total available soil N and C/N ratio, was more important for controlling N mineralization rates at the regional scale than climatic factors such as a 2°C difference in mean soil temperature and a 55% difference in mean annual precipitation. Reich et al. (1997) and Groot and Houba (1995) also highlight the influence of site characteristics (particularly soil texture) on N mineralization, with higher rates of N mineralization associated with finer-textured soils having higher percentages of silt and clay.

In the Michigan Gradient Study, temperature effects on net N mineralization in the surface 10 cm of soil were investigated during a 32-week incubation study (MacDonald et al., 1995). Results showed that net N mineralization rates increased significantly as temperatures increased from 5 to 25°C and that the highest rates of N mineralization were consistently observed for soils from site C, which had an intermediate mean annual temperature but the highest total soil N. These results are consistent with those from the Maine Gradient Study in that the highest rates of net N mineralization were observed at the most N-rich sites, regardless of temperature.

Overall, the evidence from these studies is strong that a 2 to 4°C increase in soil temperature will significantly increase N mineralization in northern U.S. forest soils. An increase in N mineralization at N-limited sites will likely lead to higher NPP (assuming that no other secondary factors become limiting), and thus a greater amount of C could be sequestered by the plant community. In contrast, increased internal generation of N at relatively N-rich sites may lead to or exacerbate N saturation, resulting in deteriorating forest health, increased leaching of N to surface waters, and a likely decrease in the amount of C sequestered by the plant–soil system. Increases in soil temperature of less than 2°C may result in minimal detectable changes in N mineralization, as has been demonstrated by

Jonasson et al. (1993) and Robinson et al. (1995) for transplant and greenhouse studies in arctic and sub-arctic soils in Sweden.

Nitrous Oxide Flux

Although atmospheric nitrous oxide (N_2O) concentrations are considerably lower than either CO_2 or CH_4 concentrations, N_2O is also an important greenhouse gas because of its long atmospheric lifetime ($\sim$120 years) and because it has the ability to trap $\sim$200 times more radiant energy than CO_2 on a molar basis (Rodhe, 1990; Houghton et al., 1996). Soils are the primary source of N_2O to the atmosphere. Nitrous oxide is produced during both nitrification and denitrification. Denitrification is the process by which denitrifying bacteria reduce NO_2^- and NO_3^- to NO, N_2O and N_2. The rate of denitrification is controlled by NO_3^-, O_2, and labile C supply. Soil moisture content is important in that it affects O_2 supply. Temperature can be important in controlling rates of denitrification once conditions become favorable for this process. Rates of denitrification may be expected to increase if future climates are warmer and wetter (Houghton et al., 1996).

Although numerous studies have investigated controls on N_2O production in agricultural soils (e.g., Bailey, 1976; Bailey and Beauchamp, 1973; Freney et al., 1979), fewer investigators have attempted to quantify N_2O production in forest soils (Goodroad and Keeney, 1985; Bowden et al., 1990, 1991; Castro et al., 1994; Groffman et al., 1993; Butterbach-Bahl et al., 1997). The production of N_2O in forest soils, as reported in these studies, is typically low (0 to $160\,\mu g$ N_2O-N m^{-2} hr^{-1}) and tends to be highly variable. Although a few studies have shown weak relationships between rates of N_2O production and soil temperature (Malhi et al., 1990; Goodroad and Keeney, 1985), other studies have shown no apparent trend (Bowden et al., 1990, 1991; Peterjohn et al., 1993).

Nitrous oxide flux was measured at the Harvard and Huntington Forest soil warming experiments. The flux of N_2O in the reference plots for both studies was low, $\sim$2 to $3\,\mu g$ N_2O-N m^{-2} hr^{-1} (Peterjohn et al., 1994; McHale et al., 1998). At the Harvard Forest, a weak positive relationship was observed between soil temperature and N_2O flux, with linear regression explaining 14% of the variation between temperature and N_2O flux (Peterjohn et al., 1994). Warming the soils by 5°C, however, had no significant effect on N_2O flux rates. At the Huntington Forest, soil temperature was not significantly correlated to N_2O flux and warming had no significant effect on N_2O flux in the first year of the study (McHale et al., 1998). In the second year of the study, the flux of N_2O was significantly correlated with soil temperature ($r^2 = 0.67$; $P < 0.0001$) and the 5.0°C increase in soil temperature resulted in significantly higher rates of N_2O flux in the heated plots compared with the controls. McHale et al. (1998) suggested that the stronger relationship between N_2O flux and soil temperature in the second year of the study may be attributed to an increased supply of NH_4^+ and NO_3^-, the precursors to both nitrification and denitrification, as a result of increased N mineralization. Alternatively, the stronger relationship could be due to the linkage between soil temperature and soil moisture during the second year of the study. In contrast to these U.S. studies, Hantschel et al. (1995) showed a 75% decline in N_2O flux over a 74-day period in response to a 3°C increase in soil temperature in an agroecosystem in Germany.

Clearly, a more thorough examination is needed of the relationship between temperature and N_2O dynamics in forest soils. On the basis of the available literature, it is likely that the effects of climate change on the flux of N_2O from northern U.S. forest soils will be predicated on warming-induced changes on the precursors of nitrification and denitrification, labile C pools, and/or soil moisture status.

Soil Solution Nitrogen

Nitrogen can be exported from forested ecosystems as both dissolved inorganic nitrogen (DIN) ($NO_3^- + NH_4^+$) and as DON. If elevated temperature increases the rate of internal N production in excess of biological demand, N concentrations in groundwater and surface waters may increase, with consequent negative impacts on downstream biota and drinking water quality.

The effect of a 5°C increase in soil temperature on soil solution export of N was evaluated at both the Harvard and Howland Forest soil warming experiments. Despite significant increases in measured rates of net N mineralization, DIN concentrations in soil solutions were typically below detection, and no significant warming effect was observed at either site (Peterjohn et al., 1994). This suggests that DIN made available via increased mineralization in response to warming at these sites was quickly immobilized by microbiota or taken up by vegetation. These results are consistent with those reported by Ineson et al. (1998), who showed either no change or a slight reduction in DIN leaching in response to a 2.8°C increase in soil temperature for a short grass ecosystem in the United Kingdom. They attributed this lack of response or even reduction in DIN leaching to increased plant uptake by the grassland community. In contrast, Lukewille and Wright (1997) reported that the flux of N in runoff increased by $\sim 12\,mmol\ N\,m^{-2}\,yr^{-1}$ in the heated catchment compared with the reference catchments during the first three years of the CLIMEX soil warming experiment in southern Norway; Van Cleve et al. (1990) reported significant increases in soil solution DIN in response to an 8 to 10°C increase in soil temperature in a black spruce (*Picea mariana* (Mill.)) ecosystem in Alaska; Joslin and Wolfe (1993) reported a 71% increase in NO_3^- concentration in soil solutions from the sun vs. the shade side of a clearing (+1.2°C) in a high-elevation red spruce stand in Virginia; and

Berg et al. (1997) reported a 55% increase in soil solution DIN leaching at sites with average mean annual temperatures of 11.0°C vs. sites with average mean annual temperatures of 8.1°C in Germany.

Results from these studies suggest that in N-saturated ecosystems (such as many sites in Europe and high-elevation sites in the northeastern U.S.), an increase in soil temperature will likely increase the export of DIN to groundwater and surface waters. In N-limited ecosystems, more typical of northern U.S. forests, increased N availability, particularly in combination with elevated CO_2, would be more likely to increase NPP and result in a greater sequestration of C in these northern temperate forest ecosystems, at least in the short-term. Over time, and if atmospheric N deposition continues at current or elevated rates, it is possible that the amount of internally generated and atmospherically deposited N could exceed the soil–plant system's capacity to take up the available N, and N could be lost from these forests. Research on the interactions of elevated temperature, N, and CO_2 will be critical to determine whether future chemical and physical environments will result in net retention or release of N from northern U.S. forested ecosystems. The effect of warming on DON is also unclear, as DON has not routinely been assessed in ecosystem warming studies. Preliminary data comparing several eastern U.S. forest sites suggest DON could range from less than 1% to greater than 80% of total N export in soil solutions and streams, possibly reflecting differences in soil drainage classes and topography. An increase in internal N cycling, due to soil warming could significantly increase soil solution DON concentrations in some forest ecosystems.

Soil Microbiology

The effect of soil warming on soil microbial communities is central to understanding soil system responses to warming. Soil microorganisms play a dual role in terrestrial nutrient cycles: (1) they are directly responsible for the release of C and nutrients from soil organic matter and (2) they serve as a net sink for nutrients during periods of increasing microbial biomass and a source for nutrients during periods of net turnover and mineralization. Most of the microorganisms commonly inhabiting forest soils are mesophiles, with temperature optima in the range of 25 to 35°C (Alexander, 1977). Thus, an increase in soil temperature in the range of 0 to 35°C will likely result in an increase in microbial activity, assuming nonlimiting moisture, oxygen (O_2), and nutrient conditions.

The response of the microbial community to warming was evaluated at the Howland and Huntington Forest soil warming experiments. At the Howland site, researchers reported that Oa horizon microbial biomass was significantly lower ($P < 0.05$) and microbial activity was numerically lower in the heated plots compared with the controls. They attribute these declines to the concomitant decline in soil moisture. In a related study, Christ et al. (1997) concluded that warming had no effect on microbial C, N, and phosphorus (P) pools at the Howland Forest, whereas, at the Huntington Forest, warming caused a decline of 0.05 and 0.007 g kg^{-1} $^{\circ}$C^{-1} ($P < 0.01$) in microbial pools of C and N, respectively. They hypothesized that the increased respiratory losses of C in response to warming exceeded the microbial community's ability to utilize additional extracellular C, resulting in a net loss of microbial biomass. This hypothesis is supported by studies demonstrating both the increase in microbial respiration with increasing temperature (see review by Kirschbaum et al., 1995; Winkler et al., 1996) as well as declines in microbial biomass (Nicolardot et al., 1994; Pohhacker and Zech, 1995).

In the Michigan Gradient Study, MacDonald et al. (1995) evaluated temperature effects on microbial respiration and biomass. Results showed that microbial respiration increased with temperature at all sites, and that both microbial biomass and the magnitude of the microbial respiration response to temperature increased from north to south, paralleling increases in percentage of soil organic C and increases in mean annual temperature. Contrary to expectations, results also indicated that microbial respiration rate constants (K_{resp}) were not significantly related to temperature, whereas the estimated pool size of substrate C available for microbial respiration was highly temperature-dependent, with increasing apparent pool size with increasing temperature. They hypothesized that these changes in apparent respirable C pool size could be caused by shifts in the microbial community, changes in transport processes such as diffusion, or lower microbial efficiency at higher temperatures. Zogg et al. (1997) later used molecular techniques of phospholipid fatty acid (PLFA) and lipopolysaccharide fatty acid (LPS-OHFA) analysis to demonstrate significant shifts in microbial community composition in response to different temperature treatments (5 to 25°C).

Taken together, these results suggest that increases in mean annual temperature may increase rates of microbial respiration, decrease soil microbial biomass (particularly if moisture is concurrently limiting), and, perhaps most significantly, result in shifts in microbial community composition toward populations with a greater ability to utilize soil organic C.

Conclusions

Results from soil warming and gradient studies in the northeastern and north central U.S. as well as elsewhere support the hypothesis that a 1.0 to 3.5°C increase in ambient soil temperature (the predicted increase in mean global temperature) can have marked effects on belowground C and N cycling in northern U.S. forests. Some processes, such as soil respiration and N mineralization, showed significant and consistent increases in response to warming, regardless of site or treatment. Other processes, such as CH_4 oxidation, N_2O flux, and litter decomposition, exhibited more variable responses, reflecting differences in litter quality, N availability, and soil moisture. Although there is great interest in ascertaining whether northern temperate forest soils will be a net source or sink for atmospheric C in coming decades, the answer remains elusive. Increased soil respiration and litter decomposition, as observed in the soil warming studies, together with geographically based observations of decreasing soil organic matter with increasing mean annual temperature, suggest that global warming will result in a net efflux of C from the soil to the atmosphere. However, increases in N availability, due either to increased internal N mineralization in response to warming or to atmospheric deposition, may increase NPP, particularly in N-limited ecosystems characteristic of northern U.S. forests, and thereby increase the storage rate of C in plant biomass. Whether this would result in a long-term increase in C in biomass is not known. This sequestration of C could be further enhanced by elevated atmospheric CO_2.

Unfortunately, the research community has not yet performed the definitive experiments on ecosystem response to an increase in mean annual temperature. Such an experiment must include interactions between different vectors of global change (including temperature, CO_2, moisture, and N), and must be performed on whole, intact ecosystems over multiple years. Particular attention must also be focused on phenological response as the simple extension of the growing season may be equal in importance to specific temperature-dependant rate increases. Although we believe that soil-only warming experiments, particularly in combination with gradient studies and laboratory experiments, have been worthwhile in evaluating in situ responses of soil processes to controlled increases in temperature, soil-only warming experiments decouple the above- and belowground components of the ecosystem, negating our ability to evaluate the balance of photosynthesis and respiration for the entire ecosystem, or possible feedbacks and linkages between above- and belowground components of nutrient cycles.

It is also important to consider that the effect of small changes in mean annual temperature on ecosystem structure and function may be trivial relative to the dramatic effects of increased fire frequency; increased frequency and intensity of severe weather events, such as droughts, floods, hurricanes, and ice storms; or increased frequency of insect and disease outbreaks. Changes in land use, such as urbanization, deforestation, conversion of grasslands and no-till croplands to intensive agriculture, and species migration, such as the predicted extension of the northern hardwood forest into regions previously occupied by boreal coniferous forests (Pastor and Post, 1988), could all result in far more dramatic changes in the landscape than the effects of a 1 to 2°C increases in temperature.

Our ability to assess the impact of global warming on northern U.S. forests remains limited. A better mechanistic understanding of the direct effects of temperature on both above- and belowground ecosystem physiology is still needed, as is more research on the interactive effects of temperature with elevated CO_2, water availability, and N availability. Results from research to date needs to continue to be incorporated into new or existing process-based models of ecosystem dynamics. These improved models can then be cautiously used to extrapolate results from site-specific studies to the local, regional, and ultimately, continental scale, as well as define future process-level research on the consequences of climate change.

References

Aber JD, Melillo JM (1982) Nitrogen immobilization in decaying hardwood leaf litter as a function of initial nitrogen and lignin content. Can J Bot 60: 2263–2269.

Aber J, Nadelhoffer KJ, Steudler P, Melillo JM (1989) Nitrogen saturation in northern forest ecosystems. BioScience 39:378–386.

Adamsen APS, King GM (1993) Methane consumption in temperate and subarctic forest soils: rates, vertical zonation, and responses to water and nitrogen. Appl Environ Microbiol 59:485–490.

Alexander M (1977) *Introduction to Soil Microbiology*. John Wiley, New York.

Bailey LD (1976) Effects of temperature and roots on denitrification in a soil. Can J Soil Sci 56:79–87.

Bailey LD, Beauchamp EG (1973) Effects of temperatures on NO_3^- and NO_2 reduction, nitrogenous gas production, and redox potential in a saturated soil. Can J Soil Sci 53:213–218.

Berg MP, Verhoef HA, Bolger T, Anderson JM, Beese F, Couteaux MM, Ineson P, McCarthy F, Palka L, Raubuch M, Splatt P, Willison T (1997) Effects of air pollutant–temperature interactions on mineral–N dynamics and cation leaching in reciprocal forest soil transplantation experiments. Biogeochem 39: 295–326.

Bonan GB, Van Cleve K (1991) Soil temperature, nitrogen mineralization, and carbon source–sink relationships in boreal forests. Can J For Res 22:629–639.

Born M, Dorn H, Levin I (1990) Methane concentration in aerated soils of the temperate zone. Tellus 42:2–8.

Bowden RD, Melillo JM, Steudler PA, Aber JD (1991) Effects of nitrogen additions on annual nitrous oxide fluxes from temperate forest soils in the northeastern United States. J Geophys Res 96:9321–9328.

Bowden RD, Steudler PA, Melillo JM, Aber JD (1990) Annual nitrous oxide fluxes from temperate forest soils in the northeastern United States. J Geophys Res 95:13997–14005.

Briggs RD, Lemin RC (1992) Delineation of climatic regions in Maine. Can J For Res 22:801–811.

Burton AJ, Pregitzer KS, Zogg GP, Zak DR (1998) Drought reduces root respiration in sugar maple forests. Ecol Applic 8:771–778.

Butterbach-Bahl GR, Breuer L, Papen H (1997) Fluxes of NO and N_2O from temperate forest soils: impact of forest type, N deposition, and of liming on the NO and N_2O emissions. Nutr Cycl in Agroecosyst 48:79–90.

Castro MS, Melillo JM, Steudler PA, Chapman JW (1994) Soil moisture as a predictor of methane uptake by temperate forest soils. Can J For Res 24:1805–1810.

Castro MS, Steudler PA, Melillo JM, Aber JD, Bowden RD (1995) Factors controlling atmospheric methane consumption by temperate forest soils. Global Biogeochem Cycl 9:1–10.

Christ MJ, David MB, McHale PI, McLaughlin JW, Mitchell MJ, Rustad LE, Fernandez IJ (1997) Microclimate control of microbial C, N, and P pools in Spodosol Oa horizons. Can J For Res 27:1914–1921.

Christensen TR, Michelsen A, Jonasson S, Schmidt IK (1997) Carbon dioxide and methane exchange of a subarctic heath in response to climate change related environmental manipulations. Oikos 79:34–44.

Cicerone RJ, Ormeland RS (1988) Biogeochemical aspects of atmospheric methane. Global Biogeochem Cycl 2:299–327.

Crill PM (1991) Seasonal patterns of methane uptake and carbon dioxide release by a temperate woodland soil. Global Biogeochem Cycl 5:319–334.

Crill PM, Bartlett KB, Hariss RC, Gorham E, Verry ES, Sebacher DI, Madzar L, Sanner W (1988) Methane flux from Minnesota peatlands. Global Biogeochem Cycl 2:371–384.

Crill PM, Martikainen PJ, Nykanen H, Silvola J (1994) Temperature and N fertilization effects on methane oxidation in a drained peatland soil. Soil Biol Biochem 26:1331–1339.

Cronan CS, Aiken GR (1985) Chemistry and transport of soluble humic substances in forested watersheds of the Adirondack Park, NY. Geochim Cosmochim Acta 49:1697–1705.

Dalva M, Moore TR (1991) Sources and sinks of dissolved organic carbon in a forested swamp catchment. Biogeochem 15:1–19.

Daubenmire R, Prusson D (1963) Studies on the decomposition of tree litter. Ecology 44:589–592.

de Jong E, Schappert H, MacDonald K (1974) Carbon dioxide evolution from virgin and cultivated soil as affected by management practices and climate. Can J For Res 54:299–307.

Delaney MT, Fernandez IJ, Simmons JA, Briggs RD (1996) *Red Maple and White Pine Litter Quality and Changes with Decomposition*. Tech Bull 162. Maine Agricultural and Forest Experiment Station, Orono, Maine.

Edwards NT (1975) Effects of temperature and moisture on carbon dioxide evolution in a mixed deciduous forest floor. Soil Sci Soc Am Proc 39:361–365.

Emmer IM, Tietema A (1990) Temperature-dependent nitrogen transformation in acid oak–beach forest litter in the Netherlands. Plant Soil 122:193–196.

Eno CF (1960) Nitrate production in the field by incubating the soil in polyethylene bags. Soil Sci Soc Am J 24:277–279.

Fenn ME, Poth MA, Aber JD, Baron JS, Bormann BT, Johnson DW, Lemly AD, McNulty SG, Ryan DF, Stottlemyer R (1998) Nitrogen excess in North American ecosystems: predisposing factors, ecosystem responses, and management strategies. Ecol Applic 8:706–733.

Fernandez IJ, Son Y, Kraske CR, Rustad LE, David MB (1993) Soil carbon dioxide characteristics under different forest types and after harvest. Soil Sci Soc Am J 57:1115–1121.

Freney JR, Denmead OT, Simpson JR (1979) Nitrous oxide emissions from soils at low moisture contents. Soil Biol Biochem 11:167–173.

Goodroad LL, Keeney DR (1985) Site of nitrous oxide productions in field soils. Biol Fertil Soils 1:1–7.

Goncalves JLM, Caryle JC (1994) Modeling the influence of moisture and temperature on net nitrogen mineralization in a forested sandy soil. Soil Biol Biochem 26:1557–1564.

Goulden ML, Wofsy SC, Harden JW, Trumbore SE, Crill PM, Gower ST, Fries T, Daube BC, Fan SM, Sutton DJ, Bazzaz A, Munger JW (1998) Sensitivity of boreal forest carbon balance to soil thaw. Science 279:214–218.

Groffman PM, Zak DR, Christensen S, Mosier AR, Tiedje JM (1993) Early spring nitrogen dynamics in a temperate forest landscape. Ecology 74:1579–1585.

Groot JJR, Houba VJG (1995) A comparison of different indices for nitrogen mineralization. Biol Fertil Soils 19:1–9.

Gunderson P, Bashkin VN (1994) Nitrogen cycling. In: Moldan B, Cerny J (eds) *Biogeochemistry of Small Catchments. SCOPE 51.* John Wiley, Chichester, United Kingdom, pp 255–283.

Hantschel RE, Kamp T, Beese F (1995) Increasing soil temperature to study global warming effects on the soil nitrogen cycle in agroecosystems. J Biogeogr 22:375–380.

Harte J, Torn M, Chang F (1995) Global warming and soil microclimate: results from a meadow warming experiment. Ecol Applic 5:132–150.

Hartley AE, Neill C, Melillo JM, Crabtree R, Bowles FP (1999) Plant and microbial responses to simulated climate change in subarctic dwarf shrub tundra, northern Sweden. Oikos 86:331–344.

Hobbie SE (1996) Temperature and plant species control over litter decomposition in Alaskan tundra. Ecological Monographs 66:503–522.

Houghton JT, Meira Filho LG, Callander BA, Harris N, Kattenberg A, Maskell K (eds) IPCC (Intergovernmental Panel on Climate Change) (1996) *Climate Change 1995: The Science of Climate Change.* Cambridge University Press, Cambridge, United Kingdom.

Ineson P, Benham DG, Poskitt J, Harrison AF, Taylor K, Woods C (1998) Effects of climate change on nitrogen dynamics in upland soils. 2. A soil warming study. Global Change Biol 4:153–161.

Jansson PE, Berg B (1985) Temporal variation of litter decomposition in relation to simulated soil climate. Long-term decomposition in a Scots pine forest. V. Can J Bot 63:1008–1016.

Jenkinson DS, Adams DE, Wild A (1991) Model estimates of CO_2 emissions from soil in response to global warming. Nature 351:304–306.

Jonasson S, Havstrom M, Jensen M, Callaghan TV (1993) *In situ* mineralization of nitrogen and phosphorus of arctic soils after perturbations simulating climate change. Oecologia 95:179–186.

Joslin JD, Wolfe MH (1993) Temperature increase accelerates nitrate release from high elevation red spruce soils. Can J For Res 23:756–759.

Kahl JS, Norton SA, MacRae RK, Haines TA, Davis RB (1989) The influence of organic acidity on the acid–base chemistry of surface waters in Maine, USA. Water Air Soil Pollut 46:221–233.

Keller M, Goreau TJ, Wofsy SC, Kaplan WA, McElroy MB (1983) Production of nitrous oxide and consumption of methane by forest soils. Geophys Res Lett 10:1156–1159.

King GM (1997) Responses of atmospheric methane consumption by soils to global change. Global Change Biol 3:351–362.

King GM, Adamsen APS (1992) Effects of temperature on methane consumption in a forest soil and pure cultures of the methanotroph *Methylomonas rebra*. Appl Environ Microbiol 58:2758–2763.

Kirschbaum M (1995) The temperature dependence of soil organic matter decomposition, and the effect of global warming on soil organic C storage. Soil Biol Biochem 27:753–760.

Koschorreck M, Conrad R (1993) Oxidation of atmospheric methane in soil: measurements in the field, in soil cores and in soil samples. Global Biogeochem Cycl 7:109–121.

Lal R, Kimble J, Levine E, Stewart BA (1995) *Soil Management and the Greenhouse Effect*. CRC Press Inc., London, pp 385.

Liechty HO, Kuuseoks E, Mroz GD (1995) Dissolved organic carbon in northern hardwood stands with differing acidic inputs and temperature regimes. J Environ Qual 24:927–933.

Lukewille A, Wright RF (1997) Experimentally increased soil temperature causes release of nitrogen at a boreal forest catchment in southern Norway. Global Change Biol 3:13–21.

MacDonald NW, Zak DR, Pregitzer KS (1995) Temperature effects on kinetics of microbial respiration and net nitrogen and sulfur mineralization. Soil Sci Soc Am J 59:233–240.

Malhi SS, McGrill WB, Nyborg N (1990) Nitrate losses in soils: effects of temperature, moisture, and substrate concentration. Soil Biol Biochem 22: 917–927.

McColl JG, Pohlman AA (1986) Soluble organic acids and their chelating influence on Al and other metal dissolution from forest soils. Water Air Soil Pollut 31:917–927.

McHale PJ, Mitchell MJ, Bowles FP (1998) Soil warming in a northern hardwood forest: trace gas fluxes and leaf litter decomposition. Can J For Res 28: 1365–1372.

McNulty SD, Aber JD, McLellan TM, Katt SM (1990) Nitrogen cycling in high elevation forests of the northeastern United States in relation to nitrogen deposition. Ambio 19:30–40.

McNulty SD, Aber JD, Newman SD (1996) Nitrogen saturation in a high elevation spruce–fir stand. For Ecol Manage 84:109–121.

Meentemyer V (1978) Macroclimate and lignin control of litter decomposition rates. Ecology 59:465–472.

Meentemyer V, Box EO (1987) Scale effects in landscape studies. In: Turner MG (ed) *Landscape Heterogeneity and Disturbance*. Ecological Studies Vol. 64. Springer-Verlag, New York, pp 15–36.

Melillo JM, Aber JD, Muratore JF (1982) Nitrogen and lignin control of hardwood leaf litter decomposition dynamics. Ecology 63:621–626.

Melillo JM, Kicklighter DW, McGuire AD, Peterjohn WT, Newkirk KM (1995) Global change and its effects on soil organic carbon stocks. In: Zepp RG, Sonntag C (eds) *Role of Nonliving Organic Matter in the Earth's Carbon Cycle*. John Wiley, New York, pp 176–189.

Moore AM (1986) Temperature and moisture dependence of decomposition rates of hardwood and coniferous leaf litter. Soil Biol Biochem 18:427–435.

Musselman RC, Fox DG (1991) A review of the role of temperate forests in the global CO_2 balance. J Air Waste Manage Assoc 41:798–807.

Nicolardot B, Fauvet G, Cheneby D (1994) Carbon and nitrogen cycling through soil microbial biomass at various temperatures. Soil Biol Biochem 26:253–261.

Pastor J, Aber JD, McCaughtery CA, Melillo JM (1984) Aboveground production and N and P cycling along a nitrogen mineralization gradient on Blackhawk Island, Wisconsin. Ecology 65:256–268.

Pastor J, Post WM (1988) Response of northern forests to CO_2-induced climate change. Nature 334:55–58.

Peterjohn WT, Melillo JM, Bowles ST (1993) Soil warming and trace gas fluxes: experimental design and preliminary flux results. Oecologia 93:18–24.

Peterjohn WT, Melillo JM, Steudler PA, Newkirk KM, Bowles ST, Aber JD (1994) Responses of trace gas fluxes and N availability to experimentally elevated soil temperatures. Ecolo Applic 4:617–625.

Pinol J, Alcaniz JP, Roda F (1995) Carbon dioxide efflux and pCO_2 in soils of three *Quercus ilex* montane forests. Biogeochem 30:191–215.

Pohhacker R, Zech W (1995) Influence of temperature on CO_2 evolution, microbial biomass C and metabolic quotient during the decomposition of two humic forest horizons. Biol Fertil Soils 19:239–245.

Pohlman AA, McColl JG (1988) Soluble organics from forest litter and their role in metal dissolution. Soil Sci Soc Am J 52:265–271.

Raich JW, Potter CS (1995) Global patterns of carbon dioxide emissions from soils. Global Biogeochem Cycl 9:23–36.

Raich JW, Schlesinger WH (1992) The global carbon dioxide flux in soil respiration and its relationship to vegetation and climate. Tellus 44:81–89.

Randlett DL, Zak DR, Pregitzer KS, Curtis PS (1996) Elevated atmospheric carbon dioxide and leaf litter chemistry: influences on microbial respiration and net nitrogen mineralization. Soil Sci Soc Am J 60:1571–1577.

Reich PB, Grigal DF, Aber JD, Gower ST (1997) Nitrogen mineralization and productivity in 50 hardwood and conifer stands on diverse soils. Ecology 78:335–347.

Robinson CH, Wookey PA, Parsons AN, Potter JA, Callaghan TV, Lee JA, Press MC, Welker JM (1995) Responses of plant litter decomposition and nitrogen mineralization to simulated environmental change in a high arctic polar semi-desert and a subarctic dwarf shrub heath. Oikos 74:503–512.

Rodhe H (1990) A comparison of the contribution of various gases to the greenhouse effect. Science 248:1217–1219.

Ruark GA (1993) Modeling soil temperature effects on *in situ* decomposition rates for fine roots of Loblolly Pine. For Sci 9:118–129.

Rustad LE, Fernandez IJ (1998a) Experimental soil warming effects on CO_2 and CH_4 flux from a low elevation spruce–fir forest soil in Maine, USA. Global Change Biol 4:597–607.

Rustad LE, Fernandez IJ (1998b) Soil warming: consequences for litter decay in a spruce–fir forest ecosystem in Maine. Soil Sci Am J 62:1072–1081.

Schiff SL, Aravena R, Trumbore S, Dillon PJ (1990) Dissolved organic carbon cycling in forested watersheds: a carbon isotope approach. Water Resour Res 26:2949–2957.

Schindler DW, Curtis PJ (1997) The role of DOC in protecting freshwaters subjected to climatic warming and acidification from UV exposure. Biogeochem 36:1–8.

Schlenter RE, Van Cleve K (1985) Relationship between CO_2 evolution from soil, substrate temperature, and substrate moisture in four mature forest types in interior Alaska. Can J For Sci 15:97–106.

Schleser GH (1982) The response of CO_2 evolution from soils to global temperature changes. Z Naturfors 37:287–291.

Shortle WC, Smith KT (1988) Aluminum-induced calcium deficiency syndrome in declining red spruce. Science 240:239–240.

Simmons JA, Fernandez IJ, Briggs RD (1996) Forest floor C pools and fluxes along a regional climate gradient in Maine, USA. For Ecol Manage 84:81–95.

Singh JS, Gupta SR (1977) Plant decomposition and soil respiration in terrestrial ecosystems. Bot Rev 43:449–528.

Steudler PA, Bowden RD, Melillo JM, Aber JD (1989) Influence of nitrogen fertilization on methane uptake in temperate forest soils. Nature 341:314–316.

Swift MJ, Heal OW, Anderson JM (1979) *Decomposition in Terrestrial Ecosystems*. University of California Press, Berkeley, CA.

Toland DE, Zak DR (1994) Seasonal patterns of soil respiration in intact and clear-cut northern hardwood forests. Can J For Res 24:1711–1716.

Tamm CO (1991) Nitrogen and Terrestrial Ecosystems. Ecological Studies. Springer-Verlag, Heidelberg, p 116.

Torn MS, Harte J (1996) Methane consumption by montane soils: implications for positive and negative feedback with climate change. Biogeochemistry 32: 53–67.

Turner DP, Koerper GJ, Harmon ME, Lee JJ (1995) A carbon budget for forests of the conterminous United States. Ecol Applic 5:421–436.

Van Cleve K, Oechel WC, Hom JL (1990) Response of black spruce (*Picea mariana*) ecosystems to soil temperature modifications in interior Alaska. Can J For Res 20:1530–1535.

Van Cleve K, Oliver LK, Schlentner P, Viereck LA, Dyrness CT (1983) Productivity and nutrient cycling in taiga forest ecosystems. Can J For Res 13:747–766.

Verburg PSJ, Van Loon WKP, Lukewille A (1999) The CLIMEX soil-heating experiment: soil response after 2 years of treatment. Bio Fertil Soils 28:271–276.

Vitouske PM, Gosz JR, Grier CC, Melillo JM, Reiners WA (1982) A comparative analysis of potential nitrification and nitrate mobility in forest ecosystems. Ecol Monogr 52:155–177.

Vitousek PM, Aber J, Howarth RW, Likens GE, Matson PA, Schindler DW, Schlesinger WH, Tilman GD (1997) Human alteration of the global nitrogen cycle: causes and consequences. Ecol Applic 7:737–750.

Waring RH, Schlesinger WH (1985) Forest Ecosystems: Concepts and Management. Academic Press, Inc., New York.

Whalen SC, Reeburgh WS (1996) Moisture and temperature sensitivity of CH_4 oxidation in boreal soils. Soil Biol Biochem 28:1271–1281.

Winkler JP, Cherry RS, Schlesinger WH (1996) The Q10 relationship of microbiol respiration in a temperature forest soil. Soil Biol Biochem 28(8):1067–1072.

Witkamp M (1966) Decomposition of leaf litter in relation to environment, microflora, and microbial respiration. Ecology 47:194–201.

Wofsy SC, Goulden ML, Munger JW, Fan SM, Bakwin PS, Daube BC, Bassow SL, Bazzaz FA (1993) Net exchange of CO_2 in a midlatitude forest. Science 260:1314–1317.

Yavitt JB, Fahey TJ, Simmons JA (1995) Methane and carbon dioxide dynamics in a northern hardwood ecosystem. Soil Sci Soc Am J 59:796–804.

Zogg GP, Zak DR, Burton AJ, Pregitzer KS (1996) Fine root respiration in northern hardwood forests in relation to temperature and nitrogen availability. Tree Physiol 16:719–725.

Zogg GP, Zak DR, Ringelberg DB, MacDonald NW, Pregitzer KS, White DC (1997) Compositional and functional shifts in microbial communities due to soil warming. Soil Sci Soc Am J 61:475–481.

11. Regional Impacts of Climate Change and Elevated Carbon Dioxide on Forest Productivity

Jennifer C. Jenkins, David W. Kicklighter, and John D. Aber

Net primary production (NPP) is defined as the rate at which carbon (C) is accumulated by autotrophs and is expressed as the difference between gross photosynthesis and autotrophic respiration. NPP is the resource providing for the growth and reproduction of all heterotrophs on Earth; as a result, it determines the planet's carrying capacity (Vitousek et al., 1986). For humans, terrestrial NPP is important because it is one determinant of the available food and wood supplies, and because it drives the rates of most other processes identified as "ecosystem services" provided by terrestrial systems (Costanza et al., 1997; Daily et al., 1997). Forests store 90% of the C in terrestrial vegetation (Graham et al., 1990), so fluxes of C between forest biomass, forest soils, and the atmosphere are key components of global and regional C budgets. In the northeastern U.S., forest production is especially important because nearly 70% of the land area in the region is forested (Lathrop and Bognar, 1994).

NPP depends on the balance between rates of photosynthesis and respiration, both of which are sensitive to changing environmental conditions. As a result, terrestrial NPP is likely to change dramatically in a future marked by increasing carbon dioxide (CO_2) concentrations and greenhouse gas-induced climate change. In this chapter we first compare

383

results from two ecosystem process models, PnET-II (Aber and Federer, 1992; Aber et al., 1995, 1996) and TEM 4.0 (Raich et al., 1991; McGuire et al., 1992, 1993; Melillo et al., 1993), which are driven by scenarios of potential future climate in order to predict forest productivity in the northeastern region under changing environmental conditions. We highlight those features of the models and input data sets that contribute to differences between model predictions. We then describe state-of-the-art methods to address issues not usually considered when developing regional estimates of forest NPP. Consideration of these additional issues will enable more accurate predictions of forest NPP for the region.

Carbon Dioxide, Climate Change, and Forest Productivity

Atmospheric CO_2 concentrations have increased by nearly 30% over the last 200 years, primarily as a result of fossil fuel combustion, land use change, and cement production (Neftel et al., 1985; Vitousek, 1992; Schimel et al., 1996a). If emissions continue to climb at this rate, a doubling of atmospheric CO_2 is possible by 2100, though the adoption of mitigation strategies may slow the growth rate or stabilize CO_2 concentrations (Houghton, 1996). Also within the last two centuries, atmospheric concentrations of gases such as tropospheric ozone (O_3), methane (CH_4), and nitrous oxide (N_2O) have increased as a result of human activity. There is substantial evidence that together with CO_2, these greenhouse gases have contributed to a global surface temperature warming of 0.3 to 0.6°C over the last century, with 0.2 to 0.3°C of this warming occurring within the last 40 years (Nicholls et al., 1996; Schimel et al., 1996a). The recent greenhouse-gas–induced warming has been greatest in the northern latitudes, from 40 to 70°N (Nicholls et al., 1996) including the northeastern region of the U.S. discussed in this chapter (lat 41 to 47.5°N, long 67 to 76°W). Future temperature changes are expected to be most dramatic toward the poles (Kattenberg et al., 1996). While most climate models predict an increase in global mean precipitation, there is little agreement about precipitation trends at the regional level (Kattenberg et al., 1996). A useful review of the sources and dynamics of greenhouse gases, and of the specific changes in climate that may result from their emissions, can be found in the report of the Intergovernmental Panel on Climate Change (Houghton et al., 1996).

The physiological responses of plants to increased CO_2 and climate change have been the subject of hundreds of publications within the past several decades. Although a detailed review of the literature on this topic is beyond the scope of this chapter, some recent discussions can be found in Strain (1987), Eamus and Jarvis (1989), Bazzaz (1990), Graham et al. (1990), Mooney et al. (1991), Bazzaz and Fajer (1992), Mousseau and

Saugier (1992), Idso and Idso (1994), Bazzaz et al. (1996), Koch and Mooney (1996), Körner (1996), Wilsey (1996), and Navas (1998). Overall, increased CO_2 is thought to have a direct effect on stomatal function and carboxylation rates, changing photosynthesis and transpiration (Jarvis and McNaughton, 1986; Field et al., 1995). Increased CO_2 has been shown to impact respiration rates as well, though the direction of change appears to vary (Amthor, 1991). At the same time climate changes, which are projected to be indirect effects of CO_2 and greenhouse-gas increases, are likely to alter the rates of respiration, decomposition, and nutrient cycling in addition to photosynthesis and transpiration. Thus at the whole-ecosystem level, complex interactions between the vegetation CO_2 response, biogeochemical cycles, and water and energy fluxes are likely. The complexity of these interactions makes it difficult to extrapolate from short-term physiological measurements to prediction of long-term system responses (Mooney, 1996).

While much progress has been made toward quantifying forest response to elevated CO_2 using short-term physiological measurements, data are becoming available that illustrate potential shortcomings of the approaches used to date. For example, in addition to affecting photosynthesis, studies by Zak et al. (1993), Wood et al. (1994), Cotrufo and Ineson (1996), and Randlett et al. (1996) have shown that enhanced CO_2 and climate change can impact allocation, foliar composition, decomposition, and nutrient cycling. Bazzaz et al. (1996) have pointed out the potential danger of extrapolating from seedling experiments to mature tree responses. Eamus (1991) has suggested that transporting plants grown at ambient CO_2 directly to increased-CO_2 environments can induce potentially unrealistic physiological responses. Few researchers have explored plant acclimation or evolutionary response to elevated CO_2 although this possibility has been acknowledged in several studies (e.g., Sage et al., 1989; Field et al., 1995; Bazzaz et al., 1996). Bolker et al. (1995), Körner (1996), Lüscher and Nösberger (1997), and Lüscher et al. (1998) have suggested that shifts in competitive interactions under increased CO_2 and climate change are likely to induce species changes, which may impact stand- and regional-scale photosynthetic rates. Finally, several modeling studies (e.g., Rastetter et al., 1991, 1992; McGuire et al., 1992, 1997; Melillo et al., 1993) have highlighted potential effects of nutrient availability on long-term photosynthetic response. Free-Air CO_2 Enrichment (FACE) experiments, in which mature stands are exposed to elevated CO_2 concentrations, are a promising alternative to greenhouse and pot experiments, but currently they are limited by technological and cost constraints and they sample only a small portion of a tree's life span (Bazzaz et al., 1996). Still, results from FACE studies conducted in forests suggest that substantial increases in photosynthesis and NPP are possible (Ellsworth et al., 1995; DeLucia et al., 1999), though long-term acclimation to elevated CO_2 remains a possibility.

In the forest ecosystems of the northeastern U.S., additional factors are thought to influence productivity, such as increasing tropospheric O_3 concentrations (Ollinger et al., 1997), acid rain and cation depletion (Likens et al., 1996), and nitrogen (N) deposition (Ollinger et al., 1993; Lovett, 1994; Townsend et al., 1996), but see also Nadelhoffer et al. (1999). Under a changed climate, the frequencies of disturbances such as fire and pathogen outbreaks are also likely to change, further altering productivity patterns (Schimel et al., 1997). An attempt at accurate and complete prediction of future forest NPP in the northeastern U.S. would require consideration of each of these separate, yet potentially interacting, factors. In this chapter, our intent is not to attempt an integrated assessment of forest response to all of the stressors that affect forest productivity in the region. Such an assessment would be premature given the current state of knowledge about the interactions between these factors. Instead, we focus on the potential impacts of enhanced CO_2 and climate change on forest productivity in the northeastern U.S. We anticipate that eventually, future research will examine the integrated responses of forests to these many stressors.

Modeling Forest Productivity

Because the NPP response of forests to the interacting factors discussed above is likely to be extremely complex, single-factor experiments to determine how intact systems will respond to future perturbations are necessarily limited in their predictive ability. The inevitable shortcomings of past experimental approaches in assessing the long-term impacts of increased CO_2 on plant production make it very difficult to use an experimental approach to measure plant response to elevated CO_2, which is just one of several stressors likely to be important in the future.

When experimental measurements are difficult or impossible, models can be useful predictive tools. A modeling approach may be used to extrapolate process descriptions from site-level measurements to regional-scale estimates (Aber et al., 1993a). Models may also be used to predict forest response to conditions that do not yet exist, such as the likely convergence of several interacting stressors in the northeastern U.S. under a changed climate and increased atmospheric CO_2. Of course, uncertainty always exists in model estimates (Oreskes et al., 1994), and it is impossible to validate predictions of the future (Rastetter, 1996). However, it is to our advantage to present our best predictions, basing those estimates on models parameterized using existing data.

Many different modeling approaches have been used to develop regional to global estimates of NPP to fulfill a variety of objectives (e.g., Cramer et al., 1999). Some models estimate NPP across the globe based on general empirical relationships of NPP with temperature, precipitation, or NDVI, whereas other models estimate NPP across a limited region based on very detailed information, such as the types of tree species found in the region or the division of vegetation into many compartments (e.g., leaves, sapwood, heartwood, fine roots). Because our understanding of ecosystem dynamics is imperfect, different models may highlight the influences of different environmental factors or feedback mechanisms on NPP (e.g., Churkina et al., 1999; Ruimy et al., 1999; Schloss et al., 1999). These differences in model structure and assumptions may or may not be important when determining regional NPP estimates or the influence of enhanced CO_2 and climate change on future forest productivity. A comparison of model estimates against field-measured data can provide useful information about model accuracy under current conditions (Aber, 1997). However, models can estimate very different NPP responses to future climate change even though they may estimate similar NPP under current conditions (VEMAP Members, 1995). Comparisons between results generated by models with different underlying principles may help to indicate what differences in model assumptions are important. These differences then suggest lines for further inquiry and point to areas where more experimental data are needed.

In this chapter, we compare the NPP responses of forests in the northeastern U.S. to enhanced CO_2 and climate change as simulated by two models that were developed for very different purposes: PnET-II and version 4.0 of the Terrestrial Ecosystem Model (TEM). PnET-II is an uncalibrated, monthly time-step C and water balance model built around generalized physiological relationships. The PnET suite of models has been used to examine the influence of N deposition, climate change, O_3 exposure, and land use history on nutrient cycling and forest production in the northeastern and southeastern U.S. and Ireland. The Terrestrial Ecosystem Model is a highly aggregated process-based ecosystem model that has been developed to examine the monthly C, N, and water dynamics within terrestrial ecosystems across the globe.

Model Descriptions

PnET-II

In PnET-II (Fig. 11.1a), aboveground vegetation C is stored in four compartments (Aber and Federer, 1992; Aber et al., 1995): foliar canopy, plant (mobile) C, bud C, and wood. Atmospheric CO_2 is taken up by the canopy during photosynthesis and C is either respired or allocated to the various compartments.

A multilayered forest canopy is constructed in which available light and specific leaf weight decline with canopy depth. Light attenuation through the canopy is based on the Beers–Lambert exponential decay equation $(y = e^{-k*LAI})$. Maximum gross photosynthesis is calculated individually

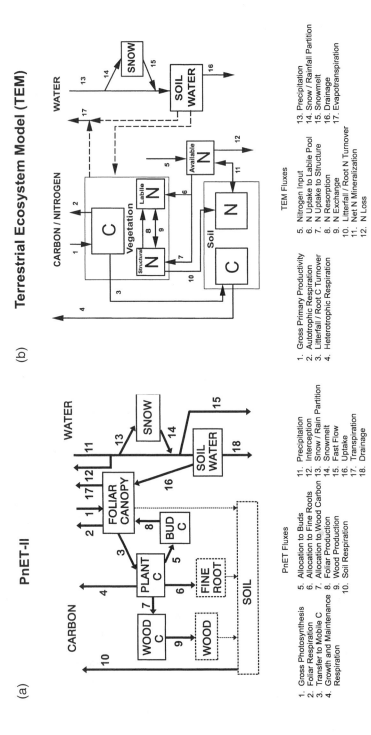

Figure 11.1. Schematic of model structures of (a) PnET-II and (b) Terrestrial Ecosystem Model (TEM) version 4.0.

for each of the 50 canopy layers in order to capture the effect of gradual light extinction on C gain. For each layer, gross photosynthesis is a function of the predicted maximum photosynthetic rate attenuated by environmental conditions, such as water availability, temperature, and daylength. Maximum photosynthetic rate is determined as a linear function of foliar N content, following the observed relationship between these two variables across species from diverse ecosystems (Field and Mooney, 1986; Reich et al., 1995). Net photosynthesis is calculated by subtracting day and night foliar respiration from gross photosynthesis. After the newly acquired C has been allocated to the wood C, plant C, and bud C compartments, NPP is calculated as the sum of wood production, foliar production, and the allocation of C to fine roots (see Fig. 11.1a).

Stomatal conductance is directly related to net photosynthetic rate, making water use efficiency (WUE) a function of vapor pressure deficit (VPD) (Tanner and Sinclair, 1983; Sinclair et al., 1984). In this way, transpiration can be predicted from canopy photosynthesis and VPD, providing a direct link between the C and water balance portions of the model. The long-term sustainability of increases in photosynthetic rate as a result of enhanced ambient CO_2 is uncertain, due to the possibility of acclimation or nutrient limitation (Bazzaz, 1990; Rastetter et al., 1991). Therefore, the atmospheric CO_2 increase is assumed to have direct effects on WUE and not on photosynthetic rate. Using PnET-II for the northeastern U.S., a doubling of CO_2 is assumed to result in a doubling of WUE (Aber et al., 1995).

While PnET-II does not calculate a complete soil C budget, it does predict some transfers between above- and belowground pools. For example, C is transferred from the mobile plant pool to roots and wood, and soil respiration (which includes both microbial and root respiration) (Aber et al., 1995) draws on the C assumed to exist belowground (see Fig. 11.1). Net ecosystem productivity (NEP) is calculated on a monthly basis as the difference between net photosynthesis and the sum of four respiration terms: foliar growth respiration, wood maintenance respiration, wood growth respiration, and soil respiration (see Fig. 11.1a). Computationally, the energy source for the aboveground vegetation respiration terms is the mobile C pool. The monthly NEP values are summed for a yearly NEP prediction.

Parameters in PnET-II are obtained on a regional basis from field data, and are not calibrated to make model results match measured output data. The model has performed well at predicting forest production and runoff at diverse locations across North America (Aber and Federer, 1992; Aber et al., 1995), and has been tested against eddy correlation CO_2 exchange measurements (Aber et al., 1996). To date, the PnET models have been applied to forests in the northeastern U.S. at spatial resolutions of 60 arcseconds (60″) (approximately 1.5 km) and 0.5° (approximately 40 to 50 km) (Aber et al., 1995, 1997; Ollinger et al., 1997, 1998; Jenkins et al.,

1999), to forests in the southeastern U.S. at spatial resolution of 0.5° (McNulty et al., 1996), and to forests in Ireland at a spatial resolution of 1 km (Goodale et al., 1997).

Terrestrial Ecosystem Model 4.0 (TEM 4.0)

Unlike PnET-II, vegetation C (both aboveground and belowground) in TEM 4.0 is simulated as a single compartment (Raich et al., 1991). Atmospheric CO_2 is taken up by plants through gross primary productivity (GPP) and C is then respired back to the atmosphere, retained as vegetation C, or transferred to the soil C compartment as litterfall (see Fig. 11.1b).

Monthly GPP is influenced by photosynthetically active radiation (PAR), leaf area, air temperatures, actual evapotranspiration, potential evapotranspiration, N availability, and atmospheric CO_2 concentration (Raich et al., 1991; McGuire et al., 1992, 1993, 1995, 1997). Monthly plant respiration depends on air temperature and vegetation C (Raich et al., 1991). NPP is calculated as the difference between GPP and plant respiration (see Fig. 11.1b).

The relationship between C assimilation and intercellular CO_2 is described by empirical functions representing limits imposed by carboxylation, light availability, synthesis, and N availability (Wullschleger, 1993; Sage, 1994; McGuire et al., 1997). Intercellular CO_2 is determined from atmospheric CO_2 by a canopy conductance term which depends on water availability (McGuire et al., 1997). For simulations of the effects of increased atmospheric CO_2 in this study, ambient CO_2 was doubled from 355 to 710 ppmv.

TEM 4.0 includes decomposition and N dynamics in its predictions. Nitrogen availability, which is determined by predicted N mineralization rate, can limit photosynthesis and NPP (McGuire et al., 1997; Pan et al., 1998; Kicklighter et al., 1999). Litterfall and root turnover are simulated as transfers from vegetation to soil C and N pools, and N is transferred between the soil N pool and the available N pool via N mineralization. In this way, N availability depends on the recycling of N from decomposing litter and soil organic matter so that rates of NPP are coupled to rates of decomposition. Thus, changes in N availability caused by climate change may also influence future NPP estimates. NEP is calculated monthly as the difference between GPP, autotrophic respiration (growth and maintenance respiration by all plant parts), and heterotrophic respiration (respiration by belowground microbes).

Although many of the vegetation-specific parameters in TEM 4.0 are defined from published information (Raich et al., 1991; McGuire et al., 1992; Melillo et al., 1993), some are determined on a biome-specific basis by calibrating the model to the fluxes and pool sizes of intensively studied field sites. To date, TEM has been applied using data sets gridded at a resolution of 0.5° latitude by 0.5° longitude for South America (Raich et al., 1991), North America (McGuire et al., 1992, 1993; VEMAP Members, 1995), and the globe (Melillo et al., 1993; McGuire et al., 1997; Xiao et al., 1997).

Equilibrium Assumptions

PnET-II and TEM 4.0 are applied in the first section of the chapter as equilibrium models. This means that they simulate a future in which CO_2 concentrations and climate have stabilized and vegetation distribution is constant. They do not include the effects of events such as changes in disturbance frequencies, reduction of C stored in biomass, changing vegetation distributions, or altered soil water-holding capacity (WHC) as a result of changing climate (Pastor and Post, 1988, 1993; Smith and Shugart, 1993). While there is growing recognition that these transient processes are potentially very important during the process of climate change (for example, see Tian et al., 1999), at present there is little consensus about the direction or magnitude of the transient responses (Melillo et al., 1996), especially at the regional level.

Influence of Model Assumptions on Net Primary Production Estimates

A comparison of the structures and assumptions of PnET-II and TEM 4.0 suggests two issues to be examined in this study. First, how does the representation of above- and belowground processes affect model predictions of NPP under enhanced CO_2 and climate change? In PnET-II (Aber and Federer, 1992; Aber et al., 1995, 1996), aboveground photosynthesis, allocation, and respiration are represented in NPP predictions (Fig. 11.1a). While a soil respiration term is included in PnET-II (Kicklighter et al., 1994), the model does not explicitly simulate biomass accumulation or belowground C and N cycling. In contrast, TEM 4.0 (Raich et al., 1991; McGuire et al., 1992, 1993; Melillo et al., 1993) simulates interactions between and among above- and below-ground lumped C and N pools, using these transfers in its predictions of NPP (Fig. 11.1b). Because PnET-II emphasizes aboveground processes while TEM 4.0 represents both above- and belowground processes, the two models represent N limitations to growth in different ways. PnET-II requires foliar %N as an input, and assumes that foliar %N represents the constant N constraints experienced by the forest. TEM 4.0 simulates soil N cycling and N uptake, and assumes the vegetation experiences dynamic N constraints.

A second major question to be asked in this study is: How does the representation of CO_2 effects on C fixation affect model predictions of NPP under enhanced CO_2 and climate change? Field research has suggested that increases in WUE are likely under enhanced CO_2 (for example, see Hollinger, 1987). Increased WUE would cause higher NPP in those biomes that experience water stress. However, a WUE increase can be caused by increased C assimilation, decreased stomatal conductance, or both (Eamus, 1991; Field et al., 1995; Bazzaz et al., 1996). If conductance decreases while C assimilation increases, WUE will increase even further. Thus WUE is a lumped measure of the end result of CO_2 increase. In PnET-II, CO_2 doubling is modeled as doubled WUE; thus PnET-II assumes that increases in C assimilation and decreases in conductance will sum to a 100% increase in WUE. In TEM 4.0, ambient CO_2 controls internal leaf CO_2, which drives C assimilation (McGuire et al., 1997). In this way, TEM 4.0 includes a physiologically based mechanism for predicting the impacts of enhanced ambient CO_2 on C assimilation.

Input Data Set Descriptions

For regional extrapolation, both PnET-II and TEM 4.0 require spatially explicit data sets of vegetation type and climate: monthly temperatures (maximum and minimum for PnET-II, average for TEM 4.0), monthly total precipitation, and monthly mean of total daily solar radiation. In addition, TEM 4.0 also requires a spatially explicit data set of soil texture. Although the models have similar input requirements, they have used different input data sets and different vegetation classification schemes to develop regional estimates. To minimize this source of variation in NPP estimates (cf. Pan et al., 1996) between the two models, we developed a common database and adapted our parameterizations to a common vegetation classification scheme.

For this study, we represented the forests of the northeastern U.S. with 115 pixels having a spatial resolution of 0.5° latitude × 0.5° longitude (see color plate Fig. 11.2). In previous analyses (Jenkins et al., 1999), we found little difference in regional NPP estimates made by these models at the 0.5° vs. the 60″ spatial resolution. The coarser spatial resolution greatly shortens the computational time required for analysis. Below, we describe how the input data sets were standardized for this study.

Land Cover Data Set

To describe the distribution of forest types in the northeastern U.S. (see Fig. 11.2), we aggregated a 30″ (roughly 1 km) data set developed from Advanced Very High Resolution Radiometer (AVHRR) data by Lathrop and Bognar (1994) to a 0.5° resolution as described by Jenkins et al. (1999). No attempt was made in the current analysis to take into account land that is not presently forested. PnET-II NPP estimates were created using parameter values for northeastern hardwood, pine, and spruce–fir forests as described by Aber et al. (1995) and Ollinger et al. (1998). To adapt the TEM 4.0 parameterizations (McGuire et al., 1992) to this vegetation classification scheme, we parameterized (1) hardwoods as temperate deciduous forests, (2) pines as temperate coniferous forests, and (3) spruce–fir as boreal forests.

For both models, hardwood/spruce–fir pixels were run assuming each pixel consisted of 40% hardwood and 60% spruce–fir, and hardwood/pine pixels were run assuming each pixel consisted of 60% hardwood and 40% pine. These forest composition estimates were generated by comparing the original AVHRR-generated vegetation map with United States Department of Argiculture (USDA) Forest Service Inventory and Analysis data (Beltz et al., 1992). For pixels containing a mosaic of hardwood/spruce–fir forests or hardwood/pine forests, TEM 4.0 estimated NPP by weighting estimates made with each of the appropriate nonmosaic calibrations (cf. McGuire et al., 1995) with the percentage cover just described for these mosaic grid cells. Similarly, PnET-II predictions for mixed pixels were created by weighting estimates made using the pine, hardwood, or spruce–fir parameterizations with the percentage cover described above for each forest type.

Soils

Both PnET-II and TEM 4.0 use the concept of soil WHC (defined as field capacity minus wilting point) to represent the maximum amount of water that can be stored in the soil and made available to plants. But while PnET-II requires soil WHC as an input variable, TEM 4.0 uses soil depth and soil texture (as %[silt + clay]) to derive total volumetric soil water at field capacity, and then uses rooting depth to predict soil WHC. Soil WHC is used as an internal variable during TEM 4.0 simulations. A constant value of 12 cm has been used for soil WHC in all northeastern U.S. PnET analyses to date (Aber and Federer, 1992; Aber et al., 1995, 1997; Ollinger et al., 1998; Jenkins et al., 1999) and soil WHC was held constant at 12 cm for this study as well. To create equivalent NPP predictions using both models, we constrained the soil WHC used internally by TEM 4.0 to 12 cm.

Climate Input Data Sets

To examine the effect of climate change on NPP, we developed common data sets for both contemporary climate and future climate, as projected from five General Circulation Model (GCM) experiments.

Contemporary Climate

For all contemporary climate data, we selected the 115 pixels that represent this part of the northeastern U.S. from the Vegetation/

Ecosystem Modeling and Analysis Project (VEMAP) (Kittel et al., 1995). To develop these contemporary climate data, minimum and maximum monthly temperatures from weather stations were adiabatically adjusted to sea level using algorithms of Marks and Dozier (1992), georeferenced to the 0.5° grid, then readjusted to grid elevations. For TEM 4.0, mean monthly temperatures were determined by averaging the maximum and minimum temperatures for each month. Mean monthly precipitation was aggregated to the 0.5° resolution from a 10-km resolution data set developed using the PRISM model (Daly et al., 1994). CLIMSIM (a simplified version of MT-CLIM for flat surfaces (Running et al., 1987; Glassy and Running, 1994) was used to estimate daily solar radiation received at the canopy level, and these daily data were averaged to obtain the monthly means.

Climate Change Scenarios

Equilibrium climate change scenarios from five GCM experiments were used to create the forest NPP predictions presented here. Three of these five were also used to create continental-scale predictions during the VEMAP exercise (VEMAP Members, 1995). These are from the Geophysical Fluid Dynamics Laboratory (GFDL) (Wetherald and Manabe, 1988; Manabe et al., 1990; Wetherald et al., 1990), Oregon State University (OSU) (Schlesinger and Zhao, 1989), and the United Kingdom Meteorological Office (UKMO) (Wilson and Mitchell, 1987).

The other two GCM scenarios come from the Hadley Centre (Mitchell et al., 1995). The first includes only the radiative forcing due to greenhouse gases (the "Hadley/gas" scenario), and the second takes into account both the warming effects of greenhouse gases and the direct radiative effect of sulphate aerosols (the "Hadley/sulphate" scenario). Thus only one set of climate data in this experiment includes the effects of sulphate aerosols, which can influence climate directly by scattering and absorbing radiation, or indirectly by modifying the optical properties of clouds (Schimel et al., 1996a). In the short term, sulphate aerosols are likely to mitigate the warming influence of greenhouse gases (Mitchell et al., 1995), though considerable uncertainty exists about the spatial distributions of sulphate and other aerosols and about their individual and combined impacts on radiative forcing of climate change (Schimel et al., 1996a).

To find monthly minimum and maximum temperature projections for use in PnET-II, the difference between the GCM and VEMAP monthly mean temperatures was found for each of the 115 pixels. This difference was then added to the VEMAP monthly minima and maxima to create an estimate of the monthly minimum and maximum temperature under climate change for each pixel. This method assumes no change in the diurnal temperature range.

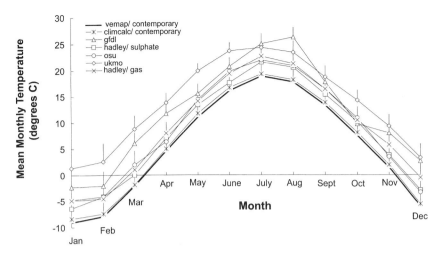

Figure 11.3. Comparison of GCM monthly mean temperature predictions for the northeastern region. Each point represents the average of 115 0.5° pixels, ±1 SD.

For the northeastern region, all of the GCM scenarios predict substantially warmer temperatures year-round. The UKMO and GFDL scenarios predict the warmest conditions under climate change, with regional yearly average temperature increases of 8.1 and 6.1°C, respectively. The Hadley/gas, OSU, and Hadley/sulphate scenarios predict smaller temperature increases, with regional yearly average increases of 3.6, 3.2, and 2.5°C, respectively (Fig. 11.3). Temperature projections made by the Climcalc model, a statistical model of contemporary climate based on latitude, longitude, and elevation, and created from weather station data collected in the region (Ollinger et al., 1995), are also plotted on Fig. 11.3 for comparison. The Climcalc and VEMAP predictions of contemporary monthly mean temperature agree closely; this agreement contrasts with the GCM projections, which differ more substantially from one another.

The difference between the VEMAP and Climcalc estimates of contemporary precipitation is greater than the corresponding difference between the contemporary temperature data sets. This difference is still small, however, compared with the projected changes in precipitation represented by the various GCM data sets (Fig. 11.4). Averaged regionally, the GCM scenarios all predict an increase in total annual precipitation for the northeastern region: +23.4% for UKMO; +19.0% for Hadley/sulphate; +18.5% for GFDL; +18.4% for Hadley/gas; and +10.9% for OSU. The GFDL scenario predicts the greatest seasonal variation in precipitation with 60 mm of precipitation occurring in August and 174 mm occurring in October.

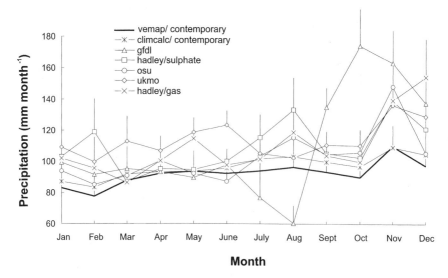

Figure 11.4. Comparison of GCM monthly precipitation predictions for the northeastern region. Each point represents the average of 115 0.5° pixels, ±1 SD.

The GCMs predict slight changes in solar radiation in the northeastern region (Fig. 11.5). During the winter, the two Hadley scenarios predict somewhat lower solar radiation than the other GCM scenarios, and the

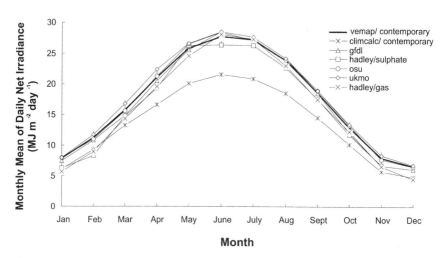

Figure 11.5. Comparison of GCM solar radiation predictions for the northeastern region. Both models use monthly mean of daily net irradiance as input. Each point represents the average of values for 115 0.5° pixels, ±1 SD.

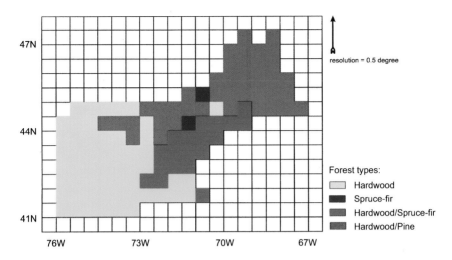

Figure 11.2. Forest classification for the portion of the northeastern region treated in this analysis. See text for details.

sulphate correction also moderately reduces the amount of radiation received during the growing season. The difference between the Climcalc and VEMAP representations of contemporary climate is larger than the differences between the predictions made by the GCM scenarios. Considerable uncertainty exists about the accuracy of the VEMAP solar radiation data at the continental and regional scales (Pan et al., 1996; Jenkins et al., 1999). Because ecosystem process models can be quite sensitive to solar radiation inputs, future research should be directed at reducing this uncertainty.

General Circulation Model Uncertainty

The 0.5° latitude by 0.5° longitude GCM estimates used in this study were interpolated from original data sets with a resolution of 5° latitude by 5° longitude. GCM predictions are less reliable at the regional scale than at larger scales (Kattenberg et al., 1996) due to the parameterizations of physical processes, which are less accurate at smaller scales (Gates, 1985; Ghan, 1992). Areas dominated by land–ocean interactions are especially prone to GCM uncertainty (Cooter et al., 1993). While regional climate projections such as those used in this analysis are necessarily uncertain, the GCM projections do represent a range of possible climate responses to radiative forcing. Analysis and comparison of model predictions of forest NPP under several of these scenarios provides us with a range of possible forest responses for a future in which the trajectory of climate change is uncertain.

Effect of Climate Change on Forest Net Primary Productivity

Regional Predictions

At the regional scale, the NPP predictions of the two models differ by only 3% for contemporary climate and by 10% for the various future climate scenarios (Table 11.1). Both models also predicted an increase in forest NPP under all climate change scenarios (see Table 11.1), with PnET-II predicting an average increase of 37.9% over the VEMAP contemporary scenario, and TEM 4.0 predicting an average increase of 30.0%. For both models, the largest increases in NPP (+58.2% for PnET-II and +54.7% for TEM 4.0) occurred under the UKMO climate. However, different scenarios induced the smallest increase in NPP for the two models. Net primary productivity estimated by PnET only increased by 30.0% under the Hadley/sulphate scenario, whereas NPP estimated by TEM 4.0 increased by only 17.7% under the GFDL scenario. PnET-II and TEM 4.0 NPP predictions increased by 2.5 and 6.2% less, respectively, under the Hadley/sulphate scenario than under the Hadley/gas scenario. Thus,

Table 11.1. NPP Predictions Made by PnET-II and TEM 4.0 under Contemporary and Enhanced CO_2 Conditions

Model	GCM Scenario	Biome-specific Averages (gOM m^{-2} yr^{-1}) (% difference from control)				Regional Total (TgOM yr^{-1})
		Hardwood[a]	Spruce–Fir[b]	Hardwood/ Spruce–Fir[c]	Hardwood/Pine[d]	
PnET-II	*VEMAP (control)*	*1367.60*	*890.40*	*1050.80*	*1186.80*	*313.14*
	GFDL	1838.41 (+34.4%)	983.56 (+10.5%)	1283.76 (+22.2%)	1819.63 (+53.3%)	418.32 (+33.6%)
	Hacdley gas only	1870.31 (+36.8%)	917.46 (+ 3.0%)	1222.24 (+16.3%)	1781.10 (+50.1%)	414.99 (+32.5%)
	Hadley sulphate	1839.26 (+34.5%)	899.30 (+ 1.0%)	1202.12 (+14.4%)	1726.24 (+45.5%)	407.06 (+30.0%)
	OSU	1907.36 (+39.5%)	910.63 (+ 2.3%)	1245.20 (+18.5%)	1813.88 (+52.8%)	422.93 (+35.1%)
	UKMO	2199.67 (+60.8%)	1124.22 (+26.3%)	1443.85 (+37.4%)	2247.13 (+89.3%)	495.38 (+58.2%)
TEM 4.0	*VEMAP (control)*	*1540.07*	*583.55*	*895.48*	*1295.35*	*324.17*
	GFDL	1757.85 (+14.1%)	709.05 (+21.5%)	1093.51 (+22.1%)	1602.62 (+23.7%)	381.68 (+17.7%)
	Hadley gas only	1941.44 (+26.1%)	744.40 (+27.6%)	1151.34 (+28.6%)	1674.89 (+29.3%)	412.38 (+27.2%)
	Hadley sulphate	1844.17 (+19.7%)	707.40 (+21.2%)	1093.40 (+22.1%)	1603.13 (+23.8%)	392.28 (+21.0%)
	OSU	1983.64 (+28.8%)	745.55 (+27.8%)	1159.77 (+29.5%)	1683.34 (+30.0%)	418.72 (+29.2%)
	UKMO	2417.64 (+57.0%)	906.45 (+55.3%)	1368.05 (+52.8%)	1943.42 (+50.0%)	501.44 (+54.7%)

[a] $n = 54$ pixels for each scenario.
[b] $n = 2$ pixels for each scenario.
[c] $n = 40$ pixels for each scenario.
[d] $n = 19$ pixels for each scenario.

while sulphate aerosol correction does impact model predictions, larger differences exist between NPP predictions made using unrelated GCM scenarios. In this analysis at the regional scale, differences between the GCM climate input data sets were more important contributors to variability in model predictions than differences between the models.

Biome-level Predictions

The biome-level differences evident in the NPP predictions for contemporary climate were emphasized when the GCM scenarios were used to drive the models. For all GCM scenarios, PnET-II predicted that NPP in the higher-productivity forests (hardwood and hardwood/pine) would increase more than in the lower-productivity forests (spruce–fir and hardwood/spruce–fir), while TEM 4.0 predicted that NPP would increase by roughly the same percentage for each biome (see Table 11.1).

To understand better the reasons behind these differences, we performed a series of experiments with both models in which we used values of one variable (i.e., atmospheric CO_2 concentration, air temperature, precipitation, or solar radiation) from the Hadley/sulphate GCM scenario together with values for all other variables from the VEMAP contemporary climate scenario as inputs to the models. The Hadley/sulphate GCM scenario was used for this exercise because it takes into account the radiation-scattering effects of sulphate aerosols, which are likely to exert a significant effect on climate in the future (Mitchell et al., 1995; Schimel et al., 1996a). We then examined how the results of these sensitivity experiments (Fig. 11.6) are related to the different conceptualizations of forest ecosystem dynamics by the two models.

PnET-II

Because doubled CO_2 is parameterized as a doubling of WUE, the forest types that are most limited by water availability in PnET-II are expected to respond the most dramatically to CO_2 increase. A previous sensitivity analysis with PnET-II (Ollinger et al., 1998) suggested that water availability was the factor most limiting to hardwood forests under the range of conditions typical of the northeastern region. Consistent with the results from the previous analysis, in this study PnET-II predicted a dramatic increase in hardwood NPP when CO_2 doubling was added to contemporary climate (see Fig. 11.6a). The forest types with a hardwood component also experienced an increase in NPP due to CO_2 doubling. In addition, hardwood NPP responded positively to the precipitation increase from the VEMAP to the Hadley/sulphate scenarios (see Fig. 11.6a). The NPP change with increased precipitation was not as dramatic as the effect of doubled CO_2, because the 19.0% increase in precipitation did not alleviate as much water stress as did the doubled WUE.

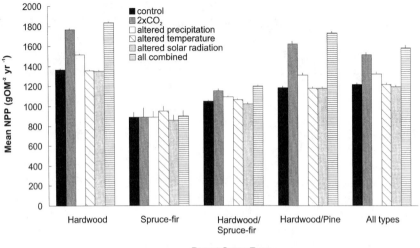

(a)

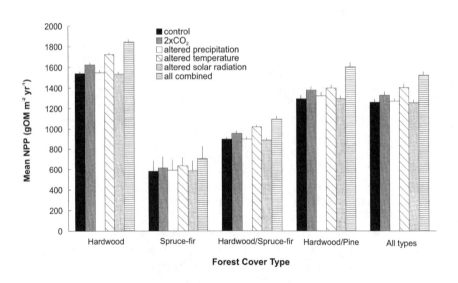

(b)

Figure 11.6. Biome-level responses to individual climate variables from Hadley/
sulphate GCM (a, PnET-II; b, TEM 4.0). Within each forest type, each bar
represents the mean ±1 SD for one run with one Hadley/sulphate climate variable
substituted for the equivalent VEMAP contemporary variable. "Altered" variables are from the Hadley/sulphate GCM; all others are from the VEMAP
contemporary climate scenario, except for the "all combined" run, in which all of
the Hadley/sulphate variables were applied at once.

For the range of conditions currently encountered by northeastern forests, PnET-II predicts that NPP in the spruce–fir biome is more limited by solar radiation and temperature than by water (Ollinger et al., 1998; Jenkins et al., 1999). Thus the 2.5°C average regional increase predicted by the Hadley/sulphate scenario caused a slight increase in predicted NPP for spruce–fir forests. Similarly, the absence of water limitation in spruce–fir forests means that spruce–fir NPP is more tightly linked to solar radiation and PAR availability. As a result, the decline in solar radiation predicted by the aerosol-corrected Hadley/sulphate scenario caused a slight decline in spruce–fir NPP (see Fig. 11.6a).

When all of the Hadley/sulphate climate variables were applied simultaneously with doubled CO_2, predicted NPP was higher in all forest types (except spruce–fir) than for any of the variables applied alone. The combined impacts of alleviated water stress and increased precipitation on hardwood NPP were important determinants of productivity in all biomes with a hardwood component (see Fig. 11.6a). However, the effects were not additive, suggesting that interactions between water and temperature may be key predictors of modeled NPP under climate change. Spruce–fir NPP under the Hadley/sulphate GCM climate combined with doubled CO_2 was approximately the same as that found under contemporary climate because the enhanced NPP caused by higher temperatures compensated for the diminished NPP caused by lower solar radiation.

TEM 4.0

In TEM 4.0, enhanced atmospheric CO_2 increases GPP in all biomes if sufficient light and N are available. TEM 4.0 does not prescribe a WUE value, but the model assumes that total canopy conductance and actual evapotranspiration do not change in response to elevated CO_2 so that the enhancement of GPP by CO_2 fertilization causes an effective increase in WUE (cf. Pan et al., 1998). Unlike PnET-II, which assumes a doubling of WUE with doubled CO_2, the effective WUE estimated by TEM increased by only 5 to 6% for all forest biomes in the northeastern U.S. with a doubling of atmospheric CO_2. The model estimates larger increases in effective WUE in warmer and drier biomes (cf. Pan et al., 1998). The precipitation increase from the VEMAP to the Hadley/sulphate scenario had little effect on the TEM 4.0 NPP predictions for the various biomes (see Fig. 11.6b), because TEM 4.0 predicts little or no water stress in the northeastern region under the VEMAP contemporary climate.

Using TEM 4.0, the temperature increase from the VEMAP to the Hadley/sulphate scenario caused a more pronounced NPP increase for all biomes than did the doubling of atmospheric CO_2 (see Fig. 11.6b). In addition to the direct effects of temperature on GPP, temperature increases also enhanced decomposition and N mineralization rates. Accelerated N mineralization increased N availability to vegetation, resulting in higher NPP (Pan et al., 1998). This result is also consistent with the continental-scale results of Schimel et al. (1997), who found a tight correlation between N mineralization and NPP in TEM 4.0 predictions under contemporary climate.

When all of the Hadley/sulphate climate variables were applied simultaneously with doubled atmospheric CO_2, predicted NPP for all forest types was higher than for any of the climate variables applied alone (see Fig. 11.6b). Using TEM 4.0 under climate change and enhanced ambient CO_2, both the increase in N availability caused by increased temperature and the increase in C availability that occurs under enhanced atmospheric CO_2 allowed larger increases of NPP to occur than under either enhanced atmospheric CO_2 or enhanced temperatures alone.

Effects of Climate Interactions on Net Primary Productivity

At this point in our analysis of forest NPP in the northeastern U.S. under climate change and enhanced CO_2, TEM 4.0 estimates appear to be primarily limited by temperature, while PnET-II estimates appear to be limited primarily by water. However, PnET-II and TEM 4.0 both estimated similar large increases in NPP associated with the large increases in temperature under the UKMO climate (see Table 11.1). In addition, the small increase in NPP estimated by TEM 4.0 under the GFDL scenario, which predicted drier summers, indicates that TEM 4.0 estimates of NPP are also sensitive to changes in precipitation. To explore this issue further, we expand our analysis to include results from model simulations using the wider range of climate variables in all five GCM scenarios.

Increased temperature can affect phenology as well as increasing photosynthesis. Myneni et al. (1997) have reported satellite evidence for a lengthening of the growing season in the northern latitudes between 1981 and 1991 as a potential result of climate change. Total annual growing degree days (GDD) is a useful measure of these combined effects and is calculated as the annual sum of temperatures for all days with average temperature greater than 0°C. Because our regional NPP estimates are based on monthly air temperature data, we can calculate GDD as:

$$GDD = \Sigma(\text{monthly average temperature}) \times (\text{days in month}) \qquad (1)$$

for all months with average temperature greater than 0°C. A comparison of predicted NPP with annual GDD for both models and all GCM scenarios (Fig. 11.7) shows the extent to which each model appears to be temperature-driven, and the comparative relationship between GDD and NPP for the different GCM scenarios as predicted by each model. For the range of climate conditions predicted by the GCM scenarios used in this analysis, GDD was a good predictor of hardwood NPP for both PnET-II

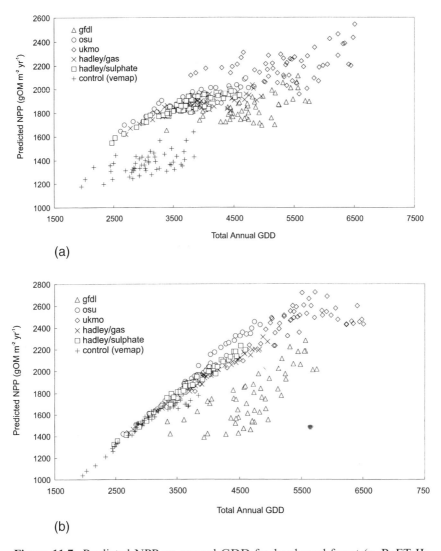

Figure 11.7. Predicted NPP vs. annual GDD for hardwood forest (a, PnET-II; b, TEM 4.0). Each point represents NPP as predicted for one pixel under one GCM scenario.

and TEM 4.0 (Table 11.2). The inclusion of six different scenarios representing both contemporary and changed climates widened the GDD range significantly for this analysis, resulting in strong correlations between GDD and predicted NPP. Annual precipitation and growing season precipitation were also good predictors of modeled NPP for both PnET-II and TEM 4.0 (see Table 11.2).

Table 11.2. R^2 Values from Linear Regression Analyses between Predicted NPP and Each of the GCM-predicted Climate Variables

Model	Forest Type	Average Yearly Temperature (°C)	GDD[a]	Net Irradiance (J m⁻² day⁻¹)	Precipitation (cm yr⁻¹)	Growing Season Precipitation (cm season⁻¹)	Suitability[b]
PnET-II	Hardwood	0.65	0.62	0.03	0.42	0.21	0.70
	Spruce–fir	0.66	0.62	0.43	0.04	0.17	0.68
	Hardwood/ spruce–fir	0.72	0.68	0.15	0.42	0.25	0.74
	Hardwood/pine	0.74	0.72	0.05	0.64	0.23	0.73
	All types	0.61	0.58	0.04	0.25	0.13	0.59
TEM 4.0	Hardwood	0.69	0.68	0.01	0.28	0.26	0.83
	Spruce–fir	0.70	0.70	0.26	0.30	0.45	0.88
	Hardwood/ spruce–fir	0.61	0.57	0.03	0.29	0.28	0.68
	Hardwood/pine	0.78	0.73	0.01	0.61	0.30	0.80
	All types	0.58	0.56	0.02	0.17	0.19	0.63

[a] Defined as the yearly sum of (monthly average temperature × days in month) for all months with average temperature >0°C.
[b] Defined as ((Growing season precipitation × GDD)/10,000).

When predicted NPP was plotted against GDD for both models (see Fig. 11.7), predictions made using the GFDL scenario were noticeably lower than predictions made using any of the other GCM scenarios. Growing season precipitation under the GFDL scenario was clearly lower than that of the other GCM scenarios (Fig. 11.4); for both models, it appeared that water stress was severely limiting productivity under this scenario. This suggests that the NPP predictions of both models under future climates may be influenced by both water availability and temperature.

To examine the interactions between growing season precipitation and GDD, we developed a diagnostic index of temperature and water suitability, defined as:

$$\text{"suitability"} = (\text{growing season precipitation} \times \text{total annual GDD})/10,000 \quad (2)$$

where growing season precipitation (cm season^{-1}) is the precipitation that occurs between May 1 and October 1. At high values of the suitability index, water and temperature are not limiting, whereas NPP is limited by water availability, temperature, or a combination of these factors at low suitability values (Fig. 11.8). Linear regression analyses indicated that this index of the interaction between precipitation and temperature was a better predictor of variability in model NPP under climate change and enhanced CO_2 than any of the climate variables alone, for all forest types and both models (see Table 11.2). The predictive power of the suitability index suggests that simultaneous accurate predictions of both temperature and precipitation are critical for NPP predictions under a changed climate.

Carbon Dioxide Fertilization and Water Use Efficiency

The stepwise increase in PnET-II hardwood NPP predictions from the VEMAP 1×CO_2 climate to the GCM 2×CO_2 scenarios (see Figs. 11.7a, 11.8a) resulted from the doubling of WUE to estimate the effects of doubled CO_2 on NPP. This approach contrasts with the approach of TEM 4.0, which causes a more continuous NPP increase from contemporary to future climate scenarios (see Figs. 11.7b, 11.8b). In experiments under doubled atmospheric CO_2, WUE increases have ranged widely though the increases are usually less than 100% (Eamus and Jarvis, 1989; Eamus, 1991). Thus, doubling WUE may overestimate the influence of doubled atmospheric CO_2 on NPP. In addition, seasonal conditions and temperature may influence the degree of CO_2 enhancement of WUE, and different patterns of C gain (and WUE increase) may be experienced by different parts of individual plants (Eamus, 1991; Bazzaz et al., 1996). The availability of nutrients such as N may also limit the benefits of CO_2 fertilization (McGuire et al.,1997; Pan et al., 1998; Kicklighter et al., 1999). These considerations potentially argue in favor of a more dynamic,

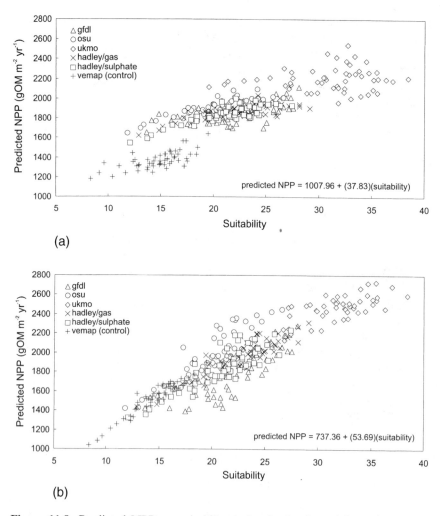

Figure 11.8. Predicted NPP vs. suitability index for hardwood forest (a, PnET-II; b, TEM 4.0). Each point represents NPP as predicted for one pixel under one GCM scenario. The best-fit linear regression equations are shown for each model (see Table 11.2 for R^2 values).

physiologically based parameterization of GPP response to enhanced CO_2, such as that found in TEM 4.0. However, the focus on GPP response to the exclusion of a stomatal response might underestimate the direct effects of enhanced CO_2 on C gain. In the absence of decisive and quantitative information about patterns of change in WUE and C assimilation under increased CO_2, doubling WUE and the more physiologically based approaches may be equally accurate.

Influence of Belowground Processes on Predicted Net Primary Productivity

Both the NPP predictions and the relative strengths of the water, temperature, and solar insolation limitations on NPP were similar for PnET-II and TEM 4.0 in this analysis despite their very different representations of the process-based limitations on NPP (see Tables 11.1 and 11.2). Specifically, TEM 4.0 includes belowground N availability in its NPP predictions by including decomposition and N mineralization routines, which are driven by climate variables, while PnET-II includes belowground N availability in its predictions via foliar %N, which is determined by forest type and remains constant. While it might be argued that including all belowground feedbacks between climate and nutrient availability offers the potential for a more accurate representation of forest ecosystem processes, there is evidence for steady-state correlations between water and nutrient limitations on NPP (Aber et al., 1991; Schimel et al., 1996b). If a model gives an accurate reflection of nutrient limitations at equilibrium (e.g., by correlating photosynthetic rate with foliar %N), then algorithms representing detailed soil processes may be extraneous. The limitation of such a model is that it does not represent conditions other than those occurring at steady state. Under transient conditions, representation of detailed belowground processes will be more important because the water, N, and C budgets are likely to respond to perturbation on different time scales (Schimel et al., 1996b), and NPP predictions will vary accordingly.

Other Issues

These two models, with their different structures and parameterization approaches, made similar predictions of forest NPP under changed climate scenarios. However, other factors likely to influence forest productivity in the northeastern U.S. were not considered here. For example, changing temperature and precipitation regimes during the period before equilibrium conditions are reached (transient conditions) are likely to cause forest NPP patterns different from those expected under equilibrium conditions (Pastor and Post, 1988, 1993; Smith and Shugart, 1993; Melillo et al., 1996). Other anthropogenically induced stresses, such as N deposition, are common in the region and may impact nutrient cycling and production rates in the future (Aber et al., 1989; Ollinger et al., 1993; Magill et al., 1996; Townsend et al., 1996; Vitousek et al., 1997; Nadelhoffer et al., 1999). Finally, past forest clearing and agriculture may have a lasting impact on current and future nutrient cycling and forest production (Cronon, 1983; Aber and Driscoll, 1997; Aber et al., 1997; Magill et al., 1997; Compton et al., 1998). Model analyses have not yet perfected techniques to predict forest NPP under each of these conditions, but progress is being made. In this section, we present some examples of our work in this area.

Transient Analyses

The variability inherent in field-measured climate data contributes substantially to variability in NPP predictions (Aber and Driscoll, 1997). Unlike field-measured climate data, GCM equilibrium scenarios capture only the long-term changes in climate that might be expected as a result of greenhouse-gas forcing. To examine the effects of climate variability on model NPP predictions, we used transient climate data spanning 1950 to 1995 as input to transient versions of both PnET-II and TEM (TEM 4.1, see Tian et al., 1999).

Climate Data Sets

The transient temperature and precipitation data were developed from the temperature anomalies of Jones et al. (1991) and the precipitation anomalies of Hulme (1995) as described by Tian et al. (1999). For the two 0.5° cells containing the Harvard Forest (HF) in north central Massachusetts and the Hubbard Brook Experimental Forest (HBEF) in central New Hampshire, the gridded data appear to damp out some of the year-to-year temperature variability and to overestimate slightly the mean annual temperature (Fig. 11.9). The gridded precipitation data capture accurately the trends in year-to-year variability of total annual precipitation (Fig. 11.10). However, the site-to-site variability in precipitation is diminished by the gridded precipitation data. For example, the drought in the mid-1960s was extremely severe at HF (see Fig. 11.10a) and less severe at HBEF (see Fig. 11.10b). The gridded data underpredict the drought's severity at HF and overpredict drought severity at HBEF, apparently smoothing drought effects across the region.

Model Comparisons

We used the gridded transient data presented in Figs. 11.9 and 11.10 to predict NPP and NEP using PnET-II and TEM 4.1 for the HF and HBEF sites. For PnET-II, foliar %N values of 2.0 and 2.4% were used in these predictions for both sites, because these foliar %N values represent the likely range of hardwood foliar %N values for the northeastern region as a result of site quality and previous land use history. The 2.4% value was measured at HBEF by Whittaker et al. (1974). The 2.0% value represents the average of two measured values in different areas of the Harvard Forest: 1.8% in the control hardwood stand used in the chronic N experiment, as measured by Aber et al. (1993b); and 2.2% in the area around the eddy correlation tower, as measured by M.E. Martin and J.D. Aber. While there is little direct evidence for reduced foliar %N in

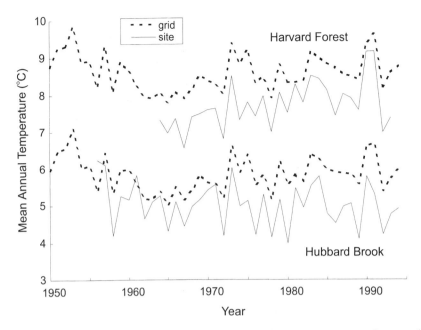

Figure 11.9. Comparison of site-measured transient temperature data and gridded temperature data for Harvard Forest and Hubbard Brook.

severely disturbed sites, there is convincing evidence for increased N uptake and foliar %N in N-rich sites (cf. Magill et al., 1997). Low soil N availability as a result of low site quality or past disturbance is likely to result in lower foliar %N on these sites, compared with undisturbed or high quality sites.

Perhaps the most obvious result from the model predictions made using transient climate data is that interannual climate variability contributes substantially to interannual variability in predicted NPP (Fig. 11.11). With PnET-II, however, the range expected as a result of climate variability was similar to or smaller than the range expected as a result of differences in foliar %N. This result is consistent with the findings of Goodale et al. (1998), who reported that PnET-predicted NPP in Ireland was more sensitive to differences in foliar %N (representing site quality and past land use history) than to climate change.

The absence of clear trends with time in this analysis suggests that more than four decades of measured data will be required in order to discern climate trends from transient data or to identify NPP and NEP trends from model predictions using the transient climate data. As suggested by Aber and Driscoll (1997), the variable characteristics of field-measured time series data should be represented in data sets representing future transient conditions.

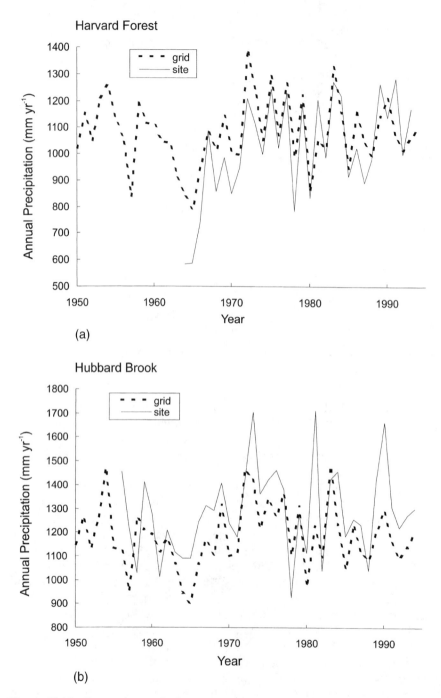

Figure 11.10. Comparison of site-measured transient precipitation data and gridded precipitation data for (a) Harvard Forest and (b) Hubbard Brook.

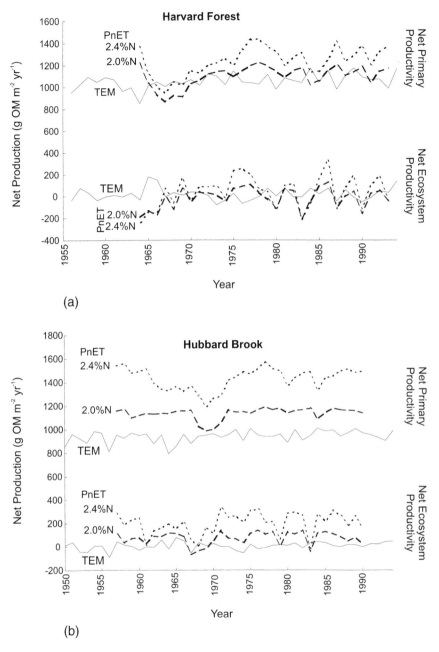

Figure 11.11. Transient predictions of NPP and NEP made by PnET-II and TEM 4.1 for (a) Harvard Forest and (b) Hubbard Brook.

At the HF site, the NPP and NEP estimates made by PnET-II (using the 2.0% foliar N value most appropriate for that site) and TEM 4.1 agreed closely (see Fig. 11.11a). When the models were applied to the HBEF site, on the other hand, the TEM 4.1 NPP predictions were lower on average, while PnET-II predictions (using the 2.4% foliar N most appropriate for that site) were higher (see Fig. 11.11b). These results reflect the model biases described earlier in the chapter.

TEM does not predict water stress for the northeastern region. As a result, temperature is most likely to be the driving variable for TEM predictions in this region under contemporary conditions. The warmer temperatures at HF, as compared with those at HBEF (see Fig. 11.9) caused TEM to predict higher NPP at HF than HBEF (see Fig. 11.11). On the other hand, PnET-II does predict water stress at HF (Aber et al., 1995). The increased precipitation at HBEF (see Fig. 11.10) reduced water stress at that site and caused PnET-II to predict higher NPP despite the cooler temperatures (see Fig. 11.11b). The cooler temperatures at HBEF also caused lower respiration rates, further contributing to the higher predicted NPP at HBEF than at HF.

Nitrogen Deposition and Land Use

Land use history and atmospheric deposition are two additional important change agents in the northeastern U.S. Interactive effects of these two can be important as land use, especially for forest harvesting or agriculture, can result in long-term reductions in N cycling in forested or reforested sites (Aber and Driscoll, 1997), which may be partially offset by increased N deposition. Nitrogen cycling rates are clearly linked with forest NPP (Pastor et al., 1984; Reich et al., 1997), suggesting that additions or extractions of N will impact C cycling rates as well.

To simulate the effects of N deposition and land use history on forest NPP at several sites in the northeast, we used PnET-CN, a model built upon PnET-II but which includes biomass storage terms for wood, roots, and soil organic matter, and which adds N pools and cycling to all compartments (Aber et al., 1997). Of critical importance for this analysis, PnET-CN does not assume a constant N availability to vegetation (via the foliar %N parameter). Rather, foliar N concentrations are predicted by the model and may change year-to-year depending on the relative availability of C and N to plants. PnET-CN also adds a "Scenario" subroutine, called at the beginning of each simulated month, which allows for the input of information on biomass removal, N deposition, and other variables such as climate change or experimental manipulations. The model has performed well when measured against streamflow and dissolved inorganic N (DIN) data at diverse sites around the northeastern U.S. (Aber and Driscoll, 1997; Aber et al., 1997), and suggests that land use events can have very long-lasting (2–3 centuries) effects on N cycling.

Several forested sites within the White Mountains of New Hampshire having contrasting land use histories were used to examine the relative effects of N deposition and land use change on predicted NPP. These include (1) the Bowl Natural Area (BNA), one of the last remaining mixed deciduous and coniferous forests in the northeastern U.S. with no history of logging, human settlement, or forest fire (Leak 1974); (2) Cone Pond, a 53-ha catchment dominated by mixed coniferous vegetation (80%), with a smaller amount of northern hardwood cover (15%) which was severely burned in 1820 and has been relatively undisturbed since (Bailey et al., 1995, 1996); (3) Watershed 6 (W6), the reference watershed at HBEF, which was logged intensively from 1910 to 1917 and experienced some salvage removals after the hurricane of 1938; (4) Watershed 4 (W4), which was commercially clearcut in the early 1970s in 25-m wide strips along elevational contours, with every third strip cut every second year over 6 years; (5) Watershed 5 (W5), which was whole-tree harvested in 1983 to 84; and (6) Watershed 2 (W2) which was experimentally devegetated for three years beginning in 1965 (see Bormann and Likens [1979] for a description of the HBEF site and the experimental manipulations performed there).

Mean atmospheric deposition of DIN (nitrate-N [NO_3^--N] plus ammonium-N [NH_4^+-N]) has been estimated at $0.87\,g\,N\,m^{-2}\,yr^{-1}$ for the HBEF, representing bulk wet deposition of $0.69\,g\,N\,m^{-2}\,yr^{-1}$ for the period 1964 to 1991 (Butler and Likens, 1991; Likens, 1992; Likens and Bormann, 1995) and dry deposition of $0.18\,g\,N\,m^{-2}\,yr^{-1}$ for the year 1989. For the simulations that included the effects of atmospheric inputs (ambient N deposition), N deposition was assumed to have increased linearly from 25% of these values to the current values beginning in 1900. Runs without N deposition effects (background N deposition) assume a constant rate of N deposition equal to 25% of current ambient values. Monthly climate data from the existing long-term records at the HBEF were used as input to the model for all sites, as described by Aber and Driscoll (1997).

Changes in NPP over time are compared for the two most extreme land use cases (BNA and W2) in Fig. 11.12. The BNA, with no prior history of N removals, showed slightly higher NPP values than W2 before the devegetation experiment, and a minor increase in NPP with ambient N deposition included. During devegetation, and for 10 years following, N availability was not limiting in W2, first due to reduced plant demand during the period of reestablishment, and then due to increased N availability because of the effects of disturbance on soil C/N ratios. During this 10-year disturbance and reestablishment period, N deposition had no effect on NPP at W2. After 1980, the long-term effects of N removals caused NPP to be more severely N-limited at W2 than BNA, with lower

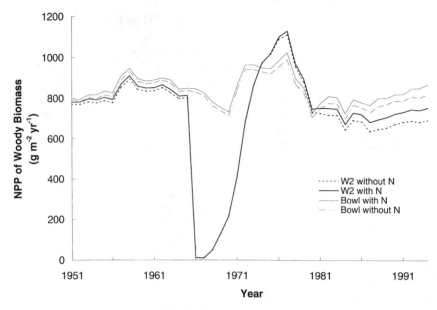

Figure 11.12. Transient PnET-CN predictions of woody NPP for two sites with different land use histories, showing the effect of N deposition on predicted NPP.

rates of NPP at W2. These results are consistent with those of Aber et al. (1997), who found from analyses using PnET-CN that while N cycling and NPP might pulse soon after disturbance, the long-term legacy of biomass and N removal is a reduction in N cycling rates. Ambient N deposition can reverse this trend, but only very slowly.

Fig. 11.13 expresses the change in NPP for each year caused by ambient vs. background N deposition for all sites. In general, increased N deposition caused small increases in NPP at all sites, with these increases due to N deposition accelerating slightly with time. Differences between sites due to disturbance histories were minor except during periods of vegetation reestablishment (e.g., W2 1966 to 1975 and W5 1984 to 1989). During these reestablishment periods, forest production was not N-limited so N deposition had no effect on predicted NPP.

The scale of current NPP differences due to contrasting land use histories is so small as to be undetectable using currently available field methods. Similarly, interannual variation in both absolute NPP values and differences between NPP predictions made using ambient and background N deposition scenarios is so high that neither could be detected by field measurement. Transient model analyses, however, can predict forward in time by decades or centuries to estimate the long-term legacies of land use history and atmospheric deposition. The transient modelling approach is especially useful for cases such as this one, in which

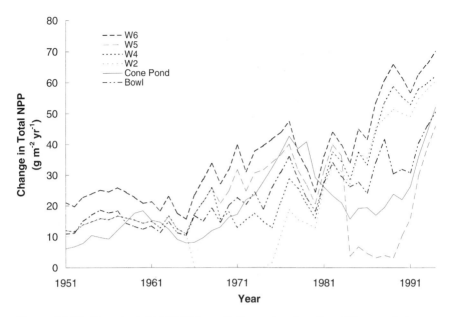

Figure 11.13. Transient PnET-CN predictions showing the difference between NPP predictions made using ambient and background levels of N deposition for six sites with different land use histories.

the temporal scales of processes such as interannual climate variability, disturbance, vegetation reestablishment, changes in C/N ratios in plant and soil compartments, and ambient N deposition overlap to produce complex patterns of C and N cycling rates.

Conclusions

PnET-II and TEM 4.0 made remarkably similar predictions of NPP under both contemporary and changed climate scenarios. That both models agreed so closely, despite their different structures and biases regarding the influence of climate variables on NPP, suggests that confidence in their predictions of forest NPP in this region should be increased, as suggested by Rastetter (1996). However, predicted responses to climate change and CO_2 enhancement varied substantially among GCM scenarios. This result suggests that further research should be directed at improving accuracy of regional-scale climate projections.

Both models predicted very large increases in forest NPP for this region under climate change and enhanced CO_2. While this is an encouraging result for policymakers attempting to balance the global C budget in the face of rising emissions, these large NPP responses are not likely to be realized because of limitations imposed by other factors. Past land management history has depleted many sites of nutrients, such as N. In these areas, nutrient limitations will reduce potential NPP in the face of climate change. In other areas, pollutants, such as acid rain or tropospheric ozone, may impose further nutrient limitations or reduce photosynthetic potential. Finally, increased variability in precipitation is likely to induce water stress at many sites, which would further reduce forest NPP.

References

Aber JD (1997) Why don't we believe the models? Bull Ecol Soc Am 78(3): 232–233.

Aber JD, Driscoll CT (1997) Effects of land use, climate variation and N deposition on N cycling and C storage in northern hardwood forests. Global Biogeochem Cycl 11:639–648.

Aber JD, Ollinger SV, Federer CA, Reich PB, Goulden ML, Kicklighter DW, Melillo JM, Lathrop RG Jr. (1995) Predicting the effects of climate change on water yield and forest production in the northeastern US. Clim Res 5:207–222.

Aber JD, Driscoll C, Federer CA, Lathrop R, Lovett G, Steudler P, Vogelmann J (1993a) A strategy for the regional analysis of the effects of physical and chemical climate change on biogeochemical cycles in northeastern (U.S.) forests. Ecol Model 67:37–47.

Aber JD, Federer CA (1992) A generalized, lumped-parameter model of photosynthesis, evapotranspiration and net primary production in temperate and boreal forest ecosystems. Oecologia 92:463–474.

Aber JD, Magill A, Boone R, Melillo JM, Steudler P, Bowden R (1993b) Plant and soil responses to chronic nitrogen additions at the Harvard Forest, Massachusetts. Ecol Applic 3:156–166.

Aber JD, Melillo JM, Nadelhoffer NJ, Pastor J, Boone RD (1991) Factors controlling nitrogen cycling and nitrogen saturation in northern temperate forest ecosystems. Ecol Applic 1:303–315.

Aber JD, Nadelhoffer KJ, Steudler P, Melillo JM (1989) Nitrogen saturation in northern forest ecosystems. BioScience 39:378–386.

Aber JD, Ollinger SV, Driscoll CT (1997) Modeling nitrogen saturation in forest ecosystems in response to land use and atmospheric deposition. Ecol Model 101:61–78.

Aber JD, Reich PB, Goulden ML (1996) Extrapolating CO_2 exchange to the canopy: a generalized model of photosynthesis validated by eddy correlation. Oecologia 106:257–265.

Amthor JS (1991) Respiration in a future, higher CO_2 world. Plant Cell Environ 14:13–20.

Bailey SW, Driscoll CT, Hornbeck JW (1995) Acid–base chemistry and aluminum transport in an acidic watershed pond in New Hampshire. Biogeochem 28: 69–91.

Bailey SW, Hornbeck JW, Driscoll CT, Gaudette HE (1996) Calcium imports and transport in a base-poor forest ecosystem as interpreted by Sr isotopes. Water Resour Res 32:707–719.

Bazzaz FA (1990) The response of natural ecosystems to the rising global CO_2 levels. Ann Rev Ecol Syst 21:167–176.

Bazzaz FA, Bassow SL, Berntson GM, Thomas SC (1996) Elevated CO_2 and terrestrial vegetation: Implications for and beyond the global carbon budget. In: Walker B, Steffen W (eds) *Global Change and Terrestrial Ecosystems*. Cambridge University Press, Cambridge, United Kingdom, pp 43–76.

Bazzaz FA, Fajer ED (1992) Plant life in a CO_2-rich world. Sci Am (Jan):68–74.

Beltz RC, Cost ND, Kingsley NP, Peters JR (1992) *Timber Volume Distribution Maps for the Eastern United States*. United States Department of Agriculture (USDA) Forest Service, Washington, DC, Gen Tech Rep WO-60.

Bolker BM, Pacala SW, Bazzaz FA, Canham CD, Levin SA (1995) Species diversity and ecosystem response to carbon fertilization: conclusions from a temperate forest model. Global Change Biol 1:373–381.

Bormann FH, Likens GE (1979) *Pattern and Process in a Forested Ecosystem*. Springer-Verlag, New York.

Butler TJ, Likens GE (1991) The impact of changing regional emissions on precipitation chemistry in the eastern United States. Atmos Environ 25A:305–315.

Churkina G, Running SW, Schloss AL, the Participants of Potsdam NPP Model Intercomparison (1999) Comparing global models of terrestrial net primary productivity (NPP): the importance of water availability. Global Change Biol 5(Suppl 1):46–55.

Compton J, Boone R, Motzkin G, Foster D (1998) Soil carbon and nitrogen in a pine–oak sand plain in central Massachusetts: role of vegetation and land-use history. Oecologia 116:536–542.

Cooter EJ, Eder BK, LeDuc SK, Truppi L (1993) *General Circulation Model Output for Forest Climate Change Research and Applications*. United States Department of Agriculture (USDA) Forest Service, Gen Tech Rep SE-85.

Costanza R, d'Arge R, de Groot R, Farber S, Grasso M, Hannon B, Limburg K, Naeem S, O'Neill RV, Paruelo J, Raskin RG, Sutton P, van den Belt M (1997) The value of the world's ecosystem services and natural capital. Nature 387:253–260.

Cotrufo MF, Ineson P (1996) Elevated CO_2 reduces field decomposition rates of *Betula pendula* (Roth.) leaf litter. Oecologia 106:525–530.

Cramer W, Kicklighter DW, Bondeau A, Moore III B, Churkina G, Nemry B, Ruimy A, Schloss A, the Participants of the Potsdam NPP Model Intercomparison (1999) Comparing global models of terrestrial net primary productivity (NPP): overview and key results. Global Change Biol 5(Suppl 1):1–15.

Cronon W (1983) *Changes in the Land: Indians, Colonists, and the Ecology of New England*. Hill and Wang, New York.

Daily GC, Alexander S, Ehrlich PR, Goulder L, Lubchenko J, Matson PA, Mooney HA, Postel S, Schneider SH, Tilman D, Woodwell GM (1997) Ecosystem services: benefits supplied to human societies by natural ecosystems. Issues Ecol 2:1–16.

Daly C, Neilson RP, Phillips DL (1994) A statistical-topographical model for mapping climatological precipitation over mountainous terrain. J Appl Meteorol 33:140–158.

DeLucia EH, Hamilton JG, Naidu SL, Thomas RB, Andrews JA, Finzi A, Lavine M, Matamala R, Mohan JE, Hendrey GH, Schlesinger WH (1999) Net primary production of a forest ecosystem with experimental CO_2 enrichment. Science 284:1177–1179.

Eamus D (1991) The interaction of rising CO_2 and temperatures with water use efficiency. Plant Cell Environ 14:843–852.

Eamus D, Jarvis PG (1989) The direct effects of increase in the global atmospheric CO_2 concentration on natural and commercial temperate trees and forests. Adv Ecol Res 19:1–55.

Ellsworth DS, Oren R, Huang C, Phillips N, Hendrey GR (1995) Leaf and canopy responses to elevated CO_2 in a pine forest under free-air CO_2 enrichment. Oecologia 104:139–146.

Field C, Mooney HA (1986) The photosynthesis–nitrogen relationships in wild plants. In: Givnish T (ed) *On the Economy of Plant Form and Function.* Cambridge University Press, Cambridge, United Kingdom, pp 25–55.

Field CB, Jackson RB, Mooney HA (1995) Stomatal responses to increased CO_2: implications from the plant to the global scale. Plant Cell Environ 18: 1214–1225.

Gates WL (1985) The use of general circulation models in the analysis of the ecosystem impacts of climatic change. Clim Change 7:267–284.

Ghan SJ (1992) The GCM credibility gap. Clim Change 21:345–346.

Glassy JM, Running SW (1994) Validating diurnal climatology logic of the MTCLIM model across a climatic gradient in Oregon. Ecol Applic 4:248–257.

Goodale CL, Aber JD, Farrell EP (1998) Predicting the relative sensitivity of forest production in Ireland to site quality and climate change. Clim Res 10:51–67.

Graham RL, Turner MG, Dale VH (1990) How increasing CO_2 and climate change affect forests. BioScience 40:575–587.

Hollinger DY (1987) Gas exchange and dry matter allocation responses to elevation of atmospheric CO_2 concentration in seedlings of three tree species. Tree Physiol 3:193–202.

Houghton JT, Meira Filho LG, Callander BA, Harris N, Kattenberg A, Maskell K (eds) IPPC (Intergovernmental Panel on Climate Change) (1996) *Climate Change 1995: The Science of Climate Change.* Cambridge University Press, Cambridge, United Kingdom.

Hulme M (1995) *A Historical Monthly Precipitation Dataset for Global Land Areas from 1900 to 1994, Gridded at 3.75 × 2.5 Resolution.* Constructed at the Climate Research Unit, University of East Anglia, Norwich, United Kingdom.

Idso KE, Idso SB (1994) Plant responses to atmospheric CO_2 enrichment in the face of environmental constraints: a review of the last 10 years' research. Agric For Meteorol 69:153–203.

Jarvis PG, McNaughton KG (1986) Stomatal control of transpiration: scaling up from leaf to region. Adv Ecol Res 15:1–49.

Jenkins JC, Kicklighter DW, Ollinger SV, Aber JD, Melillo JM (1999) Sources of variability in NPP predictions at the regional scale: a comparison using PnET-II and TEM 4.0 in northeastern U.S. forests. Ecosystems 2(6):480–496.

Jones PD, Raper SDB, Cherry BSG, Goodess CM, Wigley TML, Santer B, Kelly PM, Bradley RS, Diaz HF (1991) An updated global grid point surface air temperature anomaly data set: 1951–1990. ORNL/CDIAC-37, NDP-020/R1, Oak Ridge, TN.

Kattenberg A, Giorgi F, Grassl H, Meehl GA, Mitchell JFB, Stouffer RJ, Tokioka T, Weaver AJ, Wigley TML (1996) Climate models: Projections of future climate. In: Houghton JT, Meira Filho LG, Callander BA, Harris N, Kattenberg A, Maskell K (eds) IPCC (Intergovernmental Panel on Climate Change) *Climate Change 1995: The Science of Climate Change.* Cambridge University Press, Cambridge, United Kingdom, pp 285–358.

Kicklighter DW, Bruno M, Dönges S, Esser G, Heimann M, Helfrich J, Ift F, Joos F, Kaduk J, Kohlmaier GH, McGuire AD, Melillo JM, Meyer R, Moore B, Nadler A, Prentice IC, Sauf W, Schloss A, Sitch S, Wittenberg U, Würth G (1999) A first-order analysis of the potential role of CO_2 fertilization to affect the global carbon budget: an intercomparison study of four terrestrial biosphere models. Tellus 51B:343–366.

Kicklighter DW, Melillo JM, Peterjohn WJ, Rastetter EB, McGuire AD, Steudler PA (1994) Aspects of spatial and temporal aggregation in estimating regional carbon dioxide fluxes from temperate forest soils. J Geophys Res 99:1303–1315.

Kittel TGF, Rosenbloom NA, Painter TH, Schimel DS, VEMAP (Vegetation/ Ecosystem Modelling and Analysis Project) Modeling Participants (1995) The VEMAP integrated database for modeling United States ecosystem/vegetation sensitivity to climate change. J Biogeogr 22:857–862.

Koch GW, Mooney HA (1996) *Carbon Dioxide and Terrestrial Ecosystems.* Academic Press, San Diego, CA.

Körner C (1996) The response of complex multispecies systems to elevated CO_2. In: Walker B, Steffen W (eds) *Global Change and Terrestrial Ecosystems.* Cambridge University Press, Cambridge, United Kingdom, pp 20–42.

Lathrop RG, Bognar JA (1994) Development and validation of AVHRR-derived regional land cover data for the northeastern US region. Int J Remote Sens 15:2695–2702.

Leak WB (1974) *Some Effects of Forest Preservation.* Res Note NE-186. United States Department of Agriculture (USDA) Forest Service, Northeastern Forest Experiment Station, Radnor, PA.

Likens GE (1992) *The Ecosystem Approach: Its Use and Abuse.* The Ecology Institute, Oldendorf/Luhe, Germany.

Likens GE, Bormann FH (1995) *Biogeochemistry of a Forested Ecosystem.* 2nd ed. Springer-Verlag, New York.

Likens GE, Driscoll CT, Buso DC (1996) Long-term effects of acid rain: response and recovery of a forest ecosystem. Science 272:244–246.

Lovett GM (1994) Atmospheric deposition of nutrients and pollutants in North America: an ecological perspective. Ecol Applic 4:629–950.

Lüscher A, Hendrey GR, Nösberger J (1998) Long-term responsiveness to free air CO_2 enrichment of functional types, species, and genotypes of plants from fertile permanent grassland. Oecologia 113:37–45.

Lüscher A, Nösberger J (1997) Interspecific and intraspecific variability in the response of grasses and legumes to free air CO_2 enrichment. Acta Oecologia 18(3):269–275.

Magill AH, Aber JD, Hendricks JJ, Bowden RD, Melillo JM, Steudler PA (1997) Biogeochemical response of forest ecosystems to simulated chronic nitrogen deposition. Ecol Applic 7:402–415.

Magill AH, Downs MR, Nadelhoffer KJ, Hallett RA, Aber JD (1996) Forest ecosystem response to four years of chronic nitrate and sulfate additions at Bear Brooks Watershed, Maine, USA. For Ecol Manage 84:29–37.

Manabe S, Wetherald RT, Mitchell JFB, Meleshko V, Tokioka T (1990) Equilibrium climate change—and its implications for the future. In: Houghton JT, Jenkins GJ, Ephraums JJ (eds) IPCC (Intergovernmental Panel on Climate Change) *Climate Change: The IPCC Scientific Assessment.* Cambridge University Press, Cambridge, United Kingdom, pp 131–172.

Marks D, Dozier J (1992) Climate and energy exchange at the snow surface in the alpine region of the Sierra Nevada 2. Snow cover energy balance. Water Resour Res 28:3043–3054.

McGuire AD, Melillo JM, Kicklighter DW, Pan Y, Xiao X, Helfrich J, Moore B III, Vorosmarty CJ, Schloss AL (1997) Equilibrium responses of global net primary production and carbon storage to doubled atmospheric carbon dioxide: sensitivity to changes in vegetation nitrogen concentration. Global Biogeochem Cycl 11:173–189.

McGuire AD, Melillo JM, Joyce LA, Kicklighter DW, Grace AL, Moore B III, Vorosmarty CV (1992) Interactions between carbon and nitrogen dynamics in estimating net primary productivity for potential vegetation in North America. Global Biogeochem Cycl 6:101–124.

McGuire AD, Joyce LA, Kicklighter DW, Melillo JM, Esser G, Vorosmarty CJ (1993) Productivity response of climax temperate forests to elevated temperature and carbon dioxide: a North American comparison between two global models. Clim Change 24:287–310.

McGuire AD, Melillo JM, Kicklighter DW, Joyce LA (1995) Equilibrium responses of soil carbon to climate change: empirical and process-based estimates. J Biogeogr 22:785–796.

McNulty SG, Vose JM, Swank WT (1996) Loblolly pine hydrology and productivity across the southern United States. For Ecol Manage 86:241–251.

Melillo JM, McGuire AD, Kicklighter DW, Moore BIII, Vorosmarty CJ, Schloss AL (1993) Global climate change and terrestrial net primary production. Nature 363:234–239.

Melillo JM, Prentice IC, Farquhar GD, Schulze E-D, Sala OE (1996) Terrestrial biotic responses to environmental change and feedbacks to climate. In: Houghton JT, Meira Filho LG, Callander BA, Harris N, Kattenberg A, Maskell K (eds) IPCC (Intergovernmental Panel on Climate Change) Climate Change 1995: The Science of Climate Change. Cambridge University Press, Cambridge, United Kingdom, pp 445–482.

Mitchell JFB, Johns TC, Gregory JM, Tett SFB (1995) Climate response to increasing levels of greenhouse gases and sulphate aerosols. Nature 372:501–504.

Mitchell MJ, Driscoll CT, Kahl JS, Likens GE, Murdoch PS, Pardo LH (1996) Climatic control of nitrate loss from forested watersheds in the northeastern United States. Environ Sci Tech 30:2609–2612.

Mooney HA (1996) Ecosystem physiology: overview and synthesis. In: Walker B, Steffen W (eds) Global Change and Terrestrial Ecosystems. Cambridge University Press, Cambridge, United Kingdom, pp 13–19.

Mooney HA, Drake BG, Luxmoore RJ, Oechel WC, Pitelka LF (1991) Prediction of ecosystem responses to elevated CO_2 concentrations. BioScience 41:96–104.

Mousseau M, Saugier B (1992) The direct effect of increased CO_2 on gas exchange and growth of forest tree species. J Exp Bot 43:1121–1130.

Myneni RB, Keeling CD, Tucker CJ, Asrar G, Nemani RR (1997) Increased plant growth in the northern high latitudes from 1981 to 1991. Nature 386:698–702.

Nadelhoffer KJ, Emmett BA, Gunderson P, Kjonaas OJ, Koopmans CJ, Schleppi P, Tietema A, Wright RF (1999) Nitrogen deposition makes a minor contribution to carbon sequestration in temperate forests. Nature 398:145–148.

Navas M-L (1998) Individual species performance and response of multi-species communities to elevated CO_2: a review. Func Ecol 12:721–727.

Neftel A, Moor E, Oeschger H, Stauffer B (1985) Evidence from polar ice cores for the increase in atmospheric CO_2 in the past two centuries. Nature 315:45–47.

Nicholls N, Gruza GV, Jouzel J, Karl TR, Ogallo LA, Parker DE (1996) Observed climate variability and change. In: Houghton JT, Meira Filho LG, Callander BA, Harris N, Kattenberg A, Maskell K (eds) IPCC (Intergovernmental Panel on Climate Change) Climate Change 1995: The Science of Climate Change. Cambridge University Press, Cambridge, United Kingdom, pp 135–192.

Ollinger SV, Aber JD, Federer CA (1998) Estimating regional forest productivity and water yield using an ecosystem model linked to a GIS. Landsc Ecol 13:323–334.

Ollinger SV, Aber JD, Federer CA, Lovett GM, Ellis JM (1995) *Modeling Physical and Chemical Climate of the Northeastern United States for a Geographic Information System.* Gen Tech Rep NE-191. United States Department of Agriculture (USDA) Forest Service, Northeastern Forest Experiment Station, Radnor, PA.

Ollinger SV, Aber JD, Lovett GM, Millham SE, Lathrop RG, Ellis JM (1993) A spatial model of atmospheric deposition for the northeastern US. Ecol Applic 3:459–472.

Ollinger SV, Aber JD, Reich P (1997) Simulating ozone effects on forest productivity: interactions among leaf-, canopy-, and stand-level processes. Ecol Applic 7:1237–1251.

Oreskes N, Shrader-Frechette K, Belitz K (1994) Verification, validation, and confirmation of numerical models in the earth sciences. Science 263: 641–646.

Pan Y, Melillo JM, McGuire AD, Kicklighter DW, Pitelka LF, Hibbard K, Pierce LL, Running SW, Ojima DS, Parton WJ, Schimel DS and other VEMAP (Vegetative/Ecosystem Modeling and Analysis Project) Members (1998) Modeled responses of terrestrial ecosystems to elevated atmospheric CO_2: a comparison of simulations by the biogeochemistry models of the Vegetation/Ecosystem Modeling and Analysis Project (VEMAP). Oecologia 114:389–404.

Pan Y, McGuire AD, Kicklighter DW, Melillo JM (1996) The importance of climate and soils for estimates of net primary production: a sensitivity analysis with the terrestrial ecosystem model. Global Change Biol 2:5–23.

Pastor J, Aber JD, McClaugherty CA, Melillo JM (1984) Aboveground production and N and P cycling along a nitrogen mineralization gradient on Blackhawk Island, Wisconsin. Ecology 65:256–268.

Pastor J, Post WM (1993) Linear regressions do not predict the transient responses of eastern North American forests to CO_2-induced climate change. Clim Change 23:111–119.

Pastor J, Post WM (1988) Response of northern forests to CO_2-induced climate change. Nature 334:55–58.

Raich JW, Rastetter EB, Melillo JM, Kicklighter DW, Steudler PA, Peterson BJ (1991) Potential net primary productivity in South America: application of a global model. Ecol Applic 1:399–429.

Randlett DL, Zak DR, Pregitzer KS, Curtis PS (1996) Elevated atmospheric carbon dioxide and leaf litter chemistry: influences on microbial respiration and net nitrogen mineralization. Soil Sci Soc Am J 60:1571–1577.

Rastetter EB (1996) Validating models of ecosystem response to global change. BioScience 46(3):190–198.

Rastetter EB, Ryan MG, Shaver GR, Melillo JM, Nadelhoffer KJ, Hobbie JE, Aber JA (1991) A general biogeochemical model describing the responses of the C and N cycles in terrestrial ecosystems to changes in CO_2, climate, and N deposition. Tree Physiol 9:101–126.

Rastetter E, McKane R, Shaver G, Melillo J (1992) Changes in C storage by terrestrial ecosystems: How C–N interactions restrict responses to CO_2 and temperature. Water Air Soil Pollut 64(1–2):327–344.

Reich PB, Grigal DF, Aber JD, Gower ST (1997) Nitrogen mineralization and productivity in 50 temperate conifer and hardwood stands on diverse soils. Ecology 78:335–347.

Reich PB, Kloeppel B, Ellsworth DS, Walters MB (1995) Different photosynthesis–nitrogen relations in deciduous and evergreen coniferous tree species. Oecologia 104:24–30.

Ruimy A, Kergoat L, Bondeau A, the Participants of Potsdam NPP Model Intercomparison (1999) Comparing global models of terrestrial net primary productivity (NPP): analysis of differences in light absorption and light-use efficiency. Global Change Biol 5(Suppl 1):56–64.

Running SW, Nemani RR, Hungerford RD (1987) Extrapolation of synoptic meteorological data in mountainous terrain and its use for simulating forest evapotranspiration and photosynthesis. Can J For Res 17: 472–483.

Sage RF (1994) Acclimation of photosynthesis to increasing atmospheric CO_2: the gas exchange perspective. Photosynth Res 39:351–368.

Sage RF, Sharkey TD, Seemann JR (1989) Acclimation of photosynthesis to elevated CO_2 in five C_3 species. Plant Physiol 89:590–596.

Schimel DS, Alves D, Enting I, Heimann M (1996a) Radiative forcing of climate change. In: Houghton JT, Meira Filho LG, Callander BA, Harris N, Kattenberg A, Maskell K (eds) IPCC (Intergovernmental Panel on Climate Change) *Climate Change 1995: The Science of Climate Change.* Cambridge University Press, Cambridge, United Kingdom, pp 65–131.

Schimel DS, Braswell BH, McKeown R, Ojima DS, Parton WJ, Pulliam W (1996b) Climate and nitrogen controls on the geography and timescales of terrestrial biogeochemical cycling. Global Biogeochem Cycl 10: 677–692.

Schimel DS, VEMAP Participants, Braswell BH (1997) Continental scale variability in ecosystem processes: Models, data, and the role of disturbance. Ecol Monogr 67:251–271.

Schlesinger ME, Zhao Z-C (1989) Seasonal climate changes induced by doubled CO_2 as simulated by the OSU atmospheric GCM-mixed layer ocean model. J Clim 2:459–495.

Schloss AL, Kicklighter DW, Kaduk J, Wittenberg U, the Participants of the Potsdam NPP Model Intercomparison (1999) Comparing global models of terrestrial net primary productivity (NPP): comparison of NPP to climate and the normalized difference vegetation index. Global Change Biol 5(Suppl 1): 25–34.

Sinclair TR, Tanner CB, Bennett JM (1984) Water-use efficiency in crop production. BioScience 34:36–40.

Smith TM, Shugart HH (1993) The transient response of terrestrial carbon storage to a perturbed climate. Nature 361:523–526.

Strain BR (1987) Direct effects of increasing atmospheric CO_2 on plants and ecosystems. Tree 2:18–21.

Tanner CB, Sinclair TR (1983) Efficient water use in crop production: research or re-search. In: Taylor H (ed) *Limitations to Efficient Water Use in Crop Production.* American Society of Agronomy, Madison, Wisconsin, pp 1–28.

Tian H, Melillo JM, Kicklighter DW, McGuire AD, Helfrich J (1999) The sensitivity of terrestrial carbon storage to historical climate variability and atmospheric CO_2 in the United States. Tellus 51B:414–452.

Townsend AR, Braswell BH, Holland EA, Penner JE (1996) Spatial and temporal patterns in terrestrial carbon storage due to deposition of fossil fuel nitrogen. Ecol Applic 6:806–814.

VEMAP (Vegetation/Ecosystem Modeling and Analysis Project) Members (1995) Vegetation/Ecosystem Modeling and Analysis Project: comparing biogeography and biogeochemistry models in a continental-scale study of terrestrial ecosystem responses to climate change and CO_2 doubling. Global Biogeochem Cycl 9:407–437.

Vitousek PM (1992) Global environmental change: an introduction. Annu Rev Ecol Syst 23:1–14.

Vitousek PM, Ehrlich PR, Ehrlich AE, Matson PA (1986) Human appropriation of the products of photosynthesis. BioScience 36:368–373.

Vitousek PM, Aber J, Bayley SE, Howarth RW, Likens GE, Matson PA, Schindler DW, Schlesinger WH, Tilman GD (1997) Human alteration of the global nitrogen cycle: causes and consequences. Issues Ecol 1:1–15.

Wetherald RT, Manabe S (1988) Cloud feedback processes in a general circulation model. J Atmos Sci 45:1397–1415.

Wetherald RT, Manabe S, Cubasch U, Cess RD (1990) Processes and modelling. In: Houghton JT, Jenkins GJ, Ephraums JJ (eds) IPCC (Intergovernmental Panel on Climate Change) *Climate Change: The IPCC Scientific Assessment*. Cambridge University Press, Cambridge, United Kingdom, pp 69–91.

Whittaker RH, Bormann FH, Likens GE, Siccama TG (1974) The Hubbard Brook Ecosystem Study: forest biomass and production. Ecol Monogr 44(2):233–254.

Wilsey BJ (1996) Plant responses to elevated CO_2 among terrestrial biomes. Oikos 76:201–206.

Wilson CA, Mitchell JFB (1987) A doubled CO_2 climate sensitivity experiment with a global climate model including a simple ocean. J Geophys Res 92(D11):13315–13343.

Wood CW, Torbert HA, Rogers HH, Runion GB, Prior SA (1994) Free-air CO_2 enrichment effects on soil carbon and nitrogen. Agric For Meteorol 70:103–116.

Wullschleger SD (1993) Biochemical limitations to carbon assimilation in C_3 plants—a retrospective analysis of the A/C_i curves from 109 species. J Exp Bot 44:907–920.

Xiao X, Kicklighter DW, Melillo JM, McGuire AD, Stone PH, Sokolov AP (1997) Linking a global terrestrial biogeochemical model and a 2-D climate model: implications for the global carbon budget. Tellus Ser B 49:18–37.

Zak DR, Pregitzer KS, Teeri JA, Fogel R, Randlett DL (1993) Elevated atmospheric CO_2 and feedback between carbon and nitrogen cycles. Plant Soil 151:105–117.

12. Regional Impacts of Ozone on Forest Productivity

John A. Laurence, Scott V. Ollinger,
and Peter B. Woodbury

Coupled to potential global climate change are all the other stresses that currently affect forest health and productivity. The impacts of disease, insect infestation, nutrient limitation, drought, and atmospheric deposition are dealt with in detail in other chapters of this book. In this chapter, we examine the impact of ozone on forest productivity and the uncertainties of modeling methodologies. The common thread of our discussion is the use of predictive models to provide both scientific insight to the potential response of forests to multiple stresses and information that may be useful in the development of sound management practices and environmental policy.

The stresses that could be addressed in this chapter range from those that occur naturally, such as disease, insects, and fire, to those that are distinctly the result of human actions. We have chosen to compare two modeling methodologies that assess ozone impact on forests. We conclude with an illustration of uncertainty analysis as applied to forest modeling. A preliminary note to readers: this chapter is essentially an amalgam of our individual work, thus at the beginning of each major section, the primary author's initials appear at the end of the heading so that you may identify our interests.

Uses of Modeling in Addressing Regional Impacts

Forests are complicated systems that respond to a host of environmental factors, both biotic and abiotic, during their life cycles. In addition, since forest communities may be complex and since trees are of large stature, direct experimentation to determine the response of forests to any given stress factor is difficult and expensive. Furthermore, even if experiments are conducted at a given location, extrapolation of the results across landscapes and regions is fraught with uncertainty. To address some of the concerns, it will be necessary to perform experiments with mechanistic models to represent response at a large number of locations.

Mechanistic models are those that attempt to simulate the growth of trees or forests by addressing the fundamental mechanisms of growth rather than through a statistical analysis of empirical data. For instance, to model the growth of a tree or a stand, it is often desirable to model the process of photosynthesis and then allocate the fixed carbon to plant growth using principles of plant physiology. By using models that depend upon an understanding of the growth processes, investigators attempt to avoid the problems inherent in predicting outside the range of data used to develop a statistical model. Models based on mechanisms should be applicable across wide areas if we understand the biological and functional relationships in the models.

Statistical models may also be of great value providing the conditions that are expected to occur in the future or in a different part of the region, have occurred before and therefore can be incorporated in the model. Models developed on large data bases, for instance site index models, can used to great advantage, but will not be appropriate for predicting the response of forests under climate or pollution conditions not encountered previously.

Using models to address regional impacts of biotic or abiotic stresses also allows managers to examine "what if" scenarios, that is, they may examine the impacts of management strategies under differing stress conditions. To do so experimentally would be almost impossible, and would be too time consuming to be of value. Models may also be used to address emerging problems that may function in a similar fashion to stresses that have been observed and modeled previously. For instance, plant growth models that simulate the effects of insect injury may be appropriate to address other stresses, perhaps a fungal pathogen, which causes its effect by reducing leaf area. Thus, modeling allows investigation of emerging problems, potential problems, and the extrapolation of results across a region based on mechanisms and principles of plant science.

In the following sections, we present three examples of models applied to regional impacts of stress, and a discussion of uncertainties associated with modeling.

Estimating Regional Productivity with an
Ecosystem Model (SVO)

Description of PnET and Regional Modeling Objectives

In the New York–New England portion of the northeastern U.S., a regional forest ecosystem analysis has been underway using a series of models known collectively as PnET. The original PnET model ran at a monthly time step and was designed to predict forest carbon (C) and water balances with the aim of capturing important ecosystem processes while retaining enough simplicity to be run on relatively few data inputs (Aber and Federer, 1992). This was accomplished, in part, through the use of functionally based scaling algorithms that describe complex processes in relatively simple terms. A core equation in the model uses the nitrogen (N) content of foliage to determine maximum leaf photosynthetic rates (A_{max}). This follows years of physiological research that revealed a strong, linear relationship between the two variables across species from diverse ecosystems (Field and Mooney, 1986; Reich et al., 1995). In the model, the relationship is applied to a simulated forest canopy in which available light and specific leaf weight decline with canopy depth. Photosynthesis is calculated in a numerical integration over the multilayered canopy to account for the effects of gradual light extinction on total C gain.

Within the water balance routine, determination of canopy transpiration was simplified by relating stomatal conductance to the realized rate of net photosynthesis. This makes plant water use efficiency a function of the atmospheric vapor pressure deficit (Sinclair et al., 1984; Baldocci et al., 1987) and provides a functional link between forest C and water fluxes. Actual evapotranspiration and moisture stress are calculated as functions of plant water demand and available soil water, which is determined using equations from the BROOK model (Federer and Lash, 1978). If plant demand exceeds available soil water, drought stress ensues and canopy photosynthesis is reduced.

More recently, PnET has been expanded into a series of nested models which share these core relationships, but differ in the number of processes simulated and hence the number of input parameters required. PnET-II is similar in structure to the original PnET model, but has improved algorithms for photosynthesis and canopy structure (Aber et al., 1995). PnET-II outputs include net primary production (NPP), total ecosystem C balances, and water runoff. PnET-Day is a daily timestep version of the canopy photosynthesis routine that was designed to allow comparison against eddy correlation CO_2 flux data (Aber et al., 1996). It does not calculate site water balances and requires less input than PnET-II. PnET-CN includes all of the processes in PnET-II but adds live biomass, soil organic matter, complete cycles for C and N, and includes feedbacks between litter quality, decomposition, soil C and N fluxes, and productivity (Aber et al., 1996). With its added complexity, PnET-CN requires the greatest number of input parameters, including estimates of historical disturbance, which are necessary for predicting current soil C and N dynamics.

Although the PnET models vary in complexity and input data requirements, they are all designed to be applied without calibration. Model calibration involves setting input parameters by adjusting their values until model output comes into agreement with field observations. Although this can be a useful step when input parameters are difficult to measure directly, it can also increase the probability of obtaining accurate results through compensating errors. Because the failure of a model can provide valuable insight into the behavior of ecological systems and our understanding of them, any step that artificially reduces the chances that a model will fail also reduces the potential benefits of the model.

Regional Data Base

A geographic information system (GIS) database has been developed to facilitate regional assessments using PnET-II. The minimum data requirements of PnET-II are monthly averages of maximum and minimum daily temperature, solar radiation and vapor pressure deficit, total monthly precipitation, soil water-holding capacity (WHC), and vegetation type. Incorporating the effects of atmospheric deposition into analyses with PnET-CN will make chemical inputs an additional data requirement.

A 1 km digital elevation model (DEM) was obtained from the United States Geological Survey (USGS) and a regional land use/land cover (LULC) map was derived by Lathrop and Bognar (1994) using Advanced Very High Resolution Radiometer (AVHRR) satellite data. The map identifies forest vegetation as deciduous, spruce–fir, mixed deciduous/spruce–fir and mixed deciduous/pine, as well as a number of nonforest land cover types.

Climate data layers were derived from a statistical model developed from long-term (30-year) weather station data (Ollinger et al., 1995). Monthly temperature and precipitation were predicted using regression equations as functions of geographical position and elevation. The regressions describe regional, elevational, and temporal trends, but do not account for local factors, such as lake-effect precipitation and nighttime temperature inversions. Atmospheric vapor pressure deficit was determined as a function of daily minimum temperatures, assuming that nighttime air temperatures do not fall below the point at which dew formation begins. This assumption is generally valid in humid climates and was found to hold true for all but the highest elevations within the northeastern U.S. (Ollinger et al., 1995). Uncertainties in all climate predictions are greatest at higher elevations due to the small number of high-elevation weather stations in the region.

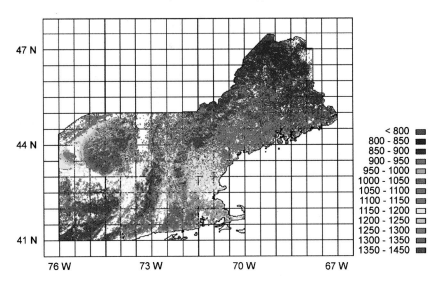

Figure 12.1. PnET-II estimated net primary production for the northeast study region. Map locations without values are nonforested.

Monthly solar radiation data planes were derived by combining potential radiation estimates with the ratio of actual to potential radiation measured at 11 field stations. This ratio was determined monthly and served as an indicator of atmospheric transmittance. A ratio of 0.5 would indicate that radiation received at the surface is half of what would be received in the absence of the atmosphere. Potential radiation is calculated for each grid cell of the DEM as a function of day of year, latitude, slope, and aspect, using algorithms from Swift (1976). The product of this and the monthly atmospheric interference index provides mapped estimates of surface radiation. The resulting maps may overestimate radiation at high elevations because no attempt was made to account for the increased cloud cover experienced at high elevations.

A plant-available soil WHC map was derived by Lathrop et al. (1995) using data from the U.S. Soil Conservation Survey's STATSGO database (USSCS, 1991), but efforts to validate it showed poor agreement with the county-level soil survey data from which the STATSGO data were derived. This was due, in part, to an incompatibility between the intended uses of STATSGO and the requirements of PnET. The large STATSGO map units contain a range of attribute values (i.e., percentage of each map unit covered by a certain soil characteristic), whereas the model required aggregation to a single WHC value for each grid cell.

As an alternative, soil hydrology equations from Clapp and Hornberger (1978) were used to examine WHCs under a range of northeastern soil types (Ollinger et al., 1998). Under the range of soil textures typically encountered in northeastern till soils, texture had a minimal effect on WHC as compared with rooting depth and stone content. Asssuming a rooting depth of 1 m and 25% large fragments, most well-drained till soils produced a plant-available WHC of 120 mm. In the absence of a well-validated WHC map, this has been used as a regional mean value.

Regional Productivity—Estimates and Uncertainties

PnET-II has been used to estimate regional forest production, total ecosystem C balances and water yield, and to make initial assessments of climate change effects (Aber et al., 1995; Ollinger et al., 1998). For regional productivity, the model was run for each pixel of the regional GIS, using climate inputs from the statistical climate model and vegetation type from the AVHRR-based land cover map. Mixed pixels were run twice, once for deciduous and once for coniferous forest (pine or spruce–fir), and the final output value was determined as a weighted average of the two, assuming a 40:60 mix for deciduous/pine and a 60:40 mix for deciduous/spruce–fir. These values were determined by comparing the LULC map with growing stock data from the United States Department of Agriculture (USDA) Forest Service Forest Inventory and Analysis (FIA) program (Kingsley, 1985).

Vegetation-specific inputs required by PnET-II include leaf traits, such as foliar N, specific leaf weight, and leaf retention time, along with growing degree day (GDD) drivers, which control the timing of leaf growth and senescence. In the absence of spatial data planes for physiological leaf traits, generalized input parameter sets were compiled from published values for the forest categories in the LULC map (Aber et al., 1995). Site-level sensitivity analyses have shown that foliar N is the most important of these parameters because it controls the maximum attainable rate of photosynthesis. After Aber et al. (1995), foliar N was given a value of 2.2% for deciduous forests, 1.2% for pines, and 0.8% for spruce–fir.

Predicted annual NPP ranged from approximately 750 to $1450 \, \text{g m}^{-2}$ yr^{-1} with a regional mean of $1084 \, \text{g m}^{-2} \, \text{yr}^{-1}$ (see color insert Fig. 12.1). Predicted NPP was higher in deciduous than coniferous forests and generally increased from north to south, although highest growth rates were predicted for mid-elevations in the southern Catskill Mountain area. This occurred because the model showed substantial water limitations in lower-elevation deciduous forests. Water limitations decreased with elevation, as precipitation increased, and were less important regionally for coniferous forests. These results are consistent with an earlier analysis of drought frequency and intensity in New England (Federer, 1982).

Variation in the importance of water limitations between forest types led to variation in the factors that influenced spatial patterns of productivity. In deciduous forests, productivity was strongly correlated with precipitation and weakly correlated with annual GDDs. This resulted from the relatively high photosynthetic rates of hardwoods, which led to high rates of transpiration and soil water consumption. In spruce–fir forests, which occur only in northern and upper-elevation areas, water limitations were rare and spatial growth patterns followed annual GDDs. Pine forests were intermediate, following patterns of precipitation in southern and low-elevation areas and GDDs across colder, wetter areas.

Uncertainties in regional PnET-II predictions have been assessed using two parallel approaches; validation and sensitivity analyses. Model validation involves comparison of model predictions against data from independent field studies, providing a direct assessment of model accuracy. Comparison of PnET-II estimates (see Fig. 12.1) with data from eight locations where annual NPP has been measured showed generally good agreement with a mean absolute error of 12.5% (Ollinger et al., 1998). Poor agreement at two sites, a high-elevation balsam fir stand in northern NY and a hardwood stand in central MA, was proportional to differences between foliar N values measured at the field sites and the values used in the regional model runs. Using the measured foliar N values in the model resulted in closer agreement at both sites and reduced the overall model error to 5.2%.

Although validation is an essential step in model testing, its utility is often limited by several factors. First, because validation agreement is affected by the combined error of model algorithms, parameters, and data inputs, it can be difficult to determine exactly why a model succeeds or fails. Second, thorough evaluation of model accuracy is often limited by the availability of appropriate validation data. This is perhaps not surprising since models are often used expressly because measurements are difficult to obtain directly. In the case reported here, data from a small number of field sites gives some indication of model performance, but are inadequate for thorough evaluation of predicted spatial trends. Hence, a second method for testing regional NPP predictions has been the use of sensitivity analyses.

The spatial trends in forest productivity predicted by PnET-II (see Fig. 12.1) result from interactions within the model between climate and the growth potential of each forest type. Important sources of uncertainty in these trends are assumptions made for soil WHC and foliar N, input variables for which reliable, spatial data planes were not available. In order to test the relative importance of these assumptions, sensitivity analyses were conducted using a Monte Carlo approach. This involved performing a large number of model runs where all relevant data inputs were determined stochastically, but were limited to the probability distributions expected to occur across the study region. Variation in temperature and precipitation was introduced by randomizing the grid cell used for each model run. Foliar N varied from 1.8 to 2.6% for deciduous forests, 0.9 to 1.5% for pines and 0.7 to 1.2% for spruce–fir. Soil WHC was allowed to vary from 6 to 18 cm, ranging from sandy soils with 50% stones to clay loam soils with no stones. The model was run a total of 1000 times for each forest type and the importance of each variable was determined by the degree to which it was correlated with the resulting growth predictions.

Results of this analysis (Table 12.1) indicate the degree to which model predictions depend on variables for which regional data layers exist vs. those that require additional data development. For all forest types, the combination of precipitation, GDDs, foliar N, and soil WHC explained between 87 and 91% of the variation in predicted annual NPP. For deciduous and pine forests, precipitation was the strongest predictor, followed by WHC. This indicates that water limitations were common in both forest types, and that variation in precipitation had a greater influence than did our simulated range of soil types. For spruce–fir, water availability was relatively unimportant and most of the variation in predicted NPP was related to foliar N.

These results suggest that currently available climate data sets can account for much of the spatial variability in deciduous and pine forest productivity, but that uncertainties remain owing to the absence of WHC and foliar N data. Although spruce–fir forests showed the lowest variation

Table 12.1. Sensitivity of Predicted NPP to Climate, Foliar N, and Soil WHC for Each Forest Type as Indicated by Correlation Coefficients (r^2) from Regressions of NPP Against Each Variable Alone and in Combination. Climate Data Are Summarized as Annual Precipitation (PPT) and Annual Growing Degree Days (GDD). Foliar N (Fol N) Is Mass-based Leaf Nitrogen Concentration and Soil WHC Is Plant-available Soil Water at Field Capacity. $n = 1000$. All Variables Significant at $P < 0.05$

	r^2			Standard Error		
	Hardwood	Pine	Spruce–fir	Hardwood	Pine	Spruce–fir
PPT	0.46	0.27	NS	83	92	78
GDD	0.02	0.02	0.10	108	106	74
Fol N	0.16	0.18	0.74	101	100	40
WHC	0.27	0.25	NS	93	93	78
PPT + GDD	0.54	0.32	0.11	81	89	73
Fol N + WHC	0.40	0.38	0.74	83	84	40
ALL	0.90	0.78	0.90	35	51	25

in predicted NPP, the important control by foliar N raises large uncertainties in current predicted spatial patterns. Recent efforts to develop spatial soils data bases (Kern et al., 1995; Iverson et al., 1996) and to estimate foliar N using remote sensing data (Martin and Aber, 1997) may help both of these problems in future analyses.

Modifications for Predicting Ozone Effects on Forest Growth

PnET-II has been modified to include ozone (O_3) effects on forest production and has been applied using ambient O_3 concentration data from across the northeast region (Ollinger et al., 1997). Because PnET is designed to simulate the structure and function of mature forest canopies, it provides an opportunity for scaling up from leaf-level physiological responses observed in seedling experiments to estimating the growth response of mature forests. The approach used was to build leaf-level uptake–response relationships into the model's photosynthesis routine and allow them to interact with factors such as light attenuation, canopy O_3 gradients, and water stress. These factors are thought to be important in mature forest stands (Pye, 1988), but their combined effects cannot be estimated from seedling exposure studies alone.

Ozone fumigation experiments have shown strong relationships between cumulative ozone exposure and reductions in net photosynthesis and plant growth (e.g., Guderian et al., 1985; Reich, 1987; Pell et al., 1992). This impact varies among species, although much of the variation is related to differences in stomatal conductance (Reich, 1987; Winner, 1994). Because under a given concentration, stomatal conductance determines O_3 uptake (Taylor et al., 1992; Munger et al., 1996), this suggests that declines in photosynthesis should be better correlated with

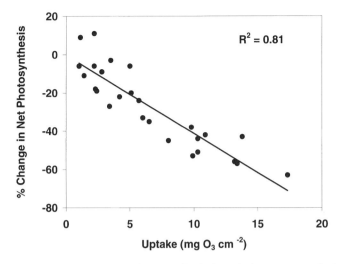

Figure 12.2. Reduction in net photosynthesis in relation to cumulative ozone uptake (from Reich, 1987) using data from independent ozone fumigation studies conducted on a variety of hardwood seedlings.

O_3 uptake than with external measures such as concentration or dose. Fig. 12.2 shows this relationship using data from a wide variety of hardwood seedlings (from Reich, 1987).

Tjoelker et al. (1995) obtained a response similar to that shown in Fig. 12.2 from mature sugar maple leaves in a partial canopy, open-air fumigation study indicating that at the leaf level, seedlings and mature trees respond similarly. Using pooled data from Fig. 12.2 and Tjoelker et al. (1995), Ollinger et al. (1997) derived the following response equation:

$$dO_3 = 1 - (2.6 \times 10^{-6} \times g \times D40) \tag{1}$$

where dO_3 is the ratio of the O_3-exposed to control photosynthetic rate, g is mean stomatal conductance (in mm s^{-1}) and D40 is the cumulative O_3 dose above a threshold concentration of 40 nmol mol^{-1}, calculated as the sum of all hourly values >40 nmol mol^{-1} after subtracting 40 from each. This threshold was used because 40 nmol mol^{-1} is approximately the level at which negative impacts appear in the data and because lower concentrations can become confused with natural, background O_3 levels.

Canopy O_3 gradients were examined using data from Munger et al. (1996). This included three years of O_3 concentrations at eight canopy positions in a mixed hardwood forest in central MA. For all months of the growing season, D40 values were calculated for each canopy level. Fig. 12.3 shows an example gradient from July 1992, during which time, O_3 below the canopy decreased to 18% of the above-canopy D40 value.

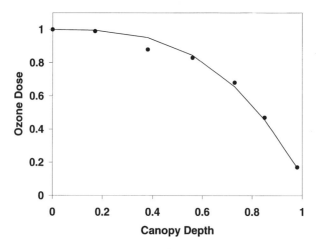

Figure 12.3. Ozone dose >40 nmol mol^{-1} in relation to canopy position at the Harvard Forest in central Massachusetts for July of 1992. Both axes have been normalized to a scale of 0 to 1 (data from Munger et al., 1996).

The shape of these curves was such that, even during months of substantial canopy depletion, O_3 levels remained high through the upper canopy. For all three years, O_3 gradients followed canopy development, increasing from spring through mid-summer and declining at the end of the growing season. These trends were described by the equation:

$$dD40_i = 1 - (i \times a)^3 \qquad (2)$$

where $dD40_i$ is the proportion of the above-canopy D40 at a given canopy level, i is the normalized canopy level from 0 at the top of the canopy to 1 at the ground, and a is an O_3 extinction coefficient, which was determined for each month. At Harvard Forest, a ranged from 0.65 to 1.05. In the model, it is calculated as a function of monthly leaf area index (LAI) following a relationship between the two observed at Harvard Forest (Ollinger et al., 1997).

For whole-forest growth predictions, these equations were built into PnET-II's photosynthesis routine for individual canopy layers. For each layer, the model tracks photosynthesis with and without O_3 in order to determine the potential, whole-canopy O_3 effect and to allow interaction with drought stress, which is calculated once for the entire canopy. Because the primary physiological response to water limitation is stomatal closure, drought stress is assumed to limit O_3 uptake and cause a reduction in the total canopy O_3 effect. This is an important assumption in that it implies no effect of O_3 on the ability of stomates to regulate water loss. If stomatal function is altered by O_3, a situation may exist in which O_3 decreases plant water use efficiency and exacerbates drought stress.

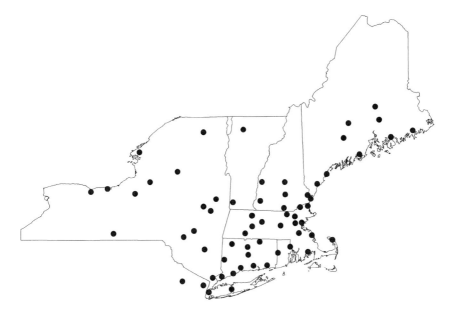

Figure 12.4. Locations of 64 EPA-sponsored ozone monitoring stations in the northeast study region.

The model was run for hardwood forests across the northeastern U.S. using ambient O_3 data obtained from the U.S. Environmental Protection Agency's (USEPA) Aerometric Information Retrieval System (AIRS) between 1987 and 1992. For each collection station, raw, hourly data were used to calculate the monthly dose above 40 nmol mol^{-1}. After screening for missing data, 64 stations were available, each containing 3 to 7 years of hourly concentrations (Fig. 12.4). Climate inputs were obtained for each site from the statistical model described above. Vegetation parameters were taken from Aber et al. (1995). To examine interactions between O_3 and drought, the model was run for each monitoring station under a range of soil WHC values, representing high to severely limited soil water retention.

Results showed decreases in predicted annual NPP of from 2 to 17% as a result of mean O_3 levels from 1987 to 1992, with greatest reductions in southern New York and New England, where O_3 levels and stomatal conductance were greatest. The predicted decrease was negatively correlated with latitude, following a trend of decreasing O_3 from south to north across the region (Fig. 12.5). Predictions varied considerably across the range of soil WHC values used (2 to 36 cm) with greater growth declines occurring on wetter soils.

Table 12.2 shows means and ranges of predicted NPP for the 64 sites under all soil WHC values used. At WHC = 36 cm, drought stress was eliminated, so these values can be used as a reference in estimating drought

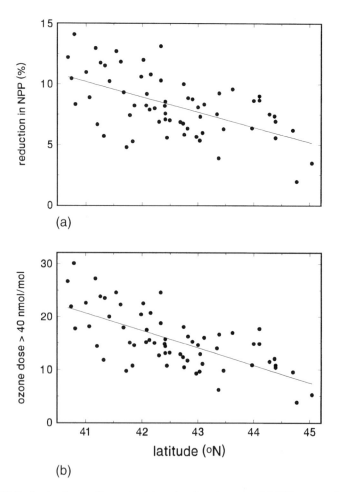

Figure 12.5. Latitude gradients in (a) reduction in annual NPP as predicted by the PnET-O_3 model and (b) mean O_3 dose >40 nmol mol^{-1} for May through October at 64 O_3 monitoring stations.

Table 12.2. Mean Predicted NPP (g m^{-2}yr^{-1}) across the 64 Study Sites With and Without O_3 Effects at 5 Levels of Soil-water Holding Capacity (cm)

Soil WHC	NPP without O_3		NPP with O_3		
	Mean	Range	Mean	Range	% Reduction
2	723	640–776	682	591–734	5.7
6	1136	991–1230	1054	915–1151	7.1
12	1354	1187–1475	1254	1190–1372	7.4
24	1732	1574–1846	1584	1411–1699	8.5
36	1840	1645–1993	1672	1527–1807	9.1

effects resulting from other moisture conditions. Under all but the two wettest conditions, drought stress limited growth to a greater extent than O_3. At WHC = 12 cm, O_3 effects on NPP were roughly 25% less than what was predicted in the absence of drought stress.

Future Assessment of Multiple Stress Interactions

Ongoing work with PnET is aimed at assessing the interactive effects of O_3, N deposition, CO_2, and climate change on regional C, N, and water cycles. Initial climate change effects have been investigated using PnET-II by Aber et al. (1995) and Jenkins et al. (see Chapter 11), and PnET-CN has been used to predict equilibrium N cycling rates across the region (Aber et al., 1997). For complete regional analyses that include multiple stress interactions, the effects of elevated CO_2 and O_3 will be included into PnET-CN. Because foliar chemistry and litter C/N ratios are not fixed in PnET-CN, but vary from year to year in response to interactions between climate, N availability, and plant demand, this will allow assessment of the ecosystem-level effects of these stressors. This is important since indirect ecosystem-level effects (e.g., effects of altered C/N ratios on soil N availability) may be as important as direct physiological effects.

Using Linked Simulation Models to Estimate Response of Forest Stands to Environmental Stress (TREGRO-ZELIG) (JAL)

At issue in estimating the impact of stress factors on the health and productivity of forests is how the interaction of one or more stresses may affect not only individuals or populations of the same species, but also mixtures of species and the composition of forest stands. In the previous section, the model used forest types (evergreen/deciduous mix), but did not address specific forest species composition. Models that simulate the development of forest stands have been developed and used for a variety of purposes, including estimating the impact of air pollutants on the growth of a mixed hardwood stand (West et al., 1980). However, the response to air pollutants, or any other stress, that is incorporated in the stand composition model is not based on the specific physiological response of the individuals in the stand, but rather on the relative tolerance to a stress, for instance drought stress, of particular species. On the other hand, detailed physiological models (Weinstein et al., 1991; Host and Isebrands, 1994; Weinstein and Yanai, 1994) that simulate the growth of individual trees under actual weather conditions and in the presence of stresses (Laurence et al., 1993; Retzlaff et al., 1997; Constable et al., 1997) address the growth of a single tree, but not of the stand, nor the composition of the stand as affected by altered competitive ability.

One way to more completely address issues of health and productivity on a larger scale is to couple models to provide the detailed physiological response to stress needed at the species level with the function, competition, and composition at the forest level. Recently, such a coupling has been accomplished (Hogsett et al., 1996) between TREGRO, a detailed physiological model used to simulate the growth of both coniferous and deciduous trees (Weinstein et al., 1991; Weinstein and Yanai, 1994), and ZELIG (Urban, 1990), a gap-succession forest stand model used to study stand development and composition. Results of the coupled model are then incorporated into a geographical information system (GIS) to provide spatial extrapolation of the simulations, allowing estimations of forest response to stress over large regions. In this section, an example of the linked-model methodology using red oak (*Quercus rubra* L.) and sugar maple (*Acer saccharum* Marsh.) is presented.

Model Descriptions

TREGRO is a physiologically-based model of a single seedling, sapling, or mature tree. TREGRO models the acquisition of C, water, and nutrients, and allocates C among competing plant parts depending upon resource availability and phenology. TREGRO operates on an hourly time step and uses real or simulated meteorological and O_3 data as an input. In addition to modeling tree growth, TREGRO has been used to asses the effects of O_3 (Laurence et al., 1993; Retzlaff et al., 1997), potential climate change (Constable et al., 1995) and drought and nutrient stress.

TREGRO uses a detailed model of photosynthesis based on the equations of Farquhar et al. (1980) and Collatz et al. (1991). Thus, growth of the tree is dependent upon the amount of C fixed, along with the availability of water and nutrients (also simulated). Since CO_2 uptake is calculated, O_3 uptake can also be calculated from gas exchange rates and hourly O_3 concentrations supplied with weather data. Each leaf class (one for deciduous trees and species-dependent for evergreens) accumulates an O_3 exposure that reduces the mesophyll conductance and/or respiration according to user-supplied exposure–response functions. Increasing CO_2 concentrations may be simulated by changing the gradient from atmosphere to the inside of the leaf. Nutrient and water stresses are modeled by changing soil conditions or altering input precipitation.

Because the model operates on a short time step, uses real meteorological and O_3 data, and allocates C and nutrients using a method that incorporates both resource availability and phenology, the modeled response to multiple stresses, or to a different temporal pattern of stress, may be quite different than to a single stress. For instance, an O_3 exposure applied before or after leaf flush may cause quite a different response than one that occurs during leafout (Constable and Retzlaff, 1997). The model output also provides the opportunity to examine the distribution of effects among different plant parts, thus giving insight to the mechanisms behind altered productivity.

For the purposes of the work reported here, parameter files were developed for red oak and sugar maple based on allometry and harvest information reported in the literature. The parameterization methodology was reported by Retzlaff et al. (1997). Simulations were conducted using weather and O_3 data from three locations: Ithaca, New York; Edmonson, Kentucky; and Sheboygan, Wisconsin. In each case, hourly meteorological and O_3 data for 1988 and 1989 were provided by James Weber, U.S. Environmental Protection Agency. The base parameterization, which includes phenology, was adjusted slightly at each site to account for differences in degree-day progression and precipitation. Three exposure–response experiments were then conducted for each species. In each case, five 3-year hourly O_3 regimes, ranging from sub-ambient to about twice ambient, and three soil N concentrations, ranging from the base parameterization (soil solution N concentration not limiting to growth) to 50% below the base were used. The model run was for three years using weather and O_3 data from 1988 (one year) and 1989 (two years). Output from the model was used to construct exposure–response surfaces that could be used by themselves, or to modify ZELIG input parameters.

ZELIG (Urban, 1990) is a gap-succession model (Botkin et al., 1972; Shugart, 1984) used to simulate succession in mixed stands typical of eastern and northern forests. It has also been used in the west (Hansen et al., 1995). ZELIG operates on an annual time step and uses stochastically generated weather based on 30-year means (with variances) for temperature and precipitation. In addition, for each species included in the stand, a growth curve and a number of other characteristics, such as maximum size, maximum age, tolerance, nutrient sensitivity, mortality, and seeding, are specified. The output from the model includes the basal area distribution by species and size class and a description of canopy structure. Data to parameterize ZELIG consisted of weather data from the three locations specified above and from published stand composition and species allometry from nearby forests. Seven species (red maple [*Acer rubrum* L.], sugar maple, white ash [*Fraxinus americana* L.], black cherry [*Pruneus serotina* Ehrh.], red oak, white oak [*Quercus alba* L.], and basswood [*Tilia americana* L.]) were included in the ZELIG simulations, but only red oak and sugar maple (the dominant species) and black cherry (an O_3-sensitive species) had growth rates modified by O_3 exposure.

In ZELIG, the expected growth rates of the species are used to predict competition among the individuals of the species modeled. Access to nutrients and light is restricted by proximity to neighboring trees, and these variables, along with the environmental conditions, are used in the calculation of growth and competition. Seedling establishment and tree death are based on species-specific parameters, environment, and the recent rate of growth. In the experiments reported here, productivity and composition were based on the mean of 100 simulated plots.

Model Linkages

In order to use the detailed simulation results from exposure–response experiments, three calculations are passed from the TREGRO to ZELIG where they modify the growth relationships in the stand model. The total biomass, leaf mass, and the fine root/leaf mass ratio, averaged over 3-year TREGRO simulations and made relative to ambient O_3, were used to modify the growth response in ZELIG. Thus, for each of the 15 (5 O_3 by 3 N levels) simulations in TREGRO, a set of modifiers was produced for use with ZELIG. The results of 100-year ZELIG runs were then used to produce exposure–response surfaces that could be incorporated in a geographical information system or used at individual locations to investigate the effects of ambient O_3 and N on stand composition and structure.

To estimate the effects of O_3 on the growth of trees across the northern region of the country, monitoring data from 1991 (a moderate O_3 year) were acquired from the USEPA AIRS database. Depending on location of the monitoring site, an exposure–response model was chosen to represent the response of trees in that area. For instance, locations east of Cleveland, Ohio and north of Pittsburgh, Pennsylvania were modeled using the function generated from simulations using Ithaca, New York, data. Locations south of about 40°N were modeled using the functions developed at the Kentucky location, and locations west of Cleveland and north of 40°N used the Sheboygan, Wisconsin, functions. While these geographic divisions are coarse, they are suitable for a first inspection of the results. A more detailed spatial model using cluster analysis to group geographical areas with similar weather conditions is under development.

Ozone Impacts on Red Oak and Sugar Maple in the Northeast

While both O_3 and N availability were used in the simulations, only the results of the O_3 experiments are presented here. In general, ambient O_3 caused a small reduction in the growth of red oak across the region (Table 12.3). The range of the predicted response included a few very small predicted growth increases, but was mostly a reduction of 2 to 4% (basal area) over the course of the 100-year simulation. The response was least in the northwestern part of the region, due to generally lower O_3 concentrations, and greatest in the more southern locations due to greater O_3 and O_3 uptake.

Sugar maple response varied widely with geographical region, but the growth response was almost always positive, indicating that the species was able to take advantage of the decrease in red oak growth (and in other minor species in the stand mix) caused by O_3. In the northern-most parts of the regions, the sugar maple growth increase about offset the red oak

Table 12.3. Summary of Change in Basal Area (Compared with That Simulated under Ambient O_3) Caused by a 100-year Exposure to 5-month SUM06 O_3 Exposures at 332 Locations in the Northern U.S. Response Functions and Weather Data Are Based on the States Named in the Table, but O_3 Exposures Are from the Actual Locations. Assignment to a Particular Data Set/Response Function Is Based on Geographical Location of the Monitoring Site

| | Change in Basal Area (%) | | | | | |
| | Wisconsin | | Kentucky | | New York | |
	Red Oak	Sugar Maple	Red Oak	Sugar Maple	Red Oak	Sugar Maple
Mean	−1.28	1.60	−2.56	11.93	−3.12	−0.72
Standard deviation	−1.25	0.63	0.46	1.80	0.32	0.16
Maximum	1.74	2.04	−2.27	17.59	−2.31	−0.21
Minimum	−4.57	−1.52	−5.42	7.41	−24.13	−1.14

reduction—generally a 3% or less increase in growth. In the southern part of the region, sugar maple growth increased up to 12% or more, due perhaps to warmer temperatures, but also due to the reduction of black cherry, a minor component of the stand that is very sensitive to O_3.

The simulations do not indicate widespread changes in forest stand productivity and composition due to ambient O_3, or even to levels that might be expected to occur in the foreseeable future. They do predict that, in some parts of the region, forest composition might change slightly due to O_3 exposure, particularly if very sensitive species are removed from the forest mix.

The predictions obtained from the linked-model methodology are consistent with those presented earlier in this chapter for the northern part of the region, particularly since the linked models predict basal area growth and not NPP.

The advantage of this method is that the response of individual species may be used to investigate potential changes in forest composition in response to stress. It is also possible to examine the effects of co-occurring stresses and to determine how the plant integrates the environment with its own characteristics of growth. Unraveling the interactions of the stresses is complicated however, and subject to a number of uncertainties that must be resolved during the interpretation.

Uncertainties in the Linked-Model Methodology

Any modeling methodology has associated uncertainties that affect the outcome and, potentially, the interpretation of the results. In the case of the linked-model methodology, there are uncertainties associated with each of the simulation models, with the linkage, and with the spatial scaling, including the estimation of O_3 exposures at a distance from the monitoring locations. New technology for predicting O_3 exposures (Phillips et al., 1997) will greatly enhance the understanding and estimation of O_3 exposures in forested areas. In addition, comparing different modeling strategies, such as is done here, will increase the confidence in forest growth estimates generated with simulation techniques.

The greatest strength of this methodology is that it allows a prediction of changes in forest composition due to differential sensitivity to environmental conditions. On the other hand, as the linkage currently stands, there is no feedback from the stand level to the individual, and this should be considered an important issue to address in the future.

Future Applications of Linked-Model Methodology

In a cooperative project with investigators at the USEPA, the University of Nevada, Reno, and Boyce Thompson Institute, we have modeled the growth and composition of red oak–sugar maple, loblolly pine–yellow poplar, red spruce–balsam fir, and ponderosa pine–white fir forests. In two cases (oak–maple and spruce–fir), N was modeled as a competing stress and in the others, drought was a competing stress. Results of the simulations are being analyzed and will be incorporated into a GIS to provide an estimate of forest response across North America.

Uncertainty in Scaling and Regional Modeling Efforts (PBW)

As discussed above, assessment of the regional impacts of environmental stresses on forest health and productivity requires the use of models. Model predictions always contain uncertainty due to (1) incomplete regional data, (2) incomplete knowledge of how forests respond to particular stresses, and (3) uncertainty about which processes and parameters should be included in the model. There is increasing awareness of the need to quantify all of these types of uncertainties when assessing risks to the environment. For example, an approach called Ecological Risk Assessment has recently been adopted by the USEPA to evaluate risks to the environment (USEPA, 1992). This approach emphasizes that the scope of the problem and analysis endpoints should be clearly defined, and that uncertainties in the analysis of pollutant exposure and effects be analyzed. A better understanding of uncertainties in data and models can improve decision making by providing more information about various management options, or demonstrating the cost-effectiveness of additional data collection (Morgan and Henrion, 1990).

The term "uncertainty" is used broadly to include natural variability, errors in data and model parameters, and lack of scientific understanding of processes. A useful distinction seems to be whether the variation may be reducible by further observation or research—henceforth uncertainty—or whether it is inherent in the system (e.g., a chaotic system or unknowable value)—henceforth variability. This section of the chapter discusses some of the important uncertainties in assessing the effects of stresses on forest growth and productivity, and presents approaches for quantitatively analyzing these uncertainties in order to improve regional assessments.

Uncertainty in Regional Data

All of the modeling approaches described earlier in this chapter require spatially explicit regional data sets. Awareness of the importance of uncertainty in spatial data has increased greatly over the last decade, concomitant with the surge in the availability of GIS software, image processing software, and digital data sets. Various techniques have been used to assess map accuracy and error propagation in a GIS context (for example, see Mowrer et al., 1996). Such analysis is particularly important when spatially and temporally interpolated data are used as inputs to regional models. As examples of regional model input data, we will briefly discuss uncertainty in historical data on forest condition in the U.S., and uncertainty in climate change scenarios.

Spatially explicit models require spatially explicit data sets, often including historical forest growth rates. The USDA Forest Service FIA produces estimates of forest growth throughout the U.S., with statistical estimates of uncertainty (in the form of confidence limits for mean values) at the scale of counties, states and regions. For smaller areas, such as counties, uncertainty about average growth rates is much larger, since data from only a few plots are available. Efforts have been made to assess the effect of such uncertainties on modeled predictions of the effect of disturbances. For example, Dale and others (1988) found that such errors are most important for commonly occurring, rapidly growing species. Their analysis showed that uncertainty in the model results could be most effectively reduced by increasing the precision of the forest data by means of additional sampling.

Spatial statistical analyses, including conditional simulation, can be used to estimate the uncertainties in species distribution based on FIA data (Riemann, 1996). Such information can also be used to indicate regions that may benefit from additional sampling. Forest inventory data are becoming available in digital form, and uncertainties in the distribution and mortality or dominant species in the northeastern U.S. are being analyzed by foresters in the USDA Forest Service's northeastern station (Riemann, 1996; see Chapter 1). For probabilistic assessments of the effects of climate change and other stresses on forests in the northeastern U.S., such quantification of uncertainty in regional forest data will be very useful. In addition, assessments of spatial uncertainty in forest growth rates and environmental conditions such as soil WHC will be required, and need to be developed. For example, as discussed above in an assessment of the effects of O_3 on forests in the northeastern U.S., reliable spatial data on soil WHC were not available in digital form, constraining the analysis. An effort should be made to develop such regional data sets along with quantitative estimates of the uncertainty in these data.

In addition to regional forest and soil data, regional assessments generally require climate data for all locations. Such data must be interpolated among climate stations (e.g., Phillips et al., 1992; Ollinger et al., 1995). Assessments of climate change require not just historical climate data, but also scenarios of future climate. In order to meet the needs of forest growth models, regional climate change scenarios must meet several criteria. Estimates of precipitation, temperature, incident solar radiation, and other climatic variables may be required for all locations in the region. These estimates should account for topographical effects, and physical consistency should be maintained among climatic variables. All data should be interpolated to a single grid cell size, including climate change projections from general circulation models (GCMs) that often operate at different spatial resolutions. One example of an effort to build a consistent climate change scenario data base for the U.S. is the Vegetation/Ecosystem Modeling and Analysis Project (VE-MAP) (VEMAP Members, 1995). Data sets meeting the above criteria were developed so that different vegetation models using the same future climate scenario could be compared. In VEMAP, several scenarios were derived from different GCMs. A probabilistic climate change scenario for the southeastern U.S. was developed by sampling among these scenarios at each analysis cell (Woodbury et al., 1996; Woodbury et al., 1998; Smith et al., 1998). This approach permitted the influence of uncertainties in predicted climatic variables to be compared with uncertainties in other aspects of a modeling effort to assess the effects of climate change on forest growth.

Spatial Interpolation

When models are used to interpolate among locations where ground data are collected, the uncertainty of the prediction will increase. While it is difficult to quantify this spatial uncertainty, several useful approaches have been developed, including kriging. Kriging is a statistical interpolation technique that can produce estimates of uncertainty in interpolated values based on fitting a spatial autocorrelation model to the data. This technique has been used primarily for interpolation of mean values, which may then be used in regional modeling efforts. Several investigators have used kriging to estimate O_3 exposure between monitoring stations (Lefohn et al., 1988; Phillips et al., 1997). The use of auxiliary data can reduce the prediction uncertainty (Phillips et al., 1997). Cross-validation of the interpolated values can produce more credible estimates of uncertainty in the predicted values because predicted values are compared directly with measured values, not just with the model that is fitted to the data. Such cross-validation has been used to estimate the uncertainty in interpolated O_3 concentrations in rural forested areas throughout the southeastern U.S. (Phillips et al., 1997). Deposition of air pollutants such as O_3 and sulfur dioxide (SO_2) to forests can be also estimated using long-range transport models. Some efforts have been made to quantify the uncertainties in these model predictions. For example, for selected source–receptor combinations, isolines of sulfate deposition were found to have uncertainties in their locations up to hundreds of kilometers (Alcamo and Bartnicki, 1987).

Aggregation and Scaling

Just as uncertainty increases when extrapolating across space, it also increases when scaling up from measurements made at the seedling or tree scale to larger scales. To predict regional effects from such measurements, objects and processes must be aggregated. Such aggregation will cause bias if responses to stresses are nonlinear (e.g., King, 1991). The first step in reducing such uncertainty is to identify which processes contribute most to the aggregation uncertainty. This task can be accomplished using either deterministic and probabilistic techniques (Rastetter et al., 1992). While a rigorous deterministic approach to reducing aggregation error is rarely practical, probabilistic methods have shown some promise.

Most model builders and model users prefer simple models—they are easier to use, modify, and examine. However, it is crucial to maintain the functional responses in the model that are most important in influencing the predictions. Probabilistic (uncertainty) analysis can help to guide such decisions by identifying the parameters that are most influential. Luxmoore et al. (1990) showed that when a tree physiological model (UTM) was linked to a forest succession model (FORET), the recruitment and mortality functions in the succession model swamped any physiological responses from the tree model. Hence, in this example, addition of physiological complexity did not alter the long-term predictions of the succession model. For predictions of stand growth, however, uncertainty about physiological parameters, such as maximal photosynthetic rate, was shown to introduce more uncertainty in the stand model FORGRO than did most stand inputs (van der Voet and Mohren, 1994), suggesting that such physiological parameters are quite important. Hence the conclusion about what type of aggregation is useful and which factors are most influential is likely to be dependent on the scale and the goals of the modeling effort.

When creating models to assess the regional impacts of stresses on forests, the goal is to generate unbiased estimates with a minimum of uncertainty. This goal is difficult to achieve due to the many interacting factors that influence forest processes, often in a nonlinear fashion. One way to examine the influence of model structure is to compare different models using the same data set. While complex models may represent more processes and be more intellectually appealing, they can also propagate errors and produce less precise predictions. For example, a comparison of two forest growth models applied to pure aspen stands in the central Rocky Mountains found that the more complex model estimated the same average stand growth with less precision than did the simpler model (Mowrer et al., 1988). The VEMAP project mentioned above used multiple models, increasing the credibility of the predictions where different models predicted similar results (VEMAP members, 1995).

Calibration and Validation of Models

Validation is the process by which a model is tested against an independent data set, that is, data that were not used in its development. While validation appears superficially to be straightforward, rarely is it clear what the appropriate domain of the model may be. For stresses such as climate change, it may be not be possible to truly validate a model until after the change has occurred; we are forced to extrapolate beyond current conditions. As discussed previously, with purely statistical models, there is awareness of the danger such extrapolation. Ecological assessment "process" models should be more reliable to the degree that they actually represent important processes in forests. Because all models are by definition simplifications of reality, they often contain empirical, that is, statistical, relationships, and may not include processes that may become important under future environmental conditions. The more aggregated a model becomes, the more it represents a statistical approach, rather than a process approach. It is difficult to evaluate the extent to which a model truly represents underlying processes versus statistical associations. Nevertheless, it is important to try to quantify the uncertainties in a model, in particular to determine whether additional effort in defining a model is worthwhile or whether more data must be collected in order to reduce uncertainties in model predictions.

One approach for estimating bias and error in regional extrapolation is to create a model based on data from one area and test its predictions against data from another area (without calibration or tuning). Such an approach requires extensive data, and may not be very useful if functional responses of ecosystem components are known to differ from one area to another. This approach was used to quantify errors in a lake acidification model applied throughout Scandinavia (Gardner et al., 1990). The investigators conclude that the use of parameters from a specific region in Finland can be used to model the response of all lakes within Finland

and much of Sweden without significant error, but that substantial error occurred for part of Sweden and much of Norway. This example demonstrates the importance of defining the spatial domain within which a regional model is useful.

Uncertainty Analysis

For issues such as climate change effects on forest growth, there are clearly large uncertainties in both the magnitude of future climate change and the magnitude of various responses to a changed climate (Dixon and Wisniewski, 1995). Regional ecological risk assessments must account for uncertainty in spatial data as well as uncertainty in the response of ecosystem components, such as forests, to stresses. While such regional assessments are difficult, it is possible to use Monte Carlo techniques to address these uncertainties quantitatively. Several investigators, notably at the U.S. Department of Energy's Oak Ridge National Laboratory, over the last two decades have demonstrated that spatial and stochastic modeling methods can be combined for such regional assessments (e.g., Graham et al., 1991). In Monte Carlo-type uncertainty analysis, probability distributions are specified for model parameters. The model is executed by sampling from these distributions. A large number of such iterations are performed in order to produce a simulated distribution of model predictions. In order to reduce the number of iterations that are required to produce stable output distributions, Latin hypercube sampling can be used to sample systematically from each distribution (McKay, 1979). Rank-order correlation analysis can be used to identify which input parameters have the greatest influence on the model output. Such techniques are gaining popularity in many disciplines, particularly risk analysis (Morgan and Henrion, 1990).

While sensitivity and uncertainty analyses are both concerned with uncertainty in model parameters, they focus on different issues. Sensitivity analysis is concerned with the propagation of errors in deterministic, but unknown, model parameters. Uncertainty analysis treats parameters as inherently uncertain, that is, they have irreducible variability, and examines the influence of parameter variability on model predictions.

Only rarely have uncertainties in both spatial data (e.g., vegetation distribution) and model functions (e.g., O_3 dose–response) been examined together. Such analyses will help to determine whether our ability to assess such effects is more limited by uncertainty in regional data availability or uncertainty in dose–response measurements. Such analysis will also improve assessments of ecological risk by identifying both processes and regions in which uncertainty in model predictions is the greatest. For example, when predicting lake acidification due to regional sulfur deposition with a complex deposition model, prediction uncertainty was found to vary with soil characteristics across the region (Hettelingh et al.,

1989). Specifically, there was much more variation in predicted acidification among watersheds that were relatively insensitive as compared with those that were relatively sensitive to acidification. In the northeastern forests, analysis of the influence of uncertainty in spatial data in Vermont determined that the best way to reduce the uncertainty of a model's predictions or forest growth was to increase the amount of forest plot data collected (Dale et al., 1988).

Recently, two of us (PBW and JAL) performed an analysis of the potential effects of climate change on loblolly pine (*Pinus taeda* L.) growth in the southeastern U.S. (Woodbury et al., 1996; Woodbury et al., 1998; Smith et al., 1998). We will present our methods and conclusions in brief as an example of an approach to simultaneously analyzing uncertainties in regional data and forest response to climate change and other interacting stresses, such as ozone.

Our modeling system is comprised of a GIS and a probabilistic forest growth model. The model produces more than a single estimate—it produces distributions of possible estimates in order to account for important sources of uncertainty in the model. This task is accomplished by defining inputs to the model and functional relationships within the model as frequency distributions, and then using Latin hypercube techniques to sample from these distributions. The GIS serves to integrate the regional data, facilitate the development of probabilistic inputs for the forest growth model, and display the probabilistic results of the forest growth model. The study area was the 12-state region where loblolly pine occurs. To produce regional probabilistic estimates of the effect of climate change, 150 simulations were run for each of 1169 30×30 km grid cells in the 12-state region.

Based on a probabilistic climate change scenario derived from the results of four GCMs, our model estimated that loblolly growth will likely decrease slightly throughout its 12-state range. However, due to large uncertainties in both climate factors and the influence of these factors on forest growth, there is a substantial chance of either a large decrease or a large increase in loblolly pine basal area growth rate under future climate conditions. The most influential factor at all sites was the relative change in C assimilation. Of climatic factors, carbon dioxide (CO_2) concentration was found to be the most influential factor at all sites throughout the region. Substantial regional variation in estimated growth was observed, and was probably due primarily to variation in historical growth rates and the importance of historical growth in our model structure. The estimate of future loblolly pine growth rate was more uncertain for stands with historically rapid growth, again due to the structure of the forest growth model.

This section of the chapter is intended to show that uncertainties in regional models are important, and should not be ignored. Techniques for quantitative analysis of such uncertainties in regional data and models are becoming established and more accessible to forest modelers. Quantification of uncertainty in regional forest data will improve the accuracy of assessments of climate change effects on forests in the northeastern U.S. Assessments of the uncertainty in regional estimates of environmental conditions such as soil WHC need to be developed. Lack of reliable regional data can constrain regional assessments. An effort should be made to develop such regional data sets along with quantitative estimates of the uncertainty in these data. We hope that uncertainty analysis will be used to guide model development, regional data collection, and to improve decisions about mitigating the effects of stresses on regional forest growth and productivity.

Conclusions

In order to understand the impacts of the various stresses that affect forests and to assess the importance of both predicted climate and pollution scenarios and potential pollution control strategies, it will be essential to use models in the future. Three different modeling methodologies have been presented—a regional simulation model, linked physiological–ecological model, and a risk assessment model—that can be used to study the response of forests and landscapes to stress. Each of the models carries with it uncertainty and variability that impose limits on the usefulness of the predictions, but each allows the interpretation of uncertainty and variability that leads to a better understanding of the biological systems.

Models do not eliminate the need for experimentation, rather they can be used to guide investigations in a time of limited research funding. Furthermore, they can be effective tools for communicating the biological consequences of management and policy to those responsible for the health and productivity of our forests.

References

Aber JD, Ollinger SV, Federer CA, Reich PB, Goulden ML, Kicklighter DW, Melillo JM, Lathrop RG (1995) Predicting the effects of climate change on water yield and forest production in the northeastern U.S. Clim Res 5:207–222.

Aber JD, Federer CA (1992) A generalized, lumped-parameter model of photosynthesis, evapotranspiration and net primary production in temperate and boreal forest ecosystems. Oecologia 92:463–474.

Aber JD, Ollinger SV, Driscoll CT (1997) Modeling nitrogen saturation in forest ecosystems in response to land use and atmospheric deposition. Ecol Model 101:61–78.

Aber JD, Reich PB, Goulden ML (1996) Extrapolating leaf CO_2 exchange to the canopy: a generalized model of forest photosynthesis validated by eddy correlation. Oecologia 106:257–265.

Alcamo J, Bartnicki J (1987) A framework for error analysis of a long-range transport model with emphasis on parameter uncertainty. Atmos Environ 21:2121–2131.

Baldocchi DD, Verma SB, Anderson DE (1987) Canopy photosynthesis and water-use efficiency in a deciduous forest. J Appl Ecol 24:251–260.

Belanger RP (1990) *Quercus falcata* Michx. var. falcata. In: Burns RM, Honkala BH (eds) *Silvics of North America. Vol. 2. Hardwoods.* Agric. Hndbk 654. United States Department of Agriculture (USDA) Forest Service, Washington, DC, pp 640–644.

Botkin DB, Janek JF, Wallis JR (1972) Some ecological consequences of a computer model of forest growth. J Ecol 60:849–873.

Clapp RB, Hornberger GM (1978) Empirical equations for some soil hydraulic properties. Water Resour Res 9:1599–1604.

Collatz GJ, Ball JT, Grivet C, Berry JA (1991) Physiological and environmental regulation of stomatal conductance, photosynthesis and transpiration: a model that includes a laminar boundary layer. Agric For Meteorol 54:107–136.

Constable JVH, Retzlaff WA (1997) Simulating the response of mature yellow poplar and loblolly pine trees to shifts in peak ozone periods during the growing season using the TREGRO model. Tree Physiol 17:627–635.

Constable JVH, Taylor GE Jr., Laurence JA, Weber JA (1995) Climatic change effects on the physiology and growth of Pinus ponderosa. Can J For Res 26:1315–1325.

Dale VH, Jager HI, Gardner RH, Rosen AE (1988) Using sensitivity and uncertainty analysis to improve predictions of broad-scale forest development. Ecol Model 42:165–178.

Dixon RK, Wisniewski J (1995) Global forest systems: an uncertain response to atmospheric pollutants and global climate change? Water Air Soil Pollut 85:101–110.

Farquhar GD, von Caemmerer S, Berry JA (1980) A biochemical model of photosynthetic CO_2 assimilation in leaves of C_3 species. Planta 149:78–90.

Federer CA (1982) Frequency and intensity of drought in New Hampshire forests: evaluation by the BROOK model. In: *Applied Modeling in Catchment Hydrology, Proceedings of the International Symposium on Rainfall–Runoff Modeling, May 1981,* Water Resources Publications, Littleton, CO, pp 459–470.

Federer CA, Lash D (1978) BROOK: *A Hydrologic Simulation Model for Eastern Forests.* Water Resour Res Ctr Rept 19. University of New Hampshire, Durham, NH.

Field C, Mooney HA (1986) The photosynthesis–nitrogen relationship in wild plants. In: Givnish T (ed) *On the Economy of Plant Form and Function.* Cambridge University Press, New York, pp 25–55.

Gardner RH, Hettelingh JP, Kaemaeri J, Bartell SM (1990) Estimating the reliability of regional predictions of aquatic effects of acid deposition. In: Kaemaeri J (ed) *Impact Models to Assess Regional Acidification.* International Institute for Applied Systems Analysis, The Netherlands. Kluwer Academic, Boston, MA, pp 185–207.

Graham RL, Hunsaker CT, O'Neill RV, Jackson BL (1991) Ecological risk assessment at the regional scale. Ecol Applic 1:196–206.

Guderian R, Tingey DT, Rabe R (1985) Effects of photochemical oxidants on plants. In: Guderian R (ed) *Air Pollution by Photochemical Oxidants: Formation, Transport, Control, and Effects on Plants.* Springer-Verlag, Berlin, Germany, pp 127–333.

Hansen AJ, Garman SL, Weigand JF, Urban DL, McComb WC, Raphael MG (1995) Alternative silvicultural regimes in the Pacific Northwest simulations of ecological and economic effects. Ecol Applic 5:535–554.

Hettelingh JP, Gardner RH, Rose KA, Brenkert AL (1989) Broad scale effects of sulfur decomposition: a response surface analysis of a complex model. In: Kanari J, Brakke DF, Jenkins A, Norton SA, Wright RF (eds) *Regional Acidification Models*. Springer-Verlag, Berlin, Germany, pp 267–277.

Hogsett WE, Herstrom A, Laurence JA, Weber JE, Lee EH, Tingey D (1996) An approach for characterizing tropospheric ozone risk to forests. Environ Manage 21:105–120.

Host GE, Isebrands JG (1994) An interregional validation of ECOPHYS, a growth process model of juvenile poplar clones. Tree Physiol 14:933–945.

Iverson LR, Prasad AMG, Scott CT (1996) Preparation of forest inventory and analysis (FIA) and state soil geographic data base (STATSGO) data for global change research in the eastern United States. In: Birsdey R, Hom J, O'Brian K (eds) *Proceedings: 1995 Meeting of the Northern Global Change Program*. Gen Tech Rep NE-214. United States Department of Agriculture (USDA) Forest Service, Northeastern Forest Experiment Station, Radnor, PA, pp 209–214.

Kattenberg A, Giorgi F, Grassl H, Meehl GA, Mitchell JFB, Stouffer RJ, Tokioka T, Weaver AJ, Wigley TML (1996) Climate models—projections of future climate. In: Houghton JT, Meira-Filho LG, Callander BA, Harris N, Kattenberg A, Maskell K (eds) IPCC (Intergovernmental Panel on Climate Change) *Climate Change 1995: The Science of Climate Change*. Cambridge University Press, Cambridge, United Kingdom, pp 285–357.

Kern JS (1995) Geographic patterns of soil water-holding capacity in the contiguous United States. Soil Sci Soc Am J 59:1126–1133.

King AW (1991) Translating models across scales in the landscape. In: Turner MG, Gardner RH (eds) *Quantitative Methods in Landscape Ecology*. Springer-Verlag, New York, pp 475–512.

Kingsley NP (1985) *A Forester's Atlas of the Northeast*. Gen Tech Rep NE-95. United States Department of Agriculture (USDA) Forest Service, Northeastern Forest Experiment Station, Radnor, PA.

Lathrop RG, Aber JD, Bognar JA (1995) Spatial variability of a digital soils map in a regional modeling context. Ecol Model 82:1–10.

Lathrop RG, Bognar JA (1994) Development and validation of AVHRR-derived regional land cover data for the northeastern U.S. Region. Int J Remote Sens 15:2695–2702.

Laurence JA, Kohut RJ, Amundson RG (1993) Use of TREGRO to simulate the effects of ozone on the growth of red spruce seedlings. For Sci 39:453–464.

Lefohn AS, Knudsen HP, McEvoy LR Jr. (1988) The use of kriging to estimate monthly ozone exposure parameters for the Southeastern USA. Environ Pollut 53:27–42.

Luxmoore RJ, Tharp ML, West DC (1990) Simulating the physiological basis of tree-ring responses to environmental changes. In: Dixon RK, Meldahl RS, Ruark GA, Warren WG (eds) *Process Modeling of Forest Growth Responses to Environmental Stress*. Timber Press, Portland, OR, pp 393–401.

Martin ME, Aber JD (1997) High spectral resolution remote sensing of forest canopy lignin, nitrogen, and ecosystem processes. Ecol Applic 7:431–443.

McKay MD, Beckman RJ, Conover WJ (1979) A comparison of three methods for selecting values of input variables in the analysis of output from a computer code. Technometrics 21:239–245.

Morgan MG, Henrion M (1990) *Uncertainty: A Guide to the Treatment of Uncertainty in Quantitative Policy and Risk Analysis*. Cambridge University Press, New York.

Mowrer TH (1988) A Monte Carlo comparison of propagated error for two types of growth models. Gen Tech Rep 120. United States Department of Agriculture (USDA) Forest Service, North Central Forest Experiment Station, St. Paul, MN, pp 778–785.

Mowrer TH, Czaplewski RL, Hamre RH (Eds) (1996) *Spatial Accuracy Assessment in Natural Resources and Environmental Sciences: Second International Symposium Proceedings*. Gen Tech Rep RM-GTR-277. United States Department of Agriculture (USDA) Forest Service, Rocky Mountain Forest and Range Experiment Station, Ft Collins, CO.

Munger JW, Wofsy SC, Bakwin PS, Fan S, Goulden ML, Daube BC, Goldstein AH, Moore K, Fitzjarrald D (1996) Atmospheric deposition of reactive nitrogen oxides and ozone in a temperate deciduous forest and a sub-arctic woodland. 1. Measurements and mechanisms. J Geophys Res 101: 12639–12657.

Ollinger SV, Aber JD, Federer CA (1998) Estimating regional forest productivity and water yield using an ecosystem model linked to a GIS. Landsc Ecol 13: 323–334.

Ollinger SV, Aber JD, Federer CA, Lovett GM, Ellis J (1995) Modeling physical and chemical climatic variables across the northeastern U.S. for a Geographic Information System. Gen Tech Rep NE-191. United States Department of Agriculture (USDA) Forest Service, Northeastern Forest Experiment Station, Radnor, PA.

Ollinger SV, Aber JD, Reich PB (1997) Simulating ozone effects on forest productivity: interactions among leaf-, canopy- and stand-level processes. Ecol Applic 7:1237–1251.

Pell EJ, Eckerdt N, Enyedi AJ (1992) Timing of ozone stress and resulting status of ribulose bisphosphate carboxylase/oxygenase and associated net photosynthesis. New Phytol 120:397–405.

Phillips DL, Dolph J, Marks D (1992) A comparison of geostatistical procedures for spatial analysis of precipitation in mountainous terrain. Agric For Meteorol 58:119–141.

Phillips DL, Lee EH, Herstrom AA, Hogsett WE, Tingey DT (1997) Use of auxiliary data for spatial interpolation of ozone exposure in southeastern forests. Environmetrics 8:43–61.

Pye JM (1988) Impact of ozone on the growth and yield of trees: a review. J Environ Qual 17:347–360.

Rastetter EB, King AW, Cosby BJ, Hornberger GM, O'Neill RV (1992) Aggregating fine-scale ecological knowledge to model coarser-scale attributes of ecosystems. Ecol Applic 2:55–70.

Reich PB (1987) Quantifying plant response to ozone: a unifying theory. Tree Physiol 3:63–91.

Reich PB, Kloeppel B, Ellsworth DS, Walters MB (1995) Different photosynthesis–nitrogen relations in deciduous and evergreen coniferous tree species. Oecologia 104:24–30.

Retzlaff WA, Weinstein DA, Laurence JA, Gollands B (1997) Simulating the growth of a 160-year-old sugar maple (*Acer saccharum* Marsh.) tree with and without ozone exposure using the TREGRO model. Can J For Res 27:783–789.

Riemann HR (1996) Understanding the spatial distribution of tree species in Pennsylvania. In: Mowrer HT, Czaplewski RL, Hamre RH (eds) *Spatial Accuracy Assessment in Natural Resources and Environmental Sciences: Second*

International Symposium Proceedings. Gen Tech Rep RM-GTR-277. United States Department of Agriculture (USDA) Forest Service, Rocky Mountain Forest and Range Experiment Station, Ft Collins, CO, pp 73–82.

Shugart HH (1984) *A Theory of Forest Dynamics: The Ecological Implications of Forest Succession Models*. Springer-Verlag, New York.

Sinclair TR, Tanner CB, Bennet JM (1984) Water-use efficiency in crop production. BioScience 34:36–40.

Smith JE, Woodbury PB, Weinstein DA, Laurence JA (1998) Integrating research on climate change effects on loblolly pine: A probabilistic regional modeling approach. In: Mickler RA, Fox S (eds) *The Productivity and Sustainability of Southern Forest Ecosystems in a Changing Environment*. Springer-Verlag, New York, pp 429–451.

United States Soil Conservation Service (USSCS) (1991) *State Soil Geographic Data Base (STATSGO) Data User's Guide*. Miscellaneous Publication 1492. United States Department of Agriculture (USDA) Soil Conservation Service, Washington, DC.

Swift LW Jr. (1976) Algorithm for solar radiation on mountainous slopes. Water Resour Res 12:108–112.

Taylor GE Jr., Hanson PJ (1992) Forest trees and tropospheric ozone: role of canopy deposition and leaf uptake in developing exposure–response relationships. Agric Ecosyst Environ 42:255–273.

Tjoelker MG, Volin JC, Oleksyn J, Reich PB (1995) Interaction of ozone pollution and light effects on photosynthesis in a forest canopy experiment. Plant Cell Environ 18:895–905.

United States Environmental Protection Agency (USEPA) (1992) *Framework for Ecological Risk Assessment. Framework Assessment Forum.* EPA/630/R-92/001. USEPA, Washington, DC.

Urban DL (1990) *A Versatile Model to Simulate Forest Pattern: A User's Guide to ZELIG Version 1.0.* Environmental Sciences Department, University of Virginia, Charlotesville, VA.

van der Voet H, Mohren GMJ (1994) An uncertainty analysis of the process-based growth model FORGRO. For Ecol Manage 69:157–166.

VEMAP (Vegetation/Ecosystem Modeling and Analysis Project) Members (1995) Vegetation/Ecosystem Modeling and Analysis Project: comparing biogeography and biogeochemistry models in a continental-scale study of terrestrial ecosystem responses to climate change and CO_2 doubling. Global Biogeochem Cycl 9:407–437.

Weinstein DA, Beloin RM, Yanai RD (1991) Modeling changes in red spruce carbon balance and allocation in response to interacting ozone and nutrient stresses. Tree Physiol 9:127–146.

Weinstein DA, Yanai RD (1994) Integrating the effects of simultaneous multiple stresses on plants using the simulation model TREGRO. J Environ Qual 23:418–428.

West DC, McLaughlin SB, Shugart HH (1980) Simulated forest response to chronic air pollution stress. J Environ Qual 9:43–49.

Winner WE (1994) Mechanistic analysis of plant response to air pollution. Ecol Applic 4:651–661.

Woodbury PB, Smith JE, Weinstein DA, Laurence JA (1998) Assessing potential climate change effects on loblolly pine growth: a probabilistic regional modeling approach. For Ecol Manage 107:99–116.

Woodbury PB, Smith JE, Weinstein DA, Laurence JA (1996) Towards an ecological risk assessment of the potential effects of global change on loblolly pine forests throughout the USA. (abstr) Bull Ecol Soc Am 77(Suppl):489.

13. Effects of Climate Change on Forest Insect and Disease Outbreaks

David W. Williams, Robert P. Long, Philip M. Wargo, and Andrew M. Liebhold

General circulation models (GCMs) predict dramatic future changes in climate for the northeastern and north central United States under doubled carbon dioxide (CO_2) levels (Hansen et al., 1984; Manabe and Wetherald, 1987; Wilson and Mitchell, 1987; Cubasch and Cess, 1990; Mitchell et al., 1990). January temperatures are projected to rise as much as 12°C and July temperatures as much as 9°C over temperatures simulated at ambient CO_2 (Kittel et al., 1997). Projections of precipitation are quite variable over the region, ranging from 71 to 177% of ambient levels in January and 29 to 153% of ambient in July among several GCMs (Kittel et al., 1997). Such climate changes clearly may affect the growth and species composition of our northern forests directly in ways discussed in previous chapters. In contrast with the discussions in previous chapters, this chapter steps up one trophic level to consider the effects of climate change on the populations of microorganisms, fungi, and insects that feed in and on forest trees.

Interactions at the community level may greatly complicate predictions of the effects of climate change on the dynamics of consumer populations and forests. Climate change may influence the population processes of microorganisms, fungi, and insects directly through the physiology of individual organisms or indirectly through physiological effects on their host plants, competitors, and natural enemies. The resulting population dynamics are difficult to predict because of the complexity and nonlinearities in the responses of physiological processes to climatic factors and interspecies interactions. Feeding by various consumer guilds is just another stress for forest trees, along with climatic and other abiotic stresses, that alters their growth processes. In addition, stresses resulting from climate change may alter the susceptibility or vulnerability of trees to consumers, exacerbating or ameliorating the effects of their feeding unpredictably.

Throughout most of the chapter, the focus is on the direct and indirect effects of climate change and increasing CO_2 levels on microorganisms, fungi, and insects. Such organisms are often referred to as "pests" and "pathogens," reflecting our anthropocentric view as plant pathologists, entomologists, and foresters. At the end of the chapter, the larger question of the effects of changing outbreak dynamics of pest organisms on forests is briefly considered. We have chosen to limit our discussion to the effects of changing climatic variables and CO_2 concentration, ignoring the potential impacts of anthropogenic disturbances. Nevertheless, land use changes by humans and changing forest harvesting regimes may also ultimately affect the frequency and severity of pest outbreaks.

Forest Pests and Climate

Distribution and Abundance of Insects

Climate is an important factor in defining the ranges of most insect species of temperate regions (Andrewartha and Birch, 1954). Climatic variables, notably temperature and precipitation (or another index of moisture), have been used historically in developing climograms that define the limits of favorability for population growth for insect species (Schwerdtfeger, 1935; Gutierrez, 1987). Climatic patterns are a major part of the abiotic setting for the many population and community interactions of a species (Huffaker and Messenger, 1964). Over a species' range, climate may play different roles from the center of the distribution to its margins (Rogers and Randolph, 1986; Hoffman and Blows, 1994). At the poleward margin, climatic factors, in particular low temperature, often define the zone in which recruitment just equals mortality and in which the survival of a population is risky. In the interior of a distribution, climatic factors provide conditions for optimum population growth and mediate interactions with hosts, competitors, and natural enemies. Reasons for limitation at the opposite margin (e.g., the southern margin in the northern hemisphere) are less explored, although they undoubtedly result from biotic interactions (Gaston, 1990).

Across a species' range, climatic factors are important determinants of its abundance through their direct and indirect effects on population increase. As poikilotherms, insects grow as a function of temperature: their growth rates, generation times, fecundity, and intrinsic survivorship are primarily temperature-dependent (Uvarov, 1931; Andrewartha and Birch, 1954). Likewise, temperature and moisture are critical factors in the growth of host plants of herbivorous insects and of their invertebrate predators and parasitoids and pathogens. Indeed, the range of a plant species often determines the distributional limits for a herbivore, and plant abundance provides limits for the herbivore's abundance. Regulation by natural enemies may also change over the climatic gradient of a species' range. Natural enemy activity, coupled with conditions less favorable for intrinsic population growth, may reduce populations to very low levels so as to define that margin of the range. Climatic factors play identical roles, albeit over smaller areas, in species distributed over ranges in elevation (Randall, 1982).

The distribution and abundance of a species are related: local abundance increases directly with range size (Gaston, 1990; Lawton, 1995). In addition, variation in abundance increases with range size. Thus, some of the most common and prevalent insect species also have the highest local populations. Many serious forest insect pests fit this description: they occur over large areas and can produce large, damaging populations. Many also display large variation in abundance, undergoing wide oscillations that result in populations increasing from virtually undetectable levels to outbreak levels over much of their ranges in only a few generations (Berryman, 1987). Such species are the primary subject of this review.

Pest Outbreaks

In the most general ecological sense, an outbreak may be defined as "an explosive increase in the abundance of a particular species that occurs over a relatively short period of time" (Berryman, 1987). Another important characteristic is that outbreaks occur episodically (Myers, 1988)—if not as population cycles, then as approximately periodic oscillations. In addition, outbreaks of some species appear to be synchronized spatially, so that numerous local populations explode almost simultaneously over wide geographical regions. A good example is the eastern spruce budworm, *Choristoneura fumiferana* (Clemens), whose populations may exist at low levels for up to 20 years and then increase through the spruce–fir forests of eastern North America virtually simultaneously in outbreaks that last over a decade (Royama, 1984; Mattson et al., 1988).

Numerous hypotheses have been proposed to explain the onset of outbreaks. They have been thought to result from changes in the physical environment, most typically from changes in weather; intrinsic changes in the genetic or physiological makeup of populations; intrinsic life history characteristics of "pest" species, such as *r* strategies; interactions with higher or lower trophic levels in the case of herbivores, which result in the cycling most often demonstrated in mathematical predator–prey models; changes in host plant physiology or biochemistry as a result of environmental stresses; and escape from regulation by natural enemies (Berryman, 1987; Myers, 1988). In the following, we consider briefly just three hypotheses—specifically, those most influenced by weather and, thus, most likely to be invoked under climate change. They are the plant stress hypothesis, the theory of climatic release, and the Moran effect.

Berryman (1987) broadly classified insect outbreaks into two types: eruptive and gradient. Eruptive outbreaks arise rapidly and, once underway, continue as a self-perpetuating process in space and time that is generally impervious to environmental changes. Eruptive outbreaks are driven by intrinsic population processes and trophic-level factors that operate under time lags. In contrast, gradient outbreaks are driven by changing environmental factors that provide more favorable conditions for population growth. Increases in abundance during gradient outbreaks reflect the opportunistic response of insect populations to increases in their resources. Clearly, gradient outbreaks are very likely to be caused directly by climate change, whereas eruptive outbreaks are much less likely to be affected by it.

Roles of Weather in Pest Outbreaks

Weather As a Stressor: Herbivore–Plant Interactions

Mattson and Haack (1987a,b) proposed several general ways in which plant water stress might increase the likelihood of herbivore outbreaks, including (1) effects on the behavior of herbivores and on their microhabitats, (2) effects on the nutritional quality of plants, and (3) effects on the abilities of plants to regulate herbivores. First, stressed plants may stimulate herbivore feeding through chemical or physical changes. In addition, water stress reduces leaf transpiration, resulting in higher leaf temperatures, which may provide more favorable microhabitats for herbivore growth and development. As a second mechanism, the nutritional hypothesis suggests generally that stress enhances the nutritional quality of plants to insect herbivores (Mattson and Haack, 1987a,b), perhaps because of increased concentrations of nitrogen (N) in stressed tissues (White, 1984). Increased nutritional quality enhances growth and development rates and fecundity of herbivores, increasing the likelihood of outbreaks. Mattson and Haack (1987a) also speculated that plant stress may increase the ability of herbivores to detoxify plant defensive chemicals. Third, water stress may impair a plant's ability to produce defensive compounds, allowing herbivores to feed more readily. In plants that are generally tolerant of insect feeding, particularly by defoliators, water stress reduces growth of new tissues, rendering plants more susceptible to loss (Mattson and Haack, 1987a).

Among forest insects, bark beetles have been frequently cited as good examples of the positive effects of plant water stress on herbivores (Mattson and Haack, 1987b; Waring and Cobb, 1992; Ayres, 1993). In an extensive review of the literature on the impact of plant stress on herbivore dynamics, Waring and Cobb (1992) reported that wood borers, which were usually associated with conifers, responded positively to drought stress in all cases reviewed. However, in a critique of the plant stress hypothesis, Larsson (1989) argued that host water stress may have different effects on different insect herbivore feeding guilds. Similar to reported results for bark beetles, he observed that water stress may have positive effects on sucking insects, such as aphids, presumably due to higher plant nitrogen concentrations under stress. On the other hand, he speculated that gall-forming insects may suffer from host stress, which may reduce the size of the galls that they depend upon for development. He also felt that stress might have negative effects or no effects on leaf chewers. Similarly, Waring and Cobb (1992) enumerated many cases of negative effects of plant stress on herbivores. As an alternative to the plant stress hypothesis, Price (1991) made a case for the positive effects of plant vigor on herbivore attack although his examples emphasized gall formers, which may be special cases among herbivore feeding guilds. Given the multitude of apparently contradictory results in the literature, it clearly will be important to consider the feeding guilds to which particular pest species belong in evaluating potential effects of climate change on them as mediated through their hosts.

The plant stress hypothesis is too simplistic even in the apparently clear-cut situation of southern pine beetle, *Dendroctonus frontalis* Zimmerman, and pine, however. Reeve et al. (1995) suggested a more physiologically realistic model of the interaction based on differential effects of water availability on growth and photosynthetic rates in pine trees. Under that model, higher than normal water levels, that is, beyond those associated with any water stress, may also make trees more susceptible to pine beetles by facilitating rapid shoot growth at the expense of allelochemical production. A similar hypothesis has been proposed for the association of western spruce budworm outbreaks with periods of increased rainfall in chronically dry regions of the southwestern U.S. (Swetnam and Lynch, 1993).

Weather As a Trigger of Outbreaks: The Theory of Climatic Release

The theory of climatic release proposes that weather causes outbreaks through direct effects on insect populations (Greenbank, 1956). Outbreaks are thought to be triggered by several successive years with weather conditions favorable to population growth. The classic case for which the theory was developed were outbreaks of the eastern spruce budworm in Canada (Wellington et al., 1950; Greenbank, 1956). Favorable weather conditions included dry, sunny, or warm weather, or some combination of those factors, during the spring period of larval development. Such conditions favored the survival and rapid growth of larvae, which resulted in increased fecundity. Outbreaks were thought to result from several years of enhanced population growth. Ultimately, the favorable local weather patterns were linked to synoptic, regional storm systems (Wellington et al., 1952).

Martinat (1987) was critical of the theory of climatic release because of the precise timing required between weather events and outbreaks. He stressed that most insect outbreaks display a fairly regular periodicity. Thus, the theory requires that weather phenomena show similar periodicity and that the critical weather variables remain in favorable ranges over several successive years. Martinat (1987) indicated the unlikelihood of such a scenario given current knowledge of long-term weather patterns.

Weather As a Synchronizer of Outbreaks: The Moran Effect

Moran (1953b) proposed an alternate hypothesis for the role of weather in regional outbreaks, which has been called the Moran effect (Royama, 1992). Under this theory, weather does not cause outbreaks directly. Instead, oscillations in abundance result from intrinsic population processes. Within the time series model that Moran proposed, the oscillations are driven by second-order autoregressive processes (Moran, 1953a), which have been interpreted as delayed density-dependent factors, such as the activity of specific parasitoids (Turchin, 1990), or as first-order autoregressive factors that act in a direct density-dependent fashion, such as the activity of generalist predators (Williams and Liebhold, 1995a). An extrinsic factor, such as weather, acts only as a stochastic perturbation in the model. Given similar intrinsic biotic processes to produce oscillations in local populations across a region, exposure to common weather keeps the populations oscillating in synchrony. Thus, weather acts in the Moran effect not to cause outbreaks, but simply to synchronize the oscillations of local populations so as to produce regional outbreaks.

Williams and Liebhold (1995b) applied Moran's technique to explain historical outbreaks of the gypsy moth, *Lymantria dispar* (L.), in the New England states of the U.S. They reported that outbreaks were synchronous among states over a 55-year period. Using time series techniques, they then identified three weather variables that were strongly correlated with gypsy moth defoliation in all four states. The variables, daily minimum temperature and daily rainfall in mid-December during the current gypsy moth generation and daily minimum temperature in mid-July during the previous gypsy moth generation, were also strongly correlated among the states, suggesting that they were common regional factors that synchronized gypsy moth populations.

Direct Responses of Insects and Pathogens to Changing Climate

Direct Effects on Insect Population Processes

Climate change will have both direct and indirect effects on insect populations (Porter et al., 1991; Cammell and Knight, 1992). Direct effects are those mediated through the physiology of individuals, including changes in development, survival, reproduction, behavior, and movement. Although effects of variable temperature and humidity are well documented at the individual level, extrapolating them to changes in dynamics at the population level under large-scale climate changes is the difficult, but essential step in understanding the effects of climate change (Kingsolver, 1989). Indirect effects on herbivores are mediated through the trophic levels surrounding them, in particular, effects on their host plants and natural enemies (Cammell and Knight, 1992).

Several direct effects on herbivores are expected, particularly as a result of rising temperatures. For multivoltine herbivores, more time will be available at the beginning and end of the growing season for population development, especially as winter temperatures are generally expected to increase more than summer temperatures (Porter et al., 1991). With increased accumulation of heat units through the growing season, generation times will decrease and more generations will be possible per season for multivoltine herbivores (Porter et al., 1991; Cammell and Knight, 1992). Overwintering survival is expected to increase, and hence, to allow increased overwintering at higher latitudes. Among migratory species, migration is expected to increase (Sutherst, 1991). At the regional scale, increasing temperature is expected to allow expansion of herbivore populations to higher latitudes and higher elevations (Williams and Liebhold, 1995c; Parmesan, 1996).

Increasing temperatures may alter overwintering patterns of species through changing diapause requirements and increasing overwinter survival. The aphid, *Elatobium abietinum* (Walker), the major defoliator of Sitka spruce in the United Kingdom (UK), produces overwintering eggs on the European continent and in Scandinavia (Straw, 1995). However, it is able to develop continuously in the UK, not producing an overwintering stage. With warmer mean temperatures and fewer extreme cold events, survival is likely to increase, populations will be able to build, and increasing pest problems are anticipated in Sitka spruce plantations (Straw, 1995).

For multivoltine herbivores with the capacity for several generations per year, a relatively modest increase in temperature may permit an additional generation over much of a species range. In a modeling study of European corn borer populations across Europe based on degree day accumulations, Porter et al. (1991) found that, with a 1°C increase, not only were univoltine populations able to shift northward by 165 to 665 km, but also multivoltine populations were able to increase generations incrementally at a given location. Similarly, some generally univoltine bark beetle species can become bivoltine under the variability of current weather conditions, suggesting the potential for multivoltine populations under actual climate change (Heliövaara et al., 1991).

Direct Effects on Pathogen Dynamics

There has been considerable speculation, but little direct experimental evidence, concerning the effects of climate change on the incidence and/or severity of forest pathogens. However, there is much information available about weather and climatic effects on specific host–pathogen systems and this can provide insights regarding possible changes in pathogen dynamics and alteration of geographical ranges.

Increased concentrations of atmospheric CO_2 alone are unlikely to directly affect pathogen life cycles substantially within the range of the predicted doubling of CO_2 (Manning and Tiedemann, 1995). In particular, many root pathogens are accustomed to elevated CO_2 levels commonly found in soils, and most aboveground pathogens, primarily fungi and bacteria, will be unaffected by increases in CO_2 from 0.03 to 0.07% (Manning and Tiedemann, 1995). However, one recent study showed a 20-fold increase in spore production by *Colletotrichum gloeosporioides*, causal agent of anthracnose on a tropical legume, under twice ambient CO_2 (Lupton et al., 1995), so the assumption of no direct effects must be viewed cautiously. Additionally, airborne fungal spores from litter and soil were shown to increase 4-fold under twice ambient CO_2 concentrations in a study conducted with *Populus tremuloides* Michx. grown in open-top chambers (Klironomos et al., 1997). While most of the spores included the common genera of fungal decomposers, *Aspergillus, Penicillium,* and *Cladosporium*, there also was an increase in spores produced by species of *Fusarium* and *Alternaria*. Both of these genera have species that can act as decomposers, weak parasites, or pathogens (Klironomos et al., 1997).

Projected increases in CO_2 may alter temperature regimes and rainfall patterns and thus have significant effects on pathogen survival, reproduction, dissemination, and present geographical distribution. Disease is the result of interactions among a host, a pathogen, and the environment (the disease triangle), since all components are necessary for disease to develop. Pathogen dynamics often vary in response to weather and climatic events. From this knowledge, various predictions about the impacts of global change have been suggested. For instance, the poplar leaf rust pathogen, *Melampsora allii-populina* Kleb., is presently largely confined to southern and central Europe, while *M. larici-populina* Kleb. dominates throughout most parts of Europe. However because it is more thermophilic in some portions of its life cycle than *M. larici-populina* (Somda and Pinon, 1981), and because *M. allii-populina* is wind-dispersed, it may become more common in northern Europe during years with warmer than average temperatures (Lonsdale and Gibbs, 1996).

A widely distributed pathogen that may significantly change its range under altered temperature and climate regimes is *Phytophthora cinnamomi* Rands the cause of Phytophthora root rot on a wide range of tree species (Lonsdale and Gibbs, 1996; Brasier and Scott, 1994). This oomycete, a soil borne root pathogen, is indigenous to the Papua New Guinea region, and requires warm, wet soils to infect roots (Brasier, 1996). It has a very broad host range affecting over 900 mainly woody perennial plants (Zentmyer, 1980). The pathogen has caused extensive dieback in southwestern Australia in jarrah (*Eucalyptus marginata* Sm.) forests (Shearer and Tippett, 1989). *P. cinnamomi* also caused a major epidemic on chestnut (*Castanea dentata* [Marsh.] Borkh.) and chinkapin (*C. pumila* [L.] Mill.) in the southern U.S. resulting in significant mortality in the 1800s and in the 1920s and 1930s though this has only been weakly related to climatic anomalies (Woods, 1953; Crandall et al., 1945). Since the early 1980s, severe decline of oaks across the Mediterranean has involved *P. cinnamomi* as an important factor predisposing trees to drought stress and affecting declines of *Quercus suber* L. and *Q. ilex* L. in southern Iberia (Brasier, 1996; Brasier and Scott, 1994). Besides the interaction of *P. cinnamomi* and drought, this decline involves interactions with changing land use, site factors, and attacks by other organisms (Brasier and Scott, 1994).

Brasier (1996) made several generalizations about *P. cinnamomi* responses to altered climate. First, climatic warming may move the range of the fungus northward and increase the incidence of periods favoring inoculum production and infection. Warmer seasons will increase the rate of spread of the pathogen, increase inoculum production and increase secondary infection rates. Additionally, warmer winters will permit greater pathogen survival while climatic extremes such as drought or waterlogging will reduce host resistance (Brasier, 1996). Because *P. cinnamomi* and drought are major predisposing factors in Iberian oak decline, this predicted alteration in *P. cinnamomi* distribution could increase infection incidence and extend oak decline to new regions. Host stress caused by climatic extremes may further increase the incidence of *P. cinnamomi* and increase attacks by secondary organisms (Brasier, 1996).

Enhanced moisture conditions will favor many fungi and increase spore production and dispersal by rain splash (Lonsdale and Gibbs, 1996). Increased precipitation in some regions will increase the incidence and severity of foliar and root pathogens that are dependent on moisture for development and dispersal (Sutherst, 1996). Alterations in rainfall patterns could also affect fungal development if rainfall events decrease throughout the growing season, or during particular portions of the growing season. This would affect dispersal and decrease incidence of leaf spot fungi such as *Marrsonina* spp. on poplars (Lonsdale and Gibbs, 1996) or *Cristulariella pyramidalis* Wat. & Marsh. on black walnut (*Juglans nigra* L.) (Neely et al., 1976).

Most climate models predict that temperature increases associated with global change will be greatest in the winter (Barron, 1995). Increasing mean winter temperatures will also affect dormant season pathogen activity. Areas in the temperate latitudes where winter dormancy of woody plants occurs simultaneously with periods when temperatures are sufficient for fungal activity are termed "asynchronous dormancy zones" (Lonsdale and Gibbs, 1996). Particularly with perennial species, many pathogens can only breach host defenses and invade host tissues during the dormant season before temperatures become too cold for pathogen growth. Milder winters may allow greater host tissue colonization for perennial stem canker fungi whose growth is temperature-limited. In addition, climatic parameters may affect low-temperature fungi, such as snow molds. These pathogens require a very humid environment that is common under snow cover and that insulates against extreme cold. These pathogens, such as *Phacidium infestans* Karst., would decrease in their incidence in regions where both the amount and persistence of snow cover decrease (Lonsdale and Gibbs, 1996). Other temperature-sensitive pathogens that can only colonize hosts in the dormant season may also be affected. *Gremmeniella abietina*, the cause of Scleroderris shoot blight of conifers, has been shown to have a "conducive" temperature range from −6 to +5°C for disease expression in tree seedlings (Marosy et al., 1989). Insulating snow cover facilitates infection by protecting the pathogen from exposure to cold temperatures. Changes in snow cover patterns would tend to decrease disease incidence of this pathogen.

The ability of pathogens to infect hosts can be controlled by the duration of leaf wetness. For example, the apple scab fungus, *Venturia inaequalis* (Cke.) Wint., requires a long period of leaf wetness and warm temperatures to infect hosts (Weltzien, 1983). Other forest pathogens may have similar requirements, and changes in the duration of leaf wetness will alter infection dynamics. A notable example of this is the defoliation of sugar maple by forest tent caterpillar (*Malacosoma disstria* Hübner) in northern Pennsylvania in 1994. Defoliation was followed by an unusually moist period from July through October. During portions of this period defoliated trees tried to refoliate and young succulent leaves were colonized by the anthracnose fungus *Discula campestris* (Pass.) von Arx, which prevented refoliation (Hall, 1995). High levels of mortality and extensive crown dieback were observed the following spring.

An increase in mean winter temperature in the northern hemisphere could increase inoculum survival over winter and provide higher levels of initial inoculum in the spring (Coakley, 1995). However, the incidence of frost damage may be expected to decrease along with the canker fungi that invade damaged tissues after injury from frost (Lonsdale and Gibbs, 1996). For agronomic crops, disease incidence of rusts and mildews (obligate parasites) are predicted to increase as the assimilate content of host plants increases. It is unknown whether a similar phenomenon will affect the incidence of rusts and mildews in forests. More vigorous growth of host foliage will lead to denser canopies accompanied by a microclimate more favorable for disease development (Coakley, 1995). With increased production of foliage and woody twigs, there will be a potential, depending on the harvesting method, for more residue after harvest and greater survival of saprophytic organisms. Some of these saprophytes are likely to be facultative parasites, and their incidence may increase.

A high degree of uncertainty is associated with predictions about pathogen dynamics because we do not know the magnitude or rapidity with which climate may be altered and because there is little empirical data about how specific hosts and pathogens will interact. Furthermore, climatic change may have a positive effect on one part of the disease triangle and a negative effect on another (Coakley, 1995). A variety of feedback mechanisms may limit effects associated with increasing CO_2 (Coakley, 1995). For instance, while increasing CO_2 may stimulate host growth, this growth could also be limited by increased competition from competing species or by increased weed or woody shrub competition. Similarly, increased temperature could be beneficial to some plant processes, but could also increase evapotranspiration and water loss. A stressed host plant may then be more vulnerable to the effects of infection by primary pathogens or susceptible to secondary pathogens.

Geographical Range Changes of Insects and Pathogens

The expansion of geographical ranges of many insect species to higher latitudes and elevations is highly likely as a result of the combined effects of the previously cited factors (Porter et al., 1991; Cammell and Knight, 1992). That is, overwinter survival will be enhanced, the capacity of increase will be augmented by more and earlier generations in some species, and dispersal may increase (Porter et al., 1991). For herbivores, presence of the host species is also a requisite of range expansion. Agricultural pests may expand distributions quickly with climate change as growers plant crops at higher latitudes and elevations (Porter, 1995). Although a considerably slower process, even with human intervention, tree species may also migrate, facilitating the range expansions of associated forest insect species (Leverenz and Lev, 1987; Davis and Zabinsky, 1992; Franklin et al., 1992; Dyer, 1995).

Perhaps because significant climate change has not yet been definitively discerned, there have been few studies documenting actual range changes in insects. Through a recent survey of established study sites and a comparison with historical records of occurrence, Parmesan (1996) documented changes in the latitudinal distribution of Edith's checkerspot butterfly over the past 30 years in western North America. Her analyses suggested a subtle thinning out at the southern end of the species distribution and an increased occurrence at northern sites, as might be expected with a northward shift of the species. Similarly, Kozár (1991) and Stollár et al. (1993) documented the northward spread of a number of agricultural pest insects in Hungary over the past two decades under the influence of milder winter temperatures.

In the absence of a clear climate signal and unequivocal observed shifts in species ranges, the only available approach to anticipating range changes is through models. Simulation of changes over large regions often entails using models in conjunction with geographic information systems. Observed species distributions, or some approximation to them, such as detectable defoliation in the case of forest defoliators, are fit as functions of environmental variables. Typically, the best information available to characterize distributions is a map of presence or absence of a species (Williams and Liebhold, 1995c). Environmental variables usually include climatic variables, such as long-term averages of temperature and precipitation, and the distribution of the host species in the case of herbivores (Williams and Liebhold, 1995c), or another indicator of habitat suitability (Rogers and Randolph, 1993). Models are fit using ambient conditions, and then they may be used to extrapolate the effects of changes in climate or other environmental factors.

Modeling studies of forest pests have demonstrated potential for considerable changes in distribution under climate change. Using a regression approach, Virtanen et al. (1996) modeled the frequency of outbreaks by the European pine sawfly, *Neodiprion sertifer* (Geoffroy), in municipalities of Finland over the period 1961 to 1990. Outbreak distributions at two spatial scales were most significantly related to minimum winter temperatures. Under an increase in winter temperature of 3.6°C, the best guess prediction for temperature change by the year 2050 according to a Finnish government policy scenario, outbreak frequencies were projected to increase dramatically over the entire country, with the greatest changes in northern regions.

Williams and Liebhold (1995c) investigated potential changes in spatial distribution of outbreaks of the western spruce budworm, *Choristoneura occidentalis* Freeman, and the gypsy moth in the U.S. states of Oregon and Pennsylvania, respectively, using maps of historical defoliation, climate, and incidence of susceptible forest types. Relationships between outbreak incidence and the environmental variables were modeled using a linear discriminant function. With an increase in temperature alone (+2°C), the projected defoliated area decreased relative to ambient conditions for the budworm and expanded slightly for the gypsy moth. With an increase in temperature and precipitation (+0.5 mm day^{-1}), the defoliated area expanded for both species. By contrast, the defoliated area decreased for both when temperature increased and precipitation decreased. Results for GCM scenarios contrasted sharply. For the GFDL GCM, defoliation by budworm was projected to cover Oregon completely, whereas no defoliation was projected by gypsy moth in Pennsylvania. For the GISS model, defoliation disappeared completely for the budworm and slightly exceeded that under ambient conditions for the gypsy moth. Williams and Liebhold (1995c) interpreted the results as due in part to changes in distribution of the susceptible forest types.

In recent studies, Williams and Liebhold (1997a,b) investigated potential changes in range of eastern spruce budworm and gypsy moth over wide geographical regions in the U.S. Changes in distribution of the susceptible forest types were also projected as functions of climatic variables using a discriminant function model, and those forest scenarios were used in generating the respective defoliator outbreak scenarios.

Changes in outbreak distributions of eastern spruce budworm under three climatic warming scenarios were projected over much of the species' range in the northeastern and north central U.S. (Williams and Liebhold, 1997a). In general, predicted areas of defoliation and of the susceptible forests (i.e., the spruce–fir forest type group) decreased in size with increases in temperature. As temperatures increased, the distribution of defoliation exhibited a general pattern of thinning and disappearance at its southern margin, suggesting a northward shift of budworm populations.

Potential range shifts by gypsy moth and the oak forests that it inhabits were projected in the northeastern U.S. (Williams and Liebhold, 1997b). Responses of both susceptible forest (i.e., the oak–hickory and oak–pine forest type groups) and defoliation distributions to rising temperature were projected as increases in area of oak forests and in areas of those forests defoliated by gypsy moth. The gypsy moth outbreak distribution was projected to occupy nearly 29% of the study area under ambient conditions and increased rapidly under three warming scenarios, reaching 100% at +6°C (Fig. 13.1). The defoliation projections exhibited a northward shift with increasing temperature, conforming with the general expectations for latitudinal shifts under climate change.

Changes in pathogen ranges have been predicted by computer modeling or climate matching programs. One such program, CLIMEX, uses climatic data to derive an ecoclimatic index for predicting an organism's potential relative abundance and distribution (Brasier and Scott, 1994). CLIMEX was used with *P. cinnamomi* and appropriate temperature and soil moisture values to predict where the pathogen is currently distributed

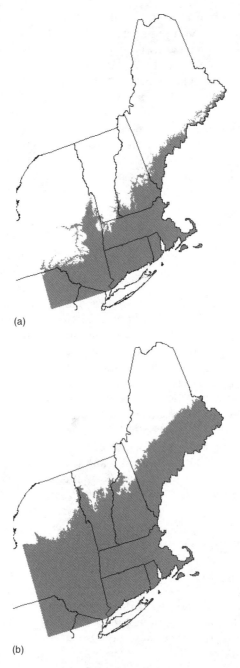

(a)

(b)

Figure 13.1. Predicted areas of gypsy moth defoliation in the northeastern United States under ambient conditions and three climate change scenarios: (a) ambient temperature, (b) 2°C increase, (c) 4°C increase, and (d) 6°C increase.

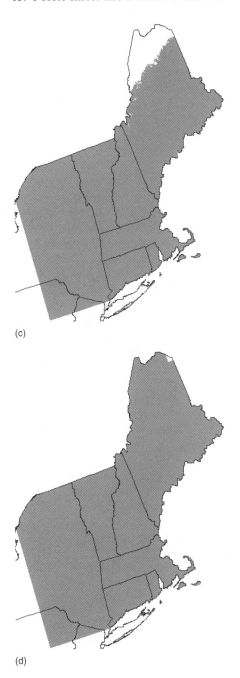

(c)

(d)

Figure 13.1. (*Continued*)

in Europe, thus allowing some verification of the matching process. Various predictions of change in the pathogen's distribution in response to changes in precipitation or temperature inputs can be examined (Brasier and Scott, 1994). The CLIMEX model suggests that given a warming of 1.5 to 3°C, *P. cinnamomi* will increase significantly within its present locations, in southern Europe and western coastal regions, accompanied by some extension of the pathogen's activity into the European continent, but no extension into regions with very cold winters (Brasier and Scott, 1994).

Indirect Effects of Climate Change on Community Processes

Seasonal Synchrony of Insect Herbivores with Host Plants

For insect herbivores, synchrony of hatch or emergence from the overwintering stages with growth and development of host plant tissue may have important consequences for early season growth and survival. Ayres (1993) noted that many forest insect species are specialists on immature plant tissues early in the season. Developing plant tissues, such as expanding leaves, tend to be low in fiber, high in nutritional value, and low in defensive compounds. Higher temperatures narrow the window of availability of such tissues, potentially altering herbivore growth rates and survival. For example, hatch of first instar leaf-eating caterpillars significantly before budbreak may result in starvation or in stunted development due to a delay in feeding (Hunter, 1993). Conversely, delayed hatch may result in caterpillars feeding on older plant tissues, which may reduce growth rates significantly even with a delay of only a few days (Ayres, 1993). Three studies, one using models and the other two controlled field experiments, illustrate this point. Dewar and Watt (1992) compared predictions of temperature-based models for the timing of budbreak in Sitka spruce and larval emergence in the winter moth, *Operophtera brumata* L. In the host model, budbreak timing depended on the accumulation of "chilling time" below a threshold, whereas larval emergence time was based on degree days above a threshold. When daily temperature increases of 0.5 to 2.0°C were simulated, warmer temperatures resulted in the predicted date of budbreak staying about the same, while the date of emergence was advanced. With a 2°C increase, larvae were predicted to emerge 20 days before budbreak in the host, potentially jeopardizing their survival. Alternatively, a recent experimental field study found that an increase of 3°C did not alter the synchrony of winter moth emergence and budbreak in pedunculate oak, a common host of winter moth in the UK (Buse and Good, 1996).

Hill and Hodkinson (1992) compared the synchrony of two psyllid species that occupy separate ranges in a latitudinal gradient. Both species have a common host, dwarf willow, and feed on its catkins, which are available for only a short window in the summer. In field experiments, the researchers increased temperatures by 0.6 to 1.4°C by enclosing trees in clear polyethylene cages. For both psyllid species, synchrony between catkin and nymph development inside the cages decreased as compared with controls. In one case, the asynchrony resulted in significantly smaller adults, while in the other, nymph survival increased and adults were significantly larger.

Using similar methods, Ayres (1993) compared developmental rates of *Epirrita autumnata* (Borkhausen) caterpillars on leaves of mountain birch in the field under ambient conditions and inside greenhouses constructed around trees. Higher temperatures ($\sim+1°C$) inside the greenhouses increased caterpillar growth rates, decreased developmental times, and resulted in larger pupae as compared with controls. The overall effect was a rise in the rate of population increase by as much as 2.9 times with a relatively small rise in temperature. The temperature increase apparently affected growth rates of caterpillars much more than those of leaves, allowing caterpillars to benefit from younger leaf tissue longer in the greenhouse treatment.

Over the long run, it is difficult to assess how important seasonal asynchrony resulting from climate change may be. As pointed out by Dewar and Watt (1992), there is considerable natural variability in the timing of budbreak and emergence, and because greenhouse warming likely will be a gradual process, genetic selection by the herbivore may quickly remedy transient asynchrony.

Insect–Plant Interactions under Increasing Carbon Dioxide

Increases in atmospheric CO_2 are likely to alter insect herbivore–plant interactions considerably, primarily through effects on plant growth and development (Ayres, 1993). The projected doubling of CO_2 concentrations over the next century (Houghton et al., 1996) probably will not have direct effects on insect herbivore populations, as insects are known to be able to tolerate CO_2 concentrations far in excess of projections (Graves and Reavey, 1996). Instead, effects will be mediated through changes in the nutritional quality of their host plants. Plants may be affected in several ways by increased availability of CO_2. In general, faster growth rates with increased CO_2 will increase the C/N ratio, resulting in lower tissue concentrations of N. Higher CO_2 levels are hypothesized to increase the potential for producing carbon (C)-based allelochemical defenses (Lincoln et al., 1993; Lindroth, 1996). Higher CO_2 also results in higher starch content of leaves and in increased tissue water content because of increased water use efficiency. Finally, plants grown in higher CO_2 conditions are hypothesized to have leaves with higher specific weight and greater toughness, although the latter phenomenon has been reported in only one experimental study (Lindroth et al., 1993).

Such changes in host plants may have several effects on insect herbivores. Decreasing N concentrations in plant tissues are generally detrimental, decreasing growth rates, increasing developmental time, and decreasing fecundity, all of which involve processes dependent on amino acids (Lincoln et al., 1993; Lindroth, 1996). Increased levels of allelo-chemicals often may have deleterious effects on growth and survival. In response to lower N and increased starch, herbivores may increase consumption rate, presumably to maintain a relatively constant N intake (Lincoln et al., 1993). In later larval stages, increased consumption may allow a relatively normal growth rate. However, earlier stages may not be able to compensate for lower nutritional quality in that manner, may not be able to eat tougher leaves, and, therefore, may suffer slower growth and increased likelihood of mortality.

Relatively few studies to date have focused on forest insects (Lindroth et al., 1993; Roth and Lindroth, 1994; Herms et al., 1996; Lindroth et al., 1996; Watt et al., 1996; Williams et al., 1997a,b; Williams et al., 1998). Lindroth et al. (1993) investigated changes in plant growth rate and leaf chemistry for three tree species, quaking aspen, red oak, and sugar maple, grown under enriched CO_2. In addition, they investigated consumption and growth parameters for late instar gypsy moth and forest tent caterpillar, *Malacosoma disstria*, fed leaves from the experimental seed-lings. Overall, the effects of increased CO_2 were most pronounced on aspen, the fastest growing tree species, and followed the hypothetical patterns just outlined: for both herbivore species, consumption rates were higher, stage durations were longer, and growth rates were lower for larvae fed leaves grown under enriched CO_2. These patterns probably resulted from changes in leaf N concentration and levels of allelochem-icals. Results were less clear for the other host species, perhaps reflecting the rather idiosyncratic natures of the differently coevolved plant–herbivore systems. Relative growth rates of forest tent caterpillar were lower on high-CO_2 maple leaves, whereas gypsy moth rates did not differ significantly between treatments and controls on maple. The negative effects on forest tent caterpillar growth were likely due to higher leaf tannin levels, although this effect was ameliorated somewhat under high CO_2. Alternatively, relative growth rates of gypsy moth were higher for high-CO_2 than for ambient CO_2 oak leaves, while forest tent caterpillar rates did not differ significantly on oak. The better performance of gypsy moth resulted from increased levels of starches and decreased levels of tannins in high-CO_2 oak leaves.

Herms et al. (1996) investigated effects of three aspen clones grown under increased CO_2, increased ozone (O_3), and an increase of both gases on first and fourth instars of four forest defoliator species: gypsy moth; forest tent caterpillar; large aspen tortrix, *Choristoneura conflictana*

(Walker); and whitemarked tussock moth, *Orgyia leucostigma* J.E. Smith. The aspen clones differed in their sensitivity to O_3, with one clone susceptible, another resistant, and the third intermediate. Growth and consumption rates and other derived parameters were estimated after feeding caterpillars excised leaves for 48 hours. No effects on larval survival were observed under increased CO_2. Similar to results from previous studies and hypothetical expectations, growth rates were generally lower for all species and instars under enhanced CO_2 (except for first instar forest tent caterpillars grown on the O_3-resistant aspen clone, in which case growth rate increased). Efficiency of conversion of digested food varied with aspen clone, showing a significant increase overall on the O_3-resistant clone under enhanced CO_2 and a significant decrease on the other clones. Unlike results from previous studies and hypothetical expectations, consumption rates were unaffected by increased CO_2, except for the large aspen tortrix, in which consumption rate decreased. Although the effects or air pollutants are beyond the scope of this review, it is interesting to note that Herms et al. (1996) reported significantly increased growth rates for all defoliator species on leaves produced under increased O_3. This increase apparently resulted from an increased efficiency of conversion, probably due to a decrease in concentration of tremulacin, an aspen allelochemical, under the O_3 treatment.

It is difficult to use studies of the effects of increased CO_2 on host plants and herbivores to draw general conclusions about effects at the population level and over long periods. Most studies were carried out in the laboratory and involved effects on individual insect stadia, ignoring effects over the entire life cycle and on fecundity in particular (Lawton, 1995; Watt et al., 1995; Awmack et al., 1997a). Thus, it is difficult to speculate how such effects may affect population dynamics. Potential effects of increased CO_2 on interactions with higher trophic levels are virtually unexplored. The only published study in this area suggests that herbivores may be more vulnerable to their natural enemies under doubled CO_2 concentrations (Awmack et al., 1997b). With chronic exposure to increased CO_2, plants may acclimate to higher levels, potentially altering their interactions with herbivores further over the long term (Watt et al., 1995). A final deficiency of extant studies is that they have tended to look at only one feeding guild, leaf chewers, and generally ignored sucking insects, leaf miners, gall-forming insects, and borers (Watt et al., 1995). Under current theories of the effects of increased CO_2 on plants, the potential effects on these feeding guilds are uncertain.

Pathogen–Plant Interactions under Increasing Carbon Dioxide

Host responses to elevated CO_2 and alterations in host growth, biomass, physiology, and nutrition can be expected to affect interactions with pathogens and diseases. From studies conducted under controlled conditions, only a few generalizations are possible concerning host responses. For instance, most studies have shown an increase in total leaf area due to larger leaves or greater numbers of leaves produced (Ceulemans and Mousseau, 1994). However with yellow poplar, *Liriodendron tulipifera* L., there was a reduction of individual leaf area in response to elevated CO_2 and an increase in fine root production (Norby et al., 1992). Most studies examining woody plant responses to elevated CO_2 have shown that dry weight per unit of leaf area increases and is likely related to increased starch content or additional cell layers (Eamus and Jarvis, 1989). Additional leaf starch may enable some foliar pathogens to increase their colonization of host tissues and increase the area of leaf necrosis, though this has not been demonstrated experimentally. Conversely, under elevated CO_2 leaf N concentrations decrease and secondary compounds, such as tannins and phenolics, increase, which may make leaves less palatable to some insect pests, while the effects on pathogens are unknown (Lindroth et al., 1993).

The shift in production from aboveground components to belowground fine roots indicates that significant C inputs may induce an increase in soil microbial biomass and activity (Zak et al., 1993). This may favor growth of some soil-borne fungal pathogens or mycorrhizae, but, again, experimental evidence is lacking. Pathogenesis in root systems is exacerbated by stresses induced by weather extremes, especially drought during a growing season (Lonsdale and Gibbs, 1996). Root diseases caused by the root-rotting fungi in the genus *Armillaria* can be expected to increase with increasing host stress due to drought or other causes (Wargo and Harrington, 1991). Other diseases are linked to drought or other stressors that may increase in temperate forests under global warming scenarios, though empirical data are lacking. Many of these involve secondary pathogens that usually become pathogenic only when the tree is weakened. Many decline diseases fall into this category and they are discussed in more detail in Chapter 4.

Interactions of Mycorrhizal Fungi and Plants under Increasing Carbon Dioxide

While mycorrhizas are clearly not pathogens, they form mutualistic associations with tree roots that enhance nutrient and water uptake and protect trees from some soil-borne pathogens. These associations are likely to be altered by elevated CO_2. The mycorrhizal association is more efficient since it requires less energy for the host to produce and maintain the mycorrhizal network than it does for the host to produce plant roots (Marshall and Perry, 1987). The fungus provides the host with nutrients, while the host provides the fungus with C (Harley and Smith, 1983). Mycorrhizal fungi protect trees from certain root pathogens (Marx, 1970) and some mycorrhizal species are more effective than others in providing this protection (Malajczuk, 1988). Interspecific competition among mycorrhizal fungi may therefore affect disease incidence (Lonsdale and Gibbs, 1996). Increases in drought frequency will enable other competing fungi to replace established species in mycorrhizal communities due to the death and subsequent regeneration of nonwoody roots (Lonsdale and Gibbs, 1996). These altered mycorrhizal communities may affect disease susceptibility depending on the dominant fungi in the mycorrhizal communities.

Interactions among hosts and mycorrhizal fungi in response to elevated CO_2 may result in a more mutualistic, a less mutualistic, or even a parasitic association (Sanders, 1996). Most recent research has shown inconsistent effects associated with growth in elevated CO_2. The work of Tingey et al. (1995) with ponderosa pine (*Pinus ponderosa* Laws.) and Norby et al. (1992) with yellow poplar show that elevated CO_2 stimulated belowground carbon allocation. This resulted in an increase in mycorrhizal occurrence and a consistent, though statistically nonsignificant increase in root area. With yellow poplar, no significant growth response to elevated CO_2 was found after three years, but there was a sustained increase in leaf-level photosynthesis and lower rates of foliar respiration for trees grown in elevated CO_2 (Norby et al., 1992). There was a consistent trend of a decrease in leaf production and leaf area and an increase in fine root production for trees growing in elevated CO_2. This implies that as CO_2 stimulates aboveground production at least some of the C is allocated to belowground components, such as mycorrhizal production or extraradicle hyphae, in order to enable support of the aboveground production. Similarly, O'Neill et al. (1987) found an increase in mycorrhizal density in response to elevated CO_2 for shortleaf pine (*Pinus echinata* Mill.) and white oak (*Quercus alba* L.).

In addition to the stimulation of belowground C allocation from elevated CO_2, Tingey et al. (1995) found that soil temperature was a key factor influencing seasonal changes in root area density and new root area density. In the summer, when soil temperatures were 15 to 22°C (at 15 cm), new root flushes were produced in ponderosa pine seedlings. Thus, while changes in air temperature will not directly influence root growth, as the soil warms increased root biomass may be produced. However at soil temperatures >25°C ponderosa pine root growth declined significantly (Lopushinsky and Max, 1990). Both the magnitude and duration of soil warming will influence mycorrhizal root growth just as it will affect soil-borne pathogens. Soil warming will have a significant influence on biological activity, including root growth and decomposition, both of which are likely to be accelerated as soil temperature increases (Luxmoore et al., 1993).

At the ecosystem scale, elevated CO_2 could affect the interactions among trees, shrubs, and herbaceous vegetation. The complexity of these interactions is suggested by research in a calcareous grassland (Leadley and Korner, 1996). The percentage of root length in the total plant community that was colonized by mycorrhizal fungi was not significantly affected by elevated CO_2 concentrations of 500 or 650 ppm as compared with ambient levels. However, a significant change in community structure was observed as a result of detrimental growth effects (decreased biomass) on *Prunella vulgaris* L. In elevated CO_2 there was a significant increase in the proportion of the root length being occupied by mycorrhizal fungi for *P. vulgaris*, thus providing a large C sink that may have inhibited growth. A coexisting species, *P. grandiflora* L., showed a significant decrease in mycorrhizal colonization, but no significant change in biomass in response to elevated CO_2. Response of extraradicle mycorrhizal hyphae were not evaluated in this study, but may also account for some of the observed differences. These differential mycorrhizal responses could, over time, lead to a reduction of *P. vulgaris* in the grassland community and a change in the community structure (Sanders, 1996). The effects of elevated CO_2 on interactions among diverse tree and mycorrhizal species have not been studied in forest communities and, therefore, few predictions can be made with any certainty.

Relationships between Insect Herbivores and Their Natural Enemies

Climate change may alter the interactions of herbivores and their natural enemies, resulting potentially in changes in levels of natural control and dynamics of the herbivore (Cammell and Knight, 1992). As with the host plant–herbivore interactions, changes in synchrony between herbivores and their natural enemies during the early season may affect the subsequent levels of control. Dynamics of the interaction during the season may also be affected because critical population parameters, such as the rates of population increase and the functional responses of predators and parasitoids, may vary with temperature. Lengthened growing seasons under a warming climate may increase the time over which overwintering stages of the herbivore are exposed to predation or parasitism. More generally, some natural enemies may be critically limited by specific climatic factors, such that regional climate changes may alter their capacity to attack hosts over large areas.

The potential for seasonal asynchrony between herbivores and natural enemies is apparent in a study by Campbell et al. (1974). They reviewed temperature requirements for several aphid species and their parasitoids, including both the lower thresholds for development and the degree-day sums, or thermal constants, required to complete development. In most cases, hosts had lower thresholds and shorter developmental times, suggesting the potential for varying levels of synchrony under different temperature regimes. Effects of phenological asynchrony were explored by Hassell et al. (1993) and Godfray et al. (1994) using a discrete time host–parasitoid model. The researchers investigated the effects of varying the day of parasitoid emergence from overwintering relative to the host and the level of density dependence on the stability of the two-species system. Increasingly later parasitoid emergence provided a temporal host refuge, permitting host populations to increase successively. At the latest emergence dates, stability was not possible. Simulations run with variable asynchrony, as might be expected with weather variability under climate change, were generally unstable (Hassell et al., 1993).

The potential for increased exposure of overwintering hosts to parasitism with rising temperature is seen with *Ooencyrtus kuvanae* (Howard), an egg parasitoid of the gypsy moth. Gypsy moth overwinters in diapause during the egg stage from about July until April each year (Brown, 1984). The parasitoid attacks host egg masses during the fall and early spring, often inflicting high levels of parasitism depending upon the size of an egg mass (Williams et al., 1990). *Ooencyrtus kuvanae* females are not generally active during the winter months, but are not in diapause. As a result, numbers of fall generations of the parasitoid have been reported to vary from two to five over the its range in North America, apparently in response to climatic differences (Brown, 1984). Increases in fall and winter temperatures clearly may permit additional generations in a specific location, resulting in higher levels of parasitism.

Climatic limitations to natural control of herbivores are generally appreciated (Messenger, 1971), but some natural enemies, such as insect pathogens, may be more definitively limited by critical dependence of their life cycles on temperature and humidity. An example is *Entomophaga maimaiga* Humber, Shimazu & Soper, a fungal pathogen of gypsy moth larvae, whose epizootics decimated host populations over wide areas of the northeastern U.S. in 1989 and 1990 (Hajek et al., 1990). Those epizootics were associated with higher than average rainfall in May, and infection levels of the fungus increased with additional moisture in controlled studies (Hajek et al., 1996). The fungus continues to increase its range, but its ability to produce epizootics under changed climatic conditions likely will depend upon relative levels of temperature and rainfall.

Potential Effects of Changing Climate on Major Forest Insect Pests of the Northeastern U.S.

Because sufficient distribution data, historical time series, and basic ecological knowledge may not be available, it is generally impossible to develop reliable models and make quantitative predictions of the effects of climate change on population dynamics and range changes for even common pest insects. Nevertheless, it may be useful to speculate on relative positive or negative effects of environmental change based simply on qualitative characteristics of insect herbivores and their interactions with trophic levels above and below. Landsberg and Stafford Smith (1992) proposed such a "functional scheme" for predicting the effects of climate change on the likelihood of outbreaks by herbivorous insects. As a conceptual model, they used a tritrophic-level system consisting of host plant, herbivore, and natural enemy complex. They enumerated the various critical trophic-level interactions that may produce density-dependent regulation of the herbivore and then identified the interactions most likely to be sensitive to climate change. For specific systems, they then attempted to identify the interactions most critical for regulation from above or below and predict how climate change may enhance or ameliorate the likelihood of herbivore outbreaks. For herbivores, they considered the plant tissues fed upon, number of generations per season, cues to initiate development in the spring and halt activity in the fall, and shelter locations during winter (e.g., in litter), and speculated how those attributes may be affected directly by change in CO_2 level or climate. Using the larch bud moth, *Zeiraphera diniana* Guenée, as an example, they proposed that the critical factor is the cue to initiate activity in the spring. Warm spring weather may enhance synchrony with host budbreak and increase the likelihood of outbreaks, whereas warm winters may increase overwintering mortality and decrease the likelihood of outbreaks. An obvious shortcoming of this scheme is that a particular system may have multiple critical interactions with conflicting responses to environmental change that confound definitive predictions.

How may climate change affect the most prevalent and damaging forest pests in the northern U.S.? Table 13.1 lists a selection of such species identified as significant forest pests in a recent report by the United States Department of Agriculture (USDA) Forest Service Forest Health Monitoring Program (USDA Forest Service, 1994). Using ideas similar to those of Landsberg and Smith (1992), we speculate on the likelihood of increased outbreak activity for individual species based on the biological and ecological characteristics listed in Table 13.1.

Several species show potential for changing activity given the host tissues on which they feed. The two beetle species may be affected in different ways if precipitation patterns change. If precipitation decreases, spruce beetle, an inner bark feeder, may be favored by the weakening effects of water stress on its host (Ayres, 1993). Alternatively, as a phloem feeder, the white pine weevil is more productive with vigorous growth of host shoots (Hamid et al., 1995), which will be enhanced under an increase in precipitation. Among the leaf feeders, spruce budworm is a specialist on young leaf tissue, and its dynamics may change if increasing temperature alters its synchrony with budbreak of its host (Fleming, 1996).

An increase in the number of generations per season is likely to be relevant for only two species, spruce beetle and hemlock woolly adelgid. Spruce beetle has a biennial life cycle over much of its range, but may

Table 13.1. Ecological Characteristics of Selected Forest Insect Species of the Northern United States

Common name	Species (Order)	Generations per Year	Overwintering Stage	Diapause Physiology	Outbreak Type	Origin	Primary Hosts	Host Tissue
Eastern Spruce Budworm[a]	Choristoneura fumiferana (Clemens) (Lepidoptera)	1	Larva	+	Eruptive	North America	Balsam fir	Leaf
Gypsy Moth[b]	Lymantria dispar (L.) (Lepidoptera)	1	Embryonated egg	+	Eruptive	Europe	Oak species	Leaf
White Pine Weevil[c,d]	Pissodes strobi (Peck) (Coleoptera)	1	Adult	–	Gradient	North America	Eastern White Pine	Phloem
Spruce Beetle[e,f]	Dendroctonus rufipennis (Kirby) (Coleoptera)	1, 2, or 3 yr life cycle (2 yr most common)	Larva, Adult	–	Eruptive	North America	Spruce species (mature stands)	Inner bark &phloem
Hemlock[g] Woolly Adelgid	Adelges tsugae Annand (Homoptera)	3	Nymph	–	Gradient	Asia	Eastern Hemlock	Parenchyma

a Mattson et al. (1987).
b Montgomery and Wallner (1987).
c Hamid et al. (1995).
d Retnakaran and Harris (1995).
e Holsten et al. (1991).
f Safranyik (1995).
g McClure (1996).

complete development in one year or extend to three years dependent on regional temperatures in its range (Holsten et al., 1991). As an example, spruce beetle uncharacteristically completed an entire life cycle throughout Alaska during the unusually warm summer of 1993. Given this flexible life cycle, it seems likely that many populations will reduce to a univoltine condition under a warming climate, allowing more rapid development of outbreaks. Hemlock woolly adelgid produces three asexual generations per year in Connecticut, one during summer and two during fall, winter, and spring (McClure, 1996). Milder winters will favor population growth if they permit another generation. Under the current climate, development time of overwinter generations can vary greatly with weather conditions, suggesting the possibility of further generations under a warmer climate (McClure, 1996).

Three species are endemic (see Table 13.1), with ranges presumably co-evolved with their hosts' ranges. Warmer temperatures will increase the potential for northward extension of their ranges, although this is likely to be a long process involving simultaneous movement of insect and host populations (Williams and Liebhold, 1995c). The two exotic species, gypsy moth and hemlock woolly adelgid, are still expanding their ranges (Liebhold et al., 1992, 1995; McClure, 1996). Both appear to be limited currently by cold temperatures at the northern extremes of their distributions, and, thus, warmer winters will favor increased movement into unexploited ranges of their hosts. However, Allen et al. (1993) predicted that gypsy moth will not do well in Florida under that state's warm climatic conditions, suggesting that it soon may encounter the southern limit of its distribution in the U.S.

Several species, such as gypsy moth and spruce budworm, display eruptive outbreaks. Because such dynamics involve complex interactions with hosts and natural enemies (Turchin, 1990; Williams and Liebhold, 1995a), the effects of climate change on, for example, changing outbreak frequency are impossible to predict. Under the Moran effect, however, regional synchrony of outbreaks may be altered if regional climatic patterns change (Moran, 1953b; Williams and Liebhold 1995b).

Overwintering may be variably affected by milder winters. In general, survival will increase with fewer periods of lethal low temperatures. However, in species that overwinter as eggs in obligate diapause, such as gypsy moth, exposure to warmer temperatures may prolong diapause development, resulting in later emergence and possible asynchrony with host development (Williams et al., 1990).

Overall, hemlock woolly adelgid seems most likely to be affected positively by climate change. It may increase numbers of generations, survive better during milder winters, begin activity earlier in spring, and disperse farther north as North America becomes warmer. Spruce beetle is also likely to benefit from warmer temperatures by altering its life cycle to univoltine in wider areas and by responding to host stress conditions if precipitation decreases.

Beyond the effects of climate change on a few serious pest species, there is a real possibility of the development of new pests. In a review of nonoutbreak species of lepidopteran defoliators of mixed fir forests in North America, Mason (1987) noted that for each major outbreak defoliator there are at least nine nonoutbreak species, whose roles in the forest ecosystem are not well known. Under significant climate change and accompanying changes in forest composition and disturbance levels, such defoliators may be released from control and become pests. Another likelihood is the introduction of new exotic pests. Although not a result of climate change per se, such introductions become more likely as a result of the wider scope of global change and increasing levels of international trade in forest products (Liebhold et al., 1995). Invasions of nonnative plant pathogens and herbivorous insects are another component of global change that may alter forest ecosystem function and reduce biodiversity (Vitousek et al., 1996). It is conceivable that climate change may open up new niches for invasion by exotic species after they are established and facilitate their spread.

Consequences of Changing Insect Outbreak Patterns for Forests

Climate change may have profound effects on forests of the northern U.S., as well as those of Europe and Asia (Fanta, 1992; Kobak et al., 1996). Not the least of the influences of a changing climate will be its effects on pest and fire disturbances (Kurz et al., 1995b). In North America, pest disturbances are likely to be more serious than fire in the wetter eastern half of the continent. However, the effects of pest disturbances are more difficult to assess because they often result in growth reduction, rather than mortality, they affect individual species selectively, and they may produce stand replacing mortality (Kurz et al., 1995b).

Despite the general opinion that forests should be net C sinks (Kauppi et al., 1992), the short-term negative effects of disturbances on C storage are likely to be dramatic even in the near future. In trying to assess the effects of climate change on disturbance, Kurz et al. (1995b) emphasized the importance of "asymmetry in the rate of change" of forest processes. That is, forest area and C in forests accumulate through regeneration at a much slower rate than they are lost through disturbance. Thus, forest dynamics are likely to be highly variable in the coming century. Kurz et al. (1995a) noted that aging boreal forests in Canada have become increasingly susceptible to disturbances in recent decades. As a result of a modeling study, Kurz and Apps (1995) reported that increased historical disturbances by fire and insects during the period 1970 to 1990 resulted in none of their scenarios being able to produce a C sink during the decade of the 1990s. Positive C fluxes were projected to resume only after the year 2000 and were not projected to approach recent historical levels until well into the next century (Kurz and Apps, 1995).

Another anticipated effect of climate change is the migration of forests over the long run. How they do so will determine equilibrium forest type compositions and affect the level of transient disturbances. The simplest assumption is that forest ecosystems will shift northward as entire integrated units (Fleming, 1996). Because species compositions and herbivore and natural enemy guilds do not change under such an assumption, the basic community regulatory mechanisms remain constant, and little change is to be expected with climate change beyond a simple relocation of the unit. Obviously, this assumption is too simplistic. More realistically, tree and insect species are likely to move individually (Fleming, 1996), according to their particular dispersion mechanisms and tolerance for conditions in their original ranges (Davis and Zabinski, 1992; Dyer, 1995). Distributions of individual tree species will shift as the result of different population processes over their current range. At the southern end, seedlings are expected to be less tolerant of warmer temperatures and only northward dispersers are expected to survive (Fleming, 1996). Senescence rates of mature trees are expected to increase, and that, coupled with possibility of stress due to climate change, may lead to catastrophic losses to insect herbivores. At the northern end of the range, seedling establishment will be possible increasingly farther north, but slow due to low temperatures (Fleming, 1996). Kurz et al. (1995b) note that population change processes will be asymmetrical at latitudinal extremes. At the northern extreme, change is expected to be slow, whereas at the southern end, change may be very rapid, giving the potential for quite variable transient distributions along the path to equilibrium.

Effects of Climate Change on Forest Decline Diseases

In a seminal paper, Hepting (1963) first summarized the impact of climate and climatic variation on forest diseases. Hepting presented the first comprehensive summary of evidence about climate change and its effects on pathogens and forest diseases. What Hepting termed "physiogenic" diseases are now generally referred to as decline diseases. Manion (1991) defined decline diseases as those "...which are caused by the interaction of a number of interchangeable, specifically ordered abiotic and biotic factors to produce a gradual general deterioration, often ending in the death of trees." In the early 1960s, the importance of decline diseases and the role of climatic fluctuation was just beginning to be appreciated. Pole blight of western white pine (*Pinus monticola* Dougl.) was known to have an edaphic–physiological cause involving fine root deterioration and soil physical properties (Leaphart, 1958). Climatic records indicated that in areas with pole blight there was a period of low precipitation and high temperature from 1917 to 1940 (Leaphart, 1958). Subsequent research (Leaphart and Stage, 1971) indicated that pole blight was triggered by the severe drought of 1936, though more recent research has disputed this claim and proposed that winter thaw–freeze events were the likely triggering factors for the decline (Auclair et al., 1990).

The theory that climatic perturbation is the mechanism that drives widespread forest dieback has been presented for northern hardwoods (Auclair et al., 1992). Region-wide declines of birch (*Betula allegheniensis* Brit., *B. papyrifera* Marsh.) from the 1930s to the early 1950s, and sugar maple (*Acer saccharum* L.) in the 1980s have been correlated with thaw–freeze events (Auclair et al., 1992). Both declines were preceded by a major thaw–freeze event that included two to four weeks of intense cold in the winter and exceptional episodes of early warm weather followed by severe frost in late winter or early spring (Auclair et al., 1992). These coincided with the years 1937 for birch and 1981 for sugar maple. However, this does not account for other maple declines that occurred in Wisconsin in the 1950s and in New England in the 1960s (Giese et al., 1964; Mader and Thompson, 1969).

Five different mechanisms have been proposed to account for changes in tree crown condition in response to these fluctuating climatic factors (Auclair et al., 1992). These include (1) direct injury from exceeding frost resistance by severe cold, (2) soil freezing and root injury due to scant snow cover or open winters, (3) exceeding frost resistance limits for rehydrated tissues following a warm period, (4) xylem cavitation induced by cold temperatures that followed an anomalous warm period, and (5) root injury from deep soil freezing that occurred with saturated soils following the warm period. Because the thaw resulted in saturation of the soil profile from snow melt, this enhanced the vulnerability to deep soil freezing since thermal conductivity of wet soils is 50 times the conductivity of dry soils of the same texture (Auclair et al., 1992). Other research in Quebec has shown that deep soil freezing induced root injury and resulted in increased crown dieback and increased crown foliage transparency for injured sugar maple (Robitaille et al., 1995). More mechanistic studies will be necessary to determine the importance of extreme climatic events on hypothesized region-wide decline scenarios.

While extreme freeze–thaw events are well documented and coincidental with major decline events, they do not account for the differential effects on specific tree species. It is difficult to account for the effects of such events on birch in the 1930s and 1940s and no effects on sugar maple until the 1980s. One current theory is that dieback susceptibility is related to tree maturation (Auclair et al., 1997). Specifically, regional forest dieback is synchronized with maturation of birch and maple populations in northern hardwood forests. The greater susceptibility of mature trees to predisposing stressors makes them more vulnerable to injury from extreme climatic events or other inciting factors (Auclair et al., 1997; Manion, 1991). These ideas are examined in Chapter 4.

The role of climate change, not just variability, is difficult to assess for decline diseases. Perhaps the most suggestive link is decline of yellow cedar (*Chamaecyparis nootkatensis* [D. Don] Spach) in southeast Alaska. For the high latitudes, some GCM models predict that early changes in mean temperatures will be greatest in the late fall/winter and eventually warming will be greatest in midwinter (Kräuchi, 1993). Changes in mean annual temperature are notable in Alaska (Juday, 1984). At Sitka, Alaska, mean annual temperature increased at a rate of 2.08°C in the first half of the century, but moderated to just 0.60°C in the period through 1981 (Juday, 1984). While few pre-1900 climatic records are available for Alaska, there are observations of glacial recessions which report continuous recession in southeast Alaska since 1850 (Heusser, 1952). The initiation of both the warming period and the extensive mortality associated with the decline occur around 1880 (Hennon and Shaw, 1994). Other research implicates the role of extreme freeze–thaw events in this decline (Auclair et al., 1990). Dieback is usually associated with extreme episodes of winter drying, unseasonable frost, summer drought, and heat stress (Auclair et al., 1990). No biotic agents have been found associated with the decline and the primary cause is probably abiotic (Hennon and Shaw, 1994). Climatic warming has likely reduced snowpack, altered decomposition dynamics, and induced soil freezing and root injury, though the exact mechanism of the decline has not been conclusively established (Hennon and Shaw, 1994). In particular, a long-lived species, such as yellow cedar, that does not reproduce often may be especially vulnerable to a moderate climatic shift since it has limited ability to adapt to a changing environment (Hennon and Shaw, 1997).

Conclusions

Even if the extreme GCM climate change projections for the northern U.S. come to fruition, the effects of such changes and accompanying rises in CO_2 on forest insect species are uncertain. The direct effects are likely to be most obvious. Most species will survive more successfully over winter with regional warming. Many will shift their ranges toward higher latitudes and elevations into hitherto unexploited ranges of their hosts. Species with flexible life histories will produce more generations per unit time, increasing their rates of intrinsic population increase. Such direct effects hold the potential to increase the frequency, level, and geographical extent of disturbance by current forest insect pests. The indirect effects are likely to be more subtle to detect and difficult to predict because they involve complex community interactions at multiple trophic levels. Species of defoliators exhibiting eruptive outbreaks, such as gypsy moth and spruce budworm, exemplify the difficulties and uncertainties of indirect effects. There is no consensus among ecologists as to the causes of their population oscillations, although they are likely to arise from complex interactions with their hosts and natural enemies. Anticipating the effects of changing climate on such systems is impossible. Similar difficulties apply to predicting the impacts of changing insect outbreak patterns on forest growth and dynamics. Such uncertainties suggest the need for increased monitoring efforts in future decades as the climate signal becomes stronger.

It is apparent that significant gaps already exist in our understanding of how global change will affect host–pathogen dynamics. The complexity of these interactions involves the many factors mentioned in this review: direct effects of increased CO_2 on hosts and pathogens, indirect effects of altered climate on hosts and pathogens, and interactions among hosts and pathogens in a climate-altered environment. Similar, though larger, gaps exist for ecosystem processes that include the role of pathogenesis in stand dynamics and succession, the role of decomposing organisms in cycling nutrients, and the alteration of C storage dynamics in the ecosystem.

If scenarios associated with increasing CO_2 are correct, then drought frequency will increase in temperate forests and trees and forests may become chronically stressed. Such stresses may lead to progressive deterioration in tree health, increasing susceptibility to secondary organisms (Sutherst, 1996), and perhaps an increase in incidence of decline diseases. Soil-borne pathogens sensitive to soil temperature will spread into new geographical areas from which they are presently excluded due to low temperatures. Alterations in pathogen occurrence and host distribution will alter patterns of disease occurrence both in type and amount (Sutherst, 1996). New disease complexes could emerge with different pathogen and host assemblages. The reproduction rate and number of disease cycles per year may increase under warmer conditions, increasing disease severity and accelerating the evolution of new pathotypes (Sutherst, 1996). The difficulty of studying these numerous interacting factors may account for the paucity of direct experimental data available in the literature. Hopefully, future research will address these issues.

References

Allen JC, Foltz JL, Dixon WN, Liebhold AM, Colbert JJ, Regniere J, Gray DR, Wilder JW, Christie I (1993) Will the gypsy moth become a pest in Florida? Florida Entomol 76:102–113.

Andrewartha HG, Birch LC (1954) *The Distribution and Abundance of Animals*. University of Chicago Press, Chicago, IL.

Auclair AND, Eglinton PD, Minnemeyer SL (1997) Principal forest dieback episodes in northern hardwoods: development of numeric indices of areal extent and severity. Water Air Soil Pollut 93:175–198.

Auclair AND, Lill JT, Revenga C (1996) The role of climate variability and global warming in the dieback of northern hardwoods. Water Air Soil Pollut 91: 163–186.

Auclair AND, Martin HC, Walker SL (1990) A case study of forest decline in western Canada and the adjacent United States. Water Air Soil Pollut 53: 13–31.

Auclair AND, Worrest RC, Lachance D, Martin HC (1992) Climatic perturbation as a general mechanism of forest dieback. In: Manion PD, Lachance D (eds) Forest Decline Concepts. APS Press, St Paul, MN, pp 38–58.

Awmack CS, Harrington R, Leather SR (1997a) Host plant effects on the performance of the aphid Aulacorthum solani at ambient and elevated CO_2. Global Change Biol 3:545–549.

Awmack CS, Harrington R, Leather SR (1997b) Climate change may increase vulnerability of aphids to natural enemies. Ecol Entomol 22:366–368.

Ayres MP (1993) Plant defense, herbivory, and climate change. In: Kareiva PM, Kingsolver JG, Huey RB (eds) Biotic Interactions and Global Change. Sinauer Associates, Sunderland, MA, pp 75–94.

Barron EJ (1995) Climate models: how reliable are their predictions? Consequences 1:17–27.

Berryman AA (1987) The theory and classification of outbreaks. In: Barbosa P, Schultz J (eds) Insect Outbreaks. Academic Press, San Diego, CA, pp 3–30.

Brasier CM (1996) Phytophthora cinnamomi and oak decline in southern Europe. Environmental constraints including climate change. Ann Sci For 53:347–358.

Brasier CM, Scott JK (1994) European oak declines and global warming: a theoretical assessment with special reference to the activity of Phytophthora cinnamomi. OEPP/EPPO Bull 24:221–232.

Brown MW (1984) Literature review of Ooencyrtus kuvanae, an egg parasite of Lymantria dispar. Entomophaga 29:249–265.

Buse A, Good JE (1996) Synchronization of larval emergence in winter moth (Operophtera brumata L.) and budburst in pedunculate oak (Quercus robur L.) under simulated climate change. Ecol Entomol 21:335–343.

Cammell ME, Knight JD (1992) Effects of climatic change on the population dynamics of crop pests. Adv Ecol Res 22:117–162.

Campbell A, Frazer BD, Gilbert N, Gutierrez AP, Mackauer M (1974) Temperature requirements of some aphids and their parasites. J Appl Ecol 11:431–438.

Coakley SM (1995) Biospheric change: will it matter in plant pathology? Can J Plant Pathol 17:147–153.

Crandall BS, Gravatt GF, Ryan MM (1945) Root disease of Castanea species and some coniferous and broadleaf nursery stocks caused by Phytophthora cinnamomi. Phytopathology 35:162–180.

Cubasch U, Cess RD (1990) Processes and modeling. In: Houghton JT, Jenkins GJ, Ephraums JJ (eds) IPCC (Intergovernmental Panel on Climate Change) Climate Change: The IPCC Scientific Statement. Cambridge University Press, Cambridge, UK, pp 69–91.

Cuelemans R, Mousseau M (1994) Tansley Review No. 71. Effects of elevated atmospheric CO_2 on woody plants. New Phytol 127:425–446.

Davis MB, Zabinski C (1992) Changes in geographical range resulting from greenhouse warming: effects on biodiversity in forests. In: Peters RL, Lovejoy

TE (eds) *Global Warming and Biological Diversity*. Yale University Press, New Haven, CT, pp 297–308.

Dewar RC, Watt AD (1992) Predicted changes in the synchrony of larval emergence and budburst under climatic warming. Oecologia 89:557–559.

Dyer J (1995) Assessment of climatic warming using a model of forest species migration. Ecol Model 79:199–219.

Eamus D, Jarvis PG (1989) The direct effects of increase in the global atmospheric CO_2 concentration on natural and commercial temperate forest trees and forests. Adv Ecol Res 19:1–55.

Fleming RA (1996) A mechanistic perspective of possible influences of climate change on defoliating insects in North America's boreal forests. Silva Fenn 30:281–294.

Franklin J, Swanson F, Harmon M, Perry D, Spies T, Dale V, McKee A, Ferrell W, Means J, Gregory S, Lattin J, Schowalter T, Walter D (1992) Effects of global climatic change on forests in northwestern North America. In: Peters RL, Lovejoy TE (eds) *Global Warming and Biological Diversity*. Yale University Press, New Haven, CT, pp 244–257.

Gaston KJ (1990) Patterns in the geographical ranges of species. Biol Rev 65: 105–129.

Giese RL, Houston DR, Benjamin DM, Kuntz JE, Kapler JE, Skilling DD (1964) *Studies of Maple Blight*. Res Bull 250. University of Wisconsin, Madison, WI.

Godfray HCJ, Hassell MP, Holt RD (1994) The population dynamic consequences of phenological asynchrony between parasitoids and their hosts. J Anim Ecol 63:1–10.

Graves J, Reavey D (1996) *Global Environmental Change. Plants, Animals and Communities*. Longman, Essex, UK.

Greenbank DO (1956) The role of climate and dispersal in the initiation of outbreaks of the spruce budworm in New Brunswick. I. The role of climate. Can J Zool 34:453–476.

Gutierrez AP (1987) Analyzing the effects of climate and weather on pests. In: Prodi F, Rossi F, Crisoferi G (eds) *Proceedings, International Conference on Agrometeorology*. 6–8 October 1987, Fondazione Cesena Agricultura, Cesena, Italy, pp 203–223.

Hajek AE, Elkinton JS, Witcosky JJ (1996) Introduction and spread of the fungal pathogen, *Entomophaga maimaiga*, along the leading edge of gypsy moth spread. Environ Entomol 25:1235–1247.

Hajek AE, Humber RA, Elkinton JS, May B, Walsh SRA, Silver JC (1990) Allozyme and RFLP analyses confirm *Entomophaga maimaiga* responsible for 1989 epizootics in North American gypsy moth populations. Proc Natl Acad Sci USA 87:6979–6982.

Hall TJ (1995) Effect of forest tent caterpillar and *Dicsula campestris* on sugar maple in Pennsylvania. Phytopathology 85:1129.

Hamid A, Odell TM, Katovich S (1995) White Pine Weevil. For Ins Dis Leaflet 21. United States Department of Argiculture (USDA) Forest Service, Washington, DC.

Hansen J, Lacis A, Rind D, Russell G, Stone P, Fung I, Ruedy R, Lerner J (1984) Climate sensitivity: analysis of feedback mechanisms. In: Hansen JE, Takahashi T (eds) Climate Processes and Climate Sensitivity. Geophysical Monograph 29. American Geophysical Union, Washington, DC, pp 130–163.

Harley JL, Smith SE (1983) *Mycorrhizal Symbiosis*. Academic Press, New York.

Hassell MP, Godfray HCJ, Comins HN (1993) Effects of global change on the dynamics of insect host–parasitoid interactions. In: Kareiva PM, Kingsolver JG,

488   D.W. Williams et al.

Huey RB (eds) *Biotic Interactions and Global Change.* Sinauer Associates, Sunderland, MA, pp 402–423.

Heliövaara K, Väisänen R, Immonen A (1991) Quantitative biogeography of the bark beetles in northern Europe. Acta For Fenn 219:1–35.

Hennon PE, Shaw CG III (1997) The enigma of yellow-cedar decline: what is killing these long-lived, defensive trees? J For 95(12):4–10.

Hennon PE, Shaw CG III (1994) Did climatic warming trigger the onset and development of yellow-cedar decline in southeast Alaska? Eur J For Pathol 24:399–418.

Hepting GH (1963) Climate and forest diseases. Annu Rev Phytopathol 1:31–50.

Herms DA, Mattson WJ, Karowe DN, Coleman MD, Trier TM, Birr BA, Isebrands JG (1996) Variable performance of outbreak defoliators on aspen clones exposed to elevated CO_2 and O_3. In: Hom J, Birdsey R, O'Brian K (eds) *Proceedings, 1995 Meeting of the Northern Global Change Program.* GTR-NE-214. United States Department of Agriculture, (USDA) Forest Service, Northeastern Forest Experiment Station, 14–16 March 1995, Radnor, PA, pp 43–55.

Heusser CJ (1952) Pollen profiles from Southeastern Alaska. Ecol Monogr 22: 331–352.

Hill JK, Hodkinson ID (1995) Effects of temperature on phenological synchrony and altitudinal distribution of jumping plant lice on dwarf willow in Norway. Ecol Entomol 20:237–244.

Hoffman AA, Blows MW (1994) Species borders: ecological and evolutionary perspectives. Trends Ecol Evol 9:223–227.

Holsten EH, Thier RW, Schmid J (1991) *The Spruce Beetle.* For Ins Dis Leaflet 127. United States Department of Agriculture (USDA) Forest Service, Washington, DC.

Houghton JT, Meiro Filho LG, Callander BA, Harris N, Kattenberg A, Maskell K (eds) IPCC (Intergovernmental Panel on Climate Change) (1996) *Climate Change 1995. The Science of Climate Change.* Cambridge University Press, Cambridge, UK.

Huffaker CB, Messenger PS (1964) The concept and significance of natural control. In: Debach P (ed) *Biological Control of Insect Pests and Weeds.* Chapman and Hall, London, UK, pp 74–117.

Hunter A (1993) Gypsy moth population size and the window of opportunity in spring. Oikos 68:531–538.

Juday GP (1984) Temperature trends in the Alaska climate record: problems update and prospects. In: McBeah JH (ed) *The Potential Effects of Carbon Dioxide-induced Climatic Changes in Alaska.* Misc Pub 83-1. University of Alaska, Fairbanks, AK, pp 76–91.

Kauppi PE, Mielikäinen K, Kuusela K (1992) Biomass and carbon budget of European forests, 1971 to 1990. Science 256:70–79.

Kingsolver JG (1989) Weather and the population dynamics of insects: integrating physiological and population ecology. Physiol Zool 62:314–324.

Kirschbaum MUF, Fischlin A (1996) Climate change impacts on forests. In: Watson RT, Zinyowera MC, Moss RH (eds) *Climate Change 1995. Impact, Adaptations and Mitigation of Climate Change: Scientific–Technical Analyses.* Cambridge University Press, Cambridge, UK, pp 95–129.

Kittel TGF, Rosenbloom NA, Painter, TH, Schimel DS, Fisher HH, Grimsdell A, VEMAP (Vegetation/Ecosystem Modeling and Analysis Project) Participants (1997) *The VEMAP Phase I Database: An Integrated Input Dataset for Ecosystem and Vegetation Modeling for the Conterminous United States.* CD-ROM. Climate and Global Dynamics Division, National Center for

Atmospheric Research and Climate System Modeling Program, University Corporation for Atmospheric Research, Boulder, CO.

Klironomos JN, Rillig MC, Allen MF, Zak DR, Pregitzer KS, Kubiske ME (1997) Increased levels of airborne fungal spores in response to *Populus tremuloides* grown under elevated CO_2. Can J Bot 75:1670–1673.

Kozár F (1991) Recent changes in the distribution of insects and the global warming. *Proceedings of the Fourth European Congress of Entomology and the Thirteen International Symposium for the Entomofauna of Central Europe.* Godollo, Hungary, pp 406–413.

Kräuchi N (1993) Potential impacts of a climate change on forest ecosystems. Eur J For Pathol 23:28–50.

Kurz WA, Apps MJ (1995) An analysis of future carbon budgets of Canadian boreal forests. Water Air Soil Pollut 82:321–331.

Kurz WA, Apps MJ, Beukema SJ, Lekstrum T (1995a) 20th century carbon budget of Canadian forests. Tellus 47B:170–177.

Kurz WA, Apps MJ, Stocks BJ, Volney WJA (1995b) Global climate change: disturbance regimes and biospheric feedbacks of temperate and boreal forests. In: Woodwell GM, Mackenzie FT (eds) *Biotic feedbacks in the global climatic system. Will the warming feed the warming?* Oxford University Press, New York, pp 119–133.

Landsberg J, Stafford Smith M (1992) A functional scheme for predicting the outbreak potential of herbivorous insects under global atmospheric change. Austral J Bot 40:565–577.

Larsson S (1989) Stressful times for the plant stress—insect performance hypothesis. Oikos 56:277–283.

Lawton JH (1995) The response of insects to climate change. In: Harrington R, Stork NE (eds) *Insects in a Changing Environment.* Academic Press, London, UK, pp 3–26.

Leadley PW, Korner CH (1996) Effects of elevated CO_2 on plant species dominance in a highly diverse calcareous grassland. In: Korner C, Bazzaz FA (eds) *Carbon Dioxide Populations and Communities.* Academic Press, San Diego, CA, pp 159–176.

Leaphart CD (1958) Pole blight—how it may influence western white pine management in light of current knowledge. J For 56:746–751.

Leaphart CD, Stage AR (1971) Climate: a factor in the origin of the pole blight disease of *Pinus monticola* Dougl. Ecology 52:229–239.

Leverenz JW, Lev DJ (1987) Effects of carbon dioxide–induced climate changes on the natural ranges of six major commercial tree species in the western United States. In: Shands WE, Hoffman JS (eds) *The Greenhouse Effect, Climate Change and U.S. Forests.* Conservation Foundation, Washington, DC, pp 123–155.

Liebhold AM, Halverson JA, Elmes GA (1992) Gypsy moth invasion in North America: a quantitative analysis. J Biogeogr 19:513–520.

Liebhold AM, MacDonald WL, Bergdahl D, Mastro VC (1995) *Invasion by Exotic Forest Pests: A Threat to Forest Ecosystems.* For Sci Monogr 30. Society of American Foresters, Washington, DC.

Lincoln DE, Fajer ED, Johnson RH (1993) Plant-insect herbivore interactions in elevated CO_2 environments. Trends Ecol Evol 8:64–68.

Lindroth RL (1996) Consequences of elevated atmospheric CO_2 for forest insects. In: Korner C, Bazzaz FA (eds) *Carbon Dioxide, Populations, and Communities.* Academic Press, San Diego, CA, pp 347–361.

Lindroth RL, Arteel GE, Kinney KK (1995) Responses of three saturniid species to paper birch grown under enriched CO_2 atmospheres. Func Ecol 9:306–311.

Lindroth RL, Kinney KK, Platz CL (1993) Responses of deciduous trees to elevated atmospheric CO_2: productivity, phytochemistry, and insect performance. Ecology 74:763–777.

Lonsdale D, Gibbs JN (1996) Effects of climate change on fungal diseases of trees. In: *Proceedings, Fungi and Environmental Change Symposium of the British Mycological Society, March 1994, Cranfield University, Cambridge*. Cambridge University Press, Cambridge, UK, pp 1–19.

Lopushinsky W, Max TA (1990) Effect of soil temperature on root and shoot growth and on budburst timing in conifer seedling transplants New For 4: 107–124.

Lupton J, Chakraborty S, Dale M, Sutherst RW (1995) Assessment of the enhanced greenhouse effect on plant diseases—a case study of Stylosanthes anthracnose. In: *Proceedings of the Tenth Biennial Australasian Plant Pathology Society Conference, Lincoln University*, New Zealand, p 108.

Luxmoore RJ, Wullschleger SD, Hanson PJ (1993) Forest responses to CO_2 enrichment and climate warming. Water Air Soil Pollut 70:309–323.

Mader DL, Thompson BW (1969) Foliar and soil nutrients in relation to sugar maple decline. Soil Sci Soc Am Proc 33:794–800.

McClure MS (1996) Biology of *Adelges tsugae* and its potential for spread in the northeastern United States. In: *Proceedings of the First Hemlock Woolly Adelgid Review*. FHTET 96-10. United States Department of Agriculture (USDA) Forest Service, Washington, DC, pp 16–25.

Malajczuk N (1988) Interaction between *Phytophthora cinnamomi* zoospores and microorganisms on non-mycorrhizal and ectomycorrhizal roots of *Eucalyptus marginata*. Trans Brit Mycol Soc 90:375–382.

Manabe S, Wetherald RT (1987) Large-scale changes in soil wetness induced by an increase in carbon dioxide. J Atmos Sci 232:626–628.

Manion PD (1991) *Tree Disease Concepts*. Prentice-Hall, Englewood Cliffs, NJ.

Manning WJ, Tiedemann AV (1995) Climate change: potential effects of increased atmospheric carbon dioxide (CO_2), ozone (O_3) and ultraviolet-B (UV-B) radiation on plant diseases. Environ Pollut 88:219–245.

Marosy M, Patton RF, Upper CD (1989) A conducive day concept to explain the effect of low temperature on the development of Scleroderris shoot blight. Phytopathology 79:1293–1301.

Marshall JD, Perry DA (1987) Basal and maintenance respiration of mycorrhizal and nonmycorrhizal root systems of conifers. Can J For Res 17: 872–977.

Martinat PJ (1987) The role of climatic variation and weather in forest insect outbreaks. In: Barbosa P, Schultz JC (eds) *Insect Outbreaks*. Academic Press, San Diego, CA, pp 241–268.

Marx DH (1970) The influence of ectotrophic mycorrhizal fungi on the resistance of pine roots to pathogenic infections. V. Resistance of mycorrhizae to infection by vegetative mycelium on *Phytophthora cinnamomi*. Phytopathology 60:1472–1473.

Mason RR (1987) Nonoutbreak species of forest Lepidoptera. In: Barbosa P, Schultz J (eds) *Insect Outbreaks*. Academic Press, San Diego, CA, pp 31–58.

Mattson WJ, Haack RA (1987a) The role of drought stress in provoking outbreaks of phytophagous insects. In: Barbosa P, Schultz J (eds) *Insect Outbreaks*. Academic Press, San Diego, CA, pp 365–407.

Mattson WJ, Haack RA (1987b) The role of drought in outbreaks of plant-eating insects. BioScience 37:110–118.

Mattson WJ, Simmons GA, Witter JA (1988) The spruce budworm in eastern North America. In: Berryman AA (ed) *Dynamics of Forest Insect Populations. Patterns, Causes and Implications.* Plenum Press, New York, pp 310–331.

Messenger PS (1971) Climatic limitations to biological control. In: *Proceedings, Tall Timbers Conference on Ecological Animal Control by Habitat Management,* Number 3. pp 97–114. 25–27 February 1971, Tall Timbers Research Station, Tallahassee, FL.

Mitchell JFB, Manabe S, Meleshko V, Tokioka T (1990) Equilibrium climate change and it implications for the future. In: Houghton JT, Jenkins GJ, Ephraums JJ (eds) IPCC (Intergovernmental Panel on Climate Change) *Climate Change: The IPCC Scientific Statement.* Cambridge University Press, Cambridge, UK, pp 131–172.

Montgomery ME, Wallner WE (1988) Gypsy moth: a westward migrant. In: Berryman AA (ed) *Dynamics of Forest Insect Populations. Patterns, Causes and Implications.* Plenum Press, New York, pp 353–375.

Moran PAP (1953a) The statistical analysis of the Canadian lynx cycle. I. Structure and prediction. Aust J Zool 1:163–173.

Moran PAP (1953b) The statistical analysis of the Canadian lynx cycle. II. Synchronization and meteorology. Aust J Zool 1:291–298.

Myers JH (1988) Can a general hypothesis explain population cycles in forest Lepidoptera? Adv Ecol Res 18:179–242.

Neely D, Phares R, Weber B (1976) Cristulariella leaf spot associated with defoliation of black walnut plantations in Illinois. Plant Dis Rep 60:587–590.

Norby RJ, Gunderson CA, Wullschleger SD, O'Neill EG, McCracken MK (1992) Productivity and compensatory responses of yellow-poplar trees in elevated CO_2. Nature 357:322–324.

O'Neill EG, Luxmoore RJ, Norby RJ (1987) Increase in mycorrhizal colonization and seedling growth in *Pinus echinata* and *Quercus alba* in an enriched CO_2 environment. Can J For Res 17:878–883.

Parmesan C (1996) Climate and species' range. Nature 382:765–766.

Porter J (1995) The effects of climate change on the agricultural environment for crop insect pests with particular reference to the European corn borer and grain maize. In: Harrington R, Stork NE (eds) *Insects in a Changing Environment.* Academic Press, London, UK, pp 93–123.

Porter JH, Parry ML, Carter TR (1991) The potential effects of climatic change on agricultural insect pests. Agric For Meteorol 57:221–240.

Price PW (1991) The plant vigor hypothesis and herbivore attack. Oikos 62: 244–251.

Randall MGM (1982) The dynamics of an insect population throughout its altitudinal distribution: *Coleophora alticolella* in northern England. J Anim Ecol 51:993–1016.

Reeve JD, Ayres MP, Lorio PL (1995) Host suitability, predation, and bark beetle population dynamics. In: Cappuccino N, Price PW (eds) *Population Dynamics: New Approaches and Synthesis.* Academic Press, San Diego, CA, pp 339–357.

Retnakaran A, Harris JWE (1995) Terminal weevils. In: Armstrong JA, Ives WGH (eds) *Forest Insect Pests in Canada.* Canadian Forestry Service, Sci Sust Dev Dir, pp 233–240.

Robitaille G, Boutin R, Lachance D (1995) Effects of soil freezing stress on sap flow and sugar content of mature sugar maples (*Acer saccharum*). Can J For Res 25:577–587.

Rogers DJ, Randolph SE (1986) Distribution and abundance of tsetse flies. J Anim Ecol 55:1007–1025.

Rogers DJ, Randolph SE (1993) Distribution of tsetse and ticks in Africa: past present and future. Parasitol Today 9:266–271.

Roth SK, Lindroth RL (1994) Effects of CO_2-mediated changes in paper birch and white pine chemistry on gypsy moth performance. Oecologia 98: 133–138.

Royama T (1984) Population dynamics of the spruce budworm Choristoneura fumiferana. Ecol Monogr 54:429–462.

Royama T (1992) Analytical Population Dynamics. Chapman and Hall, London, UK.

Safranyik L (1995) Bark beetles. In: Armstrong JA, Ives WGH (eds) Forest Insect Pests in Canada. Canadian Forestry Service, Sci Sust Dev Dir, pp 155–163.

Sanders IR (1996) Plant-fungal interactions in a CO_2-rich world. In: Korner C, Bazzaz FA (eds) Carbon Dioxide Populations and Communities. Academic Press, San Diego, CA, pp 265–272.

Schwerdtfeger F (1935) Studien über den Massenwechsel einiger Forstschädlinge. 1. Das Klima der Schadgebiete von Bupalus piniarius L., Panolis flammea Schiff. und Dendrolimus pini L. in Deutschland. Z Forst Jadgw 67:15–38, 85–104, 449–482, 513–540.

Shearer BL, Tippett JT (1989) Jarrah dieback: the dynamics and management of Phytophthora cinnamomi in the Jarrah (Eucalyptus marginata) Forest of southwestern Australia. Res Bull 3. Department of Conservation and Land Management, Western Australia.

Somda B, Pinon J (1981) Ecophysiologie du stade uredien de Melampsora laricipopulina Kleb et de M alli-populina Kleb. Eur J For Pathol 11:243–254.

Stollár A, Dunkel Z, Kozár F, Sheble DAF (1993) The effects of winter temperature on the migration of insects. Időjáris 97:113–120.

Straw NA (1995) Climate change and the impact of green spruce aphid, Elatobium abietinum (Walker), in the U.K. Scot For 49:134–145.

Sutherst RW (ed) (1996) Impacts of Climate Change on Pests Diseases and Weeds in Australia. Report of an International Workshop, 9–12 October 1995, Brisbane, Australia. CSIRO Division of Entomology, Canberra, Australia.

Sutherst RW (1991) Pest risk analysis and the greenhouse effect. Rev Agric Entomol 79:1177–1187.

Swetnam TW, Lynch AM (1993) Multicentury, regional-scale patterns of western spruce budworm outbreaks. Ecol Monogr 63:399–424.

Tingey DT, Johnson MG, Phillips DL, Storm MJ (1995) Effects of elevated CO_2 and nitrogen on ponderosa pine fine roots and associated fungal components. J Biogeogr 22:281–287.

Turchin P (1990) Rarity of density dependence or population regulation with lags? Nature 344:660–663.

USDA (United States Department of Agriculture) Forest Service (1994) Northeastern Area Forest Health Report 1992. NA-TP-01-94. USDA Forest Service, Northeastern Forest Experiment Station, Radnor, PA.

Uvarov BP (1931) Insects and climate. Trans Entomol Soc London 79:1–247.

Virtanen T, Neuvonen S, Nikula A, Varama M, Niemelä P (1996) Climate change and the risks of Neodiprion sertifer outbreaks on Scots pine. Silva Fenn 30:169–177.

Vitousek PM, D'Antonio CM, Loope LL, Westbrooks R (1996) Biological invasions as global environmental change. Am Sci 84:468–478.

Wargo PM, Harrington TC (1991) Host stress and susceptibility. In: Shaw CG III, Kile GA (eds) Armillaria Root Disease. Agric Hndbk 691. United States Department of Agriculture (USDA), pp 88–101.

Waring GL, Cobb NS (1992) The impact of plant stress on herbivore population dynamics. In: Bernays E (ed) *Insect-Plant interactions.* Vol. IV. CRC Press, Boca Raton, FL, pp 167–226.

Watt AD, Lindsay E, Leith ID, Fraser SM, Docherty M, Hurst DK, Hartley SE, Kerslake J (1996) The effects of climate change on the winter moth, *Operophtera brumata,* and its status as a pest of broadleaved trees, Sitka spruce and heather. Aspects Appl Biol 45:307–316.

Watt AD, Whittaker JB, Docherty M, Brooks G, Lindsay E, Salt DT (1995) The impact of elevated atmospheric CO_2 on insect herbivores. In: Harrington R, Stork NE (eds) *Insects in a Changing Environment.* Academic Press, London, UK, pp 197–217.

Wellington WG (1952) Air-mass climatology in Ontario north of Lake Huron and Lake Superior before outbreaks of the spruce budworm *Choristoneura fumiferana* (Clem.) and the forest tent caterpillar *Malacosoma disstria* Hbn. Can J Zool 30:114–127.

Wellington WG, Fettes JJ, Turner KB, Belyea RM (1950) Physical and biological indicators of the development of outbreaks of the spruce budworm, *Choristoneura fumiferana* (Clem.). Can J Res Sect D Zool Sci 28:308–331.

Weltzien HC (1983) Climatic zoning and plant disease potential—examples from the Near and Middle East. EPPO Bull 13:69–73.

White TCR (1984) The abundance of invertebrate herbivores in relation to the availability of nitrogen in stressed food plants. Oecologia 63:90–105.

Williams DW, Fuester RW, Metterhouse WW, Balaam RJ, Bullock RH, Chianese RJ, Reardon RC (1990) Density, size, and mortality of egg masses in New Jersey populations of the gypsy moth. Environ Entomol 19:943–948.

Williams DW, Liebhold AM (1997a) Latitudinal shifts in spruce budworm (Lepidoptera: Tortricidae) outbreaks and spruce–fir forest distributions with climate change. Acta Phytopathol Entomol Hungar 32:205–215.

Williams DW, Liebhold AM (1997b) Range shifts in gypsy moth outbreaks and oak forest distributions in the northeastern United States under climate change. In: Fosbroke SLC, Gottschalk KW (eds) *Proceedings, U.S. Department of Agriculture Interagency Gypsy Moth Research Forum 1997.* Gen Tech Rep NE-240. United States Department of Agriculture (USDA) Forest Service, Northeastern Forest Experiment Station, Radnor, PA.

Williams DW, Liebhold AM (1995a) Detection of delayed density dependence: effects of autocorrelation in an exogenous factor. Ecology 76:1005–1008.

Williams DW, Liebhold AM (1995b) Influence of weather on the synchrony of gypsy moth outbreaks in New England. Environ Entomol 24:987–995.

Williams DW, Liebhold AM (1995c) Forest defoliators and climatic change: potential changes in spatial distribution of outbreaks of western spruce budworm and gypsy moth. Environ Entomol 24:1–9.

Williams RS, Lincoln DE, Norby RJ (1998) Leaf age effects of elevated CO_2-grown white oak leaves on spring-feeding lepidopterans. Global Change Biol 4:235–246.

Williams RS, Lincoln DE, Thomas RB (1997a) Effects of elevated CO_2-grown loblolly pine needles on the growth, consumption, development, and pupal weight of red-headed pine sawfly larvae reared within open-topped chambers. Global Change Biol 3:501–511.

Williams RS, Thomas RB, Strain RB, Lincoln DE (1997b) Effects of elevated CO_2, soil nutrient levels, and foliage age on performance of two generations of *Neodiprion lecontei* feeding on loblolly pine. Environ Entomol 26:1312–1322.

Wilson CA, Mitchell JFB (1987) A doubled CO_2 climate sensitivity experiment with a global climate model including a simple ocean. J Geophys Res (D11):13315–13343.

Woods FW (1953) Disease as a factor in the evolution of forest composition. J For 51:871–873.

Zak DR, Pregitzer KS, Curtis PS, Teeri JA, Fogel R, Randlett DL (1993) Elevated atmospheric CO_2 and feedback between carbon and nitrogen cycles in forested ecosystems. Plant Soil 151:105–117.

Zentmyer GA (1980) *Phytophthora cinnamomi* and the diseases it causes. Am Phytopathol Soc Monogr 10.

14. Forest Responses to Changing Climate: Lessons from the Past and Uncertainty for the Future

Donald H. DeHayes, George L. Jacobson Jr.,
Paul G. Schaberg, Bruce Bongarten, Louis Iverson,
and Ann C. Dieffenbacher-Krall

The earth's climate has undergone dramatic and long-term changes through natural processes many millennia before humans influenced global climate. Considerable evidence indicates that increasing concentrations of carbon dioxide and other greenhouse gases in the earth's atmosphere will lead to near-term warming, perhaps as much as 2 to 4°C in northeastern North America. Given that the distribution of vegetation on earth has varied with past climate change, it is reasonable to expect that future climate change will affect forest composition and distribution. For reasons both ecological and economic, it is desirable to understand and be able to predict the extent and nature of the changes that might be expected in northern forests in response to climate warming.

This chapter offers insights as to how one might analyze potential future changes in forest composition and also elucidates the numerous complicating biological and anthropogenic factors that create uncertainty about forest responses to changing environments in the future. In particular, we explore vegetation–climate interactions of the past using paleoecological and paleoclimatic information to reveal patterns that have implications for the future. Furthermore, we examine predictive models of vegetation change in response to climate warming that are based largely on our understanding of current forest–climate relationships and of physiological characteristics of vegetation. We also explore specific biological factors, such as survival, reproductive capacity, and rate of dispersal, that influence how individual species respond to changing climate conditions, and we assess the potential ramifications of interspecific and intraspecific variation in forest responses to changing environments. Finally, we conceptualize anthropogenic stressors as a "new" set of driving variables that can either shape or constrain evolutionary responses to new environmental situations. To that end, we explore the potential "costs" of pollution as a selective agent, the implications of forest fragmentation, and even intensive forest management on both forest ecosystem stability and our ability to anticipate future forest responses.

Paleoecological Evidence of Past Climates and Vegetation Patterns

Holocene Climates of the Northeast

The climate of the Quaternary Period, which is roughly the last two and a half million years, has fluctuated widely. It is characterized by regular cycles consisting of ice ages followed by brief warm intervals known as interglacials, of which the Holocene is the most recent. The regularity of the ice-age cycles has been linked convincingly to three orbital parameters that affect the amount of energy reaching the earth's surface. These "Milankovitch cycles" include changes in (1) the precession of the equinoxes with a periodicity of approximately 19,000 to 22,000 years, (2) the tilt of the axis of the earth with a periodicity of approximately 42,000 years, and (3) the eccentricity of the earth's orbit around the sun with a periodicity of approximately 100,000 years (Imbrie and Imbrie, 1979). Hays et al. (1976) demonstrated that these three cycles were strongly evident in the geochemistry of ocean sediments deposited during the Quaternary and argued that the changing energy flux set the timing of long-term glacial–interglacial sequences.

Of the three cycles, the precessional cycle has most affected Holocene climates because it has changed the date of the year that the earth comes closest to the sun (perigee). At present, perigee occurs on January 5, in the heart of the northern winter. Approximately 9,000 to 11,000 years ago in the early Holocene, perigee occurred in the summer. As a result, seasonality in the early Holocene was increased, that is, summers were relatively warmer and winters relatively cooler than in the Northern Hemisphere today.

A record of climate change in the Holocene consistent with such changes in insolation and reduction in seasonality can be found in sedimentary deposits that have remained in place over many centuries. For thousands of years, lakes and peat bogs have passively monitored vegetation shifts caused by climate change, human activities, and other related phenomena in the study area. Cores of lake sediments or peat deposits resemble strip-charts that have recorded past vegetation change and have provided a chronology through radiocarbon dating (Jacobson, 1988). Such stratigraphic evidence has been used to infer that temperatures during much of the previous 10,000 years were as much as 2°C warmer and that net moisture was considerably lower than today (COHMAP, 1988; Webb et al., 1994). The climate 6,000 years ago in the upper Midwest was characterized by mean July temperatures approximately 2°C warmer than at present (Bartlein and Webb, 1985). The fact that mean July temperatures in central Europe were also 2°C warmer at that time (Huntley and Prentice, 1988; Prentice et al., 1996) reinforces the notion that hemispheric or global phenomena were involved.

Changes in moisture balance per se (precipitation minus evaporation) in the Holocene have been the focus of studies of past lake levels in Minnesota (Digerfeldt et al., 1992), Wisconsin (Winkler et al., 1986; Winkler, 1988), and Cape Cod (Winkler, 1985). Additional data relating to past water-level changes have been compiled from throughout the northeastern United States (Webb et al., 1993) and eastern North America (Harrison, 1989; Webb et al., 1994). Paleohydrologic research of this kind has consistently shown that lake levels were relatively low during the early to middle Holocene and that they have risen to present levels only in the past few millennia (Fig. 14.1). This is consistent with warmer and drier conditions early in the Holocene changing to cooler and moister conditions in its latter stages.

Forest Responses to Holocene Climate Change in the Northeast

Paleoecologists have used pollen analysis of lake and peat deposits to reconstruct the distribution and abundance of plant taxa during the Holocene. Data from hundreds of such studies completed during the last several decades have been used to create isopoll maps that show temporal changes in distribution and abundance of selected plant taxa, including forests of the northeast (Jacobson et al., 1987; Webb et al., 1994). The North America Pollen Database, an outgrowth of Webb's COHMAP database, is now available to the public at the National Oceanic and Atmospheric Administration (NOAA) National Geophysical Data Center in Boulder, Colorado.

Northeastern forests have been strongly influenced by changing temperature and moisture regimes during the present interglacial. Synoptic maps of pollen data spanning the past 20,000 years reveal that most modern vegetation assemblages, even at the biome scale, have been in their present configuration for no more than 6,000 to 8,000 years (Webb, 1987; Jacobson et al., 1987; Overpeck et al., 1991). Also, these results have shown that continent-wide changes in distribution and abundance of plant taxa are species-specific, consistent with Gleason's (1926) individualistic

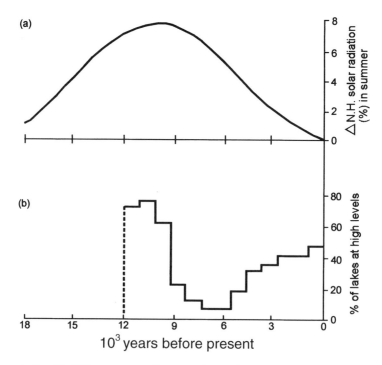

Figure 14.1. (a) Holocene variation in solar radiation during summers at mid-latitudes in the Northern Hemisphere. The pattern results from precession of the equinoxes. (b) Summary data for Holocene lake levels in eastern North America, showing the proportion of sites with relatively high levels through time (by implication, times of moist water-balance) (after Webb et al., 1994).

concept of plant–species responses (Davis, 1983; Jacobson et al., 1987). Contrary to popular belief, modern communities are not highly organized, finely tuned units representing long periods of co-evolution among species. Rather, present communities are merely transitory combinations of taxa that have been responding individualistically to continual and sometimes major climate changes (Hunter et al., 1988).

Eastern White Pine Case Study

Eastern white pine (*Pinus strobus* L.) is just one of many species whose Holocene distribution has been studied using pollen analysis (Jacobson, 1979; Jacobson and Dieffenbacher-Krall, 1995). The data show that changes in eastern white pine populations accompanied major changes in Holocene climate (Fig. 14.2). Eastern white pine was widely distributed in the early to middle Holocene (10,000 to 6,000 years ago), a time of warmer and drier climate than occurs today. As was noted earlier, this was a period of lowered lake levels that, along with other sedimentary data,

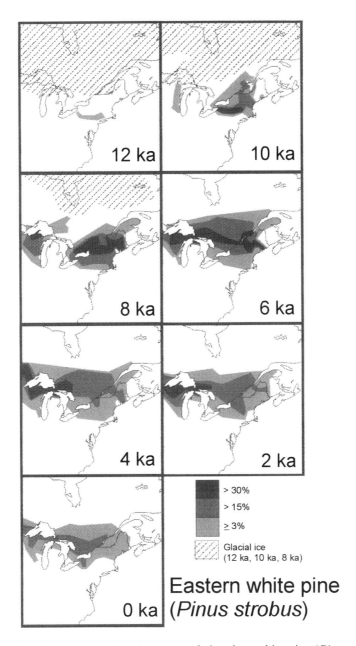

Figure 14.2. Isopoll maps showing areas of abundant white pine (*Pinus strobus*) during the past 12,000 ^{14}C years (estimated from ^{14}C pollen). Fossil pollen data from approximately 140 sites are summarized by lines of equal proportional representation across northeastern North America. Time designated in kiloanni (ka).

indicate that relatively dry conditions prevailed throughout the early Holocene in the Great Lakes–New England region (Webb et al., 1993).

Paleoecological evidence shows that eastern white pine made its first post-glacial appearance in Virginia (Craig, 1969), perhaps moving in from a full glacial location on the exposed continental shelf. It reached northern New England by 10,000 years ago (Davis and Jacobson, 1985), the central Great Lakes region by 9,000 years ago (Brubaker, 1975), and Minnesota and western Ontario by 7,000 years ago (Jacobson, 1979; Björck, 1985).

In the early Holocene, large concentrations of eastern white pine occurred in both the eastern Great Lakes–New England region and an area west of Lake Michigan. In general, regions of high abundance of eastern white pine were also areas in which forest fires were frequent and precipitation was probably not much greater than evapotranspiration. The relatively high fire frequency of the early to middle Holocene would have helped to create conditions favorable for seedling establishment (Jacobson and Dieffenbacher-Krall, 1995). Early Holocene sediments in many New England lakes have high proportions of pollen from eastern white pine and oak (*Quercus* spp.) (Fig. 14.3), along with many charred particles from past fires (Anderson et al., 1986; Patterson and Barkman, 1988; Anderson et al., 1992). Similar evidence was found in Nova Scotia (Green, 1982). The close relationship between climate and fire frequency has been well-established in detailed studies by Clark (1988, 1989).

Eastern white pine reached its northernmost extent about 4,000 years ago, with areas of high abundance shrinking substantially and shifting southward thereafter. This coincides with climate cooling that has allowed boreal taxa to move southward. Another factor in the late-Holocene decline in eastern white pine is the decrease in frequency of fire. Further details of these late-Holocene changes may be found in Jacobson and Dieffenbacher-Krall (1995).

The western range limit of eastern white pine occurs today where precipitation equals evapotranspiration (Transeau, 1905). Unless its habitat is manipulated by human activity, white pine does not thrive when conditions become too cool or moist, for example, at the southern margins of the boreal forest in northern New England and adjacent Canada where disturbance by fire may be too infrequent for widespread establishment of seedlings. The current abundance of eastern white pine in the Northeast results largely from abandonment of farmland during the last 150 years.

Isopoll maps of other taxa (Jacobson et al., 1987; Webb et al., 1994) provide a useful context for evaluating regional changes in eastern white pine. Other data show that in the early to middle Holocene, both eastern white pine and eastern hemlock (*Tsuga canadensis* [L.] Carr) were present at elevations as much as 300 to 400 m higher than their present upper limit in the White Mountains of New Hampshire (Davis et al., 1980). Jackson and Whitehead (1991) documented similar elevational patterns for tree

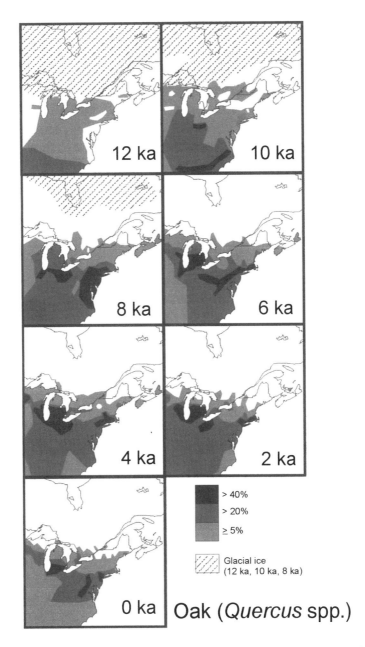

Figure 14.3. Isopoll maps showing areas of abundant oak (*Quercus* spp.) during the past 12,000 ^{14}C years (estimated from ^{14}C pollen). Fossil-pollen data from approximately 600 sites are summarized by lines of equal proportional representation across northeastern North America. Time designated in kiloanni (ka).

taxa growing in the Adirondack Mountains of New York. Paleoecological studies of the later Holocene show that the boreal forest of eastern Canada developed only in the past 6,000 years (Webb, 1987) and that hemlock has been abundant in the forests of the eastern Great Lakes–New England region for that same period of time (Fig. 14.4). Other data show that southern populations of spruce (*Picea* spp.) shifted from Canada into the northern tier of states from Maine to Minnesota in the past 1,000 to 1,500 years, accompanied by a general decrease in abundance of eastern white pine that has continued to the present (see Fig. 14.2). Small populations of balsam fir (*Abies balsamea* [L.] Mill.) were scattered throughout the northeast during most of the Holocene, but they, too, expanded recently to form the spruce–fir forests of today. The spatial array of changes has been influenced by variations in importance of fire (Foster, 1983) and other disturbances.

Spruce Case Study

Acadian forests of Maine and the adjacent Canadian provinces are characterized by abundant spruce and fir. These constitute a major resource for both the forest products industry and the millions of recreational users who enjoy the north woods. The health of these forests is so important that any threat to their productivity or aesthetics is of immediate public concern. During the last few decades, factors such as spruce budworm outbreaks, cutting practices, and various other land use issues have been the subject of major public discussions in the Northeast. However, spruce–fir forests may be as vulnerable to future changes in climate as to any of these often-debated factors.

Isopoll maps demonstrate that dramatic changes occurred in the abundance of spruce in Maine and adjacent areas, especially during the last few centuries. Spruce and, to some extent, fir had relatively little presence in the northeastern forest during most of the last 9,000 years (Fig. 14.5). Only in the last few centuries have spruce and fir covered the Acadian region as densely as they do today (Schauffler, 1998). Another way to think about this is that the older spruce trees living today in northern Maine, some 250 years or more old, have been living for roughly half the time in which recent spruce forests have densely covered the Acadian region.

Populations of spruce and fir are closely linked to climate and have responded to cooling in the late Holocene. Several lines of evidence indicate that the climate of New England became cooler starting several thousand years ago—and that it turned dramatically cooler just a few hundred years ago. A similar trend is well-documented for many areas of Europe and even New Zealand, where the recent cooling is often referred to as the "Little Ice Age." The best current estimates of the cooling in Maine and along the southern margin of the boreal forest in Canada indicate a reduction of perhaps 1°C in mean July temperature during the

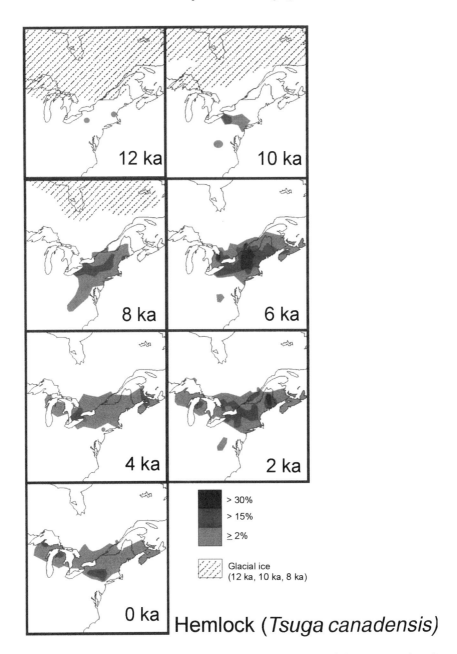

Figure 14.4. Isopoll maps showing areas of abundant hemlock (*Tsuga canadensis*) during the past 12,000 ^{14}C years (estimated from ^{14}C pollen). Fossil-pollen data from approximately 450 sites are summarized by lines of equal proportional representation across northeastern North America. Time designated in kiloanni (ka).

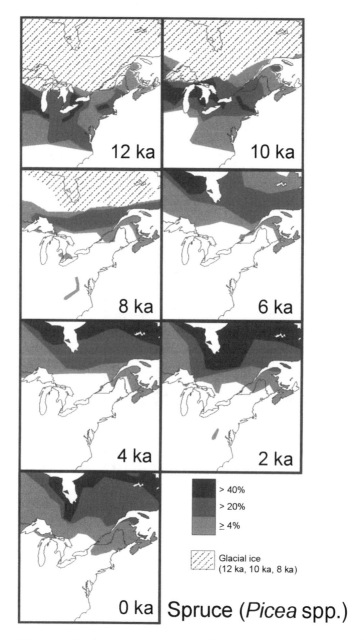

Figure 14.5. Isopoll maps showing areas of abundant spruce (*Picea* spp.) during the past 12,000 ^{14}C years (estimated from ^{14}C pollen). Fossil-pollen data from approximately 600 sites are summarized by lines of equal proportional representation across northeastern North America. Time designated in kiloanni (ka).

"Little Ice Age" (Gajewski, 1988). This change is ecologically significant because spruce survival along the southern margin of its range is thought to be limited by summer heat and drought. The examples of spruce and eastern white pine migrations of the past demonstrate the independent nature of species responses to changing climate and provide insights into the dynamic nature of species/environment relationships.

Modeling Species Responses to Changing Climate

Evaluation of paleoecological evidence is one approach for predicting future forest composition in response to climate change. The second general approach involves using mathematical models to predict how changes in climate will affect the distribution of vegetation. Increasing evidence exists of a general warming trend on the planet (MacCracken, 1995; Wigley, 1995), and various global circulation models predict a further 1 to 4.5°C temperature increase over the next century (Kattenberg et al., 1996). Major changes may occur in the earth's living systems, including temperate forests. Using models to predict vegetation response may greatly help in understanding potential impacts of global climate change.

Several approaches have been used to model possible species responses, each of which has merit. These models use modern calibrations of various types and at various spatial scales. For example, Nielsen (1995) and Kittel et al. (1995) have developed climate–vegetation models based essentially on continent-scale calibration of biomes. Other efforts have included stand-scale modeling approaches to estimate forest responses to climate changes (Pastor and Post, 1986; Post and Pastor, 1990, among others). Still another type of model, Prentice's BIOME model, was derived from a physiology-based analysis of how plant life-forms relate to climate regions of the world (Prentice et al., 1996; Haxeltine and Prentice, 1996). Indeed, because these models cannot be truly validated (Rastetter, 1996), multiple avenues of research are encouraged, with the hope that results may eventually converge (Hobbs, 1994; VEMAP members, 1995; Lauenroth, 1996). This section provides a brief overview of some of the models that have been used. It then explains in more detail one empirical approach to modeling specific species responses and finally uses that approach to present two possible scenarios of tree biodiversity effects using multiple species overlays.

Overview of Approaches

Two major approaches have been used to model potential responses of vegetation to climate change: the empirical approach using correlative/statistical models and the mechanistic approach using biogeography and biogeochemistry models. The first uses empirical relationships of current vegetation–climate patterns to predict potential vegetation distribution following climate change, while the latter incorporates physiological characteristics of the vegetation. Because modeling is an active area of research, all approaches are becoming more sophisticated and intertwined.

Empirical Approach Using Correlative/Statistical Models

Many researchers have successfully recreated current vegetation patterns via climate–vegetation analysis and using specific regression relationships (Booth, 1990; Bonan and Sirois, 1992; Box et al., 1993; Huntley et al., 1995). Iverson and Prasad (1998) used regression tree analysis (described in detail in a following section) that uses climate as well as soil, topographic, and land cover information to derive vegetation–environment relationships. Once relationships to current vegetation are established, the climate is changed according to various global change scenarios to derive potential future vegetation distribution maps. These models must assume that species are bounded by modeled characters, so that factors such as changes in species competition and CO_2-derived water use efficiency cannot be incorporated (Loehle and LeBlanc, 1996). Still, they provide valuable insights into potential species shifts under various global climate change scenarios.

Mechanistic Approach Using Biogeography and Biogeochemistry Models

At least five biogeography models and 20 biogeochemistry models now exist (Neilson et al., 1998); they use physiological characteristics of vegetation to predict future vegetative distribution from climate change. Two primary biogeography models are the Mapped Atmosphere-Plant-Soil System (MAPPS), developed by Neilson and associates (Neilson, 1995; Neilson and Marks, 1994), and the BIOME3 model (Haxeltine et al., 1996). Both models calculate the potential vegetation type (up to 45 vegetation types globally for MAPPS and 18 for BIOME3) and leaf area that a site can support (at local, regional, or global scales), as constrained by local vegetation and hydrologic process and the physiological properties of plants (Neilson et al., 1998). These two models have been used extensively by the Intergovernmental Panel on Climate Change to better assess potential regional vegetation changes according to various climate change scenarios (Watson et al., 1998). Efforts are now underway to incorporate continuous feedback from vegetation effects into dynamic models of global vegetation change (Foley et al., 1996; Neilson and Running, 1996).

Biogeochemistry models simulate carbon and nutrient cycles of ecosystems, but tend to lack the ability to predict vegetation types at a given location. In an exercise to compare among three of the primary biogeochemical models as well as three biogeography models, the Vegetation/Ecosystem Modeling and Analysis Project (VEMAP) was devised to assess model capabilities and the potential impacts of global warming on U.S. ecosystems (VEMAP members, 1995). They used TEM (Raich et al., 1991), CENTURY (Parton et al., 1988), and BIOME-BGC (Running and Hunt, 1993). The VEMAP process determined that all models can adequately simulate vegetation under the current environment, but that alternative climate scenarios produce divergences, even to the level of producing vegetation responses of opposite sign. The biogeochemical model PnET, used thus far primarily in the eastern part of the U.S., has done an excellent job of producing physiologically based estimates of primary productivity, annual drainage, and photosynthesis (Aber and Federer, 1992; McNulty et al., 1994).

All of the above modelling efforts have shown that environmental drivers, as modified by disturbance processes, generally control the distribution of tree species. These relationships are increasingly being scrutinized and verified. Within a region, species vary primarily due to regional climatic factors, whereas variations in terrain, soil, and land use history factor principally in more local studies. Geographical information systems (GIS) allow predictive mapping of vegetation based on the species-environment relationships.

Response of Species Using Regression Tree Analysis Models

We describe here an empirical modeling approach called regression tree analysis (RTA) to evaluate potential future habitat that is suitable for species to occupy given any potential future climate scenario. If we assume the species will be able to migrate to the new habitat, it will represent potential future distributions for the species. RTA is a relatively new technique in the ecological sciences that uses repeated resampling of the data to develop empirical relationships between response and predictor variables, rather than the more restrictive distributional assumptions in classical regression functions. This alternative modeling approach creates models that are fitted by binary recursive partitioning whereby a data set is successively split into increasingly homogeneous subsets to elucidate relationships between predictor and response variables (Clark and Pregibon, 1992).

The RTA approach seems appropriate for predicting landscape-level distributions of species from environmental data. Its use has grown with that of geographical information systems that allow model outputs to be readily mapped across landscapes. There are few, but increasing, ecological examples of the use of RTA (e.g., Davis and Goetz, 1990; Michaelsen et al., 1994; Reichard and Hamilton, 1997; Flemming, 1997). For this chapter, RTA was used to evaluate the relationship of 33 environmental variables to 80 eastern tree species importance values, and then used the derived relationships to predict their present and potential future importance values. The methods and assumptions involved in this effort are described in detail in Iverson and Prasad (1998), and were used to produce an associated atlas of all 80 species (Iverson et al., 1999).

The intention was to create RTA models that best match current distribution of species importance values, then project potential future distributions after climate change. For this, we substituted current climate variables with the projected outputs from two scenarios of equilibrium climate under $2\times CO_2$: the GFDL (Geophysical Fluid Dynamics Laboratory) (Wetherald and Manabe, 1988), and GISS (Goddard Institute of Space Studies) (Hansen et al., 1988). Though $2\times CO_2$ will likely occur by the year 2100, the longevity of trees, genetic variation, and the presence of remnant refugia would dictate that the possible outcomes shown in these models would take centuries to take full effect.

Regression trees were generated for 80 tree species in the U.S. Department of Agriculture (USDA) Forest Service's Forest Inventory and Analysis (FIA) database. Species importance value (based on basal area and number of stems) was the response variable, along with the 33 predictor variables mentioned above. The regression trees were generated and used to predict the importance value for each species by county.

The tree diagram produced for each species provided information on the primary driving variables for the species. Predicted current distributions matched current FIA data quite well for most species. For example, the primary variable controlling the importance value of eastern hemlock was July temperatures followed by soil characteristics, such that the modeled current distribution and abundance of eastern hemlock matches actual data reasonably well (see color insert Fig. 14.6). Eastern hemlock grows well on moist, cooler sites with good drainage, but will grow under a variety of soil conditions (Godman and Lancaster, 1990). The county-level of resolution is therefore adequate to capture the major environmental variables driving its distribution, and conditions represented by county averages are adequate to model the species.

In general, the more specialized the species is with respect to edaphic conditions, the less accuracy in the RTA model predictions because county-resolution data would not be expected to consistently capture the appropriate information for the RTA process. Projected species distributions, following equilibrium of predicted climate changes, show major shifts for many species. The GISS and GFDL scenarios would reduce area of eastern hemlock by 14% in the U.S., with over 40% loss in area-weighted importance value (see Fig. 14.6). The difference maps show the loss of importance to be primarily in New York and southern New England, while increases could occur in the upper Midwest (see Fig. 14.6). Because of differential shifts among species in area and importance, forest communities would also be expected to differentially change in composition. Regression tree analysis can be considered a valuable tool to understand species–environment relations. By operating at a species level, RTA gives an indication of potential changes in community dynamics and biodiversity.

Multiple Species Assessments

Using regression tree analysis, we can begin to make overall assessments regarding possible changes in forest composition under global climate change scenarios. Of course, overlaying multiple species projections means that, in addition to the assumptions listed by Iverson and Prasad (1998), we must assume that competition among species will not significantly alter the result of linearly combining the 80 single species models created by RTA.

Species Richness

By combining the maps for all 80 species, we can count the number of species projected to occur in each county and map the result (see color insert Fig. 14.7). Doing so, we see that the overall species richness is not projected to change wildly. However, a general homogenization occurs whereby those locations in the western part of the study area with currently the lowest species diversity are projected to gain some species, and some of the areas of rich diversity in the southern part of the country (e.g., Mississippi) are projected to lose some species. Florida also shows a possible gain in species, although diversity is currently artificially low there because its many endemic species did not meet the criteria for entering into the RTA modeling. As shown in Fig. 14.7, Minnesota and Wisconsin would lose some species, while areas that are now highly diverse, such as along the Ohio River and Mississippi River, would remain diverse.

Dominant Genera

In an effort to begin assessing possible changes in forest type, we combined the information for all 80 species and assigned a genera code to the species that had the highest importance value for each county. This process was done for current climate as well as the GISS and GFDL climate scenarios. Maps were then generated showing genera with the highest importance value from current and projected future conditions (see color insert Fig. 14.8). To aid in visual comparisons, only those 14 genera were mapped that had at least 15 counties in which they were dominant under current climate conditions, and the same legend was used for the two future scenarios. A comparison of the three maps shows the loss of all fir (mostly balsam fir) from Maine and most of the *Populus* spp. (mostly quaking aspen) from the northern states. Eastern cottonwood (*Populus deltoides* Bartr.) remains important in the western part of the study region. Maple (*Acer* spp.) also is greatly reduced in the future scenarios; red maple (*Acer rubrum* L.) is mostly eliminated from dominant position, while sugar maple (*Acer saccharum* Marsh.) appears infrequently in the Midwest and boxelder (*Acer negundo* L.) occupies a prominent position in the western counties.

The primary increase is with pine (*Pinus* spp.), primarily loblolly pine (*Pinus taeda* L.). This species is shown to move quite dramatically northward, yet, along with longleaf pine (*Pinus palustris* Mill.), retain the current area but diminish in importance in the Southeast (see Fig. 14.8). In New England, again pine is prominent, but here it is white pine, which is projected to replace the fir forests in Maine. Ashes (*Fraxinus* spp.) and elms (*Ulmus* spp.) would gain prominence in the midwest under the future climate change scenarios. The oaks (*Quercus* spp.) would also expand under future (warmer and drier) conditions, especially in the GFDL scenario in which primarily post oak (*Quercus stellata* Wang.) would take over loblolly pine in overall importance in many instances.

Both paleoecological studies and modeling efforts have clearly shown that communities are *ad hoc* mixtures of species and cannot be expected to move together as intact communities if future conditions change. Modeling exercises such as these are laden with assumptions, but they do begin to give a picture as to how species and forests could respond if the climate changes. The model outputs for spruce–fir and eastern white pine match historical patterns described in the case study presented earlier in the chapter. Efforts such as these are needed to begin to learn what the nature of our nation's future forests might look like under a globally changed climate.

Factors Contributing to Uncertainty in Predicting Future Forest Composition

Paleoecological studies and modeling efforts provide strong evidence that global warming will eventually change the geographical distribution of vegetation. However, forecasting the details of species migration presents a significantly more difficult challenge because a number of important factors are not well known. Accurately predicting future forest composition requires both an understanding of how the environment will change and how individual species will respond to that change. Considerable effort has been expended to predict future climate change, but many questions still remain, particularly about future precipitation patterns and the frequency of disturbance events. Also not yet clear is the extent and nature of predicted change in such factors as ambient air temperature increases, atmospheric ozone enrichment, and the concentration of acid inputs.

The response of tree species to complex environmental changes will be of fundamental importance to the continued adaptation, health, and sustainability of northern forest ecosystems, yet species responses to environmental change have been treated only in simple terms. For the most part, the determinants of the boundaries of any particular species are poorly understood. Many models attempting to forecast species distribution in response to global warming begin with the assumption that species can exist only within their present temperature and/or moisture regimes (Botkin and Nisbet, 1992; Davis and Zabinski, 1992). While perhaps initially appealing, there is substantial evidence that many other factors are also influential in delimiting species boundaries and distributions (Woodward, 1987). Without accurately addressing the variety of factors that influence species distribution and responses to changing environments, predictions of species migration in response to changing climate will be erroneous, perhaps greatly so. Consideration must certainly be given to life history and competition factors that influence species establishment and survival, including the potential responses and influence of exotic plant species and pathogens to climate change. Anthropogenic factors and the differential response of species and populations within species to such factors represent a new complicating element in predicting future forest composition. Anthropogenic stresses may transform some species through selective action and modify the extent, nature, and ability of populations or species to adapt or respond to climate changes.

Impact of Life History and Competition Factors

Despite many uncertainties, we do know the general categories of factors that influence species distribution. For a species to maintain a presence at a particular location, there must be (1) a physical environment (including climatic, edaphic, and fire-history factors) that permits successful completion of the plant's life stages, (2) a supply of seeds (or an effective vegetative propagation method), (3) an absence of decimating herbivores and pathogens, and (4) an ability to successfully compete with other species. If a species is absent from a particular location, it may be because the species' physiological tolerances have been exceeded at one or more stages in its life cycle. Alternatively, it may also be because past events have excluded the species from the area, or the species is not competitive with others that are present, or fires or pests preclude survival. If a species exists at a particular location, it is because that species has traits that enable it to outcompete other species on that site. As such, for a species to successfully occupy a site, it must be able to access that area, successfully colonize the site and remain competitive, and replace itself. Present ecological configurations certainly reflect competitive successes, but do not necessarily reflect the only situations in which a species could be successful.

We can generalize about the relative importance of environmental factors on species distribution. In the boreal and temperate regions, low temperatures are frequently thought to limit poleward expansion of species ranges. Where forests abut steppe, drought and/or fire are probably limiting agents. In most other situations, interspecific competition is likely to be the dominant factor limiting species distribution. Examples of effective interspecific competition strategies may include more aggressive reproduction, phenological synchrony with physical and/ or biological components of the environment, or more effective methods to withstand insect or pathogen attack. By Woodward's (1987) analysis, the equator-ward margins of species ranges tend to be established by competitive forces. Unfortunately, though, the precise factors that determine species ranges are myriad and idiosyncratic. The number of factors and paucity of information about them led Franklin and others (1992) to conclude that, aside from the general tendency for species to migrate poleward and uphill as climates warm, "... surprises (exceptions to our logical predictions) will probably be the rule."

Robustness of Natural Populations

Where trees exist at their physiological limits, climate changes could induce rapid death. While some express concern about this possibility (Botkin and Nesbit, 1992), others have noted the ability of mature trees to withstand altered climates for considerable lengths of time (Brubaker, 1986). Support for the latter view comes from historical data indicating that during the past 10,000 years North American spruce trees existed in climates that were significantly warmer than in any part of their current ranges (Davis, 1978). There is general recognition from paleoecological data that the distribution of vegetation was nonanalogous with climate prior to about 6,000 to 7,000 years ago (Woodward, 1987; Overpeck et al., 1991). This suggests that factors other than physiological tolerance were determining the southern portions of the range of spruce. It also suggests that the phenotypic plasticity of spruce is greater than that evident from the range of environments inhabited by spruce today.

The degree to which climate change can be tolerated before mortality is accelerated is not well known for any tree species, but provenance tests indicate that it is probably substantial. In provenance tests, trees of a single species representing a wide range of habitats are planted together at one or more locations for comparison. In most cases, such tests reveal considerable genetic differentiation among populations from disparate parts of the species' range. As a general rule, trees from northern populations are slower-growing than their southern relatives (except when low temperatures injure trees from southern provenances) when grown in common garden, and they suffer almost no mortality when grown hundreds of miles south of their origin. Clearly, northern populations of most temperate and boreal zone species have no difficulty tolerating climates more than 5°C warmer on an annual basis. The only exception may be with trees having substantial winter chilling requirements. For example, red maples from Massachusetts exhibit sporadic and delayed spring budbreak and have poor survival when grown in Florida (Perry and Wang, 1960). This possibility should be examined carefully because some climate models project that much of the future warming will be experienced in the winter (Woodward, 1992). Observations of horticultural plantings also suggest that species can be grown in climates far warmer than any place in their natural range.

While increasing warmth, per se, will probably not be injurious to most established trees, results could be different if accompanied by drought. The anticipated warming, of course, will be accompanied by increased atmospheric carbon dioxide (CO_2) concentration. A number of studies have shown that increased CO_2 concentration should increase the growth of trees under well-watered conditions (Teskey et al., 1998). More importantly, perhaps, other studies have shown that increased CO_2 will reduce transpiration by allowing more CO_2 to enter through partially closed stomata (Jarvis and McNaughton, 1986). The reduction in transpiration that is expected to accompany increased CO_2 concentration may well offset the effects of increased drought that could occur. Indeed, Woodward (1992) projects that the world's forested area will increase despite increased dryness if atmospheric CO_2 concentration doubles. These projections are based on extrapolations from small plot studies and, thus, are subject to considerable error when applied on a global scale. Nevertheless, it is expected that higher CO_2 concentrations will partially offset the effects of decreased moisture availability, but the degree of compensation is not firmly established.

Provenance tests also indicate that trees from moist areas generally suffer greater mortality than trees from drier sites, especially when planted on sites in the drier portions of the species' range. For example, loblolly pine from North Carolina (with wetter growing seasons) grew much faster than native populations in Arkansas (with drier growing seasons). However, after 25 years, the North Carolina trees show pronounced mortality (Wells and Lambeth, 1983). The exact cause of death was not known, but it was correlated with drought. Plasticity in drought tolerance has not been quantified for any species, but appears to be smaller than for tolerance of high temperatures.

Empirical information from provenance tests and horticultural plantings, albeit based on trees in largely noncompetitive growing environments, demonstrates the great resiliency of forest tree species, especially relative to high temperatures. Coupled with the relatively rapid rate of predicted climate warming, these data and experiences highlight the possibility that many species distributions may not simply shift northward or upward, but may actually remain competitive in their current locations and actually expand their distributions.

Regeneration Success in Changed Climates

The inherent robustness of natural populations will likely enable existing trees to withstand warmer temperatures currently predicted to accompany higher atmospheric CO_2 levels. Concerns have been raised, however, about the ability of these trees to regenerate in a warmer and perhaps drier climate (Franklin et al., 1992).

The ability to regenerate depends on the ability of existing trees to produce seeds and for those seeds to germinate and survive. Various climatic parameters can influence seed production. Cold temperatures have been shown to restrict seed production in some species, such as little-leaf linden (*Tilia cordata* Mill.), at the northern extremities of their range (Pigott, 1981). On the other hand, viable seed production in American beech (*Fagus grandifolia* Ehrh.) is an inexplicably uncommon event in the southern portion of its range, exclusive of Mexico. Physiological stress has been shown to increase seed production in many species, at least in the short term. Conversely, stress that depletes carbon reserves can result in poor seed production, and severe drought stress can lead to flower and fruit abortion. Numerous other climatic factors could potentially affect seed production as well. However, no generalizations can be made at this time. Some species will undoubtedly show increased seed production, and others less. In most cases, though, there is no reason to expect seed production will be inhibited following warming.

A greater concern is that a warmer and drier environment may reduce germination and seedling survival. The most sensitive stage of a tree's life is the beginning. Losses in this period are very high, owing mostly to conditions that are inhospitable at this vulnerable stage. On sites that are prone to drought or lethal temperatures, such as south-facing slopes, higher temperatures would exacerbate losses. The area affected by these lethal agents would increase to some degree. Furthermore, germination could be reduced or unfavorably delayed in species with unmet cold stratification requirements. These unknown factors related to seed germination and success of the seedling stage in forest trees all contribute to uncertainty in predicting future forest composition.

Uncertain Rates of Migration

Paleoecological studies have indicated that the ranges of many species have migrated throughout history in response to changing environmental conditions. As climate warmed, species ranges typically shifted to higher latitudes and higher altitudes; as climate cooled, the reverse occurred.

Other, more subtle, changes in species ranges could also occur to adjust for climate changes, such as a species inhabiting cool, moist microsites if the climate becomes warmer and drier. The ability to adjust their distribution to changing environmental conditions ensured continued representation of many forest tree species after such upheavals as the Quaternary's glacial periods.

Of course, individual trees did not change location; rather, a shift occurred in the distribution of the range of the overall species at historical rates of approximately 10 to 50 km per century (Schwartz, 1993). Consider the example of climate cooling. More seedlings on the southern portion of the species range would likely become established than on the northern edge of the species range. In time, more seedlings would germinate southward than northward, and the species range as a whole would shift southward. If a topographical obstacle, such as a high mountain range or water body happens to get in the way of this southward migration, the species may be unable to migrate far enough southward to withstand a significant climatic cooling. In fact, this scenario may have occurred for a number of northeastern forest tree species during glacial periods; many species found a "glacial refugium" in the southern Appalachian region. If the Appalachian Mountains were aligned east–west instead of north–south, perhaps many of these species would have been unable to migrate far enough southward to endure the climatic cooling experienced during glaciation.

Uncertainty exists about the nature and extent of future climatic change and its effect on migration of forest tree species. Although it is tempting to speculate that climate change may be too rapid for forest tree species to successfully migrate or that large gaps may be created that can restrict species dispersal, historical evidence indicates that climate changes in the past have been more rapid than changes projected for the next few centuries without any great restriction to species movement. It is possible, however, that species rates of migration could be so different that some species may be "overrun" and obliterated by more competitive species. Furthermore, it is unclear to what extent species migration might be influenced by human activities, such as being restricted by forest fragmentation or aided by the planting of trees. Migration is an important component that needs to be addressed in predicting future forest composition.

Competition from Exotic Species

The competitive influence of exotic species, including both nonnative plant competitors and exotic pathogens (insects and diseases), also contributes uncertainty to predictions of how northern forests will respond to environmental change. Whether part of the planned extension of a species range or the result of an uncoordinated release, the addition of nonnative tree species to northern forests would likely interact with the competitive dynamics already influenced by environmental change and further alter forest community structure. In fact, because exotic species often aggressively fill new or undeveloped niches within ecosystems, there is reason to believe that some exotics could thrive in an environment in which traditional niches are redefined in response to environmental change. Empirical observations of the survival, growth rate, and successful establishment of exotic species in northern forests, such as Norway spruce (*Picea abies* [L.] Karst.), common buckthorn (*Rhamnus cathartica* L.), common lilac (*Syringa vulgaris* L.), and black locust (*Robinia pseudoacacia* L.), demonstrate that aggressive exotic species can modify the composition of forested landscapes, especially in disturbed areas. The widespread existence of mature forest plantations, including both non–North American and off-site native species, and abundant ornamental plantings provide a ready source of exotic germ plasm for dissemination. Likely variation in the net physiological responses and resulting competitive fitness of native species, combined with uncertainties about the possible interplay of introduced species, raises substantive questions about our ability to accurately predict the outcomes of complex and potentially dynamic environmental change.

Perhaps of greater concern are the potential impacts of exotic insects and diseases. Exotic pathogens, such as chestnut blight, Dutch elm disease, and gypsy moth, have already modified the dynamics and composition of north temperate hardwood forests. If warm temperatures permit the spread of insects and diseases that are currently held at bay by cold, or if drought increases the acreage affected by fire or the susceptibility of some trees to secondary stresses, the impact on forest health and future composition would be substantial. Although the effects of fire may be mitigated by human fire-control activity, it is likely that unpredictable insect and disease epidemics will materialize. Abundant present day empirical information supports this contention. Recent epidemics of Scleroderris canker, pear thrips, the Asian longhorn beetle, and the apparent aggressive northern migration of the hemlock wooly adelgid and hemlock looper illustrate the potential influence of exotic pathogens on native forests. The globalization of forest industry and frequency of long-distance travel and transport has enhanced pathways for exotic introductions. Indeed, this element of uncertainty can have truly frightening implications.

Indeed, the potential for significant climate-driven changes in the distributions of north temperate forest tree species will likely blur the distinction between some species thought to be "native" vs. "exotic" to a specific region or habitat. In the same way that some very adaptable Eurasian exotic plant or pathogen species may have become widespread and competitive in the northeast, the potential exists for some currently native northeastern species to become dominant as an "exotic" of sorts in some new environment. For example, recent data documents a dramatic ecological expansion in the distribution of red maple throughout eastern forests, which is expected to shift wildlife populations and alter biodiversity of these forests (Abrams, 1999). The expansion is attributed to a combination of the species being an ecological generalist and changes in habitat disturbance patterns, including the suppression of forest fires (Abrams, 1999). Under new disturbance regimes associated with changing climate, other species may emerge as dominant generalists to alter the distribution and composition of future forests.

Impact of Anthropogenic Factors

Anthropogenic factors also need to be incorporated into future environment scenarios because human influences on the earth's ecosystems are substantial and growing (Vitousek et al., 1997). Through a variety of enterprises (e.g., industry, agriculture, recreation, and international commerce), humans are transforming fundamental natural processes such as climate, biogeochemical cycling, and even the biological diversity upon which evolutionary change depends (Vitousek et al., 1997). Although a dominant component of this global transformation is climate change, there are many other environmental factors that could affect the health and sustainability of forests. Prominent among these is a diverse pool of pollutant inputs. The list of airborne chemicals that northern forests receive is long, and probably restricted only by current monitoring limitations. Principal contaminants include acid deposition (which is dominated by four ionic constituents: hydrogen $[H^+]$, nitrate $[NO_3^-]$, sulfate $[SO_4^{2-}]$, and ammonium $[NH_4^+]$) and gaseous pollutants (primarily ozone $[O_3]$, sulfur dioxide $[SO_2]$, and oxides of nitrogen $[NO_x]$ in addition to elevated CO_2) (Mohnen, 1992). Trace additions of heavy metals (e.g., lead, mercury, and arsenic) and numerous man-made toxins (e.g., polychlorinated biphenyl [PCB] congeners, polycyclic aromatic hydrocarbons [PAHs], dichlorodiphenyltrichloroethane [DDT]), and derivatives also occur (Baisden et al., 1995; Hoff et al., 1996; Golomb et al., 1997a,b).

   Although climate change and pollution are often emphasized, northern forests also contend with a host of other anthropogenic influences, including changes in land use and associated forest fragmentation and the possible impacts of intensified forest management. Alone and in combination, these and other anthropogenic forces are changing forest environments and the selection pressures that shape forest communities and will affect the future distribution of species.

Direct Threats to Tree Health and Survival

It is increasingly evident that some anthropogenic factors directly alter tree physiology and influence the health and survival of forests. The impact of gaseous air pollutants on eastern white pine and acid deposition on red spruce (*Picea rubens* Sarg.) provide vivid examples.

Eastern white pine, although a robust species from both ecological and paleoecological perspectives (Jacobson and Dieffenbacher-Krall, 1995), may very well be injured by air pollution more than any other tree species in eastern North America and is particularly sensitive to O_3 and SO_2 pollution (Gerhold, 1977). Morphological and physiological symptoms of air pollution injury in eastern white pine include direct foliar injury, growth reductions, depressed photosynthetic rates, and mortality (Karnosky and Houston, 1979). It has been demonstrated that greater than 10% of the sensitive trees and 5% of trees intermediate in sensitivity in a population died as a result of air pollution–induced viability selection (i.e., reduced growth rate and competitive ability for light, water, etc.) over several years (Karnosky, 1980). Most estimates suggest that the proportion of trees sensitive to air pollution in a population may vary between about 5 and 25%, and many of the sensitive individuals have been lost from natural populations. In northern Ohio, it is estimated that 40% or more of a native eastern white pine population may have been lost as a result of air pollution–induced selection (Kriebel and Leben, 1981). Although the population is adapting to changing pollution levels, such selection for pollution resistance results in an immediate loss of forest productivity. Also, pollution-induced mortality would be expected to reduce genetic diversity that may be needed for future adaptation to anthropogenic stress, such as climate warming.

The response of red spruce to acid deposition is more indirect and physiologically complex, but further illustrates an anthropogenic threat to forest health, survival, and potential evolutionary development. It has recently been documented that acid inputs directly leach calcium from the membranes of leaf cells, decrease cell stability, reduce cold tolerance, and enhance the potential for secondary freezing injury (DeHayes et al., 1999; see Chapter 6). In fact, acid-induced reductions in cold tolerance are of sufficient magnitude (3 to 10°C) to explain the now common and widespread freezing injury to red spruce observed in northern montane forests over the past 40 years (Johnson et al., 1988; DeHayes, 1992; Johnson et al., 1996) and growth and vigor losses typical of red spruce decline (Wilkinson, 1990; Tobi et al., 1995). Most northern evergreen species develop a tolerance to low temperature that protects leaves from freezing injury even at temperatures well below the lowest ambient temperatures experienced. In contrast, the current-year foliage of red spruce achieves a maximum cold tolerance that is barely sufficient to prevent freezing injury during most winters (DeHayes, 1992; Schaberg et al., 2000). As a consequence of this low baseline hardiness, acid-induced reductions in cold tolerance make red spruce uniquely vulnerable to freezing injury (DeHayes et al., 1999; see Chapter 6). The study of acid deposition's influence on red spruce cold tolerance has provided a detailed mechanistic model of how a pollutant can alter tree physiology and forest health. This case study also exemplifies how variation in stress response can have profound ecological consequences.

Substantive differences in survival following exposure of eastern white pine to gaseous air pollution and red spruce to acid deposition compared with other apparently less sensitive species highlight the possibility that variation in species responses to anthropogenic change could be an important force in shaping the composition, form, and function of tomorrow's forests.

Heterogeneity of Species Responses to Environmental Change

To further complicate predictions of future forest composition, the response of trees to any given environmental change will not be uniform. If existing research is any indication, a major source of variability will likely be at the species level.

Numerous studies have evaluated potential differences in the physiology of tree species following perturbations in temperature, water availability, atmospheric gases, or pollutant loading that may accompany climate change. And, almost uniformly, these studies detect significant differences in species responses to environmental perturbation. Table 14.1 provides examples of studies that examined the impact of these perturbations on a broad range of physiological traits for eastern North American tree species. Collectively, these examples include evaluations of eight different climate change factors, three crossed factors, 24 tree species, and a range of physiological parameters including seed germination, growth, foliar injury, cold tolerance, carbohydrate storage, water transport capacity, cation nutrition and mortality (see Table 14.1). Yet, despite the diversity exemplified here, one commonality is clear: in each study, meaningful differences in response were detected among the species evaluated (see Table 14.1).

If trees in forests showed the same heterogeneity in response depicted in Table 14.1, one consequence would be great uncertainty in predicting the overall impacts of environmental change. This uncertainty is fueled, in part, by the many levels of response that require integration in order to determine "net responses." For example, even for a single climate change factor (e.g., acid rain) and a single tree species (e.g., sugar maple), the physiological impact could be seemingly positive for one trait (e.g., increased extension growth and leaf weight under low soil nutrient conditions, Raynal et al., 1982b), neutral for a different trait (e.g., seed germination, Table 14.1, Raynal et al., 1982a), and potentially negative for yet another (e.g., increased foliar cation leaching, Table 14.1, Lovett and Hubbell, 1991). So, to understand the net impact of even a solitary climate change factor on a single species, one has to integrate impacts on all physiology throughout the life and reproductive cycle of that species. Once this was accomplished for each species, this information would have

Table 14.1. Examples of Differential Responses of Eastern North American Tree Species to Individual or Combined Environmental Change Factors

Factor	Species Compared	Response	Reference
CO$_2$	Acer rubrum L. Acer saccharum Marsh. Betula papyrifera Marsh. Fagus grandifolia Ehrh. Pinus strobus L. Prunus serotina Ehrh. Tsuga canadensis (L.) Carr	Relative to plants receiving 400 µl l^{-1} CO$_2$, shade tolerant A. saccharum, F. grandifolia, and T. canadensis seedlings had the greatest stimulation of biomass accumulation in response to elevated CO$_2$ (700 µl l^{-1}), P. serotina and B. papyrifera showed intermediate increases, whereas A. rubrum and P. strobus experienced insignificant increases in biomass growth.	Bazzaz et al., 1990
Temperature	Abies balsamea (L.) Mill. Picea rubens Sarg.	Mature montane P. rubens trees dehardened 3 to 14°C during a natural midwinter thaw, whereas sympatric A. balsamea trees showed no evidence of dehardening.	Strimbeck et al., 1995
O$_3$	A. rubrum Betula alleghaniensis Britton Fraxinus americana L. Liquidambar styraciflua L. Liriodendron tulipifera L. P. serotina Quercus alba L. Quercus rubra L.	Following 12 weeks of experimental exposure to 0, 0.075 or 0.15 ml l^{-1} O$_3$ seedlings showed a range of leaf stippling and defoliation responses. P. serotina was the most sensitive to O$_3$, followed by L. styraciflua, L. tulipifera, F. americana, A. rubrum and B. alleghaniensis. Q. spp. exhibited no foliar injury.	Davis and Skelly, 1992
UV-B	Picea glauca (Moench) Voss Picea mariana (Mill.) B.S.P. Pinus banksiana Lamb.	All seedlings experienced reductions in dry weight following 16 weeks of high supplemental UV-B (150 W cm^{-2}) relative to low additions (1.1 W cm^{-2}). Shade tolerant P. glauca and P. mariana showed greater reductions (45 and 40%, respectively) than P. banksiana (29%).	Yakimchuk and Hoddinott, 1994
Acid precipitation	A. rubrum A. saccharum B. alleghaniensis P. strobus T. canadensis	The impact of pH 3.0, 4.0 or 5.6 treatment solutions on seed germination was tested. Compared with pH 5.6, germination was inhibited at pH 4.0 and 3.0 for A. rubrum and at pH 3.0 for B. alleghaniensis. No inhibition was observed for A. saccharum or T. canadensis. Germination was stimulated by decreasing pH for P. strobus.	Raynal et al., 1982

Nitrogen deposition	*A. balsamea* *A. rubrum* *Acer spicatum* Lam. *Betula* spp. *P. rubens*	Following 7 years of low level nitrogen additions (0 to 31.4 kg N ha^{-1} yr^{-1}) to montane forest plots. *P. rubens* had the highest rates of annual mortality (% trees that die per species), followed by *A. balsamea* and *Betula* spp. No *Acer* spp. mortality was noted.	McNulty et al., 1996
Water stress	*Populus deltoides* Bartr. *Thuja occidentalis* L. *A. saccharum* *Juniperus virginia* L.	Xylem tensions causing a 50% loss in hydraulic conductivity were lower (a more positive Ψ) for *P. deltoides*, intermediate for *T. occidentalis* and *A. saccharum*, and greater for *J. virginiana*.	Tyree et al., 1994
Ca depletion	*P. strobus* *A. saccharum*	For both pH 3.8 and 5.0 mist treatments, *A. saccharum* canopies lost approximately 6 times more Ca (μg m^{-2} leaf area) than *P. strobus* canopies.	Lovett and Hubbell, 1991
$CO_2 \times$ N	*Populus tremuloides* Michx. *Q. rubra* *A. saccharum*	Seedlings were exposed to 2 levels of atmospheric CO_2 (355 or 650 μl l^{-1}) and 2 soil NO_3 treatments (1.25 or 7.5 mM). Significant $CO_2 \times NO_3 \times$ species interactions were found for shoot length ($P \leq 0.05$). *P. tremuloides* shoot growth increases were almost solely associated with NO_3 treatment, whereas *Q. rubra* and *A. saccharum* had growth increases responsive to both CO_2 and NO_3.	Kinney and Lindroth, 1997
$CO_2 \times$ UV-B	*P. banksiana* *P. mariana* *P. glauca*	Seedlings were exposed to CO_2 (350 or 370 μl l^{-1}) and UV-B radiation (1.1 μW cm^{-2} or 150 μW cm^{-2}). A $CO_2 \times$ UV-B interaction ($P \leq 0.05$) for height growth was found for only *P. banksiana*. For this species, UV-B treatment fully offset increases in growth associated with CO_2.	Yakimchuk and Hoddinott, 1994
$CO_2 \times$ water stress	*P. tremuloides* *A. saccharum*	Seedlings were grown under ambient or elevated CO_2 (350 or 700 μl l^{-1}) and under well-watered or drought conditions. A $CO_2 \times H_2O \times$ species interaction for foliar hexose levels ($P \leq 0.06$) was found. For *P. tremuloides*, drought-induced reductions in hexose concentration were partially offset by elevated CO_2. In contrast, *A. saccharum* experienced increases in foliar hexoses with drought, but these increases were reduced by elevated CO_2.	Roth et al., 1997

to be compiled and tested to determine how these physiological changes could impact competition and survival among species. In addition, the complexity in response would only increase as the mix of environmental change factors is augmented to more closely resemble field conditions. Even limited study of differences in response to combinations of environmental factors has identified significant interactions with tree species (see Table 14.1). Considering the probability that multi-factor interactions with tree species occur, the interplay of species responses to complex mixes of environmental change factors in the field are likely to be different from results of simplified experimental simulations. These differences could have great importance if they influence the dynamics of species competition that have helped establish and sustain existing forest communities.

The models that have been used so far to predict future forest composition (examples described earlier in this chapter) have been valuable first steps in the ever-evolving prediction process. Refinements can made after results of these models are evaluated. Heterogenetity of species response and other factors affecting species distribution will need to be addressed in future prediction models.

Genetic Variation in Response to Changing Environments

Forest tree species clearly have the potential to respond in dramatically different ways to changing environmental conditions, including climate warming and elevated levels of CO_2 and other greenhouse gases. That is, species × environment interactions, a measure of the differential response of species to varying environmental conditions, are expected to be substantial. Similarly, although perhaps less intuitive, different populations and individuals within a single species also exhibit considerable genetic variation in response to changing environmental conditions. Such genotype × environment interaction may reflect an adaptive response of populations to anthropogenic stressors, adding uncertainty about how complex forest systems may respond to global environmental changes. Because forest tree species are dynamic and exist as an aggregate of potentially genetically variable and divergent populations, it is not reasonable to expect individuals of a given species to uniformly respond to changing environmental conditions.

Genetic variability has been documented in the response of eastern forest tree species to an array of anthropogenic stressors, including tropospheric O_3, acid deposition, elevated CO_2, elevated nitrogen, and temperature stress (Table 14.2 on pages 526–527). The findings highlight both the resiliency and adaptability of forest tree species and also the potential for complex responses to interactive stresses in environments where competition for limited resources is prevalent. For example, Berrang et al. (1986, 1989) demonstrated genetic variation in ozone sensitivity among populations of quaking aspen (*Populus tremuloides* Michx.). Interestingly, their data show that high ambient levels of ozone in certain locations have differentially selected against sensitive aspen genotypes resulting in populations with greater average resistance to both artificial ozone fumigations (Berrang et al., 1986) and exposure under ambient field conditions. Trees originating from locations with low ambient O_3 levels were generally most sensitive to O_3 exposure. Thus, it appears that natural selection for O_3 tolerance has occurred relatively quickly in this species because anthropogenically elevated levels of O_3 had existed for only about 60 to 70 years prior to the study. Important, however, is the fact that more than a 3-fold difference in O_3 injury was observed among clones within some populations, including the highly selected populations that had evolved greater O_3 tolerance. As such, it seems that a relatively short period (perhaps a single generation) of ozone selection was enough to enhance tolerance with little evidence of a reduction in genetic variation.

Recent evidence has suggested that plant species may also exhibit genetic variation in response to global change factors, such as CO_2-enriched environments. As such, it might appear that rising CO_2 levels would result in selection favoring genotypes and populations that exhibit increased productivity in CO_2-enriched atmospheres leading to enhanced carbon sequestration in terrestrial ecosystems and a concomitant reduction in future atmospheric CO_2 levels. However, empirical data suggests a more complex form of selection pressure and population response. Bazzaz et al. (1995) have shown that experimental populations of yellow birch (*Betula alleghaniensis* Britt.) exhibited responses to enriched CO_2 that were both genotype-specific and density-dependent (see Table 14.2). That is, under conditions of minimal competition, all half-sib families of yellow birch that were tested exhibited a growth enhancement in response to elevated CO_2. However, the magnitude of the family response varied genetically. For instance, the fastest-growing family at ambient CO_2 levels was the slowest growing in the CO_2-enriched atmosphere. Under high density, maximum competition conditions, variation among genotypes in CO_2 responsiveness was greater, but the overall magnitude of the response was considerably less than in minimal competition environments. An enriched CO_2 atmosphere appeared to result in pronounced shifts in genetic composition, even though overall productivity enhancements were small. The authors concluded that CO_2 enrichment intensified interplant competition and that selection favored genotypes with a greater ability to compete for resources other than CO_2. While this may result in the evolution of populations with a greater competitive ability on the one hand, there is little evidence to suggest the evolution of increased CO_2 responsiveness or carbon sequestering on the other. Furthermore, the increased selection intensity evident in CO_2-enriched environments would be expected to contribute to a loss of genetic diversity in these populations and perhaps to influence population responses to other localized or global-scale environmental changes.

These examples illustrate the dynamic nature, potential evolutionary responsiveness, and overall resiliency of forest tree populations in the face of anthropogenically induced environmental change. However, they also highlight the challenge in predicting species-level responses to changing climatic conditions because forest tree populations have been, and will continued to be, faced with a potentially new set of environmental changes that are not well understood. Such changes may lead to the evolution of resistance or population elimination, greater or lesser competitive abilities, reductions in overall levels of genetic diversity and responsiveness, and/or individuals and populations that exhibit greater levels of fitness in new environments. There clearly is uncertainty. In fact, the only certainty is that environments and selection pressures and forest tree populations themselves will be different from those of the past. As such, extensive variability among and within species will make accurate prediction of overall responses to anthropogenic change difficult at best. However, because genetic and phenotypic variability is fundamentally important to biological survival, the presence and persistence of such variability as the raw material for future evolutionary adaptation will impart the greatest likelihood of continued health and stability within northern forests.

Potential Threats to Evolutionary Adaptation

Forest trees possess and depend upon extensive levels of genetic diversity for their adaptive potential and for overall forest ecosystem stability (Gregorius, 1996). Anthropogenic disturbances, such as air pollution selection, forest fragmentation, and intensive timber harvesting or management regimes, and perhaps even mutagenic effects of pollution (Grant, 1998), could alter the size and structure of tree populations, levels of genetic diversity, and gene flow, which are the foundations of species adaptation and evolution. The genetic foundation of forest ecosystem stability lies at the population level, which is the unit of adaptation and evolution (Gregorius, 1996). As such, loss of genetic variability within populations poses a potential threat to evolutionary adaptation.

Fitness vs. Flexibility: Potential "Cost" of Resistance to Anthropogenic Factors

Forest tree species generally possess very high levels of genetic diversity, and much of the genetic polymorphism is concentrated among individuals within populations. The ubiquitous presence of genetic variation in forest trees implies that selection for resistance to stressors (e.g., pollution) may be occurring in natural populations. The previously cited example with trembling aspen (Berrang et al., 1986, 1989) illustrates this relationship.

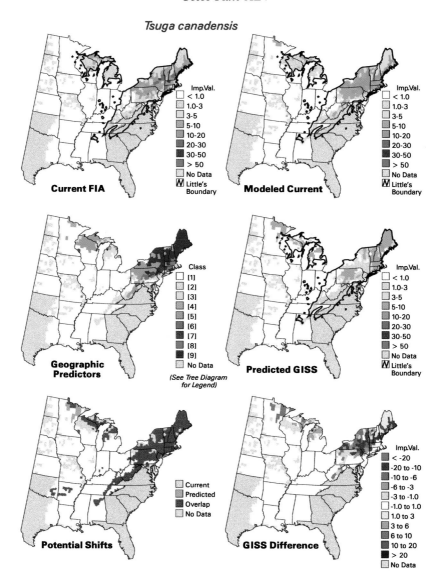

Figure 14.6. Example model outputs for eastern hemlock (*Tsuga canadensis*), including (UL) actual county importance value (IV) as calculated from FIA data; (UR) modeled current IV from the RTA model; (ML) predicted future IV after climate change according to the GFDL GCM; (MR) predicted future IV after climate change according to the GISS GCM; (LL) difference between modeled current and predicted future IV according to the GFDL model; and (LR) difference between modeled current and predicted future IV according to the GISS model.

Color Plate XLVI

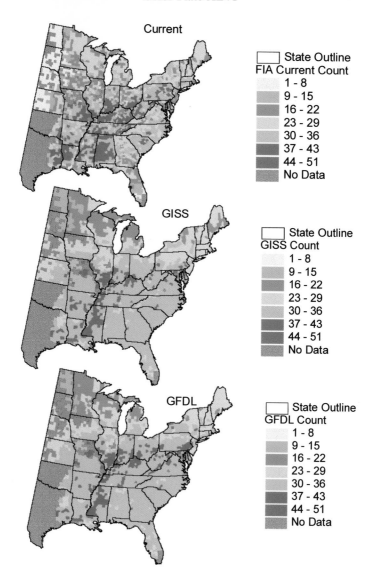

Figure 14.7. Estimated species richness, of the possible 80 common tree species in the eastern U.S., for (top) current times according to FIA data; (middle) future climate according to the GISS scenario; and (bottom) future climate according to the GFDL scenario.

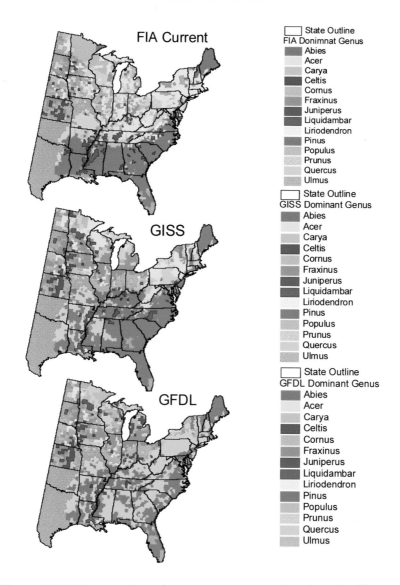

Figure 14.8. Estimated dominant genera, by county, based on the 80 common tree species in the eastern U.S., for (top) current times according to FIA data; (middle) future climate according to the GISS scenario; and (bottom) future climate according to the GFDL scenario.

The review by Pitelka (1988) provides additional examples. The extent of a physiological or evolutionary "cost" of resistance is not clear. Resistant individuals clearly have the benefit of increased fitness in the presence of changing environmental conditions. However, such fitness may come with a price: loss of genetic variation that could lead to decreased flexibility of the population to respond to future environmental change.

Competition experiments indicate that resistant plants are less able to compete and survive in environments lacking the pollution stress and that there can be rapid selection against tolerant individuals once pollution stress declines (Pitelka, 1988; Winner et al., 1992). Although "costs" of resistance may indeed be real, they are expected to vary with the physiological mechanism(s) involved and, therefore, might be expected to vary among co-occurring resistant species exposed to the same pollutant (Parsons and Pitelka, 1992). As such, competitive balances and species composition may be altered.

From an evolutionary standpoint, the "cost" of pollution selection in terms of loss of genetic diversity will be influenced by numerous factors, such as the selection intensity, the magnitude of resistance and pattern of its inheritance, and the life cycle of the population/species under selection. Given the episodic nature of most air pollution (acid deposition, ozone, sulfur dioxide) exposures, high levels of genetic diversity in trees, and the relatively long life cycles of tree species, one might expect a relatively slow rate of loss of genetic diversity in response to air pollution selection in most tree species. The persistence of genetic diversity in ozone resistance among trembling aspen clones from relatively resistant populations is consistent with this expectation. If, however, pollution-induced mortality is high, as Kriebel and Leben (1981) have indicated for eastern white pine populations from Ohio, then one would expect the residual population to effectively experience a bottleneck, genetic drift, and a reduction in genetic diversity. That is, genetic diversity and the potential for evolutionary change may be reduced as a result of the random loss of alleles (including alleles unrelated to the actual cause of mortality) associated with reduced population size as well as the byproduct of intensive pollution selection. Pollution-induced changes in both the genetic structure and levels of genetic diversity of forest tree populations will influence the evolutionary opportunities and pathways possible in the future in response to new selection pressures and intensities.

Forest Fragmentation

Fragmentation of forest ecosystems has occurred as a result of agriculture, urban and suburban development, and road construction. Fragmentation serves as a potential migration barrier that can inhibit species or population dispersal in response to environmental change, especially latitudinal migration expected in response to climate warming. Certainly,

Table 14.2. Examples of Genetic Variation in Responses of Eastern North American Tree Species to Individual or Combined Environmental Change Factors

Factor	Species Compared	Response	Reference
Temperature	*Picea rubens* Sarg.	Freezing injury of trees from 12 provenances growing in a NH plantation was assessed for 4 winters. Trees from Quebec, NY, and New Brunswick were consistently among the least injured, whereas trees from MA and NH were consistently among the most injured.	DeHayes et al., 1990
O_3	*Populus tremuloides* Michx.	Ozone sensitivity of offspring derived from populations in 5 national parks with differing ambient O_3 levels was evaluated. Populations varied significantly in ozone-induced foliar injury resulting from both fumigation and ambient exposure. This provides evidence of natural selection for ozone tolerance.	Berrang et al., 1986, 1989
Acid precipitation	*Acer rubum* L.	Trees from 3 seed sources grown in plantation were sprayed with simulated rain solutions of pH 5.2, 4.2 or 3.2 on 5 dates over the growing season. Evaluations of throughfall chemistry revealed small but significant differences among provenances ($P \leq 0.05$) in the concentrations of Ca, Fe, and Zn leached from canopies.	Schier, 1987
N	*Betula papyrifera* Marsh.	Seedlings from 4 populations were grown with either low or high soil N treatments (20 or 80 N supplied at NH_4NO_3) for 3 months. Populations differed in average root biomass and root/shoot ratios following treatment ($P \leq 0.01$).	Wang et al., 1998

Water stress	*A. rubrum*	Seedlings from 4 seed sources (2 wet sites and 2 dry sites) were subjected to 3 levels of water stress. Wet site seed sources wilted sooner in response to water stress than dry site sources. However, at all levels of water stress, growth was greatest for wet site seedlings.	Townsend and Roberts, 1973
CO_2	*B. alleghaniensis*	Experimental populations were exposed to CO_2-enriched atmosphere. CO_2 enhanced biomass in noncompetitive situations. In high density stands, enriched CO_2 resulted in significant shifts in genetic composition and only small productivity enhancements.	Bazzaz et al., 1995
CO_2 and temperature	*Pinus banksiana* Lamb.	Seedlings from 15 maternal families were grown in growth chambers with either 390 µl l^{-1} CO_2 and simulated ambient temperatures or 700 µl l^{-1} CO_2 and ambient + 4°C temperatures. Families differed ($P \leq 0.05$) in their instantaneous water use efficiency (WUE) following treatment.	Cantin et al., 1997
CO_2 and temperature × N	*P. banksiana*	Seedlings in the above study were also divided into subplots that received either 5 or 100 µg g^{-1} N fertilization. A CO_2 and temperature × N × family interaction ($P \leq 0.05$) was found for WUE.	Cantin et al., 1997
$CO_2 \times O_3$	*P. tremuloides*	Rooted cuttings from one O_3-tolerant and one O_3-sensitive clone were exposed to either O_3-reduced air, 0.10 µl l^{-1} O_3, 150 µmol mol^{-1}CO_2, or elevated O_3 and CO_2. Simultaneous $O_3 + CO_2$ exposure decreased photosynthetic carboxylation efficiency more than the O_3 treatment alone for the O_3 tolerant clone only.	Kull et al., 1996

paleoecological lessons have shown that forest tree species have survived changing environmental conditions by dispersing to more suitable environments. Such dispersal is expected to be constrained in today's increasingly fragmented forests throughout the north temperate zone. Historical rates of migration (10 to 50 km per century) are not likely to occur within a fragmented habitat (Schwartz, 1993). Realizing this, current investigators are assessing more realistic migration scenarios based on species regeneration characteristics and actual landscapes (Schwartz et al., 1996; Iverson et al., 1999).

Equally as important, however, fragmentation subdivides populations into small subunits that could become prone to inbreeding or, more importantly, subject to localized extinction as a result of random demographic uncertainties, such as catastrophic events or periodic reproductive failure (Lande, 1988). As noted by Ledig (1992), such partial extinctions (i.e., loss of populations) represent a loss of genetic diversity and threat to evolutionary development because they may result in the breakup of a complex structure of locally adapted populations.

Small, fragmented populations are expected to experience inbreeding, genetic drift, and an erosion of genetic variability and evolutionary potential within subpopulations, especially if isolation and small population size persists for multiple generations. Under such conditions, genetic divergence among populations is expected to increase, while genetic diversity within populations becomes eroded. While individual populations perhaps become vulnerable to environmental stresses because of genetic uniformity, the genetic distance and isolation of surviving populations could conceivably lead to further evolutionary division and separation. Empirical data from modern-day marginal populations reinforces such theoretical predictions. Table 14.3 shows that relatively small, isolated marginal populations of pitch pine and red spruce exhibit reduced heterozygosity and genetic polymorphism compared with large

Table 14.3. Estimates of Genetic Variability for Central and Marginal Populations of Pitch Pine and Red Spruce

Species	No. of Populations	Polymorphic Loci (%)	Heterozygosity (%)	Reference
Pitch pine				
Central	3	71 a[a]	24.6 a	Misenti,
Marginal	3	56 b	20.9 b	1990
Red spruce				
Central	6	40 a	8.9 a	Hawley and
Marginal	5	36 a	6.2 b	DeHayes, 1994

[a] Means within species followed by the same letter are not significantly different by analysis of variance.

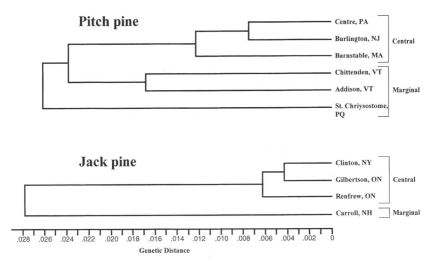

Figure 14.9. Phenogram of genetic distance estimates depicting the degree of genetic differentiation among marginal and central populations of pitch and jack pines.

central populations, and genetic distance estimates among marginal populations of pitch and jack pines are at the extremes for most conifer species (Fig. 14.9). The latter presumably have resulted from genetic drift. Indeed, fragmentation does alter the genetic structure of forest tree populations. Although population subdivision in response to changing environmental conditions is not an entirely new phenomenon, the prevalence and expected increasing rate of fragmentation in today's north temperate forests is unprecedented and a potential threat to evolutionary adaptation.

Intensive Forest Management

Forest management and harvesting practices that include selective logging can conceivably alter the genetic structure of forest tree populations and the evolutionary potential of forest tree species, especially when the intensity of selection is high. Selective cutting could also increase inbreeding, which would be expected to reduce reproductive output and viability (Ledig, 1992). However, very little empirical information has been generated about the influence of intensive forest management practices on the genetic structure, diversity, and mating systems within forest ecosystems. Neale (1985) and Neale and Adams (1985) found that shelterwood harvesting had little impact on the genetic structure of Douglas fir (*Pseudotsuga menziesii* [Mirb.] Franco) forests and attributed population resiliency to exceedingly high levels of genetic diversity in that species and the almost pure, high density Douglas-fir stands. In contrast, Cheliak et al. (1988) reported that phenotypically selected white spruce (*Picea glauca* [Moench] Voss) trees possessed on average only 75% of the genes found among randomly selected trees from the same population. Thus, a loss of some genes and changes in gene frequencies would be expected in response to selective cutting.

Indeed, recent studies have demonstrated that selective logging can alter the genetic structure of residual populations and presumably the potential for populations to adapt to future environmental changes. Buchert et al. (1997) contrasted genetic diversity parameters between preharvest and postharvest gene pools in virgin, old-growth stands of eastern white pine in central Ontario. Following harvesting-driven reductions in tree density of about 75%, the following changes in population genetic structure were observed: 25% reduction in number of alleles, 33% reduction in polymorphic loci, and an 80 and 40% loss of rare and low frequency alleles, respectively. The latent genetic potential of these stands was reduced by about 50%, and the ability of these gene pools to adapt to changing environmental conditions was likely compromised. Given the selective logging history of eastern white pine over the past century or so, one might conclude that harvesting has caused genetic erosion in this species with some loss of evolutionary potential.

The influence of selective logging on the genetic structure of eastern hemlock is more complex, but also demonstrates a dysgenic influence of selective logging on the population. Hawley et al. (1998) compared genetic diversity parameters between an unmanaged control stand and stands that had undergone fixed diameter limit cuts or selection cuts three times over the past 40 years (years 0, 20, and 40) as part of a long-term replicated silvicultural experiment at the Penobscot Experimental Forest in eastern Maine. Residual stands that had undergone intensive and repeated diameter limit cuts, leaving only the very worst trees in terms of size and form, had significantly higher levels of heterozygosity, polymorphic loci, and effective number of alleles per locus. These seemingly counterintuitive results reflect an apparent association between rare alleles and defective phenotypes in eastern hemlock. Several alleles that occurred at very low frequencies ($P < 0.03$) in the natural unmanaged stand occurred at a higher frequency in the diameter limit cut because the defective residual trees preferentially possessed these rare alleles. That is, rare alleles conferred a negative fitness impact (i.e., slow growth and poor form) on eastern hemlock trees. In fact, 68 and 24% of the residual trees from the diameter limit cut possessed at least one or two rare alleles, respectively, compared with 26 and 6% and 32 and 2% for the unmanaged and selection cut stands, respectively. Although these results contrast dramatically with the harvest-induced loss of such alleles and potential positive evolutionary implications of rare alleles suggested for eastern white pine, they nonetheless illustrate that intensive human-induced manipulations of forest tree populations have altered the genetic structure, long-term adaptability, and productivity of north temperate forest ecosystems.

Summary: The Net Effect

One fact is certain: responses of tree species to changing global environments are exceedingly complex. As such, meaningful predictions of future forest composition are challenging and inherently uncertain. Our best understanding of future forest composition from contemporary models based on species relationships to climate and site characteristics indicates substantial geographical shifts for many species in response to climate warming. In addition, forest communities are expected to change in composition. Both the paleoecological record and predictive models indicate that individual species rather than intact communities will disperse and that forest communities will transition to a new mixture of species able to co-exist on the same site for a period of time.

Issues of substantial uncertainty include the rate of change in forest species composition, migration pathways, and the implications of inter-specific competition. One scenario is that most tree populations will be maintained in place for an extended period, despite a warming climate. The ability of existing trees to survive, produce seeds, and adapt should allow such maintenance. If the physiological tolerances of existing species are not exceeded, they should remain in place until other, more competitive, invaders become established. At that point, they will be eliminated. And by the time more successful invaders enter the site, species fading from one location will have migrated to more favorable locations. However, a more likely scenario is that of global climate change causing the demise of some species because better adapted species will outcompete them. Disease, insects, and drought are the primary agents that are expected to create these situations, and disturbance frequency will strongly influence the rate of change in forest composition. That is, rapid change will likely follow disturbance. In the absence, of disturbance living trees will likely persist and only be replaced as gaps occur.

Anthropogenic factors, including pollution exposure, exotic plants and pathogens, and a diverse array of other direct and indirect human influences on forests, represent a new "wild card" that can profoundly influence species responses to global climate change. Neither the paleo-ecological record nor modern day models reflect this sort of biotic and abiotic disturbance. Extrapolations of existing research regarding the impact of single or simple combinations of anthropogenic stressors suggest that the health and survival of individual trees, specific tree populations, and perhaps even some species, will likely be degraded in the years ahead if predicted levels of climate change and pollutant exposures are realized. However, because actual stress exposure combinations will probably be more complex than mixes already tested, predictions based on existing data can only produce generalized potential outcomes. Our modern-day experiences with the impacts of aggressive exotic species, especially exotic insects and diseases, indicate potential for a dramatic influence on the migration, establishment, and reproductive success of forest tree species growing in altered environments.

Whatever the response to individual or multiple stressors, evidence of extensive variation in stress response both within and among species suggests that North American trees currently have a broad capacity for differential response and potential adaptation to environmental change. However, climate change and pollutant additions are not occurring in isolation from other anthropogenic changes. Other factors, including increased forest fragmentation and intensive silvicultural management, could alter the genetic structure of and diversity within northern forests. Collectively, these activities have probably altered the level and distribution of genetic diversity that confers the adaptability necessary for trees to persist across environments that vary spatially and over time. This may have important consequences for the response of forests to global climate warming.

Because of existing and ever-increasing forest fragmentation, migration in response to climate warming may be impeded and evolutionary change may be the only viable means for some species to respond to and cope with changing climates over the long term. Forest tree species depend on high levels of genetic diversity for their adaptive potential and for overall forest ecosystem stability. If the reproductive and/or genetic structure within species or populations is altered due to population reduction and isolation, as is evident in marginal populations of many species today, or the influence of anthropogenic selective forces, then the adaptability and stability of many forest ecosystems in response to changing climates may also be compromised. In this context, perhaps the greatest anthropogenic threat to northern forests lies in the combined dangers of unprecedented climate change, pollution exposure, and competition from exotic plants and pathogens concomitant with a degradation of the genetic diversity needed for adaptation and survival. Indeed, our understanding of and ability to predict forest responses to changing climates may now be influenced as much by uncertainty as by the well-documented lessons from the past.

References

Aber JD, Federer CA (1992) A generalized, lumped-parameter model of photosynthesis, evapotranspiration and net primary production in temperate and boreal forest ecosystems. Oecologia 92:463–474.
Abrams MD (1999) Red maple taking over eastern forests. J For 97:6.

Anderson RS, Davis RB, Miller NG, Stuckenrath R (1986) History of late- and post-glacial vegetation and disturbance around Upper South Branch Pond, northern Maine. Can J Bot 64:1977–1986.

Anderson RS, Jacobson GL Jr., Davis RB, Stuckenrath R (1992) Gould Pond, Maine: late-glacial transition from marine to upland environments. Boreas 21:359–371.

Baisden WT, Blum JD, Miller EK, Friedland AJ (1995) Elemental concentrations in fresh snowfall across a regional transect in the northeastern U.S.: apparent sources and contribution to acidity. Water Air Soil Pollut 84:269–286.

Bartlein PJ, Webb T III (1985) Mean July temperatures at 6000 yr B.P. in eastern North America: Regression equations for estimates from fossil pollen data. Sylloogeus 55:301–342.

Bazzaz FA, Coleman JS, Morse SR (1990) Growth responses of seven major co-occurring tree species of the northeastern United States to elevated CO_2. Can J For Res 20:1479–1484.

Bazzaz FA, Jasienski M, Thomas SC, Wayne P (1995) Microevolutionary responses in experimental populations of plants to CO_2-enriched environments: parallel results from two model systems. Proc Natl Acad Sci USA 92:8161–8165.

Berrang P, Karnosky DF, Bennett JP (1989) Natural selection for ozone tolerance in *Populus tremuloides*: field verification. Can J For Res 19:519–522.

Berrang P, Karnosky DF, Mickler RA, Bennet JP (1986) Natural selection for ozone tolerance in *Populus tremuloides*. Can J For Res 16:1214–1216.

Björck S (1985) Deglaciation chronology and revegetation in northwestern Ontario. Can J Earth Sci 22:850–871.

Bonan GB, Sirois L (1992) Air temperature, tree growth, and the northern and southern range limits to *Picea mariana*. J Vegetat Sci 3:495–506.

Booth TH (1990) Mapping regions climatically suitable for particular tree species at the global scale. For Ecol Manage 36:47–60.

Botkin DB, Nisbet RA (1992) Projecting the effects of climate change on biological diversity in forests. In: Peters RL, Lovejoy TE (eds) *Global Warming and Biologic Diversity*. Yale University Press, New Haven, CT, pp 277–296.

Box EO, Crumpacker DW, Hardin ED (1993) A climatic model for location of plant species in Florida, USA. J Biogeogr 20:629–644.

Brubaker LB (1986) Responses of tree populations to climatic change. Vegetation 67:119.

Brubaker LB (1975) Postglacial forest patterns associated with till and outwash in north-central Upper Michigan. Quaternary Res 5:499–527.

Buchert GP, Rajora OP, Hood JV, Dancik BP (1997) Effects of harvesting on genetic diversity in old-growth eastern white pine in Ontario, Canada. Cons Biol 11:747–758.

Cantin D, Tremblay MF, Lechowicz MJ, Povtin C (1997) Effects of CO_2 enrichment, elevated temperature, and nitrogen availability on the growth and gas exchange of different families of jack pine seedlings. Can J For Res 27:510–520.

Cheliak WM, Murray G, Pitel JA (1988) Genetic effects of phenotypic selection in white spruce. For Ecol Manage 24:139–149.

Clark JS (1989) Effects of long-term water balances on fire regime, northwestern Minnesota. J Ecology 77:989–1004.

Clark JS (1988) Effect of climate change on fire regimes in northwestern Minnesota. Nature 334:233–235.

Clark LA, Pergibon D (1992) Tree-based models. In: Chambers JM, Hastie TJ (eds) *Statistical Models in S*. Wadsworth, Pacific Grove, CA, pp 377–419.

COHMAP Project Members (1988) Climatic change of the last 18,000 years: observations and model simulations. Science 241:1043–1052.

Craig AJ (1969) Vegetational history of the Shenandoah Valley, Virginia. Geol Soc Am Special Paper 123:283–296.

Davis DD, Skelly JM (1992) Foliar sensitivity of eight eastern hardwood trees species to ozone. Water Air Soil Pollut 62:269–277.

Davis FW, Goetz S (1990) Modeling vegetation pattern using digital terrain data. Landsc Ecol 4:69–80.

Davis MB (1983) Quaternary history of deciduous forests of eastern North America and Europe. Ann Missouri Bot Gard 70:550–563.

Davis MB (1978) Climatic interpretation of pollen in quaternary sediments. In: Walker D, Guppy JC (eds) *Biology and Quaternary Environments*. Australian Academy of Science, Canberra, Australia, pp 35–51.

Davis MB, Spear RW, Shane CLD (1980) Holocene climate of New England. Quaternary Res 14:240–250.

Davis MB, Zabinski C (1992) Changes in geographical range resulting from greenhouse warming: effects on biodiversity in forests. In: Peters RL, Lovejoy TE (eds) *Global Warming and Biologic Diversity*. Yale University Press, New Haven, CT, pp 297–308.

Davis RB, Jacobson GL Jr. (1985) Late glacial and early Holocene landscapes in northern New England and adjacent areas of Canada. Quaternary Res 23: 341–368.

DeHayes DH (1992) Winter injury and developmental cold tolerance of red spruce. In: Eager C, Adams MB (eds) *The Ecology and Decline of Red Spruce in the Eastern United States*. Springer-Verlag, New York, pp 295–337.

DeHayes DH, Schaberg PG, Hawley GJ, Strimbeck GR (1999) Acid rain impacts calcium nutrition and forest health. BioScience 49:789–800.

DeHayes DH, Waite CE, Ingle MA, Williams MW (1990) Winter injury susceptibility and cold tolerance of current and year-old needles of red spruce trees from several provenances. For Sci 36:982–994.

Digerfeldt G, Almendinger JE, Björck S (1992) Reconstruction of past lake levels and their relation to groundwater hydrology in the Parkers Prairie sandplain, west-central Minnesota. Palaeogeogr Palaeoclimatol Palaeoecol 94:99–118.

Environmental Systems Research Institute (1992) ArcUSA 1:2M, User's guide and data reference. Environmental Systems Research Institute, Redlands, CA.

Flemming S (1997) Stand and neighbourhood parameters as determinants of plant species richness in a managed forest. J Vegetat Sci 8:573–578.

Foley JA, Prentice IC, Ramankutty N, Levis S, Pollard D, Sitch S, Haxeltine A (1996) An integrated biosphere model of land surface processes, terrestrial carbon balance and vegetation dynamics. Global Biogeochem Cycl 10: 603–628.

Foster DR (1983) The history and pattern of fire in the boreal forest of southeastern Labrador. Can J Bot 61:2459–2471.

Franklin JF, Swanson FJ, Harmon ME, Perry DA, Spies TA, Dale VH, McKee A, Ferrell WK, Means JE, Gregory SV, Lattin JD, Schowalter TD, Larsen D (1992) Effects of global climate change on forests in northwestern North America. In: Peters RL, Lovejoy TE (eds) *Global Warming and Biologic Diversity*. Yale University Press, New Haven, CT, pp 244–258.

Gajewski K (1988) Late Holocene climate changes in eastern North America estimated from pollen data. Quaternary Res 29:255–262.

Gerhold H (1977) *Effect of Air Pollution on Pinus strobus* L. *and Genetic Resistance—A Literature Review*. Tech Rep EPA-600/3-77-002. Environmental Protection Agency (EPA), Corvallis Environmental Research Laboratory, Corvallis, Oregon.

Gleason HC (1926) The individualistic concept of the plant association. Bull Torrey Bot Club 53:7–26.

Godman RM, Lancaster K (1990) *Tsuga canadensis* (L.) Carr. Eastern hemlock. In: Burns RM, Honkala BH (eds) *Silvics of North America. Vol. 1. Conifers.* Agric Hndbk 654. United States Department of Agriculture (USDA) Forest Service, Washington, DC, pp 604–612.

Golomb D, Ryan D, Eby N, Underhill J, Zemba S (1997a) Atmospheric deposition of toxins onto Massachusetts Bay. I. Metals. Atmos Environ 31:1349–1359.

Golomb D, Ryan D, Eby N, Underhill J, Zemba S (1997b) Atmospheric deposition of toxins onto Massachusetts Bay. II. Polycyclic aromatic hydrocarbons. Atmos Environ 31:1361–1368.

Grant WF (1998) Higher plant assays for the detection of genotoxicity in air polluted environments. Ecosy Health 4:210–229.

Green DG (1982) Time series and postglacial forests of southwest Nova Scotia. J Biogeogr 9:29–40.

Gregorius HR (1996) The contribution of the genetics of populations to ecosystem stability. Silv Genet 45:267–271.

Hansen J, Fung I, Lacis A, Rind D, Lebedeff S, Ruedy R (1988) Global climate changes as forecast by Goddard Institute for Space Studies three-dimensional model. J Geophys Res 93:9341–9364.

Harrison S (1989) Lake level and climate changes in eastern North America. Clim Dynam 3:157–167.

Hawley GJ, DeHayes DH (1994) Genetic diversity and population structure of red spruce (*Picea rubens*). Can J Bot 72:1778–1786.

Hawley GJ, DeHayes DH, Brissette J (1998) *Changes in the genetic structure and diversity of forest ecosystems as a result of forest management practices. Final Report: Global Change Program.* United States Department of Agriculture (USDA) Forest Service, Northeastern Forest Experiment Station, Radnor, PA.

Haxeltine A, Prentice IC (1996) BIOME3: an equilibrium biosphere model based on ecophysiological constraints, resource availability and competition among plant functional types. Global Biogeochem Cycl 10:693–709.

Haxeltine A, Prentice IC, Creswell ID (1996) A coupled carbon and water flux model to predict vegetation structure. J Vegetat Sci 7:651–666.

Hays JD, Imbrie J, Shackleton NJ (1976) Variations in the earth's orbit: pacemaker of the ice ages. Science 194:1121–1132.

Hobbs RJ (1994) Dynamics of vegetation mosaics: can we predict responses to global change? Ecoscience 1:346–356.

Hoff RM, Strachan WMJ, Sweet CW, Chan CH, Shackleton M, Bidleman TF, Brice KA, Burniston DA, Cussion S, Gatz DF, Harlin K, Schroeder WH (1996) Atmospheric deposition of toxic chemicals to the Great Lakes: a review of data through 1994. Atmos Environ 30:3505–3527.

Hunter ML Jr., Jacobson GL Jr., Webb T III (1988) Paleoecology and the coarse-filter approach to maintaining biological diversity. Conserv Bio 2:375–385.

Huntley B, Berry P, Cramer W, McDonald AP (1995) Modelling present and potential future ranges of some European higher plants using climate response surfaces. J Biogeogr 22:967–1001.

Huntley B, Prentice IC (1988) July temperatures in Europe from pollen data, 6000 years before present. Science 241:687–690.

Imbrie J, Imbrie KP (1979) Ice Ages. Solving the Mystery. *Enslow*, Short Hills, NJ.

Iverson LR, Prasad AM (1998) Predicting potential future abundance of 80 tree species following climate change in the eastern United States. Ecol Monogr 68:465–485.

Iverson LR, Prasad AM, Hale BJ, Sutherland EK (1999) *An Atlas of Current and Potential Future Distributions of Common Trees of the Eastern United States.* Gen Tech Rep. NE-265, United States Department of Agriculture (USDA) Forest Service, Northeastern Forest Experiment Station, Radnor, PA.

Iverson LR, Prasad AM, Schwartz MW (1999) Modeling potential future individual tree-species distributions in the Eastern United States under a climate change scenario: a case study with *Pinus virginiana*. Ecological Modelling 115:77–93.

Jackson ST, Whitehead DR (1991) Holocene vegetation patterns in the Adirondack Mountains. Ecology 72:641–653.

Jacobson GL Jr. (1988) Ancient permanent plots: sampling in paleovegetational studies. In: Huntley B, Webb T III (eds) *Vegetation History*. Vol. 7. In: Leith H (ed) *Handbook of Vegetation Science*. Kluwer Academic, Dordrecht, The Netherlands, pp 3–16.

Jacobson GL Jr. (1979) The palaeoecology of white pine (*Pinus strobus*) in Minnesota. J Ecology 67:697–726.

Jacobson GL Jr., Dieffenbacher-Krall A (1995) White pine and climate change: insights from the past. J For 93:39–42.

Jacobson GL Jr., Webb T III, Grimm EC (1987) Patterns and rates of vegetation change during the deglaciation of eastern North America. In: Ruddiman WF, Wright HE Jr. (eds) *North America and Adjacent Oceans During the Last Deglaciation*. DNAG Vol. K-3. Geological Society of America, Boulder, CO, pp 277–288.

Jarvis PG, McNaughton KG (1986) Stomatal control of transpiration: scaling up from leaf to region. Adv Ecol Res 15:1.

Johnson AH, Cook ER, Siccama TG (1988) Climate and red spruce growth and decline in the northern Appalachians. Proc Natl Acad Sci USA 85: 5369–5373.

Johnson AH, DeHayes DH, Siccama TG (1996) Role of acid deposition in the decline of red spruce (*Picea rubens* Sarg) in the montane forests of Northeastern USA. In: Raychudhuri SP, Maramorosch K (eds) *Forest Trees and Palms: Disease and Control*. Oxford and IBH, New Delhi, India, pp 49–71.

Karnosky DF (1980) Changes in southern Wisconsin white pine stands related to air pollution sensitivity. In: Miller PR (ed) *Proceedings of the Symposium on the Effects of Air Pollutants on Mediterranean and Temperate Forest Ecosystems*. 22–27 June 1980, Riverside, CA, Gen Tech Rep PSW-49. United States Department of Agriculture (USDA) Forest Service, Pacific Southwest Forest and Range Experiment Station, Albany, CA. p 238.

Karnosky DF, Houston D (1979) Genetics of air pollution tolerance of trees in the northeastern United States. In: *Proceedings of the 26th Northeastern Forest Tree Improvement Conference*, 25–26 July 1979, University Park, PA, pp 161–178.

Kattenberg A, Giorgi F, Grassl H, Meehl GA, Mitchell JFB, Stouffer RJ, Tokioka T, Weaver AJ, Wigley ML (1996) Climate models—projections of future climate. In: Houghton JT, Meira-Filho LG, Callander B, Harris N, Kattenberg A, Maskell K (eds) IPCC (Intergovernmental Panel on Climate Change) *Climate Change 1995: The Science of Climate Change*. Cambridge University Press, Cambridge, UK, pp 285–357.

Kinney KK, Lindroth RL (1997) Responses of three deciduous tree species to atmospheric CO_2 and soil NO_3^- availability. Can J For Res 27:1–10.

Kittel TGF, Rosenbloom NA, Painter TH, Schimel DS, VEMAP (Vegetation/ Ecosystem Modeling and Assessment Project) modeling participants (1995) The VEMAP integrated database for modeling United States. Ecosystem/vegetation sensitivity to climate change. J Biogeogr 22:857–862.

Kriebel HB, Leben C (1981) The impact of air pollution on the gene pool of eastern white pine. In: Proceedings of the 17th International Union of Forestry Research Organizations (IUFRO) World Congress, Division 2, 6–17 September 1981, Kyoto, Japan, pp 185–189.

Kull O, Sober A, Coleman MD, Dickson RE, Isebrands JG, Gagnon Z, Karnosky DF (1996) Photosynthetic responses of aspen clones to simultaneous exposures of ozone and CO_2. Can J For Res 26:639–648.

Lande R (1988) Genetics and demography in biological conservation. Science 241:1455–1460.

Lauenroth WK (1996) Application of patch models to examine regional sensitivity to climate change. Clim Change 34:155–160.

Ledig FT (1992) Human impacts on genetic diversity in forest ecosystems. Oikos 63:87–108.

Loehle C, LeBlanc D (1996) Model-based assessments of climate change effects on forests: a critical review. Ecol Model 90:1–31.

Lovett GM, Hubbell JG (1991) Effects of ozone and acid mist on foliar leaching from eastern white pine and sugar maple. Can J For Res 21:794–802.

MacCracken MC (1995) The evidence mounts up. Nature 376:645–646.

McNulty SG, Aber JD, Newman SD (1996) Nitrogen saturation in a high elevation New England spruce–fir stand. For Ecol Manage 84:109–121.

McNulty SG, Vose JM, Swank WT, Aber JD, Federer CA (1994) Regional-scale forest ecosystem modeling: database development, model predictions and validation using a Geographic Information System. Clim Res 4:223–231.

Michaelsen J, Schimel DS, Friedl MA, Davis FW, Dubayah RC (1994) Regression tree analysis of satellite and terrain data to guide vegetation sampling and surveys. J Vegetat Sci 5:673–686.

Misenti TL (1990) *Genetic Diversity of Marginal vs. Central Populations of Pitch Pine and Jack Pine.* MS Thesis, University of Vermont, Burlington, VT.

Mohnen VA (1992) Atmospheric deposition and pollutant exposure of eastern U.S. forest. In: Eager C, Adams MB (eds) *The Ecology and Decline of Red Spruce in the Eastern United States.* Springer-Verlag, New York, pp 64–124.

Neale DB (1985) Genetic implications of shelterwood regeneration of Douglas-fir in southwest Oregon. For Sci 31:995–1005.

Neale DB, Adams WT (1985) The mating system in natural and shelterwood stands in Douglas-fir. Theor Appl Genet 71:201–207.

Neilson RP (1995) A model for predicting continental-scale vegetation distribution and water balance. Ecol Applic 5(2):362–385.

Neilson RP, Marks D (1994) A global perspective of regional vegetation and hydrologic sensitivities from climatic change. J Vegetat Sci 5:715–730.

Neilson RP, Prentice IC, Smith B (1998) Simulated changes in vegetation distribution under global warming. In: Watson RT, Zinyowera MC, Moss RH (eds) *The Regional Impacts of Climate Change: An Assessment of Vulnerability.* Cambridge University Press, New York, pp 439–456.

Neilson RP, Running SW (1996) Global dynamic vegetation modelling: coupling biogeochemistry and biogeography models. In: Walker B, Steffen G (eds) *Global Change Terrestrial Ecosystems.* Cambridge University Press, New York, pp 461–465.

Olson RJ, Emerson CJ, Nungesser MK (1980) *Geoecology: A County-level Environmental Data Base for the Conterminous United States*. Environ Sci Div Publ 1537. Oak Ridge National Laboratory, Oak Ridge, TN.

Overpeck JT, Bartlien PJ, Webb T III (1991) Potential magnitude of future vegetation change in eastern North America: comparisons with the past. Science 254:692–695.

Parsons DJ, Pitelka LF (1992) Plant ecological genetics and air pollution stress: a commentary on implications for natural populations. In: Taylor GE, Pitelka LF, Clegg MT (eds) *Ecological Genetics of Air Pollution*. Springer-Verlag, New York, pp 337–343.

Parton WJ, Stewart JWB, Cole CV (1988) Dynamics of C, N, P and S in grassland soils: a model. Biogeochem 5:109–131.

Pastor J, Post WM (1986) Influences of climate, soil moisture, and succession on forest carbon and nitrogen cycles. Biogeochem 2:3–27.

Patterson WA III, Backman AE (1988) Fire and disease history of forests. In: Huntley B, Webb T III (eds) *Vegetation History*. Kluwer Academic, Boston, MA, pp 603–632.

Peart DR, Nicholas NS, Zedaker SM, Miller-Weeks MM, Siccama TG (1992) Condition and recent trends in high-elevation red spruce populations. In: Eager C, Adams MB (eds) *The Ecology and Decline of Red Spruce in the Eastern United States*. Springer-Verlag, New York, pp 125–191.

Perry TO, Wang CW (1960) Genetic variation in the winter chilling requirement for date of dormancy break for *Acer rubrum*. Ecology 41:790–794.

Pigott CD (1981) Nature of seed sterility and natural regeneration of *Tilia cordata* near its northern limit in Finland. Ann Bot Fenn 18:255–263.

Pitelka LF (1988) Evolutionary responses of plants to anthropogenic pollutants. Trends Ecol Evol 3:233–236.

Post WM, Pastor J (1990) An individual-based forest ecosystem model for projecting forest response to nutrient cycling and climate changes. In: Wensel LC, Biging GS (eds) *Forest Simulation Systems*. Bull 1927. University of California, Division of Agriculture and Natural Resources, pp 61–74.

Prentice IC, Bartlein PJ, Webb T III (1991) Vegetation and climate change in eastern North America since the last glacial maximum. Ecology 72:2038–2056.

Prentice IC, Cramer W, Harrison SP, Leemans R, Monserud RA, Solomon AM (1992). A global biome model based on plant physiology and dominance, soil properties and climate. J Biogeogr 19:17–134.

Prentice IC, Guiot J, Huntley B, Jolly B, Cheddadi R (1996) Reconstructing biomes from palaeoecological data: a general method and its application to European pollen data at 0 and 6 ka. Clim Dynam 12:185–194.

Raich JW, Rastetter EB, Melillo JM (1991) Potential net primary productivity in South America: applications of a global model. Ecol Applic 1:399–429.

Rastetter EB (1996) Validating models of ecosystem response to global change. BioScience 46:190–197.

Raynal DJ, Roman JR, Eichenlaub WM (1982a) Response of tree seedlings to acid precipitation. I. Effect of simulated acidified canopy throughfall on sugar maple growth. Environ Exp Bot 22:377–383.

Raynal DJ, Roman JR, Eichenlaub WM (1982b) Response of tree seedlings to acid precipitation. II. Effect of substrate acidity on seed germination. Environ Exp Bot 22:385–392.

Reichard SH, Hamilton CW (1997) Predicting invasion of woody plants introduced into North America. Conserv Bio 11:193–203.

Roth S, McDonald EP, Lindroth RL (1997) Atmospheric CO_2 and soil water availability: consequences for tree-insect interactions. Can J For Res 27: 1281–1290.

Running SW, Hunt ER Jr. (1993) Generalization of a forest ecosystem process model for other biomes, BIOME-BGC, and an application for global-scale models. In: Ehleringer JR, Field CB (eds) *Scaling Processes between Leaf and Landscape Levels.* Academic Press, San Diego, CA, pp 141–158.

Schaberg PG, Strimbeck GR, Hawley GJ, DeHayes DH, Shane JB, Murakami PF, Perkins TD, Donnelly JR, Wong BL (2000) Cold tolerance and photosystem function in a montane red spruce population: physiological relationships with foliar carbohydrates. J Sustain For 10:225–230.

Schauffle M (1998) *Paleoecology of Coastal and Interior Picea (Spruce) Stands in Maine.* Ph.D. dissertation, University of Maine, Orno, ME.

Schier GA (1987) Throughfall chemistry in a red maple provenance plantation sprayed with "acid rain." Can J For Res 17:660–665.

Schwartz MW (1993) Modeling the effect of habitat loss on potential rates of range change for trees in response to global warming. Biodiv Conserv 2:51–61.

Schwartz MW, Iverson LR, Prasad A (1996) Projected tree distribution shifts under global climate change in the fragmented Ohio landscape. Bull Ecol Soc Am 77:398.

Strimbeck GR, Schaberg PG, DeHayes DH, Shane JB, Hawley GJ (1995) Midwinter dehardening of montane red spruce during a natural thaw. Can J For Res 25:2040–2044.

Teskey RO, Dougherty PM, Mickler RA (1998) The influences of global change on tree physiology and growth. In: Mickler RA, Fox S (eds) *The Productivity and Sustainability of Southern Forest Ecosystems in a Changing Environment.* Springer-Verlag, New York, pp 279–290.

Tobi DR, Wargo PM, Bergdahl DR (1995) Growth response of red spruce after known periods of winter injury. Can J For Res 25:669–681.

Townsend AM, Roberts BR (1973) Effect of moisture stress on red maple seedlings from different seed sources. Can J Bot 51:1989–1995.

Transeau EN (1905) Forest centers of eastern America. Am Natur 39:875–889.

Tyree MT, Davis SD, Cochard H (1994) Biophysical perspectives of xylem evolution: is there a tradeoff of hydraulic efficiency for vulnerability to dysfunction? IAWA J 15:335–360.

VEMAP (Vegetation/Ecosystem Modeling and Analysis Project) members (1995) Vegetation/Ecosystem Modeling and Analysis Project: comparing biogeography and biogeochemistry models in a continental-scale study of terrestrial ecosystem responses to climate change and CO_2 doubling. Global Biogeochem Cycl 9:407–437.

Vitousek PM, Mooney HA, Lubchenco J, Melillo (1997) Human domination of earth's ecosystems. Science 277:494–499.

Wallace B (1991) *Fifty Years of Genetic Load: An Odyssey.* Cornell University Press, Ithaca, NY.

Wang JR, Hawkins CDB, Letchford T (1998) Relative growth rate and biomass allocation of paper birch (*Betula papyrifera*) populations under different soil moisture and nutrient regimes. Can J For Res 28:44–55.

Watson RT, Zinyowera MC, Moss RH (1998) The regional impacts of climate change: an assessment of vulnerability. Cambridge University Press, New York.

Webb RS, Anderson KH, Webb T III (1993) Pollen response-surface estimates of late-Quaternary changes in the moisture balance of the northeastern United States. Quaternary Res 40:213–227.

Webb T (1992) Past changes in vegetation and climate: lessons for the future. In: Peters RL, Lovejoy TE (eds) *Global Warming and Biologic Diversity*. Yale University Press, New Haven, CT, pp 59–75.

Webb T III (1987) The appearance and disappearance of major vegetational assemblages: long-term vegetational dynamics in eastern North America. Vegetation 69:177–187.

Webb T III, Bartlein PJ, Harrison S, Anderson KH (1994) Vegetation, lake levels, and climate in eastern United States since 18,000 yr B.P. In: Wright HE Jr., Kutzbach JE, Webb T III, Ruddiman WF, Street-Perrott FA, Bartlein PJ (eds) (1994) *Global Climates Since the Last Glacial Maximum*. University of Minnesota Press, Minneapolis, MN, pp 514–535.

Wells OO, Lambeth CC (1983) Loblolly pine provenance test in southern Arkansas: 25th year results. South J Appl For 7:71–75.

Wetherald RT, Manabe S (1988) Cloud feedback processes in a general circulation model. J Atmos Sci 45:1397–1415.

Wigley TML (1995) A successful prediction? Nature 376:463–464.

Wilkinson RC (1990) Effects of winter injury on basal area and height growth of 30-year-old red spruce from 12 provenances growing in northern New Hampshire. Can J For Res 20:1616–1622.

Winkler MG (1988) Effect of climate on development of two sphagnum bogs in south-central Wisconsin. Ecology 69:1032–1043.

Winkler MG (1985) A 12,000-year history of vegetation and climate for Cape Cod, Massachusetts. Quaternary Res 23:301–312.

Winkler MG, Swain AM, Kutzbach JE (1986) Middle Holocene dry period in the northern Midwestern United States: lake levels and pollen stratigraphy. Quaternary Res 25:235–250.

Winner WE, Coleman JS, Gillespie C, Mooney HA, Pell EJ (1992) Consequences of evolving resistance to air pollutants. In: Taylor GE, Pitelka LF, Clegg MT (eds) *Ecological Genetics of Air Pollution*. Springer-Verlag, New York, pp 177–202.

Woodward FI (1992) A review of climate change on vegetation: ranges, competion, and composition. In: Peters RL, Lovejoy TE (eds) *Global Warming and Biologic Diversity*. Yale University Press, New Haven, CT, pp 105–123.

Woodward FI (1987) *Climate and plant distribution*. Cambridge University Press, Cambridge, UK.

Yakimchuk R, Hoddinott J (1994) The influence of ultraviolet-B light and carbon dioxide enrichment on the growth and physiology of seedlings of three conifer species. Can J For Res 24:1–8.

4. Summary

15. Summary of Prospective Global Change Impacts on Northern U.S. Forest Ecosystems

Richard A. Birdsey, Robert A. Mickler,
John Hom, and Linda S. Heath

In January 1989, the President's Fiscal Year 1990 Budget to the Congress was accompanied by a report entitled, "Our Changing Planet: A U.S. Strategy for Global Change Research" (Committee on Earth Sciences, 1989). The report focused the attention of policy makers on the significant environmental issues arising from natural and human-induced changes in the global Earth system. The report announced the beginning of a research program, the U.S. Global Change Research Program, with a mission to improve our understanding of the causes, processes, and consequences of the changes affecting our planet. Interest in global change was heightened in 1990 with the publication of "Climate change: The IPCC Scientific Assessment" (Houghton et al., 1990) by the Intergovernmental Panel on Climate Change (IPCC), jointly sponsored by the World Meteorological Organization and the United Nations Environmental Programme. This assessment was updated and new technical issues were added in the second assessment volumes, "Climate Change 1995: The Science of Climate Change"(Houghton et al., 1996) and "Climate Change 1995: Impacts, Adaptations, and Mitigation of Climate Change" (Watson et al., 1996). The first IPCC assessment in 1990 and subsequent assessments have concluded that continued accumulation of anthropogenic greenhouse gases in the atmosphere will lead to climate change whose rate and magnitude are likely to have important impacts on natural and human systems. Concurrent with these reports, several scientific literature reviews and research summaries by Eamus and Jarvis (1989), Bazzaz (1990), Musselman and Fox (1991), Strain and Thomas (1992), Rogers and Runion (1994), Gunderson and Wullschleger (1994), Ceulemans and Mousseau (1994), Idso and Idso (1994), Curtis (1996), and Mickler and Fox (1998) have quantified the direct effects of rising carbon dioxide (CO_2) on plant growth and development. Unlike the statements made with scientific certainty in the 1970's which heralded the beginning of the *Waldsterben* or "forest decline" in Germany, Europe, and the U.S. but concluded with the uncertainty associated with characterizing the effects of multiple stressors on forest ecosystems, predictions of global change and its effects began in the 1980s by acknowledging many uncertainties. Today, scientific assessments of climate change are characterized by ever increasing scientific understanding and improved predictive capabilities.

This volume summarizes the progress in our understanding of how multiple interacting stresses associated with global change are affecting or are likely to affect northern U.S. forest ecosystems at multiple spatial and temporal scales. In this summary chapter we attempt to draw conclusions from the material presented, and highlight a few dominant themes from these studies and others that comprise several decades of environmental effects research. We are only beginning to understand the paleoecology of northern forest ecosystems and how these forests are likely to evolve over the next century. Our goal has been to assemble the available though incomplete knowledge of current and prospective responses of northern forest ecosystems to multiple environmental changes, to develop a scientific basis for planning future global change research, and to transfer available information to land managers and policy makers.

Characteristics of Northern Forests and Their Environment

Current Forest Conditions and Trends in Carbon Sequestration

The northern region contains 23% of the forest area of the U.S., 18% of the land area, and 45% of the population. The region is unique in that heavily forested areas are located near areas of high population density. As a result, forest fragmentation has become a major issue for the region. Private forest landowners control 80% of forest land. Most of the private owners own timberland primarily for recreation and esthetic enjoyment. However, one third of private owners managing 61% of private forest lands intend to harvest their forests in the next 10 years.

Northern U.S. forests had been cleared for agriculture or heavily logged by the mid-1800s. Harvesting of forests in the Lake States removed the climax pine-hemlock (*Pinus* spp.-*Tsuga canadensis* [L.] Carr), and agricultural land clearing and fire is responsible for the present northern conifer, hardwood, and mixed forest of New England states. Since the 1920s, reestablishment and regrowth of forests on farmed or logged land has resulted in the large areas of maturing forest common today. The region is still gaining forest land from other uses, primarily agriculture. The most recent inventories by the USDA Forest Service Forest Inventory and Analysis Program indicate a 4% increase, despite continued development and fragmentation in some areas.

A major event shaping the species composition of today's forest was the chestnut blight, which all but eliminated this important species by mid-20th century. Oaks (*Quercus* spp.) and other hardwoods replaced this once-dominant species. Oak-hickory (*Carya* spp.) and maple (*Acer* spp.)-beech (*Fagus* spp.)-birch (*Betula* spp.) forest types together account for over 60% of the forest area, with oak-hickory common in the southern part of the region and maple-beech-birch common in the northern part of the region. The aspen (*Populus* spp.)-birch type is common in the North, particularly in the Lake States. The most northern parts of Maine and the Lake States contain significant amounts of spruce (*Picea* spp.)-fir (*Abies* spp.). In terms of timber volume, the North contains 47% of the nation's hardwood volume but only 11% of the nation's softwood volume. Lack of oak regeneration following harvest is a major issue for the region, as loss of oak forests coupled with a major expansion of red maple results in lower economic value and reduction of important mast species. Areas of major pest problems include spruce-fir forests in Maine, heavily damaged by spruce budworm in the 1980s, and the Allegheny Plateau region of Pennsylvania, where numerous biotic and abiotic agents have caused damage and decline of sugar maple.

Northern forests contain large reservoirs of carbon (C) in maturing forests and forest soils. Northern forests are estimated to contain more carbon than forests of the southern and western regions of the U.S., and are projected to continue to contain more carbon (Birdsey and Heath, 1995). In 1992, northern forest ecosystems contained an estimated 13.4 Pg C, with a rate of increase of 0.16 Pg C yr^{-1}. Projections through 2040 indicate that northern forest ecosystems will contain 17.5 Pg C with a rate of increase of 0.08 Pg C yr^{-1}. Comparable estimates for the Southern U.S. are current storage of 11.5 Pg C increasing at 0.03 Pg C yr^{-1}, and projections through 2040 indicating future storage of 11.9 Pg C and an average rate of increase of 0.03 Pg C yr^{-1}. Comparable estimates for the Western U.S. are current storage of 12.9 Pg C with a rate of increase of 0.02 Pg C yr^{-1}, with projections through 2040 indicating future storage of 16.8 Pg C and an average rate of increase of 0.08 Pg C yr^{-1}. Although northern forests generally do not grow as fast as southern forests, they are not harvested as heavily which leads to the increases in carbon inventory.

In the recent past, forest growth has been more than twice forest removals. Expected increases in harvesting coupled with slower growth characteristic of aging forests will reduce the high rate of carbon sequestration over the next 50 years or so.

Northern Forest Soils and Nutrient Depletion

Soil may be the most important factor that determines health and productivity of forests. Nutrient depletion as a consequence of decades of acid deposition and intensive land use are of particular interest in the North, and the subject of much research over the last two decades. To compile complete nutrient budgets for forest ecosystems, researchers have begun to study the influence of bedrock on nutrient and water cycling, which may be minimal where thick surficial deposits cover the bedrock, and great where surficial deposits are shallow.

Soil taxonomic units provide a convenient framework to examine effects of environmental change on soil and forest resources. Taxonomic units reflect differences in age of parent material, texture, and composition, which interact with climate, topography, and vegetation to determine water and nutrients movement through the soil profile, deposition affects on water and nutrients, and differing nutrient content and flux rates.

Depletion of base cations is a critical issue for northern forest soils and is recognized as an important long-term problem in southern U.S. soils (Richter and Markewitz, 1996). If weathering of parent material is not sufficient to replace base cations lost to forest growth and leaching, then depletion is likely. Mass balance studies show that depletion of base cations is a problem at some specific sites, but spatial variability of forest soils has hindered accurate extrapolation of these observations to landscape and regional scales. Improved models of susceptibility to nutrient depletion will likely follow better understanding of spatial patterns of the mineralogic composition of soil parent materials, and better knowledge of the mechanisms and locations of weathering patterns at the landscape scale.

Climate and Atmospheric Deposition Changes

Atmospheric CO_2 concentration has increased by 25% in the last century and is expected to increase globally 1.43 μL L^{-1} each year, to a doubling or more of historical concentrations during the middle of the next century. Coupled with increases in other greenhouse gases (e.g., methane, nitrous oxide, and halocarbons), these atmospheric changes are expected to cause significant warming of the earth's surface. How such climate change would affect the northern region depends in large part on changes in the development and behavior of weather systems affecting the region.

However, projecting these changes is highly uncertain because of the extreme complexity of earth-atmosphere interactions, and an inability to account for small-scale weather phenomena in global circulation models.

Global-scale model simulations of future climate conditions suggest a global average temperature increase of between 1.0° and 4.5°C with a doubling of atmospheric CO_2 (Houghton et al., 1996). The greatest warming is expected over land and at higher latitudes. On a seasonal basis, warming is expected to be most significant in late autumn and winter. All models project an increase in global precipitation, particularly in the winter over northern and mid-latitudes. Global models currently do not have the capability to make predictions of changes in extreme weather events, although such changes are theoretically possible.

Over North America, the most significant observed temperature changes over the last 40 years have occurred from the North Central U.S. through Northwestern Canada into Alaska. Average surface temperatures have increased from 0.25° to 1.5°C over this region. On a seasonal basis, the most significant changes occurred during the winter and spring. Some areas within the Northern Region, particularly in the west and north, have experienced a large number of extreme maximum and minimum temperature events over the last 40 years. Other notable temperature events include late spring freezes and midwinter thaws. Thaws followed by rapid freezing can be particularly damaging to vegetation.

Precipitation patterns are highly variable and more unpredictable than temperature changes. Precipitation in the Northern Region has increased an average 2 to 5% per decade since 1900. Extreme precipitation events and droughts occur periodically in different parts of the region. Such extreme events are manifestations of atmospheric circulation patterns that can be associated with probabilities of event occurrence. A well-known example of such a circulation pattern is the El Nino/Southern Oscillation, driven by changes in Pacific Ocean surface temperatures. El Nino events have the largest effect on the U.S. during the winter and early spring months. In the Northern region, the risk of extreme precipitation in the winter months is particularly high in the central and southern Great Plains, based on observed precipitation patterns.

Atmospheric deposition patterns are determined by air pollution concentration gradients associated with emission sources, meteorological conditions, topography, and prevailing air transport patterns. The Northeast and North Central regions of the U.S. contain a high concentration of pollution sources, but the pattern of pollution exposure varies markedly within the region from the highest national exposure levels to unpolluted background levels.

Acid deposition patterns reflect the emissions source areas in the Ohio River Valley and the Midwest, and the prevailing winds which deposit the highest acidity to a region including eastern Ohio, northern West Virginia, western Pennsylvania, and western New York. Other areas of high acid deposition include the Adirondack Mountains and Catskill Mountains of New York. Nitrogen (N) deposition, of particular interest because of its effects on forest processes, ranges from approximately $2.5\,kg\,ha^{-1}$ in northern Maine to more than $20\,kg\,ha^{-1}$ in Pennsylvania.

Recent changes in deposition have resulted from the passage of the 1990 Clean Air Act (Stoddard et al., 1999). During the period between 1983 and 1994, deposition of sulfur (S) compounds and hydrogen ions has decreased significantly, 10 to 25% and 12%, respectively. Nitrogen deposition increased nationally for the same period.

Ozone (O_3) concentrations over urban/industrial areas of the northern U.S. average about twice background levels, or 80 ppb. The highest exposures occur in the Washington–New York corridor, and lowest levels in northern Minnesota and northern Maine. Ozone concentrations vary greatly across time and space. There is a tendency for areas of high O_3 exposure to also have high deposition of N and S compounds.

Incidence and Causes of Forest Declines

Decline diseases, linked to stress and environmental change, have increased over the past century and in particular in the last two decades. A well-developed theoretical basis explains decline diseases in terms of the interactions between predisposing factors, inciting factors, and contributing factors. If the theoretical models are correct, then increased levels of various interacting stressors in the Northeast are likely to lead to increased incidence of decline disease. Decline diseases are commonly associated with leaf anomalies (e.g., reduced leaf size, scorch, necrosis, premature coloration, and leafdrop), progressive crown dieback, reduction in radial and terminal stem growth, and reductions in root carbon reserves. Increasing environmental stress is occurring at the same time as many species reach biological maturity across much of their range, a consequence of past land use impacts. Aging forests are known to be more susceptible to decline disease.

Drought and defoliation are the most common stressors associated with decline disease in the Northeast. Other important stressors include sucking insects, defoliation from late spring frost, and fungal leaf pathogens. Examples of the occurrence of decline disease include: (1) sugar maple (*Acer saccharum* Marsh) decline in northwest Pennsylvania associated with biotic factors (defoliating insects, borers, and canker fungus), a series of droughts, and acid deposition; (2) red spruce (*Picea rubens* Sarg.) decline in the Northeast associated with winter injury induced by acidic deposition; and (3) a series of widespread and simultaneous declines associated with climatic extremes over the last century. The extent and severity of declines seems to be higher over the last 25 years compared with the previous 75 years.

Physiological Responses of Trees to Environmental Change

Interacting Effects of Multiple Stresses

Two pollutants that have substantial impacts on plant growth and are increasing in the atmosphere as a consequence of human activity are CO_2 and O_3. Plant responses to CO_2 and O_3 are complex, and become even more difficult to interpret when other known stressors are considered, including N limitation, temperature and moisture extremes, and pests. In general, increasing atmospheric CO_2 increases photosynthetic rates, height growth, and biomass production, while increasing atmospheric O_3 decreases photosynthetic rates and biomass production, and increases leaf senescence. The amount and sometimes the direction of change depends on internal plant factors, such as age and genotype. Higher CO_2 concentrations may compensate for some other environmental stresses (see review by Mickler, 1998). For example, most studies show that CO_2 enrichment increases growth even though light and/or nutrients are limiting. It is becoming clear that both increasing CO_2 and O_3 impact fundamental plant processes, which in turn affect susceptibility to plant-feeding insects.

Studies on trembling aspen (*Populus tremuloides* Michx.) show that O_3 usually decreases growth although the effect varies significantly with genotype. Similar results have been reported by Flagler et al. (1998) in shortleaf pine (*Pinus echinata* Mill.) and in other southern pine species (see research program findings by Fox and Mickler, 1996). Root growth appears particularly sensitive to O_3. In contrast, substantial increases in relative below-ground C allocation were found in response to elevated CO_2. Experiments with both elevated CO_2 and O_3 suggest that elevated CO_2 does not compensate for reduced growth caused by elevated O_3. When N limits growth, there appears to be no response to elevated CO_2. Because CO_2 and O_3 change the chemical composition of the foliage, resistance to insect attack and nutritional value of foliage are altered. Elevated O_3 appears to increase insect growth and elevated CO_2 decreases insect growth. Under field conditions, these changes in insect physiology may offset increases or decreases in biomass production with elevated CO_2 that are associated with a changing atmosphere.

Consistent growth responses of yellow poplar (*Liriodendron tulipifera* L.) to O_3 have not been reported even though the species shows visual foliar symptoms of exposure. One study reported an increase in biomass production during the first year of exposure in open-top chambers, and a decrease after two seasons. Elevated CO_2 appears to increase yellow poplar growth regardless of level of exposure to O_3. In general, research on yellow poplar suggests that under field conditions, this species will increase biomass production even when nutrients and moisture are limited and in the presence of O_3.

Experimental methods have a major impact on how results from these studies should be interpreted. Significant chamber effects are common, limiting extrapolation of many experimental results to field conditions. Because experiments are typically short-lived and employ a large step increase in simulated exposure (as opposed to a gradual increase), there is no direct evidence showing how changing atmospheric chemistry would affect plant processes over the long term. Extrapolations of experimental results to natural ecosystems using process models have concluded that the long-term effects of increasing atmospheric CO_2 are likely to decline in magnitude over time (Luo, 1999). However, field exposures of loblolly pine seedlings (Alemayehu et al., 1998), saplings (Dougherty et al., 1998; Hennessey and Harinath, 1998), and mature trees (Teskey, 1998) to elevated CO_2 have not demonstrated any acclimation. Open-air exposure experiments and field physiological studies conducted over long time periods will eventually increase our understanding of individual species and forest community responses to elevated CO_2 and global change stressors as will improvements of physiological process models.

Winter Injury to Red Spruce

There has been an increase in the incidence of freezing injury to northern mountain red spruce forests and red spruce forests at lower elevations, leading to forest decline, over the past five decades (Eagar and Adams, 1992). Red spruce winter injury has been caused by subfreezing temperatures, and the species has only sufficient midwinter cold tolerance to protect foliage from the minimal temperatures found in its northern mountain range. It is likely that the increased injury is the result of acid deposition and climate pertubations. Red spruce is much more susceptible to freezing injury than the co-occurring species balsam fir (*Abies balsamea* [L.] Miller).

Many environmental factors have been studied for their potential to reduce cold tolerance. Ozone does not appear to reduce foliar cold tolerance or increase susceptibility to freezing injury. Carbon dioxide enrichment and N nutrition may interact to influence cold tolerance but the evidence is not clear. It is virtually certain that rapid freezing stress explains localized injury concentrated on sun-exposed branches, but it is unlikely that rapid freezing explains injury on shaded foliage or region-wide injury events. Long winter thaws (4 to 5 days) cause reductions in cold tolerance of up to 14°C, a response unique to red spruce. Numerous studies have shown that short-term N additions either have no impact on freezing tolerance or may improve hardiness. Sulfur additions appear to have some impact on autumn hardiness, but no impact on winter cold tolerance. Aluminum (Al) has many effects on red spruce physiology, but no apparent effect on cold tolerance. Calcium (Ca) loss due to acid deposition affects red spruce health, and may have indirect effects on cold tolerance.

The strongest evidence for an environmental effect on freezing injury is that exposure to acid cloud water increases the risk of foliar freezing injury by reducing the cold tolerance of current-year foliage from 5° to 12°C. Of all the possible combinations of factors, interactions between acid mist and thaw have the greatest potential to increase the risk of freezing injury during winter, but this hypothesis has not been fully tested.

The specific mechanisms causing winter injury appear to include perturbations of winter temperatures on a species with limited genetic potential for cold tolerance, coupled with alterations to the structure and function of plasma membrane-associated Ca in mesophyll cells (DeHayes et al., 1999). Acid-leached calcium is primarily derived from current-year needles and increases hydrogen ion uptake 60 times in response to acidic mist. Acid mist-induced calcium losses in needles reduces membrane stability and decreases freezing tolerance 4° to 10°C. Because of the strong linkage of this mechanism to acidic precipitation, there is great concern that membrane integrity of other species may also be affected, but only red spruce exhibits visible damage symptoms because of its susceptibility to winter injury.

Acid Deposition and Tree Health

Tree rings are good measures of growth and indicators of health. Tree ring characteristics are a composite response to intrinsic and extrinsic factors, and although growth is integrative, it is restricted by the essential factor that is most limiting in supply. Trees sampled at or near the edge of their range are likely to contain a stronger common signal, especially of climate. Given these principles, the use of tree rings as a proxy record of climate or environmental disturbance depends on uniformity of linkage between external conditions and tree biology. Separating the signal of external and internal factors in the tree-ring record can be difficult.

Based on a large sample of red spruce tree rings, enrichment of Ca and magnesium (Mg) evident in stemwood formed in the 1960s is consistent with the mobilization of base cations in the soil, which also coincides with increases in the atmospheric deposition of nitrates and sulfates. Root physiology and pathology are affected indirectly by acidic deposition, through changes in soil chemistry and C allocation patterns to the roots. The Ca/Al molar ratio of the soil solution seems to be one of the prime mechanisms by which acidic deposition affects forest growth and is an important indictor of potential stress.

The availability of biological markers that can assess the current status of stress in apparently healthy trees in a forest is crucial for planning a potential treatment or management practice for mediating or removing the stress. In conjunction with soil chemistry, putricine and/or spermidine may potentially be used as early indicators of Al stress before the appearance of visual symptoms in red spruce trees. The combination of Al induced growth suppression in soils with air pollutant induced growth suppression may be a major cause of tree mortality.

Ecosystem Responses to Environmental Factors

Atmospheric Deposition Effects

Research during the 1980s yielded conclusive evidence that acidic deposition had acidified poorly buffered surface waters causing loss of fish populations and other aquatic organisms. Although acidic deposition affected soil chemistry, effects on forest health were not apparent, with the exception of stand dieback of high elevation red spruce. The National Acid Precipitation Assessment Program (NAPAP) provided much information about natural processes within aquatic and forest ecosystems, and initiated baseline monitoring of deposition rates and chemical changes in ecosystems. NAPAP also highlighted the importance of interactions between multiple stressors, which together threatened the long-term structure, function, and productivity of ecosystems by changing chemical composition and nutrient cycling. In particular, it was noted that acidic deposition could decrease nutrient retention in forest ecosystems and cause imbalances in the availability of nutrients (NAPAP, 1991).

Research in the 1990s began to address the issue of recovery following declining acid deposition rates. Researchers demonstrated the complex interactions between precipitation, soil characteristics, and downstream processes. For example, release of nitrate from watersheds is controlled by biological processes that determine N mobility within soils, not hydrologic transport of atmospherically deposited N directly through soils into surface waters. Also, the importance of the forest floor in supplying Ca for root uptake was noted, along with the regional decline of available Ca that had occurred in the second half of the 20th century. Important mechanisms causing Al mobilization were discovered. Acidic deposition lowers the pH of the mineral soil, causing mobilization of Al that can then be transported into the forest floor, reducing the amount of available Ca. There seems to be a strong connection between soil Ca availability and the health and long-term growth rates of sugar maple; however, this is part of a complex of factors rather than a single factor effect (Horsley and Long, 1999).

Deposition of N may cause ecosystem saturation and a possible shift in species composition because of differential utilization by deciduous and coniferous species. Growth responses to N additions depend not only on species but on tree health and availability of nutrients in the soil, in particular, the availabilty of Ca. Nitrogen additions cause decreases in C allocation to fine roots, with implications for uptake of other nutrients. However, N may also increase the retention of nutrients on the site through a series of processes involving litter quantity and quality, microbial activity, and water retention capacity.

Poor air quality in urban areas may affect plant biota and biogeochemical processes. Based on a study of oak stands along an urban-rural gradient, there is evidence that deposition of heavy metals affects litter fungi, but inconclusive evidence that acid deposition affects soil properties (Pouyat et al., 1995a). Ozone damage has been suggested as a possible cause of changes in litter quality along the urban-rural gradient (Pouyat et al., 1995b).

Nitrogen Saturation

In the U.S. the concept of nitrogen saturation is variously defined as (1) the absense of a growth response in vegetation to N additions, (2) the initiation of nitrate leaching, or (3) the lack of net N accumulation in ecosystems as evidenced by an equivalence between inputs and losses. The region of the U.S. that is most susceptible to N saturation, due to high inputs of N, is the Northeast and in particular high elevation sites where deposition is the greatest. Many factors affect the susceptibility of a forest site to N saturation, including vegetation type, soil characteristics, and land use history.

Watershed-scale additions of N at both Bear Brook in Maine and the Fernow Experimental Forest in West Virginia have clearly shown that the majority of deposited N is retained even in ecosystems showing symptoms of N saturation, although increased stream export of NO_3 was evident. Somewhat scanty evidence suggests that some of the N is retained in vegetation, and that more is likely retained in the soil. However, the mechanisms responsible for N retention are unknown, as are the consequences for forest management and the potential for recovery if N inputs are reduced. There is no conclusive evidence that increased N inputs have resulted in increased vegetative growth and some evidence indicates that the forest growth is still N limited.

Response of Forest Soils to Warming

Experimental evidence supports the hypothesis that an increase in soil temperature of 1.0° to 3.5°C can have significant effects on belowground C and N cycling in northern U.S. forests. Soil C and N cycling are important because of the potential feedbacks to the atmosphere which could affect the direction and magnitude of climate change, the relationship of these cycles to forest productivity and health, and the potential for nutrient export from watersheds to sensitive downstream wetlands and coastal water bodies.

The responses of soils processes to experimental warming are mixed. Soil respiration and N mineralization showed significant and consistent increases regardless of site or treatment. Oxidation of methane, nitrous oxide flux, and litter decomposition showed variable responses that depended on litter quality, N availability, and soil moisture.

Observations along a latitudinal gradient in Michigan suggest that warmer soil temperatures contribute to greater fine root mortality (Hendrick and Pregitzer, 1993). Faster root turnover is one controlling factor for below-ground carbon and nutrient cycling.

Because of the complexity of belowground processes and lack of knowledge with which to construct definitive process models, it is not possible to state whether northern U.S. forest soils will be a net source or sink of C as a consequence of atmospheric warming. However, the balance of experimental evidence and observations suggests that increased soil respiration and litter decomposition, together with decreasing soil organic matter from increasing temperature, will result in a net efflux of C from the soil to the atmosphere. Other possible responses that could counteract this effect include increased N availability and, therefore, increased net primary productivity (NPP) in N limited ecosystems, which could increase the rate of C uptake by plants. However, in N-saturated systems, which are rarer (but increasing from N deposition) than N-limited systems, C uptake could decrease as a consequence of deteriorating forest health.

If temperature changes are small, regional effects on ecosystems are likely to be insignificant relative to more pronounced effects from harvesting, insects and diseases, and other disturbances. Nonetheless, because of the ubiquitous nature of prospective warming, even a small effect spread over a large area could be significant.

Regional Impacts of Multiple Stresses on Productivity and Health

Effects of Increased Carbon Dioxide and Climate Change on Productivity

Changes in NPP of forest ecosystems in the Northeast in response to increasing atmospheric CO_2 and scenarios of climate change were estimated with two ecosystem process models, PnET-II and TEM 4.0. Models were used for this regional analysis because of complex interactions at the ecosystem scale between CO_2 response, biogeochemical cycles, and water and energy fluxes. The two models differ in their representation of above- and belowground processes, their mechanisms of response to CO_2, and their approaches to regional parameterization. To the extent possible, input data was consistent for the two models in order to illustrate similarities and differences in NPP predictions for the region in response to enhanced CO_2 and climate change.

At the regional scale, both models predicted an increase in productivity under climate change scenarios, with PnET-II predicting an average increase of 37.9% and TEM 4.0 an average increase 30.0% for contemporary climate. There was a large difference in model predictions for different climate change scenarios, with the models differing by 3% for contemporary climate and by 10% for the future climate scenarios. PnET-II predicts higher productivity in hardwood and hardwood-pine forest types whereas TEM 4.0 predicts approximately the same productivity for forest biomes. Estimates of NPP with PnET-II showed more sensitivity to different forest types. The NPP responses in TEM 4.0 appear to be limited by temperature, while PnET-II appears to be limited by water. Differences in model structures and representations of response to increased CO_2 contributed to variability in predictions of future productivity. Neither model included some factors known or suspected to influence productivity, such as transient (as opposed to step-wise) climate change, N deposition, and past land use.

The responses of NPP to doubled CO_2 alone were investigated with three biogeochemistry models in the Vegetation/Ecosystem Modeling and Analysis Project (VEMAP). For the conterminous United States, doubled atmospheric CO_2 causes NPP to increase from 5–11% according to these model results (Pan et al., 1998). The national C accounting model (FORCARB) indicates that with an average increase in productivity as predicted with the TEM model, forest trees in the northeast and north central U.S. will have sequestered 19% more C as compared with a business-as-usual scenario (Birdsey and Health, 1995).

Based on these modeling studies, and considering the experimental evidence, it is likely that the combined effects of increased atmospheric CO_2 and projected climate change will increase forest productivity. Uncertainty in this conclusion is related to the high uncertainty in the projected effects of increasing greenhouse gases on precipitation. An increase in temperature accompanied by a decrease in precipitation could cause plant stress and a reduction in NPP. The magnitude of the water stress would be ameliorated to some degree by CO_2-induced stomatal closure, which increases water use efficiency, a result found in many experimental studies (Tyree and Alexander, 1993).

Ozone Effects on Productivity

Mechanistic models simulate changes in forest structure or function by quantifying fundamental mechanisms or processes, whereas statistical models rely on analysis of empirical data. Mechanistic models can therefore be used to extrapolate forest responses beyond conditions that have already occurred, if the appropriate response mechanisms are understood and modeled correctly.

PnET-II, an ecosystem-scale model used to estimate regional forest production of hardwoods, total ecosystem C balances and water yield, and responses to climate change, was modified to include O_3 effects on productivity including interactions with other stressors. Predicted NPP ranged from 750 to 1450 $g\,m^{-2}\,yr^{-1}$ and generally increased from north to south across the region. The model was then modified to include leaf-level uptake-response relationships and allow interaction with factors such as light attenuation, canopy ozone gradients, and water stress. Results for New England and New York showed decreases in predicted annual NPP from 2 to 17% as a result of mean O_3 levels from 1987 to 1992, with greatest reductions occurring where both O_3 levels and stomatal conductance were greatest. Growth declines were greatest on sites with wetter soils.

Another approach to estimating O_3 effects was to couple a model of forest stand development and composition (ZELIG) with a model of tree-level physiological response to stress (TREGRO). TREGRO models the acquisition of C, water, and nutrients, and allocates C among competing plants parts depending upon resource availability and phenology. ZELIG is a gap-succession model used to simulate succession in mixed stands typical of eastern and northern forests, and has both mechanistic and empirical characteristics. The two models were coupled by passing calculations from TREGRO to ZELIG in order to modify the growth relationships in the stand model. Simulated response to ambient O_3 from 1991 using the TREGRO/ZELIG combination for red oak and sugar maple in the Northeast showed a 2 to 4% reduction in red oak basal area growth over a 100-year simulation and a 3 to 12% growth increase in sugar maple across the region due to reduced competition from ozone sensitive species.

Model predictions include uncertainty due to (1) incomplete regional data, (2) incomplete knowledge of how forests respond to particular stresses, and (3) uncertainty about which processes and parameters should be included in the model. Some of this uncertainty is related to spatial interpolation, aggregation, and scaling errors.

Effects of Climate Change on Forest Insect and Disease Outbreaks

There is much uncertainty in predictions of the effects of climate change on insects and diseases. Effects can act directly on the physiology of the organisms, and indirectly through interactions with physiological changes in host plants or natural enemies.

Considering only the direct effects of warming, most insect species would survive more successfully over winter. Also, many insect species would shift their ranges toward higher latitudes and elevations, but this response is partly dependent on concomitant host and enemy changes. Species with flexible life histories could produce more generations per unit of time, as has already happened with the spruce bark beetle in Alaska. In summary, direct effects of warming are likely to increase the frequency, level, and geographical extent of disturbance by insect pests.

Indirect effects are more subtle and, therefore, more difficult to predict. Population changes of defoliators, such as gypsy moth and spruce budworm, are governed by complex interactions with hosts and natural enemies. For example, the effects of increased CO_2 on insects is mediated by changes in foliar nutrition and presence of defensive chemicals in foliage.

If drought frequency changes as a function of altered precipitation, temperature, or both, trees and forests may become chronically stressed, leading to progressive deterioration in tree health and increasing susceptibility to secondary organisms and decline diseases. Pathogens could spread into new areas where they are presently limited by low temperature. Reproduction rates could increase, increasing disease severity and accelerating the evolution of new pathotypes.

Lessons from the Past and Uncertainty for the Future

Changes in the distribution of vegetation are strongly correlated with climatic change over long periods, thus it is reasonable to expect that future climate changes will affect forest composition and distribution. Specific biological factors that affect species response to environmental changes include survival, reproductive capacity, rate of migration, and response variation between or within species. Anthropogenic stressors are a relatively new set of factors that affect evolutionary responses to changing environment.

Most modern vegetation assemblages have been in their present configuration for no more than 6000 to 8000 years. Present vegetation communities are transitory combinations of species that have responded individually to environmental changes and competition. Eastern white pine is an example of a species whose distribution has closely followed climate changes of the last 10,000 years.

Regression tree analysis on trees has been used to indicate prospective responses of individual species which are then aggregated to examine potential changes in community dynamics and biodiversity. Under climate change scenarios, balsam fir and quaking aspen were mostly lost from northern U.S. forests, and maples were greatly reduced. Pine species, primarily loblolly pine (*Pinus tadea* L.) and eastern white pine (*Pinus strobus* L.), expanded their dominance in the southern and northern parts of the region, respectively. Oaks (*Quercus* spp.) also expanded in some areas.

Both paleoecological studies and modeling efforts have clearly shown that vegetation communities are unlikely to move together as intact communities. Consideration of additional factors that contribute to the uncertainty of prediction only strengthens this general conclusion. For example, the determinants of the boundaries of the current distribution of any particular species are poorly known. Anthropogenic factors, such as exotic species introductions, changes in land use, and forest management, also play a role in conjunction with changing environmental factors.

There are differential responses to environmental change between species and within species, as shown by numerous experimental studies. For example, eastern white pine is highly sensitive to air pollution, with many of the sensitive individuals already lost from natural populations. Greenwood and Hutchison (1996) showed that temperature during breeding can have effects on the genotype and phenotype of progeny even if identical parent material is used. They observed significant growth differences as a function of crossing environment.

Most natural populations of temperature and boreal species seem to be quite tolerant of climate differences as indicated by a long history of provenance testing. In the northern U.S., most species should tolerate climates as much as 5°C warmer than present, as long as other factors, such as moisture availability, do not change simultaneously. However, regeneration in the face of climate change is likely to be more difficult than survival, because the most sensitive stage of a tree's life is the beginning, when warmth and drought can have strong effects.

Species migration is dependent on regeneration success. Other factors are also important, such as nonnative tree species and their competition effect, or forest fragmentation, which may be a barrier to migration.

The Potential Effects of Global Change on Northern U.S. Forests

Ecosystem responses to changes in multiple environmental factors are exceedingly complex and cannot be predicted with certainty using current experimental results and models. Environmental factors known to cause changes in ecosystem processes include climatic variables, tropospheric O_3, N deposition, acid deposition, and anthropogenic factors such as past land use and introduction of exotic species. Experimental studies can logistically include controls for only a few of the important factors governing ecosystem processes. Observational studies lack control for establishing definitive cause and effect relationships. Models lack sufficient mechanistic detail and input data, and are often not validated for regional applications. However, there is some knowledge to be gained from each approach, and when the different approaches begin to converge by indicating similar ecosystem responses, we can be more certain about conclusions.

Some Significant Findings

Evidence from experiments that address combinations of factors has shown:

- Elevated CO_2 usually increases tree biomass growth, while exposure to elevated and ambient O_3 usually decreases biomass growth. Controlled exposure of trembling aspen to both elevated CO_2 and O_3 shows that elevated CO_2 may not compensate for reduced growth caused by elevated O_3 in sensitive genotypes. When N is limiting, there may be a smaller positive growth response to elevated CO_2.

- Because CO_2 and O_3 change the chemical composition of the foliage, resistance to insect attack and nutritional value of foliage are altered. Elevated O_3 appears to increase insect growth and elevated CO_2 decreases insect growth.
- It is likely that increased winter injury to red spruce is the result of acid deposition and altered weather patterns. The strongest evidence for an environmental effect on incidence of freezing injury is that exposure to acid cloud water increases the risk of foliar freezing injury by reducing the cold tolerance of current-year foliage by 5 to 12°C. Recent research has pointed toward a mechanism of acid-induced loss of calcium from cellular membranes.
- Watershed-scale additions of N at both Bear Brook in Maine and Fernow Experimental Forest in West Virginia have clearly shown that the majority of deposited N is retained in ecosystems that have experienced chronic N addition. There is no conclusive evidence that increased N resulted in increased vegetative growth. Although most fertilizer trials shown an increase in growth, at least two experiments with conifers show reductions in growth from chronic N additions.
- Experimental evidence supports the hypothesis that an increase in soil temperature of 1.0 to 3.5°C can have significant effects on belowground C and N cycling in northern U.S. forests. Soil respiration and N mineralization showed significant and consistent increases regardless of site or treatment. Oxidation of methane, nitrous oxide flux, and litter decomposition showed variable responses that depended on litter quality, N availability, and soil moisture.

Observational studies have shown strong correlations between some environmental factors and responses:

- Acidic deposition lowers the pH of the mineral soil, causing mobilization of Al that can then be transported into the forest floor, reducing the amount of available Ca. The Ca/Al molar ratio of the soil solution seems to be related to the mechanisms by which acidic deposition affects forest growth and is an important indicator of potential stress.
- The direct effects of warming on insect physiology are likely to increase the frequency, level, and geographical extent of disturbance by insect pests. The effects of warming may be mediated by changes in feeding behavior brought about by the exposure of foliage to elevated O_3 and CO_2, which alter the nutritional value of the foliage.
- Most natural populations of temperate and boreal tree species seem to be quite tolerant of climate differences as indicated by a long history of provenance testing. In the northern U.S., most species should tolerate

climates as much as 5°C warmer than present, as long as other factors such as moisture availability do not change simultaneously.

Integrated modeling studies and model comparisons suggest that:

- Estimates of the effects of ozone on annual net primary production of hardwood forests at the regional scale range from no effect to 17% decrease in productivity. Productivity reductions are more likely to occur where both ozone levels and stomatal conductance are greatest.
- It is likely that the combined effects of increased atmospheric CO_2 and projected climate change will increase forest productivity. Uncertainty in this conclusion is related to the high uncertainty in the projected effects of increasing greenhouse gases on precipitation. An increase in temperature accompanied by a decrease in precipitation could cause plant stress and a reduction in NPP. The magnitude of the water stress would be ameliorated to some degree by CO_2-induced stomatal closure, which increases water use efficiency.
- Regression tree analysis of composition changes suggests that under warming scenarios, balsam fir and quaking aspen were mostly lost from northern U.S. forests, and maples were greatly reduced. Pine species, primarily loblolly pine and white pine, expanded their dominance in the southern and northern parts of the region, respectively. Oaks also expanded in some areas.

Two or more lines of evidence coverage in supporting the following statements:

- Numerous experimental and observational studies have shown major differences in the response of species and genotypes to multiple interacting environmental factors.
- Both paleoecological studies and modeling efforts have clearly shown that vegetation communities are unlikely to move together as intact communities.
- Experimental research and models suggest that exposure to ambient levels of O_3 may damage sensitive plants, causing growth losses under conditions conductive to damage such as during high water uptake.
- Northern U.S. forests have sequestered a significant amount of C over the past 50 years, and are expected to continue sequestering C unless affected by natural and human-induced disturbances that far exceed recent levels.
- Because of the complexity of soil process responses, it is not possible to definitively state whether Northern U.S. forest soils will be a net source or sink of C as a consequence of atmospheric warming. However, the balance of experimental evidence and observations suggests that warming of northern temperate forests will result in a net efflux of C from the soil to the atmosphere.

The effects of many environmental changes on forest productivity and health are mediated through changes in soil process:

- Depletion of base cations has been identified as a critical issue for northern forest soils. If weathering of parent material is not sufficient to replace base cations lost to forest growth and leaching, then depletion is likely. It has been clearly established that acidic deposition can decrease nutrient retention in forest ecosystems and cause imbalances in the availability of nutrients.

Human factors underlie some of the observed changes in forest ecosystems:

- Land use history is among many factors such as vegetation type and soil characteristics that affect the susceptibility of a forest site to N saturation.
- Anthropogenic stressors are a relatively new set of factors that affect evolutionary responses of tress to a changing environment. Some influential anthropogenic factors include exotic species introductions, changes in land use (and forest fragmentation), and forest management.

Vulnerable Northern Forests

Northern forests have heretofore shown remarkable resiliency and adaptability despite high levels of environmental stress. We have documented climate trends, ozone exposure, high levels of acid and nitrogen deposition, and land use pressures, all simultaneously affecting northern forests. Yet northern forests appear healthy as a whole despite evidence of local problems, and regional inventories continue to show accumulation of biomass and low (though increasing) mortality. Evidence suggests that decades of stress may have altered long-term soil processes sufficiently to begin affecting regional indicators of health and productivity. Therefore, an increased level of monitoring, targeted to areas suspected to be sensitive to environmental change, is warranted.

Forest ecosystems that appear vulnerable to past or continued environmental stress include:

- Northern Red Spruce Forests—These forests are damaged by acid deposition and winter injury and are vulnerable to additional climatic stress. Winter injury, common when temperatures fluctuate broadly and when foilage is damaged by acid deposition, could become more common. High-elevation spruce-fir forests in the middle Appalachians are remnants of an earlier, cooler climate now found only at the highest available elevations.
- Aging Hardwoods in the Mid-Atlantic Region—If the theoretical models are correct, then increased levels of various interacting stressors

in the Northeast are likely to lead to increased incidence of decline disease. Increasing environmental stress is occuring at the same time as many species reach biological maturity across much of their range, a consequence of past land use impacts. Aging forests are known to be more susceptible to decline disease. Vulnerable types include maple-beech-birch, oak-hickory, and black cherry in Mid-Atlantic States. Mid-Atlantic forests are exposed to high levels of O_3 and acid deposition, and as they age, are showing increased mortality typical of mature forests. For example, mature sugar maple on unglaciated soils in northwestern Pennsylvania are noticeably affected by biotic factors (defoliating insects, borers, and canker fungus), a series of droughts, and acid deposition. Regeneration of some aging forests is very uncertain. Lack of oak regeneration following harvest is considered a major issue for the region, as loss of oak forests coupled with a major expansion of red maple results in lower economic value and reduction of important mast species. However, regeneration in the face of climate change is likely to be more difficult than survival, because the most sensitive stage of a tree's life is the beginning, when warmth and drought can have strong effects.

- Aspen-Birch in the Lake States—Aspen-birch forests in the U.S. grow at the southern end of their current range. Aspen is sensitive to O_3, and for some genotypes, CO_2 benefits are not sufficient to offset the negative effects of O_3. It is unknown how climate change might alter aspen-birch forests when coupled with these physiological responses. Over North America, the most significant observed temperature changes over the last 40 years have occurred from the north central U.S. through northwestern Canada into Alaska. Thus, it is possible that multiple stresses will converge on aspen-birch forests in the Lake States with unknown but potentially harmful consequences.

Protecting the Values of Northern Temperate Forests

Northern temperate forests are important components of earth's terrestrial ecosystems and help regulate the concentration of atmospheric CO_2. Maintenance of these forests in a healthy and productive state will ensure their continued role as reservoirs of C that could otherwise be released and exacerbate the global problem of increasing atmospheric CO_2. Northern forests contain large reservoirs of C, particularly in abundant maturing forests and forest soils. Northern U.S. forests contribute an estimated two-thirds of the total U.S. accumulation of C in forest ecosystems, which helps offset emissions from fossil fuels (Birdsey and Heath, 1995). Although this rate of accumulation may slow as the forests continue maturing (Fig. 15.1), it is nonetheless important to maintain healthy forests and avoid large future releases of CO_2 from forest biomass or soils.

Long-term productivity and C storage of forests in the North are threatened by acid deposition (Adams, 1999; Hornbeck et al., 1997). Acid

TgC/yr

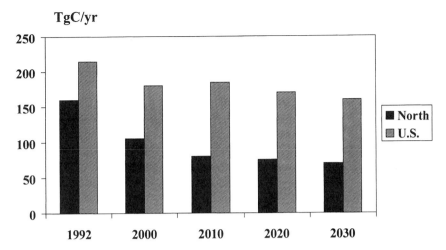

Figure 15.1. Projected periodic average annual carbon uptake (million metric tons yr^{-1}) in northern and all U.S. forests.

deposition affects cation availability and nutrient cycling in a variety of ways depending on soil parent material, management history, and other controlling factors. Forest managers will increasingly be required to monitor soil characteristics to sustain forest values over the long term.

Forests protect water supplies by keeping soil in place and reducing reservoir siltation, by absorbing pollutants and regulating the release of pollutants, and by regulating the discharge rate of water. Because of the proximity of Northern forests to great population centers, their value in protecting water quality and quantity is very significant. For example, the Catskill Mountains, nearly 100% forested, provide water for residents of New York City that is clean enough to be supplied without treatment.

Many northern forests provide timber for economic uses, recreation opportunities for residents and tourists, and wildlife habitat. Only a few of the many forest values that are potentially affected by climate change have been studied, but those that have show a variety of significant positive and negative responses. For example, a study of the potential effects of warming on brook trout in the Central Appalachians suggests that as stream temperature increases, trout growth increases in the spring and decreases in the summer; habitat is reduced; and predator abundance could increase (Ries and Perry, 1995). Knowledge of these potential effects would help sustaining forest values and maintain the stability of dependent human communities.

Genetic and phenotypic variability is fundamentally important to survival on a changing planet. Maintaining and protecting sufficient natural variability within tree populations is thus critical for future evolutionary adaptation and maintenance of health and stability within northern forests. Perhaps the greatest anthropogenic threat to northern forests is the combination of unprecedented climate change, pollutant exposure, and degradation of genetic diversity needed for adaptation and survival.

Mitigating Climate Change Through Forest Management

By signing the Kyoto Protocol in 1998, the U.S. government reaffirmed its commitment to meet the challenge of climate change by reducing greenhouse gases. Under the terms of the agreement, the U.S. must reduce net emissions 7% below 1990 levels by the first reporting period, 2008 to 2012. Although the exact terms that will be used to implement the protocol are still under discussion, there will be some opportunity for land use change and forestry activities to contribute to greenhouse gas reductions if it can be shown that these activities increase the size of the terrestrial C sink. The increased C sequestration should be relative to an average baseline that would occur in the absence of the activity.

It is relatively easy to show that conversion of nonforest land to forest land (afforestation) causes an increase in sequestered C, and that conversion of forest land to nonforest land (deforestation) causes a decrease in sequestered C. Because the North is already heavily forested and land use change affects a small proportion of the land base, it is likely that changes in forest management, which could affect large areas, would have the greatest impact on reducing atmospheric CO_2.

Forest inventory data for the North show that the region's forests currently sequester C at an average rate of $2.3\,t\,ha^{-1}\,yr^{-1}$ (Birdsey and Heath, 1995). Even unmanaged, aging forests in the North continue to accumulate C at a high rate. For example, Johnson and Strimbeck (1995) measured the change in biomass and soil C over a 33-year period for 23 stands of aggrading sugar maple in Vermont, ranging in age (at initial measurement in 1957 to 1959) from 47 to 97 years. They found that only 2 stands lost biomass whereas 21 gained biomass. The average gain in biomass for all stands was $1.8\,t\,ha^{-1}\,yr^{-1}$. There was no detectable change in soil C over this period. Another study used the eddy covariance technique to directly measure the exchange of CO_2 in a 60-year old northern hardwood ecosystem in Massachusetts (Wofsy et al., 1993). They estimated a significant net C sink of approximately $3.7\,t\,ha^{-1}\,yr^{-1}$.

Much higher rates of C sequestration are possible in forests that are managed for biomass production. The average C storage in biomass (including cut and dead trees) for 40-year old hardwoods in the Lake States under different management intensities was $331\,Mg\,ha^{-1}$ (Strong, 1997). This is equivalent to an average annual C accumulation of $8.3\,t\,ha^{-1}\,yr^{-1}$. There was a small effect of harvesting on soil C, with heavier cutting reducing soil C by 25% compared with the control (no treatment) which had $120\,\mathrm{Mg\,C\,ha}^{-1}$. Another study showed that conversion of cropland to fast-growing hybrid poplar caused a significant gain in soil C after about 10 years (Hansen, 1993). Soil C accretion beneath 12- to 18-year-old poplar plantations exceeded that of adjacent agricultural crops by $1.6\,\mathrm{Mg\,ha}^{-1}\mathrm{yr}^{-1}$.

Carbon from forests can remain stored in forest products long after forests are harvested and the wood processed into products. Carbon stored in trees harvested in the early 1900s still remains stored in wood of houses built from this wood. It is estimated that the accumulation of C in wood products and landfills from harvested northern forests totals approximately $0.015\,\mathrm{Pg\,C\,yr}^{-1}$ (Heath et al., 1996).

Urban and suburban forests may play a significant role in sequestering carbon through tree growth and energy conservation. A study of Chicago's urban forests found an average of 14 to $18\,\mathrm{t\,ha}^{-1}$ of C in tree biomass (Nowak, 1994). Urban trees also reduce the use of energy for heating and cooling, thus avoiding CO_2 emissions.

The need to protect forests from fire, pests, and other stresses is important both to secure the role of forests in mitigating climate change and for avoiding adverse impacts of climate change and air pollution. Forest managers need to be aware of threats to forest health and productivity that may not be too obvious and take adaptive actions as necessary to sustain the many values provided by northern forests.

References

Adams MB (1999) Acidic deposition and sustainable forest management in the central Appalachians, USA. *Forest Ecology and Management* 122:17–28.

Alemayehu M, Hileman DR, Huluka G, Biswas PK (1998) Effects of elevated carbon dioxide on the growth and physiology of loblolly pine. In: Mickler RA, Fox S (eds) *The Productivity and Sustainability of Southern Forest Ecosystems in a Changing Environment Ecological Studies* 128. Springer-Verlag, New York, pp 93–102.

Bazzaz FA (1990) The response of natural ecosystems to the rising global CO_2 levels. *Ann Rev Ecol Syst* 21:167–196.

Birdsey RA, Heath LS (1995) Carbon changes in U.S. forests. In: Joyce LA (ed) Productivity of America's forests and climate change. US Department of Agriculture (USDA) Forest Service, Rocky Mountain Forest and Range Experiment Station, General Technical Report RM-GTR-271, Ft. Collins, CO, pp 56–70.

Ceulemans R, Mousseau M (1994) Effects of elevated atmospheric CO_2 on woody plants. *New Phytolog* 127:425–446.

Committee on Earth Sciences (1989) Our Changing Planet. *The FY 1990 Research Plan. A Report by the Committee on Earth Sciences.* Office of Science and Technology Policy in the Executive Office of the President, Washington, DC.

Curtis PS (1996) A meta-analysis of leaf gas exchange and nitrogen in trees grown under elevated CO_2. *Plant, Cell, and Environment* 19:127–137.

DeHayes DH, Schaberg PG, Hawley GJ, Strimbeck GR (1999) Acid rain impacts on calcium nutrition and forest health. *BioScience* 49(10):789–800.

Dougherty PM, Allen HL, Kress LW, Murthy RM, Maier CA, Albaugh TJ, Sampson DA (1998) An investigation of the impacts of elevated carbon dioxide, irrigation, and fertilization on the physiology and growth of loblolly pine. In: Mickler RA, Fox S (eds) *The Productivity and Sustainability of Southern Forest Ecosystems in a Changing Environment. Ecological Studies* 128. Springer–Verlag, New York, pp 19–168.

Eagar C, Adams MB (eds) (1992) *Ecology and Decline of Red Spruce in the Eastern United States. Ecological Studies* 96. Springer-Verlag, New York.

Eamus D, Jarvis PG (1989) The direct effects of increase in the global atmospheric CO_2 concentration on natural and commercial temperate trees and forests. *Adv Ecol Res* 19:1–55.

Flagler RB, Brissette JC, Barnett JP (1998) Influence of drought stress on the response of shortleaf pine to ozone. In: Mickler RA, Fox S (eds) *The Productivity and Sustainability of Southern Forest Ecosystems in a Changing Environment. Ecological Studies* 128. Springer-Verlag, New York, pp 73–92.

Fox S, Mickler RA (eds) (1996) *Impact of Air Pollutants on Southern Pine Forests. Ecological Studies* 118. Springer-Verlag, New York.

Greenwood MS, Hutchison KW (1996) Genetic aftereffects of increased temperature in *Larix*. In: Hom J, Birdsey R, O'Brian K (eds), Proceedings of 1995 Meeting of the Northern Global Change Program Gen. Tech. Rep. NE-214, USDA Forest Service, Northeastern Forest Experiment Station. Radnor, PA, pp 169–174.

Gunderson CA, Wullschleger SD (1994) Photosynthetic acclimation in trees to rising atmospheric CO_2: a broader perspective. *Photosynthesis Res* 39:369–388.

Hansen EA (1994) Soil carbon sequestration beneath hybrid poplar plantations in the North Central United States. *Biomass and Bioenergy* 5(6):431–436.

Heath LS, Birdsey RA, Row C, Plantinga A (1996) Carbon pools and fluxes in U.S. forest products. In: Apps MJ, Prince DT (eds) *Forest Ecosystems, Forest Management and the Global Carbon Cycle*. NATO ASI Series I: Global Environmental Change, Vol. 40, Springer-Verlag, Berlin, pp 271–278.

Hendrick RL, Pregitzer KS (1993) Patterns of fine root mortality in two sugar maple forests. *Nature* 361:59–61.

Hennessey TC, Harinath VK (1998) Effects of elevated carbon dioxide, water, and nutrients on photosynthesis, stomatal conductance, and total chlorophyll content of young loblolly pine (*Pinus taeda* L.) trees. In: Mickler RA, Fox S (eds) *The Productivity and Sustainability of Southern Forest Ecosystems in a Changing Environment. Ecological Studies* 128. Springer-Verlag, New York, pp 169–184.

Horsley SB, Long RP (1999) *Sugar Maple Ecology and Health*: *Proceedings of an International Symposium*. June 2–4, 1998, Warren, PA. Gen Tech Rep NE-261. United States Department of Agriculture (USDA) Forest Service, Northeastern Research Station, Radnor, PA.

Hornbeck JW, Bailey SW, Buso DC, Shanley JB (1997) Streamwater chemistry and nutrient budgets for forested watersheds in New England: variability and management implications. *Forest Ecology and Management* 93:73–89.

Houghton JT, Kenkins GJ, Ephraums JJ (eds) (1990) *Climate Change. The IPCC Scientific Assessment*. Cambridge University Press, Cambridge, MA.

Houghton JT, Meira Filho LG, Callander BA, Harris N, Kattenberg A, Maskell K (eds) (1996) *Climate Change 1995. The Science of Climate Change*. Cambridge University Press, Cambridge, MA.

Idso KE, Idso SB (1994) Plant responses to atmospheric CO_2 enrichment in the face of environmental constraints: a review of the past 10 years' research. *Agric for Meteror* 69:153–203.

Johnson AH, Strimbeck GR (1995) Thirty-three year changes in above- and below-ground biomass in northern hardwood stands in Vermont. In: Hom J, Birdsey R, O'Brian K (eds) *Proceedings of 1995 Meeting of the Northern Global Change Program* Gen Tech Rep NE-214. United States Department of Agriculture (USDA) Forest Service, Northeastern Forest Experiment Station, Radnor, PA, pp 169–174.

Luo Y, Reynolds JF (1999) Validity of extrapolating field CO_2 experiments to predict carbon sequestration in natural ecosystems. *Ecology* 80(5): 1568–1583.

Mickler RA (1998) Southern forest ecosystems in a changing environment. In: Mickler RA, Fox S (eds) *The Productivity and Sustainability of Southern Forest Ecosystems in a Changing Environment. Ecological Studies* 128. Springer-Verlag, New York, pp 3–14.

Mickler RA, Fox S (eds) (1998) *The Productivity and Sustainability of Southern Forest Ecosystems in a Changing Environment. Ecological Studies* 128. Springer-Verlag, New York.

Musselman RC, Fox DG (1991) A review of the role of temperate forests in the global CO_2 balance. *J Air Waste Manage Assoc* 41(8):798–807.

NAPAP (1991) National Acidic Precipitation Assessment Program 1990 Integrated Assessment Report, S/N040-000-00560-9. United States Government Printing Office, Pittsburg, PA.

Nowak DJ (1994) Atmospheric carbon dioxide reduction by Chicago's urban forest. In: McPherson EG, Nowak DJ, Rowntree RA (eds) *Chicago's Urban Forest Ecosystem: Results of the Chicago Urban Forest Climate Project* Gen Tech Rep NE-186 United States Department of Agriculture (USDA) Forest Service, Northeastern Forest Experiment Station, Radnor, PA, pp 83–94.

Pan Y, Melillo JM, McGuire AD, other VEMAP members (1998) Modeled responses of terrestrial ecosystems to elevated atmospheric CO_2: a comparison of simulations by the biogeochemistry models of the Vegetation/Ecosystem Modeling and Analysis Project (VEMAP). *Oecologia* 114:389–404.

Pouyat RV, McDonnell MJ, Pickett STA (1995a) Soil characteristics along an urban-rural land-use gradient. J Environ Quality 24:516–526.

Pouyat RV, McDonnell MJ, Pickett STA, Groffman PM, Carreiro MM, Parmelee RW, Medley KE, Zipperer WC (1995b) Carbon and nitrogen dynamics in oak stands along an urban-rural gradient. In: McFee WW, Kelly JM (eds) *Carbon Forms and Functions in Forest Soils*. Soil Science Society of America, Madison, WI, pp 569–587.

Richter DD, Markewitz D (1996) Atmospheric deposition and soil resources of the southern pine forest. In: Fox S, Mickler RA (eds) *Impact of Air Pollutants on Southern Pine Forests. Ecological Studies* 118. Springer-Verlag. New York, pp 315–336.

Ries RD, Perry SA (1995) Potential effects of global climate warming on brook trout growth and prey consumption in central Appalachian streams, USA. *Climate Research* 5:197–206.

Rogers HH, Runion GB (1994) Plant responses to atmospheric CO_2 enrichment with emphasis on roots and their rhizosphere. *Environ Pollut* 83:155–189.

Stoddard JL, Jeffries DS, Lukewille A, et al. (1999) Regional trends in aquatic recovery from acidification in North America and Europe. *Nature* 401:575–578.

Strain BR, Thomas RB (1992) Field measurements of CO_2 enhancement and climate change in natural vegetation. *Water Air Soil Pollut* 64:45–60.

Strong TF (1997) Harvesting intensity influences the carbon distribution in a northern hardwood ecosystem. United States Department of Agriculture

(USDA) Forest Service. North Central Experiment Station, Research Paper 329, St. Paul, MN.

Teskey RO (1998) Effects of elevated carbon dioxide levels and air temperature on carbon assimilation of loblolly pine. In: Mickler RA, Fox S (eds) *The productivity and Sustainability of Southern Forest Ecosystems in a Changing Environment. Ecological Studies* 128. Springer-Verlag, New York, pp 131–148.

Tyree MT, Alexander JD (1993) Plant water relations and the effects of the elevated CO_2: a review and suggestions for future research. *Vegetation* 104/105:47–62.

Watson RT, Zinyowera MC, Moss RH (eds) (1996) *Climate Change 1995. Impacts, Adaptations and Mitigation of Climate Change: Scientific-Technical Analyses.* Cambridge University Press, Cambridge, MA.

Wofsy SC, Goulden ML, Munger JM, Fan SM, Backwin PS, Daube, BC, Bassow SL, Bazzaz FA (1993) Net exchange of CO_2 in a mid-latitude forest. *Science* 260:1314–1317.

Index

Ecological Studies

Volumes published since 1992

Volume 89
Plantago: A Multidisciplinary Study (1992)
P.J.C. Kuiper and M. Bos (Eds.)

Volume 90
**Biogeochemistry of a Subalpine Ecosystem:
Loch Vale Watershed** (1992)
J. Baron (Ed.)

Volume 91
**Atmospheric Deposition and Forest Nutrient
Cycling** (1992)
D.W. Johnson and S.E. Lindberg (Eds.)

Volume 92
**Landscape Boundaries: Consequences for
Biotic Diversity and Ecological Flows** (1992)
A.J. Hansen and F. di Castri (Eds.)

Volume 93
**Fire in South African Mountain Fynbos:
Ecosystem, Community, and Species
Response at Swartboskloof** (1992)
B.W. van Wilgen et al. (Eds.)

Volume 94
The Ecology of Aquatic Hyphomycetes (1992)
F. Bärlocher (Ed.)

Volume 95
Palms in Forest-Ecosystems of Amazonia
(1992)
F. Kahn and J.-J. DeGranville

Volume 96
**Ecology and Decline of Red Spruce in the
Eastern United States** (1992)
C. Eagar and M.B. Adams (Eds.)

Volume 97
**The Response of Western Forests to Air
Pollution** (1992)
R.K. Olson, D. Binkley, and M. Böhm
(Eds.)

Volume 98
Plankton Regulation Dynamics (1993)
N. Walz (Ed.)

Volume 99
Biodiversity and Ecosystem Function
(1993)
E.-D. Schulze and H.A. Mooney (Eds.)

Volume 100
Ecophysiology of Photosynthesis (1994)
E.-D. Schulze and M.M. Caldwell (Eds.)

Volume 101
**Effects of Land-Use Change on
Atmospheric CO_2 Concentrations: South
and South East Asia as a Case Study** (1993)
V.H. Dale (Ed.)

Volume 102
Coral Reef Ecology (1993)
Y.I. Sorokin (Ed.)

Volume 103
**Rocky Shores: Exploitation in Chile and
South Africa** (1993)
W.R. Siegfried (Ed.)

Volume 104
**Long-Term Experiments With Acid Rain in
Norwegian Forest Ecosystems** (1993)
G. Abrahamsen et al. (Eds.)

Volume 105
Microbial Ecology of Lake Plußsee (1993)
J. Overbeck and R.J. Chrost (Eds.)

Volume 106
Minimum Animal Populations (1994)
H. Remmert (Ed.)

Volume 107
**The Role of Fire in Mediterranean-
Type Ecosystems** (1994)
J.M. Moreno and W.C. Oechel (Eds.)

Volume 108
**Ecology and Biogeography of
Mediterranean Ecosystems in Chile,
California, and Australia** (1994)
M.T.K. Arroyo, P.H. Zedler, and
M.D. Fox (Eds.)

Volume 109
**Mediterranean Type Ecosystems: The
Function of Biodiversity** (1994)
G.W. Davis and D.M. Richardson (Eds.)